Dimensions of Ecology

Dimensions of Ecology
is one of a series of textbooks
published in connection
with E. K. Georg Landsberger

Dimensions of Ecology

Jonathan L. Richardson
Associate Professor of Biology
Franklin and Marshall College
Lancaster, Pennsylvania

The Williams & Wilkins Company · Baltimore

Made in the United States of America

Library of Congress Cataloging in Publication Data

Richardson, Jonathan L
Dimensions of ecology.

Bibliography: p.
Includes index.
1. Ecology. I. Title.
QH541.R52 574.5 76-20701
ISBN 0-683-07278-1

Composed and printed at the
Waverly Press, Inc.
Mt. Royal and Guilford Aves.
Baltimore, Md. 21202, U.S.A.

To
Alice,
Cathy
and
Mary

Preface

To him who, in the love of Nature, holds communion with her visible forms, she speaks a various language.
William Cullen Bryant (1774–1878)

The language of nature is the language of ecology, much talked about but little understood. One aim of this book is to set forth the elements of this language; another and really more basic aim is to convey the *delight* of looking at nature and at ourselves from an ecological perspective, for to do so is to begin to understand the warp and woof of Earth's living tapestry.

It is a paradox of the modern world, and one of its great dilemmas, that the average human—even the average biologist—has become increasingly ignorant of the natural world in its totality. We have lost touch with our own roots even while making great strides toward understanding life processes, eliminating disease, and efficiently channeling Earth's wealth to ourselves. Mankind everywhere is rushing toward an urban existence and the natural world seems, to most of us, another world, not our own. In reality, of course, even our cities and civilizations remain part of nature; we simply are blind to our place in the larger scheme. In many fundamental ways the question of who we are is an ecological question, and it is remarkable how ecological is the sound of historians, economists, and anthropologists when they stand back and take the long view.

This book is an introduction to ecology, not a treatise; it is designed for use in a one-semester course, and it does not assume in the reader a majestic scientific and mathematical background. Even so, this is not always an easy book, but I have tried to make it always understandable, and to infuse it with my own enthusiasm for the subject matter.

In trying to teach one-semester courses in ecology, I have learned at least one thing: that one semester is impossibly short. Ecology is at heart a biological subject, but on every side it meshes with other disciplines. In a one-semester text, how does an author handle such enormous breadth? One way is to ignore it, and to deal only with areas of specific interest to the writer. To my mind, however, this approach does a disservice to the beginning student. Another possibility, also unsatisfactory, is to try to cover everything, at a necessarily superficial level. A third approach, the one I take in this book, is to attempt to provide a sampler. "Sampler" does not here imply superficiality, but selectivity; in this book, for example, I treat several habitat types and communities in detail, while perhaps slighting others. The danger of the sampler approach is that each ecologist has his own ideas regarding what should constitute the sample. I have tried to cast my net in different directions, capturing some of the interfaces of ecology, but dealing with topics in sufficient leisure and depth to give the reader a true feeling for the subjects (not just the samples) that are covered.

I must admit here to a conviction that colors the content of the book. I believe that modern ecological theory is most useful and exciting to those who have been introduced in some detail to the real communities, populations and ecosystems with which the theory deals. I therefore have included in this text considerable descriptive material as well as theory, but the instructor easily can achieve a self-tailored balance simply by choice of lecture topics. For example, the ideas of theoretical ecology, if presented in an introductory course, need to be explained carefully in class. I believe the present text allows this, by providing sufficient background material dealing with real habitats and real species so that class time can be focused on the conceptual underpinnings of ecology.

I have tried to keep this book short enough so that supplementary readings can be assigned without overburdening the student. Such readings have real value, particularly when tailored to the local scene or to the special enthusiasms of the instructor. Except for an appendix describing indices of community diversity and similarity, this work does not try to be a methods manual, but books describing ecological methods are included in the "Suggested Readings" at the end of each chapter.

Plunge in! Although initially impatient with details, you will, I hope, find your appetite whetted as the details coalesce into patterns. There are more specialized references to move on to, but even these may not answer specific questions that develop from your own observations. If this book whets your curiosity about how nature works, its chief purpose will be satisfied, and, when all is not answered, there is opportunity to look, to test, to discover relationships for yourself. This is the ultimate delight of science generally, and of the ecological approach to the world in particular.

JONATHAN L. RICHARDSON

Acknowledgments

Those who have helped me in this endeavor are legion: my appreciation extends to many that I cannot name below, simply for reasons of space.

Sarah Allan executed many original illustrations for this book with skill, attentiveness to my desires, and patience in my changes of mind. Sara Finnegan, my editor at Williams & Wilkins, suffered through delays with understanding, and was a sure guide in the difficult later stages of preparation. No one suffered more during these labors than my wife, Alice, and I am deeply grateful for her patience and encouragement during these past three years. Daniel Livingstone, my professional mentor and colleague since student days, volunteered to read and criticize the entire manuscript, and did so to its great benefit. I much appreciate his enthusiasm. William Niering turned a keenly critical eye to several chapters, and Alice Richardson, Dorothy Richardson, Mowbray Allan, and Emily Orzack helped with criticism and preparation of the manuscript. I am grateful to Robert Schmouder and David Miner for help with photographs, and to the skillful staff illustrators at Williams & Wilkins.

Cathy and Mary, my children, assisted, too, by asking the questions about nature that adults forget to ask. Students in my ecology classes have kept me asking questions—and have even helped to answer a few. I have used some of their data in this book. My colleagues on the biology staff at Franklin and Marshall College have provided stimulation and ideas, and the college itself provided the sabbatical leave that permitted real work on this book to begin. That sabbatical was spent at the Museum of Northern Arizona, and the staff members there were helpful in every way.

Although he played no direct role in this undertaking I also remember my father, who used to detour his lawnmower around the uninvited Queen Anne's lace on his Cape Cod lawn. Occasionally he intentionally spared even a showy dandelion, with my mother's willing concurrence. Although the lawn was more ragged than our neighbors', it was more interesting. Clearly I inherit the leanings toward ecology I have always had. Probably I even inherit my perverse admiration for weeds, woodchucks, and other varmints that cheerily flourish in habitats monopolized by man.

Contents

PART I

The Dimensions of Ecology

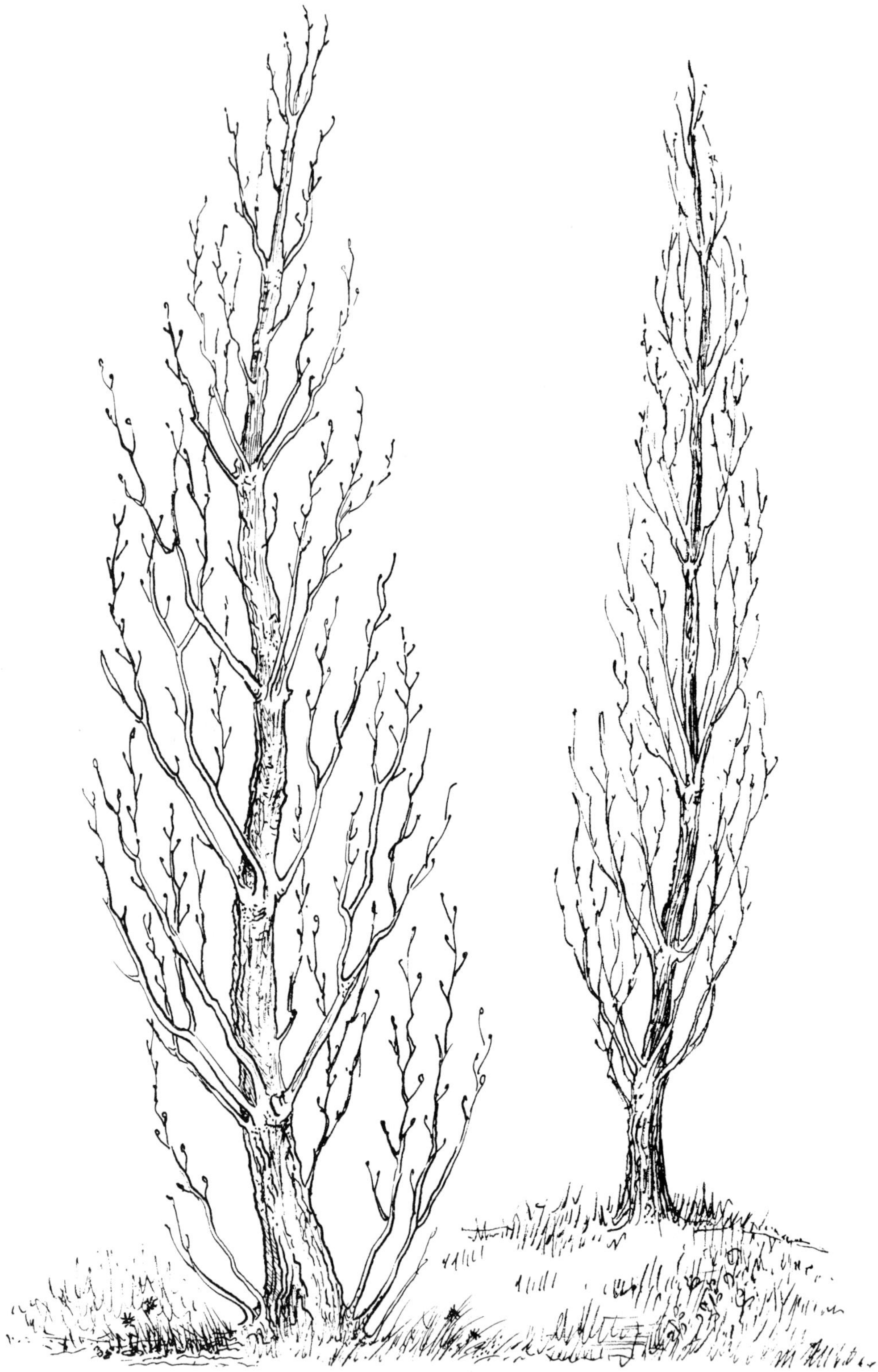

FIG. 1.1. The svelte form of lombardy poplars, shown here in winter aspect, may represent an adaptation to the climate of their Mediterranean and Near Eastern homeland. Summers there are hot and dry; a tree with a spreading crown would risk overheating since much of its foliage would be oriented perpendicular to the noonday sun. In addition there would be much evaporative water loss from the hot, exposed leaves. By contrast, the stem-hugging branches of these poplars support a foliage that is oriented laterally, minimizing solar uptake during the hot hours of the day and forming a dense foliage meshwork that retains moisture within the confines of the tree and thereby slows the rate of evaporation from the leaves. (Drawing by Sarah Allan.)

chapter 1

Ecology: Image and Reality

ECOLOGIST AND ENVIRONMENTALIST

Spirelike lombardy poplars punctuate American landscapes like exclamation points. In towns, farmyards and prairie shelterbelts these slim, familiar trees can be mistaken for no others; we know and appreciate their graceful form even if we do not know their name. But why should the shape of a lombardy poplar be so distinctive? Have you ever wondered (Fig. 1.1)?

Look closely this fall at the gray squirrels of your town or campus. Where I live, their ears whiten on top as winter approaches. We know that some Arctic animals turn completely white in winter and explain this seasonal change most easily by saying it has camouflage value. But is such an explanation reasonable for the barely discernible color change of a gray squirrel?

Why is the immature stage of 17-year cicadas ("locusts") so ridiculously long? These insects take more time to mature than do humans. Equally interesting, why are the yearly emergences of adults so unequal in size? In a "year of the locust," adults are deafening in their abundance; in a lean year one would hardly know the species existed, though beneath the earth the grubs are busy.

Why are deer populations so large in many eastern states? In Pennsylvania in 1972 more than 125,000 were killed by hunters or by automobiles, yet these losses are but a sign of their abundance. No one worries about their becoming extinct as have several other large mammals in the East.

Is Robert Ardrey correct in stating that acquisition and defense of a home turf is an instinctive component of human behavior? If this territorial imperative is as powerful and unquenchable as Ardrey claims, we might as well stop talking about world peace. But how good is Ardrey's argument?

Do some or all of these questions strike you as "unecological"? If so, doubtless you are prey to the popular media's blurred distinction between the terms "ecologist" and "environmentalist." A person may be both at once; but ecology basically is not pollution control, not recycling bottles or paper, is not in fact practicing or preaching *anything*, but like any science it is an area of inquiry and investigation.

Ecologists are spiritual kin to the great naturalists of the past from Aristotle to Darwin. They ask questions about how nature works: how species are adapted to their habitats; what they require for those habitats to be livable; how they satisfy their nutrient and energy needs; how, in so doing, they interact with other species; and finally how, within a species, the individuals are organized and function as viable populations.

Ecologists are interested in man too—because he is a species with his own needs and adaptations, and because he is such a potent factor in the environment of other species and in the whole global life system. But ecology is not basically anthropocentric; and fundamentally it is an area of inquiry, not of activism.

An environmentalist, on the other hand, is anyone who uses his pen, his voice, his investigative or managerial skills in *efforts to improve man's relationship to his environment.* Some environmentalists are trained ecologists (Paul Ehrlich, LaMont Cole, Rachel Carson). Others are scientists trained in other disciplines: Barry Commoner (plant physiology), Garrett Hardin (microbiology), Preston Cloud (geology). Environmentalists also may be aviators (Charles Lindbergh), economists (Kenneth Boulding), landscape architects (Ian McHarg), legal experts (Joseph Sax), businessmen (Laurence Rockefeller), editors (David Brower), ex-politicians (Stewart Udall), housewives, farmers, doctors, novelists. While many environmentalists are professionally trained for their activities, others are largely self-taught. Some concerns of the environmentalist—the effects of smog on human health, the radiation dangers of atomic power plants, the dwindling of mineral resources—lie outside the expertise of any traditionally trained ecologist. Even issues that *are* close to traditional ecological concerns—for example, the effects on wildlife of habitat disruption, or the merits of different approaches to insect control—require such diverse knowledge that no ecologist can be competent in all.

Does this mean that ecologists are not the environmental troubleshooters the public thinks they are? Yes and no. A stream ecologist is likely to know a great deal about stream pollution and its effects, because he or she knows about the habitat requirements of many stream species and often knows first-hand how pollutants alter stream conditions. But a stream ecologist, or any other type of ecologist, may be very uncomfortable in the role of environmental ombudsman expected by the public, and also uncomfortable about the eloquent but sometimes overstated "doomsday extremism" that everyman has come to associate with the word "ecology."

What is embarrassing about the ombudsman role? Mainly this: a trained ecologist who is honest with himself experiences the conservatism characteristic of any specialist who knows that the answers are few and predictions shaky even in his own area of expertise. Moreover, he knows he is not a specialist in all the areas in which the public assumes an ecologist to be expert.

On the other hand, the trained ecologist knows he can talk more sensibly about many environmental issues than can most nonecologists, and he often worries about the inaccurate, exaggerated or unsupported statements of emotional crusaders. What can or should the ecologist say to the public at large? Clearly, in his area of specialization he can speak with considerable authority. Even on issues outside his specialty he can at least speak from an ecological perspective, grounding his remarks in principles basic to this broad discipline. The principles of ecology do provide a coherent framework for viewing many practical and public issues. More, they constitute the basis of a fairly complete worldview. But if an ecologist is honest he will make it very clear that he is speaking from specific knowledge in some cases and from general principles in others. Rachel Carson's *Silent Spring*, which has been called one of the two most important publications in biology in the 20th century (Hardin, 1969) has also been criticized (by impartial scientists as well as pesticide manufacturers) because it fails to discriminate sufficiently between solidly based facts and alarming rumors. The rumors, based on real ecological principles, are one reason for the impact of this book, and many of Carson's fears have proven justified in the years since *Silent Spring* was published. But if the rumors had been couched in more frankly theoretical terms, the book would have retained its impact and escaped much of the criticism that marred Miss Carson's last years.

Beyond the issue of how an ecologist should speak out is the question of whether he or she has a duty to speak out at all. Some ecologists are eloquent spokesmen and writers for environmental causes. Others are awkward on a soapbox and can write only "scientese." Some ecologists *want* to turn their research in practical directions—toward elucidating problems in water pollution or discovering vulnerable points in the life cycles of insect pests. Others genuinely want to continue studying the population fluctuations of butterflies. Paul Ehrlich does both, but not many are as energetic or versatile as he.

Ecologists, then, frequently find themselves in a not wholly comfortable spotlight. As the most knowledgeable bridge between nature and unnatural modern man they are needed as spokesmen and public figures. But not all ecologists are temperamentally or intellectually inclined in this direction, and none at all can satisfy the public's expectation that they be authorities on every environmental issue. Two interesting and contrasting statements regarding this "identity problem" among ecologists are those of Bormann (1972) and of Barbour (1973). Bormann exhorts ecologists to play a major public role in designing a human existence that harmonizes with natural imperatives; Barbour advises the individual ecologist to do what his own talents and priorities urge him to, rather than necessarily to cast his life in a new and perhaps awkward public mold. Students of ecology, whatever their age, might profitably read both statements!

A SCIENCE OF INTERFACES

Ecology is a life science of enormous breadth. Because it centers on living things it is at heart a branch of biology, but one that rubs against branches of many other disciplines. It is possible to envision living systems as a nesting series of distinct levels of organization, differing in size and complexity but susceptible to the same sorts of analysis. These levels of organization include the cell, the organism, the single species population, the multi-species community, and the ecosystem—the last named simply being the interacting unit of a community and its nonliving environment. One can ask equally of cells, ecosystems, and all levels between such questions as these:

1. How are they constructed so as to be viable, self-maintaining systems?
2. By what sorts of metabolic pathways is energy obtained and distributed, and how efficient are transfers of energy from one part of the system to another?
3. How are nutrients obtained, transformed, and reused or released from the system?

Ecologists focus sharply on these questions at the levels of population, community and ecosystem. They also deal with the question of how whole species are adapted to their environments.

Adaptation, in fact, is a principle that infuses biology, and ecologists' preoccupation with it links them to other biological disciplines. There are two distinct uses of the term "adaptation": one dynamic, one static. In the dynamic sense, adaptation is a *process*, the process of adjusting to environmental pressures in ways beneficial to the organism or species. We will be using the word in this sense, as a synonym for responsive change, in later chapters relating ecology to ongoing evolution. But one also talks of organisms *being* adapted rather than *becoming* adapted, of "having adaptations" that suit them to a particular environment. In this static sense the word refers to the consequences of evolutionary response, not its process.

To learn about adaptations in this second sense, *i.e.* to learn how organisms presently are tailored to specific environments, biologists of many persuasions—anatomists, physiologists, biochemists—use a comparative approach. What can an anatomist say about the leg of a horse? among other things, that it is long and slender, that the metacarpal and metatarsal bones are reduced in number, and that the middle toe is unusually enlarged at the expense of others. All these observations imply comparisons: to make such statements the antomist must know something about leg lengths, metacarpals and toes in other organisms. Once the comparisons are made, he knows what pecularities of the horse limb need explanation, and can suggest adaptive answers if he knows something about the habitat and way of life of the animal. Knowing that the horse is an herbivorous animal, ill equipped for defense and native to grassland, the anatomist might see the elongated limb and toe as an adaptation to an open habitat, where cover is scarce and speed is one of the few possible protections against predators (Fig. 1.2). But if long legs mean speed and speed is advantageous, should not all members of the horse family have evolved equally far in this direction? What does one make of the burro, with its stubbier build? The burro is less speedy, yes, but its ancestors probably inhabited more broken terrain than did the ancestors of modern horses. In rough surroundings, sheer speed may be less advantageous than a combination of moderate speed, sure-footedness, and less fracture-prone bones.

The link between anatomy and ecology is clear in this example. Consider now a physiologist, interested in how mammals cope with reduced oxygen supply at high altitudes. If he studies *both* low and high altitude mammals he is bound to learn more about the nature of respiratory adaptation in the high altitude animals than if he were to study only the latter. He will also undoubtedly gain a clearer idea of basic mechanisms of oxygen uptake and transport in all animals. Thus a comparative approach will not only illuminate the question of how mammals adapt to high altitudes but will also add to the physiologist's understanding of basic respiratory biology. Like the anatomist, he benefits from having a point of reference, a "standard model" against which to measure the organisms that particularly interest him. August Krogh is known as one of the 20th century's greatest physiologists, rather than as an ecologist. But his comparative approach and his interest in adaptation lend a rich ecological flavor to books like *The Comparative Physiology of Respiratory Mechanisms*. One can use an adaptive approach equally at the biochemical

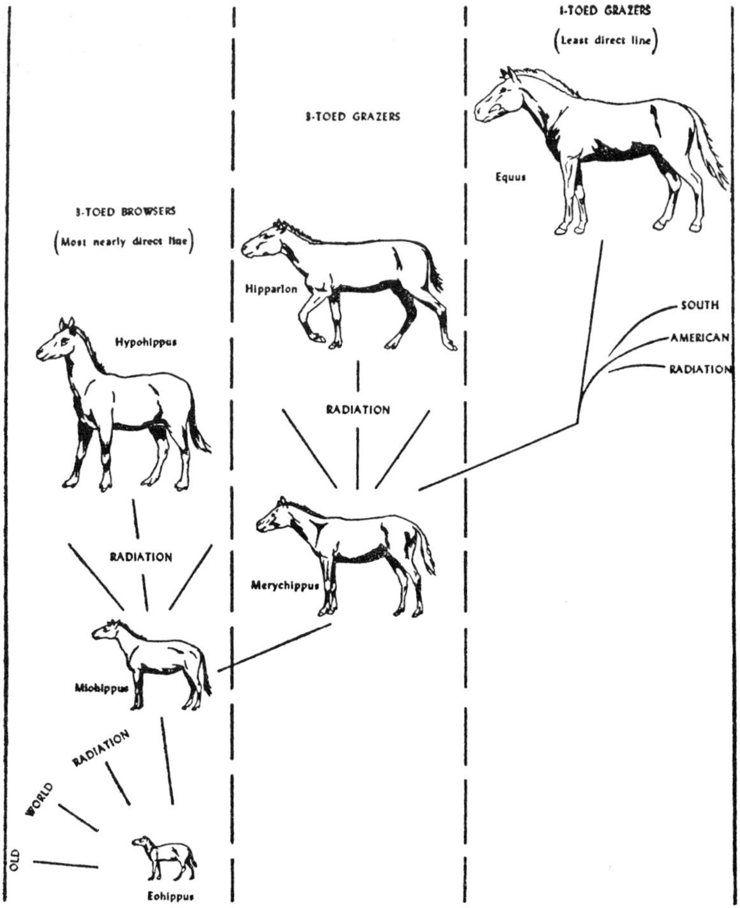

FIG. 1.2. Horse and donkey (genus *Equus*) are one-toed, grazing herbivores descended from smaller, browsing ancestors having less specialized limbs. Being browsers, these ancestral types presumably lived in habitats containing numerous small trees and shrubs. Larger size and longer, more specialized limbs in *Equus* probably are consequences of the horse family adopting an open, grassland habitat during its later evolutionary history. (From G. G. Simpson: *The Meaning of Evolution*, Ref. Ed., New Haven, Conn.: Yale University Press, 1967. Reproduced with permission.)

level, as Ernest Baldwin does in another small classic, *Comparative Biochemistry*. Krogh and Baldwin have many spiritual colleagues and descendents among modern ecologists.

Students of animal behavior, too, overlap into ecology whenever they use a comparative, adaptive approach to their subject. Many morphologists, physiologists, biochemists and behaviorists may never think of calling themselves ecologists, but in fact, whenever they study how a particular species copes with its environment they are engaging in the practice of **autecology** (the ecology of the individual species.)

Another great branch of ecology, **synecology,** dwells less on individual species than on the relationships within groups of species living together (communities). To some degree, autecologists and synecologists comprise separate camps, each often inclined to feel slightly superior to the other. This is unfortunate, because the more one knows about individual species in a community the more one knows about the community. Ideally, the two camps should merge when a community is being exhaustively studied. Philosophically, however, a difference in approach often keeps them somewhat apart: the synecologist believes he can gather much useful information about community relationships without necessarily knowing a great deal about every constituent species, whereas the autoecologist finds the specific knowledge of a particular species more satisfying than a broader but thinner knowledge of groups of species. The two ecological camps are analogous to those sometimes uncomfortable bedfellows, the cell physiologist and the whole organism physiologist. Both tell us many useful things; together they tell us even more.

We have spoken of adaptation to the environment, but what exactly is meant by environment? To ecologists, an organism's environment includes the following components: (*a*) the physical and chemical conditions (*e.g.* temperature or moisture) that permit and govern its existence; (*b*) its resources, *i.e.* the raw materials and energy necessary to sustain it; (*c*) other organisms of the same species with which it interacts; (*d*) organisms of other species with which it interacts.

Consider the ways in which the concept of environment involves ecologists with other disciplines. The need to characterize the physicochemical environment and its resources (the first two components of environment) provides a link between ecology and other natural sciences such as meteorology, geology and geochemistry, and physical oceanography. Well defined subdisciplines, often with jawbreaking hybrid names, exist at these interfaces. Biometeorology and microclimatology, for example, focus on the climates of specific habitats, *e.g.* the humidity within an animal's burrow or the light regime beneath a forest canopy. Biogeochemistry deals with the availability of inorganic nutrients to organisms, as well as their biological processing and eventual fate.

The third component of environment, other organisms of the same species, links ecology (particularly population ecology) to disciplines such as sociology, animal behavior and demography. Studying organisms of other species, the fourth component of environment, again involves ecologists with sociology and animal behavior. This last element in our concept of environment obviously blurs the distinction between autecology and synecology, since an autecologist studying the environmental relationships of a particular species must recognize other species as an important part of his subject's environment.

In the summary phrasing of Andrewartha (1961), what ecologists mean by the word environment is everything that may influence an organism's chance to survive and multiply. That is a large kettle of fish indeed!

THE NATURE OF ECOLOGICAL INQUIRY

Although enormously varied, ecological investigations can be characterized by several general sorts of approach. These approaches often complement each other, and more than one may be used to attack a single problem.

Observation and Intuitive Interpretation

The oldest form of ecological inquiry, practiced by all past and present naturalists, is simply the careful observation of organisms in their environments, the recognition of natural patterns, and the deduction of probable explanations for these patterns. Ecologists do this all the time, and the most lasting benefit of any basic ecology course is that it opens the student's eyes and mind more widely both to mysteries and to explanatory clues in the natural world around him. Every encounter with living things, roadside weeds as well as wilderness, provides potential detective stories.

What does one observe? Perhaps, that sycamore trees characteristically grow near water (Fig. 1.3); or in forests, that gypsy moth infestations get worse for several years, then disappear whether pesticides are used or not; or in a city park in spring, that flocks of migrating robins spread out quite evenly over a patch of lawn, rather than feeding together.

What might one deduce? That sycamores need more water than most other trees; that gypsy moth populations exhaust the available food in a forest, then collapse from starvation; that robins space themselves across a lawn because each finds more worms when it has an exclusive hunting territory.

The observations are all correct; the deductions may or may not be. Although logical, they may not be the only possible explanations, or even the most likely ones. Once an explanation is proposed, then, the obvious next step is to test it. Often this is easier said than done: suitable ways to test a deduction sometimes are difficult to devise, and may take great time and effort. Therefore, in many cases an ecologist may proceed no further with a particular problem, but simply draw speculative attention to it. This in itself is valuable: even if we are never sure of the answer, the act of asking ourselves

FIG. 1.3. White, scabby trunks of sycamore trees define from afar the watercourses of many American streams. This photograph was taken near Lancaster, Pennsylvania.

questions about what we observe is vastly enriching and entertaining. Experience at this basic level of inquiry makes us able to see more in nature and to interpret correctly more of what we see. Joseph Wood Krutch, theater critic-turned-naturalist, put it well:

> There is all the difference in the world between looking at something and living with it. In nature, one never really sees a thing for the first time until one has seen it for the fiftieth. It never means much until it has become part of some general configuration, until it has become not a "view" or a "sight" but an integrated world of which one is a part.
>
> *The Voice of the Desert* (1951)

Data Collection and Correlation

A further step in unravelling an ecological problem is the gathering of data and establishment of correlations. Having observed that sycamores grow near water, we suspect that they have high water requirements. This hypothesis would be strengthened if we had some objective data, say, on the percentage of sycamores that actually are found within 20 feet of open water in a study area; or better yet, on the average moisture content of the soil beneath both sycamores and other species of tree in some locale. We might end up with a convincing correlation: high soil moisture, sycamores present; low soil moisture, no sycamores. The truth of our hypothesis seems self-evident. Yet we cannot be absolutely sure, because correlation does not necessarily mean causation. Perhaps sycamores require a mineral nutrient which is most abundant in low-lying or waterlogged terrain. If that nutrient were added to drier soils, sycamores might flourish there as well. In other words, the correlation we found does not *necessarily* mean that sycamores need unusual quantities of water.

Correlations are often the best evidence one can get in a scientific investigation; this is why statistical analysis is so important in research. But the problem of going from a demonstrated correlation to a statement of causation is very real, regardless of statistics. It is the problem that has bedeviled many questions regarding pollution hazards to human health. Does cigarette smoking cause lung cancer? There is a statistical correlation between the two, but there may never be proof that smoking causes cancer. Perhaps attraction to tobacco and susceptibility to lung cancer are combined in the physiological makeup of some humans. To take a related example, it has been shown that several sorts of cancer show increased incidence in cities with high atmospheric levels of sulfur dioxide (Hickey, 1971) or soot particles (Lave and Seskin, 1970). But these cities share many other environmental features besides sulfur dioxide and soot. Again, we may suspect we have found the cause; we can't be sure.

Experiments—Natural and Artificial

To really attack the problem of causation, controlled experiments are necessary. Let us pursue our sycamore question a little further, because the most likely answer illustrates two things: the danger of accepting statistical correlation as proof of causation, and the power of critical observation when, as in this case, nature has actually performed an experiment for us.

In fact, our correlation is misleading; it is probably untrue that sycamores require more water than most other trees in the eastern United States. Asking questions of our data, or making further field observations, would have provided clues to this. Were there exceptions to the statistical pattern? A few misfits may well be obscured by a high statistical correlation, but they cannot be dismissed: the one or two sycamores perhaps found growing in dry soil, or far from water, were flourishing in the absence of the very thing this species is presumed to require! If we had extended our observations to a town, further doubts might have crept in: sycamores generally do well when planted as shade trees in yards or along streets, whether near water or not. Back in the natural setting, if we had been observant we might have noted a curious thing as we gathered our data: sycamores are very common along streams, but quite *uncommon* around ponds and lakes. If they do indeed require water, it is not just any old kind; evidently it must be running water!

What is the answer? In his delightful and wide-ranging book *Plants, Man and Life*, Edgar Anderson (1967) suggests that the critical requirements of sycamore are those of the seedling, which needs bare soil and relatively large quantities of light, not water. Bare soil is required because the tiny seeds of this tree store only a small amount of food, so the roots, once sprouted, must be able to obtain fresh nutrients from soil almost immediately. Also, of course, bare soil generally means less serious competition for light from seedlings of other plants. The water, temperature and nutrient needs of sycamore are probably not exceptional; in climates suitable for deciduous forests this species can grow wherever the requirement of exposed soil is met: on abandoned farmland, on tracts bulldozed for shopping centers, in yards where the homeowner assures sufficient light and removes competitors of the seedling. Why, in nature, do sycamores grow along streams? The erosional and depositional activities of running water, particularly of intermittent floods, provide the bare soils that are necessary. Aside from seaside beaches and landslides in hilly terrain, stream banks and sand bars are the chief natural sites of exposed soil in all but desert environments.*

In the above example, nature performed a controlled experiment for us and we observed the result: sycamores growing along streams but not around ponds. Moisture is presumably as abundant besides a pond as beside a stream, so the result of the natural experiment effectively disposes of the "water need hypothesis" of sycamore distribution, at least in its simplest form. This experiment, by itself, does not exclude the possibility that sycamores need abundant water *plus* something else that pond margins do not provide.

Laboratory experiments in ecology have the great advantage that one can conduct research under conditions more strictly controlled than those usual in nature. We might call nature's experiment with the sycamores "semi-controlled," because while soil moisture was effectively controlled, variables other than bare soil could have accounted for the differential distribution we observed. Soil texture or nutrient status, for example, might have varied between pond fringe and stream bank, but we could control these in a greenhouse experiment. Using standard soil as well as uniform moisture conditions, we might test Anderson's "bare soil hypothesis" by varying the amount of dead leaf cover or the number of competitors for light in pots sown with seeds of sycamore. Fully controlled experiments could lead to more conclusive answers to the sycamore distribution question, just as they do to many other biological questions.

Thus, not surprisingly, many ecologists work in the laboratory, or supplement field studies with laboratory experiments. Research in the field, indeed, often is considered suspect by nonecologists, to whom the rigidly controlled experiment is the only reliable sort of research. Unfortunately, many kinds of important ecological research are not amenable to laboratory investigation, even if one desired it, and much ecological research must stop at the level of correlation and inference. But ecological studies restricted to the field *can* go beyond this level, often (as in our example) by making use of the experiments nature performs for us.

Ecological research in the laboratory raises real problems also. Chief among these is the question of whether results attained under artificial laboratory conditions have any real meaning in nature. Laboratory experiments usually simplify natural conditions so drastically that one must extrapolate to the real world with caution. An ecologist who works only in the laboratory can be accused of playing games. He

* The foregoing discussion applies primarily to the most common American sycamore, *Platanus occidentalis*, of the eastern half of the country. The sycamores of western canyons certainly grow there partly because of relatively abundant moisture.

may get interesting results, but factors that are shown to be influential in a laboratory experiment may not be critical at all in nature. Ecologists have long realized this, and in recent years have sometimes gone to great trouble and expense in attempts to combine the complexities of nature with the control possible in laboratory settings. The products of such concern are often imposing and expensive: examples include phytotrons for plant growth studies, in which a multitude of climatic factors can be controlled and juggled at will; and laboratory streams, where a natural stream (or portion thereof) is channelled right through a research building where light conditions, current, or substrate size can be easily manipulated (Figs. 1.4 and 1.5).

To the ecologist, the ideal experiment is probably a strictly controlled experiment in the field, preferably using natural controls rather than artificial ones. Natural situations that lend themselves to such experiments are difficult to find, but are worth looking for. Prominent ecologists often owe their prominence to their eye for the *solvable* problem, not simply the *interesting* one, and they may spend seemingly disproportionate quantities of time and travel funds in finding the most favorable situation for a field experiment. The field researcher cannot expect freedom from harassment by man, animal and weather, even when he has found the site he wants. I once spent a summer vainly pursuing a very simple field investigation of the relationship of light intensity to algal distribution in a stream. I had found a small stream running through a long, head-high culvert beneath a highway, and realized that this presented a beautiful light gradient. If I set up, at increasing distances into the culvert, identical surfaces to which algae could attach (glass slides or bricks, for example), I would have a situation in which current, substrate and nutrient status were the same at each point, and only the light varied. I did this and waited, confident of definitive results. They never materialized. It was a summer of violent

FIG. 1.4. The phytotron at Duke University in Durham, North Carolina. The building has a total floor space of 34,000 ft^2 and contains 6 greenhouses and 40 controlled environment chambers. Artificial light is used in the latter, so that intensity and day-length can be manipulated as desired. Temperature and watering regime also are strictly controlled; in these chambers one therefore can grow plants under conditions ranging from tundra to rain forest climates, and can alter these climates in subtle ways.

FIG. 1.5. Greenhouses of the Duke phytotron make use of natural sunlight, but this may be supplemented at certain seasons with artificial illumination to increase day-length or intensity, as may be desired if tropical light conditions are to be simulated. Temperature in these greenhouses can be strictly controlled: water flows constantly across the glass roof, absorbing most of the incoming solar heat and enabling temperature control to be largely a matter of artificial heating or cooling. Phytotron photographs provided by Henry Hellmers.

thunderstorms, and despite numerous anchoring efforts, the slides and bricks were washed away time and again, before sufficient time had elapsed for algae to colonize them. It was a good idea, but the wrong stream—or the wrong summer!

The results of a similar, successful experiment using natural controls are presented in Figure 1.6. In the outflow of a hot spring, Stockner (1968) found a natural environment that was essentially constant throughout the year in temperature, water chemistry and current. Yet he found pronounced seasonal changes in the algal flora, and was thereby able to identify with considerable confidence the influence of light, the only physicochemical factor that changed seasonally.

There are, of course, so-called "experiments" which can be run in the field without controls. Into this category fall tracer studies which, since the advent of radioactive tagging methods and telemetry, have been so useful simply in discovering where nutrients go in ecosystems, what eats what, and what the movements of animals are in their habitats (Fig. 1.7). These are really not experiments at all, but simply sophisticated observations.

Historical Experiments

Some ecological questions are best answered by looking at the past. Others can only be answered in this way. Ecologists alive to this possibility often learn useful things by "coaxing history to conduct experiments" (Deevey, 1969). Paleoecology, the subdiscipline that investigates environments and ecological relationships in the past, probably developed principally because a large number of humans have antiquarian interests; some become historians or archaeologists, some paleontologists, some paleoecologists. These people are working more and more often as teams in attempts to flesh out more completely our understanding of prehistory. I have been part of such a team in the Rift Valley of Kenya, a paleoecologist joining with archeologists and geologists to elucidate prehis-

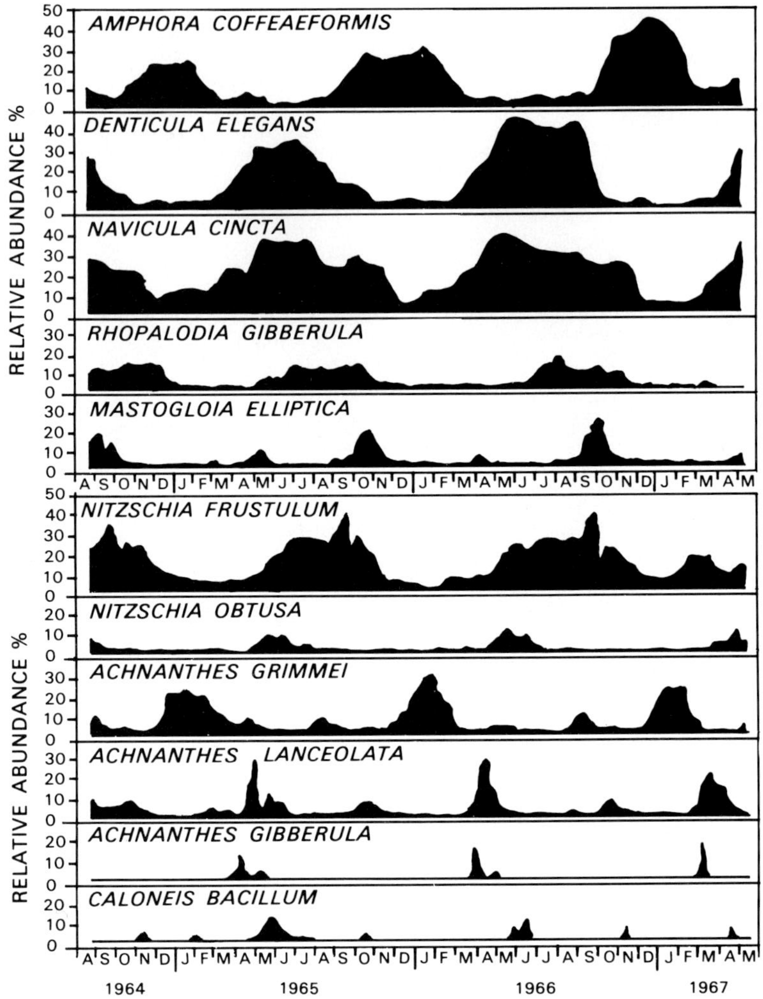

FIG. 1.6. Seasonal pulses of diatom species in the outflow of a thermal spring on Mt. Rainier, Washington. Except for seasonal changes in light, the spring represents a nearly constant environment from the physicochemical standpoint. (From J. G. Stockner: *British Phycological Bulletin 3:* 501–514, 1968. Reproduced with permission.)

toric relationships between human cultural change and environmental change. Like a true antiquarian, I found it deeply rewarding simply to discover what happened in the past.

A second impetus to paleoecology comes from questions that we ask of the *present*. Why is the modern large mammal fauna of Africa so much richer in species than that of other parts of the world? If we wish to answer this present-day question we are led almost immediately to another, specifically historical, question: has this disparity between Africa and the other continents always existed? The answer from paleontologists is no; other continents had comparably rich large mammal faunas, but these were decimated by massive extinctions during the final stages of the last ice age, 10,000 to 30,000 years ago. This information narrows our initial question considerably, and it resolves into two new ones. (*a*) What environmental pressures brought about the species extinctions on other continents? (*b*) Were the environmental pressures less severe in Africa, and if so, how? Paleoecologists presently are working on both these questions, focusing more and more on the role of prehistoric man as an extinction agent (Martin and Wright, 1967; Martin, 1973). No matter what the cause of extinction, however, ecologists working in today's world obviously must take notice of what paleoecologists tell them: that the temperate communities of today have undergone great simplification in the not too distant past, and now contain far fewer species than before. These communities may still be in a condition of recovery and readjustment, despite their longevity and stability to our contemporary eyes. Ecologists today, comparing different community types in search of common principles of organization and operation, sometimes find surprising contrasts, particularly between tropical and temperate regions. Differing histories may help to provide the explanation.

Interest in the *future* also prompts studies of the past. How will North America be affected by the next ice age? If paleoclimatologists are correct, we are living in an interglacial epoch, and today's environments thus are a poor basis for speculation about ice age conditions. Useful answers, however, may be obtained from paleoecologists who have been studying the fossil pollen preserved in bog and lake sediments. Often identifiable to species, the ancient pollen has produced detailed records from many parts of North America of the vegetational changes, and by implication, the climatic changes that occurred during the last glacial advance. If the present is often a key to the past, so too the past may be a key to the future.

Not just prehistory, but much more recent history is of value to ecologists. Controversy presently swirls over the biological consequences of building a sea-level canal across Panama. What sorts of ecological upsets may

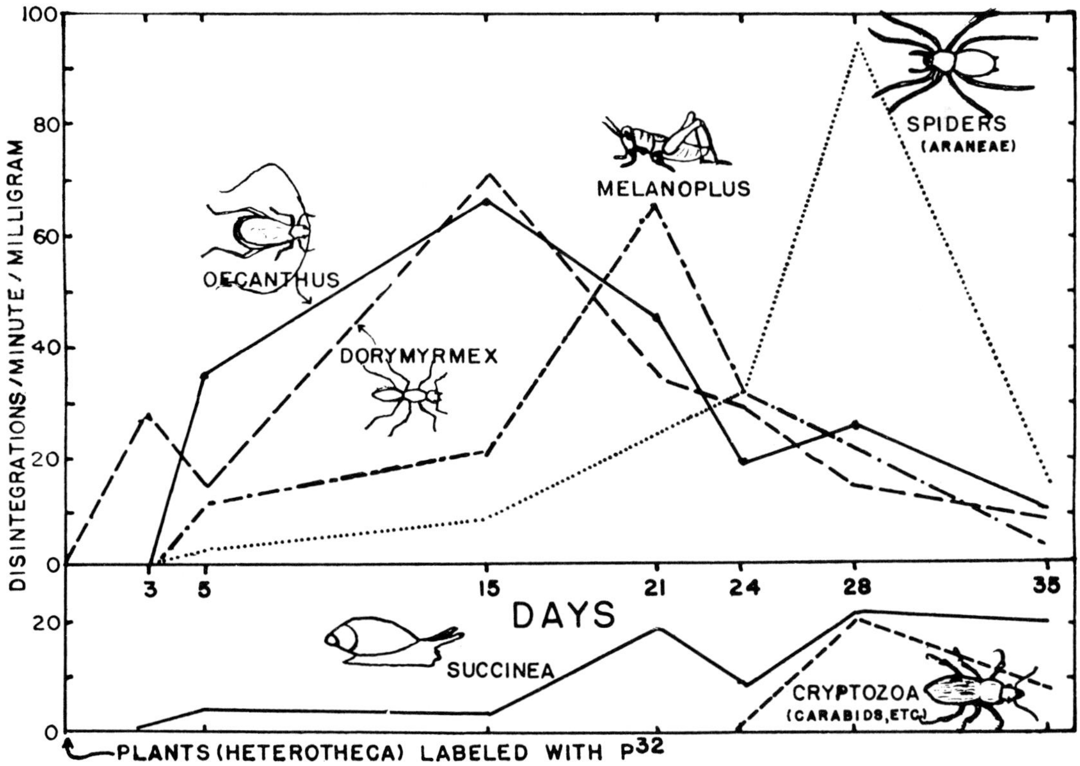

FIG. 1.7. Identification of food habits and food chain relationships of organisms in an abandoned field community. Radioactive phosphorus was applied to one of the dominant plants in the community, and various animal species were analyzed during subsequent days for the presence of this isotope. Its rapid buildup in a small cricket (*Oecanthus*) and a species of ant (*Dormyrmex*) indicate that these organisms were feeding directly on the labeled plants; so, probably, was a species of grasshopper (*Melanoplus*), although it accumulated the isotope more slowly. Spiders apparently were the next step in the food chain, the isotope building to high levels in them later than in the herbivores. The patterns of isotope buildup in snails and cryptozoa suggest that the former may feed on both living and dead plant leaves and the latter on dead leaves and/or dead herbivores. (From E. P. Odum: *Japanese Journal of Ecology 12:* 108–118, 1962.)

result from permitting exchanges between the quite dissimilar biological communities of the Caribbean and the eastern Pacific? Comparative studies of the biota on either side of the isthmus lead to general predictions (Briggs, 1969), but surely these predictions can be refined by investigating what has *actually* happened in similar cases in the past. Both Briggs (1969), and Aron and Smith (1971), point out that the Suez and Great Lakes canals constitute historical experiments for which we have over 100 years' worth of results. May we expect some Panamanian version of the story of the lamprey and alewife in the Great Lakes, where valuable native faunas were displaced by these less desirable immigrants? A careful comparison of the two situations, with the results in hand from one of them, may help answer the question *before* the new canal is built.

Models and Mathematics in Ecology

A final form of ecological investigation is the building of models. Levins (1968) defines a biological model as "a reconstruction of nature for the purpose of study," and considers models to be the basic units of theoretical investigation.

Model-building is one aspect of mathematical ecology; it stands apart from the mere mathematical treatment of data, or even from the induction of mathematical generalizations from ecological data, in that it generates theories (or even new "data") primarily by mathematical reasoning or manipulation. Pielou (1969) writes as follows in the introduction to her useful book on mathematical ecology:

> The fact that ecology is essentially a mathematical subject is becoming ever more widely accepted. Ecologists everywhere are attempting to formulate and solve their problems by mathematical reasoning, using whatever mathematical knowledge they have acquired, usually in undergraduate courses and private study.

As a mathematical ecologist herself, Pielou may be suspected of overstating the case, but perusal of current articles in ecological journals will convince us that the overstatement is not extravagant. This book operates on the assumption that the reader's mathematical training may be rather meager; the mathematical side of ecology is thereby slighted, but it cannot be completely ignored.

Let us pursue the distinction noted above between the use of mathematics in ecology and the building of ecological models. Ecologists frequently perform mathematical or statistical analyses of their data, or develop equations to express empirical relationships. For example, a well known equation of population growth, the logistic equation, describes population growth as an S curve: numbers increase slowly at first, then much more rapidly, then level off as growth ceases (see Fig. 14.6*A*). This equation is simply a mathematical description, a generalization derived from experiments and observations. It is useful because data from the study of any population can be compared to it, and such comparisons may yield insight. If the result of an ecological study can be expressed in mathematical form, there is a greater chance to arrive at generalizations that cut across species or community differences and allow fruitful comparisons. To take another example, the comparison of communities in terms of their species diversity has become a point of interest among ecologists. How do biological communities differ in terms of the number of species they contain, and in the way these species share the available resources? Are there a few superabundant species and many rare ones, or is there a more equal distribution of numbers? Several mathematical equations have been developed to answer these questions (see Appendix), and by using these equations we can compare species diversities among communities as disparate as forests and oceanic plankton. These equations are useful analytical tools, but they are not models.

What then *are* mathematical models? In ecology there are two basic kinds, the pure mathematical model and the simulation model. The first type represents ecological theory developed by mathematical logic and argument. Models of this type are relatively simple and spare; mathematically minded ecologists have been creating them for some time. The second type represents an elaborate attempt, usually involving a computer, to mimic the workings of a natural system.

What is the nature and purpose of pure mathematical models? Models of this type exclude many of the complicating factors of the real world. The mathematics may be difficult, but the essentials of such a model often are so uncomplicated as to be expressed clearly in a simple graph. Obviously one can hardly expect reality to match in close detail the predictions of such a model. But this is not the main aim of the model. As Pielou (1969) points out, simple mathematical models are poor mimics of the world because they *do not try* to imitate it in detail; rather, they provide theoretical explanations, based on a few fundamental assumptions, of how ecological relationships should work. Such models should be tested in the real world, of course, with real ecological data. If the data depart radically from a model's predictions, the theory it expresses is suspect; if the data generally bear the model out, it provides a surge of insight into real ecological processes.

Some examples will illustrate the utility of such models—sometimes even when they are "failures." Among the earliest and most famous of pure mathematical models are those of Lotka and Volterra, who in the 1920s independently developed mathematical predictions of the effect of competition or predation on the numbers of two interacting populations. These models are logical but theoretical modifications of the S-shaped logistic growth pattern, and they predict quite different patterns from the classic single population pattern. In the case of the competition model, what is predicted is the elimination of one competing population if the two species involved are very similar in their requirements. This model, in fact, is the basis of the important ecological principle of *competitive exclusion*, which we will encounter again in Chapter 3. What is interesting is that competitive exclusion was predicted by a model years before it was actually demonstrated in the famous laboratory experiments of Gause (page 67). Lotka and Volterra's predator-prey model is also successful, though unrealistically simple; it predicts interlocking fluctuations of predator and prey populations (see Fig. 14.16) similar to those that often occur in nature. These models therefore are helpful: they seemingly cut through to basic processes operating in nature and sharpen our understanding of what we see.

In Chapter 7 we will discuss a pure mathematical model that "failed," Robert MacArthur's famous "broken stick model" which attempts to explain why species in a community differ in abundance. This model makes an assumption—that differing species abundances are primarily a matter of chance, operating within the constraints of a finite pool of resources—and from this premise develops a prediction of what the actual pattern of relative abundances should be among species in a community. In this case, unfortunately, the prediction is only rarely borne out in the real world, so the model and its working hypothesis now have been largely discarded. But the broken stick model served a purpose: it drew attention to a whole series of little explored questions about how communities are organized and how they have evolved, and model builders have been fruitfully working on these questions ever since. MacArthur, in fact, has been the prime mover behind the major current thrust of ecology in an evolutionary direction. Model builders at this interface of ecology frequently think in terms of optimizing "strategies" of evolution: given that natural selection is a perfecting or optimizing instrument, they ask what possible ways of satisfying basic needs are open to a species having certain potentialities. For example, if one assumes that natural selection will act on an insectivorous bird species to increase its efficiency as an insectivore (rather than, say, lead it to become a nectar feeder), one can then ask what combination of food items and feeding pattern would be most efficient for the species. That is, what would be the optimum feeding strategy for such a species, so that it could satisfy its food needs with the least energy expenditure or least danger? Model builders like Thomas Schoener (1971) have come up with a number of theoretical optimal feeding strategies, depending on the initial constraining assumptions that are made. Turning to a real community, such a model builder can then look for resemblances between the predicted feeding strategies and the feeding patterns of real species. If a species provides a convincing analogue for the predicted strategy of the model, our ecological understanding of that species, its position in the community and the selective pressures that have molded it all are greatly enlarged.

The second basic sort of ecological model, the simulation model, does not challenge the real world as does the pure mathematical model. Simulation models are built as unashamed imitations of real ecological situations, and the more real data available during the building of the model, the better the model will be. Because real ecological systems (even simple ones) are highly complex, simulation models also must be highly complex if they are to be faithful imitations. Hence, almost of necessity, ecological simulation is a child of the computer age. Each of the many assumptions on which a computer simulation is based must be relatively accurate if the model is to be at all accurate in mimicking natural processes. Thus, in developing the model, there is need for vast amounts of reliable data for all the parameters about which assumptions must be made. In a model of forest processes, for example, we would need actual climatic data, data on the climatic tolerances and optima of the species in the ecosystem, data on the rates of consumption and transfer of nutrients by each species, numerical data on relationships of the species to one another, etc. As Watt (1966) has pointed out, the shortage of information on which such models can be based has provided a powerful impetus to the development of automatic monitoring methods which systematically and easily gather large quantities of ecological data. Ecologists can thank the simulators for much of the sophisticated instrumentation that has become available in recent years.

Ecological simulation is part of the larger realm of systems analysis, an approach now common to fields as disparate as engineering, industrial operations research and global political analysis. Ecologists involved in computer simulation often find themselves talking as much to physicists, statisticians, economists and engineers as to other ecologists. The goal of simulation modeling may be simply to understand a natural system, or it may be more practical: to understand the system *and* to manage it intelligently. Once the model is constructed, it can be used to conduct experiments in which a change is introduced at some point in the model and the consequences are observed. Sometimes there are entirely unexpected ramifications of such a change; at other times the results are predictable, but their magnitude is enlightening. Simulation models thus are a valuable tool in the development of sane environmental management policies, policies which can be continually updated if there is an ongoing program of monitoring environmental parameters in the real world. As the real data come in, they can be fed into the model to make its predictions consonant with current conditions.

CONCLUSION

The usefulness of ecological simulation for analyzing and intelligently managing man's impact on the environment brings us back to the role of the ecologist in society. The complexities of ecological systems are dauntingly great and the data we really need to understand them are still clearly inadequate. But to make order out of these complexities, ecologists today are searching hard for generalizations. In their present predicament, ecologists should find special pertinence in Mark Twain's sardonic comment:

> There is something fascinating about science. One gets such a wholesale return of conjecture out of such a trifling investment of facts.

Generalizations based on insufficient data or simple correlations may be wrong, or at least misleading. So may computer results from a simulation model based on faulty or naive assumptions. The voices of ecologists need to be heard, but if they are wrong they can do more than intellectual damage. Ecologists have made rapid recent progress in understanding how the natural world really functions; and building on these achievements, they are stepping out with increasing boldness. But they must be careful that their footing is secure.

SUGGESTED READING

Curious naturalists

DARWIN, C. The Voyage of the Beagle. New York, Bantam paperback edition, 1958. The trip around the world that epitomized the links between ecology, natural history and evolution.

KRUTCH, J. W. The Best Nature Writing of Joseph Wood Krutch. New York, Pocket Books, 1971. Selections from the work of one of the best contemporary American naturalists.

LEOPOLD, A. A Sand County Almanac. New York, Oxford University Press paperback, 1968. (Also issued as a paperback by Ballantine Books.) A classic of natural history by one of the pioneers of game management in America.

TEALE, E. W., ed. The Insect World of J. Henri Fabre. New York, Fawcett Premier paperback edition, 1956. A prominent American naturalist has chosen selections from the work of a renowned French student of the lives of insects.

WALLACE, A. R. The Malay Archipelago. New York, Dover paperback edition, 1962. An engrossing account of southeast Asian adventures in the mid-nineteenth century by the famous naturalist and biogeographer who, with Darwin, authored the Theory of Natural Selection.

Ecology and other disciplines: The link through adaptation

BALDWIN, E. An Introduction to Comparative Biochemistry. New York, Cambridge University Press paperback, 1970. A small gem, easily understood.

HOCHACHKA, P., AND G. SOMERO. Strategies of Biochemical Adaptation. Philadelphia, W. B. Saunders & Co., 1973. Requires some knowledge of biochemistry; an exciting approach to this subject.

KROGH, A. The Comparative Physiology of Respiratory Mechanisms. New York, Dover paperback edition, 1968. One of two Dover reprints of classic short books by this Nobel prize-winner. The other book is Osmotic Regulation in Aquatic Animals.

KLOPFER, P. Behavioral Aspects of Ecology. Englewood Cliffs, N. J., Prentice-Hall, Inc., 1973. The title is self-explanatory.

PROSSER, C. L., ed. Comparative Animal Physiology. Philadelphia, W. B. Saunders Co., 1973. A voluminous and most useful reference work.

Mathematical ecology and modeling

LOTKA, A. J. Elements of Mathematical Biology. New York, Dover paperback edition, 1956. A wide-ranging book, first published in 1925, that remains full of insights.

MACARTHUR, R. H. Geographical Ecology. New York, John Wiley and Sons, Inc., 1972. The final contribution of one of America's most influential theoretical and mathematical ecologists.

MEADOWS, D. H. *et al.* The Limits to Growth. New York, Universe paperback edition, 1972. The controversial computer study that predicts the global consequences of continued technological and population growth in the face of increasing global pollution and limited food and mineral resources.

PIELOU, E. C. An Introduction to Mathematical Ecology. New York, Wiley Interscience, 1969. A useful if somewhat difficult text and reference source.

SCHOENER, T. W., AND S. A. BOORMAN. Mathematical ecology and its place among the sciences. *Science* 178: 389–394 (Oct. 27, 1972). A review of MacArthur's book that is, in addition, a lucid overview of mathematics and modeling in ecology.

SMITH, J. M. Models in Ecology. Cambridge, England, Cambridge University Press, 1974. A relatively brief, highly mathematical treatment of a number of important ecological models.

WATT, K. E. F., ed. Systems Analysis in Ecology. New York, Academic Press, 1966. A collection of interesting papers describing simulation models in ecology, with a useful introduction by Watt.

WATT, K. E. F. Ecology and Resource Management. New York, McGraw-Hill Co., 1968. A text illustrating the uses of simulation models in environmental and wildlife management.

chapter 2

The Living Planet

We do not realize what we have on Earth until we leave it.

James Lovell, member of the first team of astronauts to circle the moon

INTRODUCTION

The more we know of our planetary neighbors, the clearer it becomes that Earth is the one place in the solar system where an ecologist really could ply his trade. Earth is the only living planet, or so it seems.

We were not always so sure of this. Already half forgotten is the month-long quarantine of the first moonwalkers to return to Earth, and the elaborately sterile precautions taken with everything they brought back, lest some pestilence be unleashed upon our globe. Better remembered are the excited speculations of much longer ago regarding the "canals" on Mars. Are they the work of a Martian civilization, present or past? And are the cyclic changes of the planet's appearance the seasonal greening and browning of vegetation? We no longer suspect either, though Mars remains the most likely of the sun's other planets to harbor some form of life, or to have harbored it in the past. Astronomers tell us that life undoubtedly *does* exist on planets in other solar systems, but our telescopes cannot even detect these distant planets, let alone discover the life they might support. Earth seems indeed a lonely outpost —in the words of Archibald Macleish, "small and blue and beautiful in the eternal silence where it floats."

But to call Earth the living planet betrays the bias lent to our perceptions by our immediate surroundings and by our own living nature. For Earth's realm of life is confined to a thin concentric shell, like one of the layers of an onion, bounded within and without by nonliving earthly realms vastly more voluminous. And the life that teems around us, in that thin layer known as the biosphere,* is but a minute fraction of Earth's total mass.

Nevertheless, Earth does possess a biosphere, and it is a feature of some uniqueness. Large questions regarding the biosphere will be the focus of this chapter. What makes our planet, unlike others that we know, a suitable habitat for life? Why, even on Earth, is life restricted to such a narrow zone? And within that zone, the biosphere, what physical and chemical constraints give our living realm its major patterns?

THE OUTER FRINGE OF THE BIOSPHERE

Some years ago an interest in one of Earth's anomalies, glacial lakes in the tropics, led me to the Ruwenzori Mountains, an ice-capped African range straddling the equator. As we ascended the mountains we passed through marvelously exotic and varied scenery: tree and grass savanna at the base, then a belt of rain forest, then zones of bamboo and giant heath, festooned with the weeping lichen known as Old Man's Beard; then, higher yet, the alpine zone, characterized by giant members of plant groups that are usually small and herbaceous. In the alpine zone of the Ruwenzori, everlasting flowers grew chest-high as woody shrubs; other members of the daisy family, the groundsels, grew to the size of small trees, and we encoun-

*The original use of this term, by the Austrian geologist Eduard Suess and the Russian geochemist V. I. Vernadsky, was in the sense we use it here—to describe Earth's narrow, life-supporting layer of air, water and soil. In another frequent usage "biosphere" is used as a synonym for Earth's total cargo of living material.

tered 4-m representatives of the genus *Lobelia*, best known to North Americans as the fragile cardinal flower. To someone from the temperate zone, this surely was a bizarre landscape.

The last lake we visited lay beyond the Scott-Elliott Pass, 4400 m (about 14,500 feet) above sea level. Struggling upward through dwindling vegetation I finally gained the pass and collapsed thankfully to rest. Looking about, suddenly I felt strangely at home, as if I had been here before. The realization was abrupt and forceful: once one gets high enough, all mountains begin to look alike. To me—though possibly not to a geologist—the rocks of the pass could have been those of ridges on Mt. Katahdin or Pikes Peak in the United States. Certainly high mountain landscapes differ in their shapes, sizes and snowiness, but when the distinctive plant forms of lower slopes are left below there is a sameness to the immediate environment that makes harsh mountain heights seem familiar and almost friendly.

That pass in the Ruwenzori lay at the edge of the aeolian zone, so called because wind is what makes life possible at great heights on Earth. L. W. Swan of San Francisco State College, who first recognized the biological uniqueness of the aeolian zone, has investigated it in mountains as widely separated as the Himalayas and the Mexican volcanoes. Some clouds contain an "aeroplankton" of spores and microorganisms, and birds such as the bar-headed goose pass above the Himalayan peaks when migrating, but we may consider the aeolian zone to be the outer portion of the biosphere that is inhabited by active, metabolizing communities of organisms. This zone begins at different elevations on different mountains and exists at low altitudes in the Arctic and Antarctic, but whereever it occurs the aeolian zone can be defined as that terrestrial realm of life lying beyond the limits of higher plants. Algae—including those responsible for the occasional red coloration of snow banks—are sometimes found growing in the aeolian zone, as are lichens. But unlike the **autotrophic communities** (plant-based and self-nourished) that we know so well, the communities of the aeolian zone are primarily **heterotrophic** (other-nourished); their existence depends on organic debris, blown up from below and trapped at high elevations by snow, pools or projecting rocks. Macroscopic plant fragments, seeds, and insects comprise part of this debris; embedded in the ice field capping the Ruwenzori I found a large hawk moth, its final resting place far from its normal haunts. But there is considerable microscopic debris as well—pollen, spores and the albuminoid residues of sea foam, the latter derived from dead oceanic plankton and often carried great distances before being deposited on the heights (Swan, 1963). These imported foodstuffs form the base of simple food chains, consisting of fungi and various arthropods that feed on the debris itself, on the fungi, or on each other (Fig. 2.1). Salticid spiders, the apex of one of these aeolian food chains, have been found at 6700 m (22,000 feet) on Mt. Everest, and Swan speculates that debris-based communities, or some elements of them, may occur as high as one can find snow-free areas. On Everest, this means nearly as high as the 8850-m summit.

At least three community types exist in the aeolian zone: those of snow, those of pools and glacial streams, and those of dry, rocky slopes. Where glaciers extend to lower elevations the aeolian zone may also be extended downward, interfingering with alpine communities and supporting debris-based animal communities on the glacial surfaces. It is mystifying how poikilothermic (cold-blooded) insects and arachnids can be active on snow, but Swan (1961) points out that as long as their dark bodies are raised above the surface, they can be warmed to surprisingly high temperatures by the intense, high altitude radiation; on soft snow, these creatures actually risk entombing themselves in pockets melted by the warmth of their bodies. When the sun leaves the snowfields these arthropods quickly cool to a state of torpor in which they are able to endure subfreezing temperatures.

Even at very high altitudes the sun raises air temperatures to levels that make men shed their parkas, particularly in snow-free, sheltered areas. At least on tropical mountains, the upper limit of higher plants (the boundary between the alpine and aeolian zones) may be set not by temperature but by the availability of water. In temperate and polar regions the length of the growing season may help set this boundary as well. Water is unavailable for use by plants when in either the solid or the vapor state, and under the intense radiation of high altitudes it often passes directly from solid to gas (sublimes) omitting the intermediate liquid

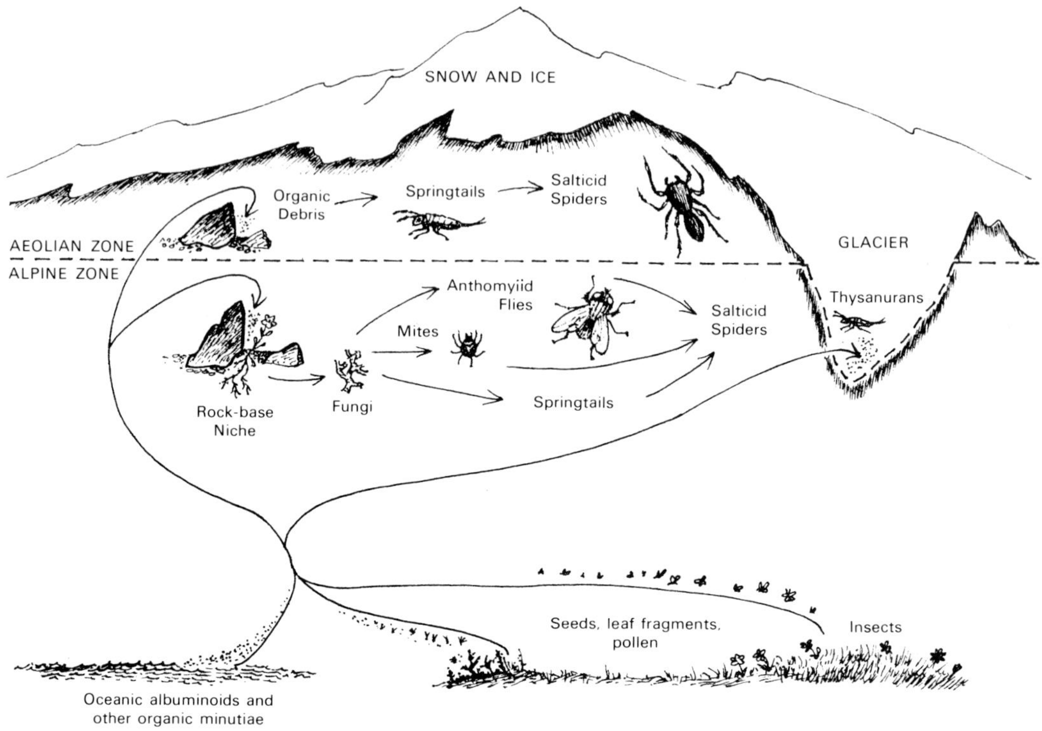

FIG. 2.1 Food chains of the aeolian zone and upper alpine zone in the Himalayas. (Redrawn from L. W. Swan: *Scientific American*, October 1961, with permission from Scientific American, Inc., All rights reserved.)

state. For plants, these conditions represent drought in the midst of plenty. Projecting rocks are efficient heat absorbers, however, and along their edges snow as well as frozen soil may be melted, creating local supplies of usable water. The highest flowering plants in the Himalayas are conspicuously restricted to such "rock base niches," where in addition to available water there is a greater opportunity for wind-blown nutrients to accumulate (Swan, 1961). In addition to the water problem at high altitudes, Hutchinson (1970) suggests that decreasing atmospheric CO_2 (needed in photosynthesis) could also set an upward limit to terrestrial plants. Thus, the existence of submerged algae within the aeolian zone, above the limits of terrestrial plants, might be permitted by the high solubility of CO_2 in water as much as by the presence of the water itself.

The arthropods, which characterize the animal life of these heights, probably extend even beyond the algae in elevation. What limits them at this outer fringe of the biosphere? Lack of oxygen? Low temperatures? Lack of water? If we were speaking about homoiothermic (warm-blooded) animals, which have much higher metabolic requirements, low oxygen might certainly be a factor, though adaptation and acclimation to the reduced oxygen of high altitudes is well known (*cf.* Hock, 1970). It is noteworthy in this regard that Swan found no birds or mammals living permanently above 5800 m (19,000 feet) in the Himalayas, although feats such as the flight of the bar-headed geese argue against an altitudinal limit set by oxygen. For poikilothermic arthropods, however, biological tolerance limits may not be exceeded even on the highest mountains; perhaps liquid water simply becomes too scarce, or perhaps aeolian arthropods are limited only by the absence of higher land surfaces. The salient ecological fact, however, is that the communities of the aeolian zone are partly or wholly parasitic; they depend on the effluvia of communities lower in the biosphere. Gravity-defying contributions of nutrients to the aeolian zone must become progressively more meager at increasing elevations, and the life that can be supported must become progressively sparser. Hence, eventually, the biosphere's limit is met.

THE INNER FRINGE

Beneath the surface of the sea the biosphere stretches some 10 km into the deepest ocean trenches, farther below sea level than the tops of the highest mountains are above it. The floor of the deep ocean is known almost solely through photographs from cameras lowered into the depths, and through the samples and soundings of men working from the surface. The oceanic abyss is as fascinatingly alien to us as the surface of the moon—to which it bears a startling resemblance (Fig. 2.2). Several recent books describe the unfolding mysteries of the ocean floor; perhaps the grandest in both picture and word is Heezen and Hollister's *The Face of the Deep* (1971), which proves that the reality of the deep sea is more enthralling than even Jules Verne imagined. In *The Sea Around Us*, Rachel Carson captures the essence of this little known realm:

> Once beyond the edge of the shelf, as we visualize the steeper declivities of the continental slope, we begin to feel the mystery and alien quality of the deep sea—the gathering darkness, the growing pressure, the starkness of the seascape in which all plant life has been left behind.

For the deep ocean floor, too, is a realm without plants, symmetrical in this respect with the biosphere's outer fringe. Light is absorbed rapidly by water, and organisms that live by photosynthesis are restricted to an upper layer of the ocean little more than 100 m deep. Here seaweeds and microscopic phytoplankton form the basis of a thin autotrophic zone in which the majority of oceanic life, animal as well as plant, is concentrated. As in the sunlit heights of the aeolian zone, life in the dark ocean depths depends on organic debris, now raining down from above rather than blowing up from below.

Both fringes of the biosphere thus support heterotrophic communities, dependent for their existence on leftovers from the plant-dominated mid-regions of the biosphere.

There are sharp contrasts as well as similarities between these fringes. One of the most

FIG. 2.2. Not the moon, but the floor of the abyss. The "meteorite craters" and other features are in reality the burrows, trails, and fecal mounds of abyssal invertebrates. (From B. Heezen and C. Hollister: *The Face of the Deep*. New York: Oxford University Press, 1971. These and other photographs of the abyss were provided by C. D. Hollister.)

striking contrasts is in extent. Only a small fraction of the biosphere encompasses the aeolian zone, but over half the earth's solid surface is deep ocean floor lying beneath at least 3 km of water. The exact boundary of the abyssal zone is somewhat arbitrary; Heezen and Hollister define it in terms of temperature (the 4° C isotherm), which is equivalent to depths of 1 to 3 km in the different oceans of the world.

The most important contrasts between the aeolian and abyssal zones relate to environmental constancy. The aeolian zone is characterized by fluctuation: light and dark, freeze and thaw, wind and calm, wetness and drought. The floor of the abyss is uniquely stable: eternally wet, dark and cold, insulated from diurnal and seasonal change by its thick, watery mantle. Some parts of the ocean floor are swept by currents, others are still; but relative constancy rather than fluctuation prevails in either case. Even during the glacial ages of the Pleistocene, when shallower regions of the ocean floor were exposed by a 100-m drop in sea level and when vast environmental changes affected the land, the abyss remained nearly inviolate.

The communities of the abyssal floor are unusual in their structure; they consist of many closely related species belonging to only a few major animal groups. Sanders and Hessler (1969) believe that the high species diversity of abyssal faunas, which contrasts sharply with the simplicity of communities in the aeolian zone, is related to the stability and relatively great age of the deep sea environment. According to recent theories of sea floor spreading, much of the abyssal floor is less than 200 million years old—far younger than heretofore believed—and the Atlantic Ocean probably originated no earlier than the Mesozoic age of dinosaurs. But even this "youthfulness" is venerable compared to the high, young mountains of the aeolian zone; if the environmental stability of even part of the ocean floor extends back to the Mesozoic, it represents an impressively long period of relative constancy.

Soft sediment mantles most of the ocean floor, and the major features in abyssal photos are the trails, feeding furrows and feces left by animals that browse on bacterial slime at the interface between sediment and water. The holes and mounds of burrowers are also prominent in photographs; only in regions where the bottom waters are stagnant and anaerobic, as in the Black Sea basin, are these "biogeomorphic features" uncommon. However, most ocean bottoms are well oxygenated by a slow, deep ocean circulation.

We all know of the bizarre fishes, many with oversized jaws and their own "miner's lights," that dwell at mid-depths below the sunlit surface layer. The animals of the abyssal floor are less familiar, though often no less bizarre (Fig. 2.3). Aside from bacteria, the pre-eminent group of abyssal organisms is the Echinodermata. Within this phylum the Holothuroidea (sea cucumbers) and Ophiuroidea (brittle starfish) are particularly abundant. Sea cucumbers are the giants of the abyssal invertebrate fauna, some approaching one-half meter in length. Many of them are more mobile than those of shallow water; they use both walking podia and a creeping sole (like that of a snail) to move across the bottom like vacuum cleaners. Other abyssal sea cucumbers are burrowers. Brittle stars, using their arms as sidewinder rattlesnakes use their bodies, loop across the bottom in seach of a variety of small creatures, and perhaps feed on the bacterial slime as well. Life is found over the whole vertical extent of the abyssal floor, amounting to 8 or 9 km. There is no distinctive trench fauna restricted to the deep gashes in the ocean floor, but sea cucumbers become even more dominant as one descends to the greatest depths (Fig. 2.4).

Particularly beneath rich surface waters, life on the deep sea floor is more abundant than in the aeolian zone, but the supply of organic debris (detritus) that reaches the ocean floor probably diminishes with increasing depths. The deeper the water, the greater the fraction of detritus that is intercepted and consumed by suspended decomposer organisms before it reaches the bottom. This presumably is the cause of the logarithmic decrease in abundance of bottom organisms that usually occurs with increasing water depths (Rowe, 1971; Smith and Teal, 1973). This progressively impoverished food supply perhaps also is responsible for the decline in diversity of sea floor animals that occurs with increasing depths (Rex, 1973). Certainly the abyssal community utilizes virtually all of the rain of food that reaches the bottom; almost no organic matter accumulates in deep sea sediments.

In addition to nutrient problems, the enormous water pressures of the abyss may consti-

FIG. 2.3. Long-legged sea cucumbers tiptoe between the arms of brittle stars on the ocean floor off California. At top is a deep sea skate. The depth is 1100 m, and the density of organisms is greater than that normally found deeper in the abyss. (From B. Heezen and C. Hollister: *The Face of the Deep*. New York: Oxford University Press, 1971. U.S. Navy photo by Eric Barham.)

tute a physical stringency. Clearly these pressures do not preclude life on the ocean floor, and many species actually occur over a considerable range of depths. *Individuals* of most species, however, live within a narrow depth range throughout their lives, and the pressure within their bodies balances that without. It is changes in this balance that pose a potential problem, and such changes are encountered only by individuals migrating vertically over a range of depths. One of the few organisms that makes such migrations is the sperm whale. This air-breathing mammal plunges from the surface to depths of more than a kilometer to plow the soft abyssal sediments for food, using its curiously narrow lower jaw as a trenching tool. Sperm whales possess many morphological specializations that help them resist the tremendous pressure changes encountered during their dives: muscular plugs that seal off the blow hole; a stiff, thickened sclera forming the outer covering of the eye; a bony rather than membranous eardrum; and cartilaginous reinforcement of all major air passages.

Most organisms of the deep ocean floor live there permanently and escape the sperm whale's problem. But the great hydrostatic pressure of the abyss influences their lives in at least one important way. In an ingenious experiment that was conducted partly from a research submarine, Smith and Teal (1973) measured *in situ* the respiratory rate of organisms of the deep sea floor, 1.85 km beneath the surface. Their results indicate that the metabolic rates of these bottom dwellers average $\frac{1}{10}$ the rates of bottom organisms on the shallow continental shelf. High pressure apparently depresses the rate of oxygen uptake in deep sea organisms, thereby depressing their overall metabolic activity. The pace of life becomes increasingly sluggish as one descends into the abyssal troughs. But there is no indication that life, even in the deepest parts of the abyss, is made impossible by pressure. Rather, the limit to the biosphere's inner fringe is imposed by food. Though sparse, enough rains into the deepest parts of the abyss to support some life at the sediment-water interface. But less than a meter beneath this interface the ooze probably is sterile; organic nutrients have been exhausted.

WHY LIFE ON EARTH ALONE?

The striking ecological symmetry of the biosphere's fringes is this: both fringes support heterotrophic, plantless communities that scrounge a living from the layer of green that lies between them. This central portion of the biosphere is the only region in which plant life can obtain both sufficient energy and sufficient water and CO_2 to exist. It is the plants, rather than the animals, that are most strictly limited; but dependent as they are on plant-derived food, the animals cannot venture much farther.

One prerequisite for the existence of life *anywhere* is a source of energy. Living organisms represent local concentrations of stored chemical energy, captured from outside sources and used to power the syntheses that result in growth and fuel the organism's activities. A second requirement for life is a supply of the elements and compounds from which living matter is constructed. A third may be a rather specialized set of physical and chemical conditions. On Earth, for example, severe constraints are set on life by temperature. At high temperatures proteins are denatured and many biological processes accelerate to uncontrolled rates or cease altogether. At low temperatures important reactions may occur too slowly to sustain the organism. Thermal control of the states of water may also set temperature limits to life on Earth, as we shall see. But we cannot be certain that life everywhere is subject to the same physicochemical constraints as it is on Earth. We can only be sure of what we know: that Earth, at least, possesses physicochemical conditions, as well as energy resources and raw materials, that permit life and perhaps, as some scientists think, even make it inevitable.

The Lunar Environment Compared to Earth's

Let us look first at Earth's moon and then at the other planets in the light of these considerations. Solar radiation, the chief energy source of earthly life, bathes the moon in quantities similar per unit area to those received by Earth. Analysis of returned moon rocks reveals a chemistry very similar to that of Earth. Not only are the abundances of elements fairly similar, but even many of the more complex minerals are the same. Oxygen is abundant on the moon but it is all tightly bound in mineral compounds; there is no free molecular oxygen as there is on Earth. Still, this is not a necessary prerequisite of life; Earth harbors anaerobic organisms to whom free oxygen is a poison, not a necessity.

The most biologically significant discovery from the analysis of moon rocks is that they contain no water in any form, either free or bound to minerals. Water may not be essential to all life, but it is uniquely important to life on Earth.

Is the moon's lack of life, then, due to lack of water? Perhaps, but there are other stringencies as well. For one thing, the moon lacks an atmosphere. The light elements and compounds comprising the moon's former atmosphere, if it ever had one, may have succumbed to the gravitational pull of the larger Earth and now reside in our own atmosphere. Evidence exists, in fact, that Earth itself has selectively lost lighter elements from its atmosphere into space. Earth's remaining atmosphere, nevertheless, is a primary reason why our planet has life. The exact chemistry of our atmosphere is probably less important than two other aspects: atmospheric movements redistribute the heat energy received from the sun; and the atmosphere serves as a protective shield for the living matter on Earth. To our list of prime prerequisites for life perhaps we should add protection from extraplanetary hazards, for biological systems are fragile entities indeed. Radiant energy sustains most life, but radiation that is too strong or of the wrong type is lethal.

This may be a major reason why life was restricted to the oceans for so long after it evolved on Earth. The atmosphere presently contains a layer of ozone that absorbs much of the incoming ultraviolet radiation. Ultraviolet is damaging to present forms of life on Earth and may well be damaging to all possible forms of life. The protective ozone layer in our atmosphere is believed to be a relatively late development in Earth's history. Before the ozone shield appeared life was probably not possible on land; but since water is a rapid absorber of ultraviolet, life could exist beneath a thin surface layer of the oceans.

Our atmosphere shields us from more than radiation, as the meteorite-pocked surface of the moon attests. The presence or absence of an atmosphere is primarily a function of a planet's mass (which determines the strength of its gravitational pull), so it is probably justifiable to conclude that a planet must have a minimum mass before it can harbor life. The moon's mass

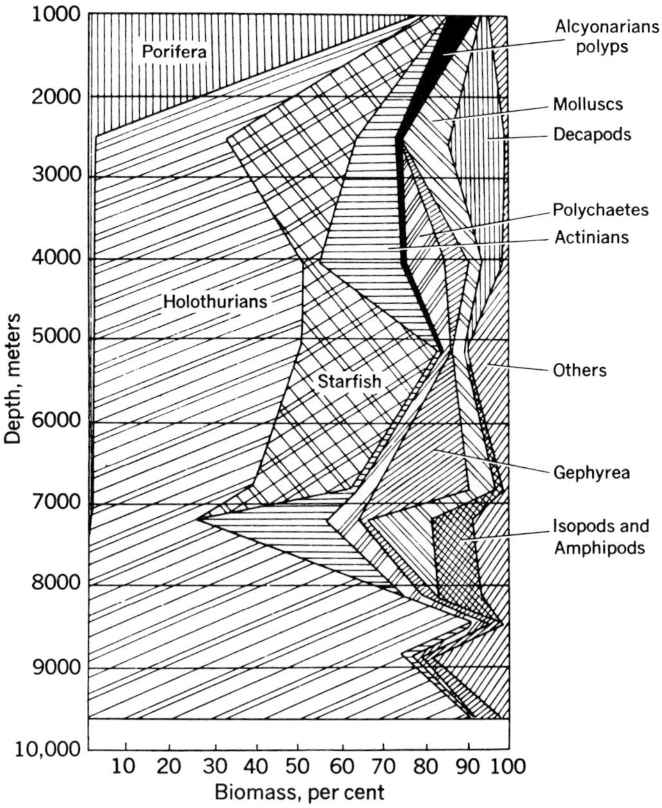

FIG. 2.4. Depth distribution of ocean floor invertebrates. Most groups occur over a wide depth range. Common names are as follows: Porifera—sponges; alcyonarians—certain corals; decapods—crabs, lobsters, shrimp; polychaetes—annelid worms; actinians—sea anemones; holothurians—sea cucumbers; Gephyrea—another class of annelids; isopods and amphipods—two orders of small crustaceans. (From L. Zenkevitch: *Biology of the Seas of the USSR*. London: George Allen and Unwin, Ltd., 1964.)

simply is too small, at least for a body so close to a planet of larger mass.

Other pertinent contrasts exist between Earth and moon. Earth rotates; the moon does not, and one side is perpetually shadowed from solar radiation while the other is perpetually bathed in it. This creates temperature extremes far greater than those on Earth. In addition, Earth is density-stratified and its dense core is at least partly molten. Electric currents set up within the liquid core create a magnetic field extending far beyond Earth's atmosphere. This magnetic field traps ionizing particles from space and captures or deflects smaller meteorites, thus serving as a second important shield against extraterrestrial hazards. It has been claimed that weakenings of Earth's magnetic field in the past frequently coincided with episodes of accelerated extinction. During such periods our magnetic shield became more porous, and the hazards to life were correspondingly increased.

By contrast to doubly protected Earth, the moon has no shield at all, either atmospheric or magnetic, against extralunar hazards. It appears to be a solid body with little or no density stratification and only the weakest of magnetic fields.

Environments of Neighboring Planets

Of all the planets, only Earth and Mars seem capable of harboring water in its liquid state. The other planets are too far from the sun, and hence too cold, or are too near it and have surface temperatures above water's boiling point. Mars, our outer neighbor, is colder than Earth but may not be too cold for water to exist there as a liquid. Mars has a thin atmosphere, which permits considerable warming radiation to reach the planetary surface, but the thin atmosphere also permits rapid night-time cooling by reradiation of heat into space. Daytime temperatures at the Martian equator may be 30 to 40° C (not unlike Earth's), but temperatures probably drop well below freezing every night (Sagan, 1970). The question of liquid water on Mars is not simply a matter of temperature. Sagan thinks that the low atmospheric pressure on that planet would result in the vaporization of most water. Night-time cooling might lead to condensation on land surfaces, however. Earthly microorganisms have been grown successfully in simulated Martian environments that were provided with small amounts of water, together with particulate matter to absorb ultraviolet radiation, and Sagan concludes that life *could* exist on Mars, perhaps in briny pools or in soil made wet by nightly dew. Recently he has suggested that higher atmospheric pressures, higher temperatures and abundant liquid water could have occurred on Mars in the past and provided an environment more favorable to life than today (Sagan *et al.*, 1973).

Is liquid water really a *sine qua non* of life, as many have believed? We have been talking as if it were, but it is time to examine the notion. The biological membranes of earthly creatures, at least, permit passage only of substances in solution; water, the closest thing known to a universal solvent, thus seems to be life's ideal if not indispensable medium. Many of water's other properties—its heat capacity and dissociation characteristics, for example—add to its suitability as a vital substance (Table 2.1). But water is not the only conceivable medium of life (liquid ammonia and mercury are others) and it is not essential to all organic syntheses. Wollin and Ericson (1971) have shown that amino acids can be formed in the complete absence of water, by the reaction of ammonia, methanol, formic acid and formaldehyde. Hence amino acids may constantly be forming in interstellar space, where these ingredients (but not water) occur.

Liquid ammonia probably exists on one or more of the larger outer planets (Jupiter, Saturn, Neptune, Uranus), several of which have atmospheres rich in ammonia vapor. These atmospheres, indeed, are nearly identical in composition to the mixtures used by Stanley Miller and others in many "origin of life" experiments; the constituent gases are quite capable of reacting to form a variety of organic molecules. But low temperatures are probable on these outer planets. Temperatures beneath their atmospheres are unknown; if heat is internally generated, the surface temperatures may be higher than the measured temperatures of their upper atmospheres. But it is questionable whether sunlight is a sufficiently strong energy source to sustain life on these planets, particularly beneath their thick, energy-absorbing atmospheres. The possibility of life on these planets cannot presently be ruled out, but their environments certainly seem less benevolent that that of Earth.

Earth, then, seems the most favored haven for life in our solar system. Its advantages include: (*a*) temperatures at which water can exist as a liquid, these temperatures being prevented by planetary spin and atmospheric heat retention abilities from becoming too extreme; (*b*) atmospheric and magnetic shields that provide protection from extraterrestrial hazards but permit useful radiation to reach the planetary surface; and (*c*) a density-layering of elements and compounds that concentrates many biologically important materials within the biosphere. We have barely touched on this last, but will expand on it in the following section

THE CHEMISTRY OF EARTH AND THE CHEMISTRY OF LIFE

When an engineer decides what metals will be used in a new aircraft, his choice is governed by suitability and economy. Only some metals possess the desired properties, and of the possible alternatives some are more expensive than others. If we translate cost into relative abundance, it is evident from the composition of protoplasm that life has evolved on earth under the same pair of constraints faced by the engineer with regard to raw materials. The elements present in living matter are principally those high in the periodic table. These light elements share two features that are essential if an element is to be a constituent of protoplasm: considerable chemical reactivity and (in some form) solubility in water. Carbon, the chief structural basis of living matter, exemplifies these properties. It is highly soluble, both as the bicarbonate ion and as CO_2; and with four bonding sites per atom it reacts readily not only with other elements but also with itself, to form the long carbon chains and benzene rings that

TABLE 2.1

Some Physical Properties of Water and Their Environmental and Biological Consequences

Modified from Sverdrup, Johnson and Fleming (1942).

Property	Comparison with Other Substances	Importance to Organisms
Heat capacity	Highest of all solids and liquids except liquid ammonia	1. Prevents extreme ranges of temperature in watery environments, and slows the rates of change in wet as opposed to dry environments. (Note that organisms themselves, being largely water, are similarly affected). 2. Movements of water or water vapor can transfer large amounts of heat.
Latent heat of fusion	Highest except for ammonia	Water is thermostatic at the freezing point, *i.e.* is "reluctant" to change from solid to liquid or vice versa.
Latent heat of evaporation	Highest of all substances	Extremely important to heat transfer capability of atmospheric water vapor; responsible for evaporative cooling of moist surfaces.
Density	Close to that of protoplasm	Organisms living in water receive much more support from the medium than those living in air. But it requires considerable energy expenditure to move through water or resist the force of flowing water.
Thermal expansion	Temperature of maximum density is 4° C; unlike most substances, density does not continue to increase with decreasing temperature.	Fresh water and dilute sea water have their maximum density at temperatures above the freezing point; ice therefore floats and forms at the surface rather than at the bottom of lakes and seas. Vertical circulation patterns in these habitats are also much affected.
Surface tension	Highest of all liquids	1. Important in cell functioning. 2. Controls formation and behavior of drops. 3. Small organisms can "walk on water."
Dissolving power	In general, dissolves more substances and greater quantities of these than any other liquid	Water is highly suitable as an "internal environment" of organisms; as an external medium it is a ready source of nutrients.
Dielectric constant (capacity to neutralize ionic charges)	Highest of all liquids when water is pure	Promotes high dissociation of acids, salts and bases, explaining their typically high solubility in water.
Electrolytic dissociation	Very small	Water is a neutral substance, yet contains both H^+ and OH^- ions.
Transparency to light	Relatively great	Photosynthetic organisms can live as much as 100 m below surface in waters low in light-absorbing contaminants.
Transparency to infrared and ultraviolet	Relatively slight	Solar heat and ultraviolet are absorbed in surface layers of water bodies rather than penetrating deeply. Atmospheric water vapor also absorbs much infrared.
Conduction of heat	Highest of all liquids	Important only on a small scale, as inside living cells. Large-scale heat transfers in water are largely by convection currents, not by conduction.

constitute the skeletons of most large organic molecules. Carbon skeletons provide an additional lesson: once formed, they are relatively stable. Chemical reactivity is important, but organic matter, especially its structural components, cannot be *too* reactive.

Evolving life was opportunistic in what it used as raw materials, primarily selecting the soluble, reactive elements that are most available. Hydrogen, oxygen, carbon and nitrogen comprise more than 99% of living matter; these elements rank 1, 3, 5 and 4 in their abundance in the universe (Table 2.2). Helium, which ranks second in cosmic abundance, is unreactive and therefore valueless as a constituent of protoplasm.

This apparent instance of evolutionary opportunism represents natural selection acting at the chemical level, rather than early organisms "choosing" their own constituents. In a fascinating paper, Pirie (1959) speculates that the exotic chemistry of some modern organisms is an evolutionary relict of past aeons, when organisms made more catholic use of the available elements. An example of "relict biochemistry" is the use made by some tunicates of vanadium, rather than iron, in oxygen-binding molecules analogous to hemoglobin. Pirie believes that the "mainstream" chemical uniformity of present-day life is the result of natural selection, acting to weed out most of those primitive organisms whose dependence on rare chemicals created problems of inorganic supply *versus* biological demand. But natural selection could only weed out chemical substitutes, where they existed, not chemical essentials. From a comparison of the relative abundance of elements in the universe and in living matter, at least four elements for which no biological substitute exists appear to be in dangerously short supply. The four are phosphorus, boron, potassium and calcium. Phosphorus is probably the chief global limiting nutrient, *i.e.* the element whose supply restricts the amount of life the biosphere can support. Apparently evolving life could find no substitute for the vital, energy-trapping capabilities of phosphorus; despite its scarcity, phosphorus is required by all organisms.

In the case of boron, calcium and potassium, the supply problem is eased by Earth's chemi-

TABLE 2.2

Relative Abundances of Elements in Universe, Earth Layers and Living Matter

All values are percentages of the total number of atoms present. Composition of universe has been deduced from spectral analysis of starlight and from chemical analysis of meteorities. Composition of Earth's lithosphere as a whole is poorly known, and percentages cannot be suggested. Authorities vary widely even on the percentage composition of Earth's crust alone. The H and C in the atmosphere occur principally as water vapor and as CO_2. Both are quite variable in concentration; the values suggested are averages. In order of relative loss, the elements preferentially lost from Earth (*i.e.* more abundant in the Universe as a whole than in Earth) are He, Ne, Kr, H, Xe, A, N.

Universe[a]	Lithosphere as a Whole[b]	Crust Only (Outer Lithosphere)[a, c]	Hydrosphere[a, c]	Atmosphere[c, d]	Plant Material (Wood)[a]	Animal Material (Man)[a]
H 91	Fe	O 47–60	H 66	N 78	H 50	H 63
He 9	O	Si 21–28	O 33	O 21	O 25	O 26
O 0.06	Mg	Al 6–8	Cl 0.3	A 0.93	C 25	C 10
N 0.04	Si	Fe 2–5	Na 0.3	H 0.5	N 0.27	N 1.4
C 0.02	Ni	Ca 2–4	Mg 0.03	C 0.03	Ca 0.07	Ca 0.3
Si 0.003	S	Na c.2.5	S 0.02		K 0.05	P 0.2
Ne 0.003	Ca	K 1.4–2.5	Ca 0.006		Si 0.03	K 0.06
Mg 0.002	Al	Mg 1.8–2.2	K 0.006		Mg 0.03	S 0.05
Fe 0.002	Co	Ti 0.3–0.5	C 0.001		P 0.03	Na 0.03
S 0.001		H 0.2–3.0	Br 0.0005		S 0.02	Cl 0.03
		C 0.16–0.19	B 0.0002		Al 0.02	Mg 0.01

[a] From Frieden, 1972.
[b] From D. A. Livingstone, unpublished.
[c] From Deevey, 1970.
[d] From Hare, 1963.

cal layering, which increases their abundance in the biosphere over that in the universe as a whole. In the case of other elements, too, the chemical layering of the earth produces patterns of biological availability quite different from the elements' cosmic abundance or even their average abundance on Earth itself. Figure 2.5 illustrates the major layers of Earth; the abundance of elements in several of these layers is shown in Table 2.2. The layers accessible to living things, and hence potential sources of their raw materials, are the atmosphere, hydrosphere and outer crust of the lithosphere.

What causes the stratification of Earth's elements? Gravity seems mainly responsible. There is a strong tendency for light elements to be concentrated in outer layers and for heavier elements to be concentrated toward the planet's center. This tendency exists even in the solid portions of Earth and provides strong evidence that the whole lithosphere was once molten, permitting the migration of elements to their preferred position in the gravitational gradient.

In the case of some of the lightest and/or least reactive elements, the centrifugal force of planetary spin has exceeded Earth's gravitational force, and these elements have been preferentially lost from Earth's outer layers into space. This conclusion may be drawn from the fact that helium, neon, krypton, hydrogen, xenon, argon and nitrogen are all less abundant constituents on Earth than they are of the universe. Helium is the most extreme in this regard; it is 10^{14} times more abundant in the universe than on Earth—where it still dominates a rarefied layer beyond what we generally consider our atmosphere. Even some of the heavier inert gases have been largely lost to space. The lightest element, hydrogen, has also suffered this fate, but to a lesser extent. By its incorporation into heavier molecules such as water and carbohydrate compounds, hydrogen has been retained in considerable quantities within the biosphere. Reactivity has been important to the retention on earth of many light, biologically important elements, and in many cases it has affected their pattern of stratification. If a light element is typically found as part of heavier compounds, it may be concentrated in a deeper layer than its own weight would imply.

The strict layering by weight of elements has

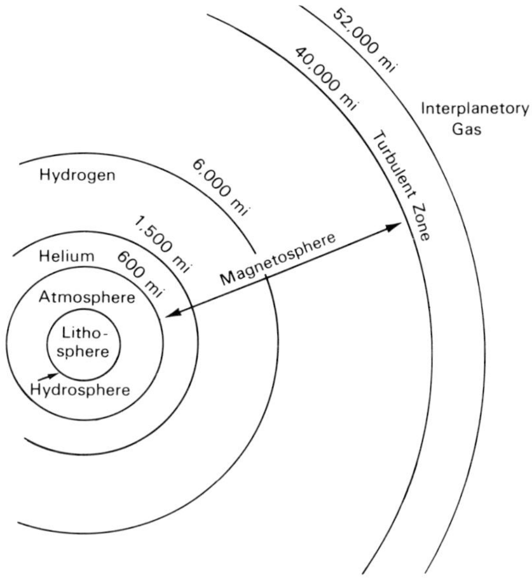

FIG. 2.5. Gross features of the Earth's layering. The magnetosphere is thought to be a region in which ionizing particles from space are caught in Earth's magnetic field. The biosphere includes the hydrosphere and the interface layers between it and the atmosphere and lithosphere. (Modified from G. R. Rumney: *Climatology and the World's Climates*. New York: The Macmillan Co., 1968.)

been further disrupted by another process. The minerals now comprising the solid lithosphere crystallized from the molten state at differing temperatures. An element such as magnesium, enormously abundant and rather light, formed early crystallization products at relatively high temperatures. These solid mineral products, heavier than the element itself, were spun down through the still molten medium toward the center of the earth; thus magnesium is now much more abundant in the deeper portion of the lithosphere than in the earth's crust. By the same process, much phosphorus has been lost to the deeper lithosphere, magnifying its potential as a limiting nutrient for the biosphere (Hutchinson, 1943).

These processes all are reflected in the ordering of elements in Table 2.2. Comparison of columns 2 and 3 reveals something of the layering of the lithosphere; elements higher in column 3 than in column 2 are concentrated in the outer crust. As might be expected, the demand for elements by protoplasm (columns 6 and 7) is geared more to the supply in layers accessible to the biosphere (columns 3 to 5) than to the supply in the earth as a whole. This statement is exemplified by magnesium, for which the biological demand is relatively low in keeping with the relative rarity of this element in accessible layers. An element such as oxygen, common in more than one sphere, usually occurs in different forms, *e.g.* primarily as oxidized ores in the lithosphere, as water in the hydrosphere, and as molecular O_2 in the atmosphere. From this table it is relatively clear where the primary reserves exist for elements needed by organisms. Plants and animals differ somewhat in the elemental demands they make on their environment: plants have a higher content of carbon, vertebrate animals of phosphorus, primarily because they rely on different materials for structural support (cellulose in plants, calcium phosphate in vertebrates).

Table 2.2 does not include all the elements found in living organisms, but only those required in relatively large quantity. Twenty-four elements are known to be essential to life (Frieden, 1972). Several more are useful to at least some organisms, and numerous additional elements occur in one or more species but have no presently known function. Many essential trace elements, such as iron and silicon, are abundant in the earth's crust but their biological availability is limited by their low solubility in water. Only three of the essential elements (molybdenum, tin and iodine) have atomic numbers higher than 34; all three are needed only in trace amounts (Frieden, 1972). Iodine, the heaviest of the essential elements, perhaps plays a biological role because it is unusually reactive for a substance so heavy. Many heavy elements act as poisons and living organisms preferentially exclude them, using instead lighter elements of the same subgroup in the periodic table (Patterson, 1965). It is not entirely clear whether this discrimination is an evolved biochemical exclusion process or whether it is simply a consequence of the lesser reactivity of heavier metals. In any case, living tissues cannot entirely exclude heavy elements, as our current problems with lead and mercury poisoning attest.

It is too simple to think only in terms of elements when considering the relationships between biochemistry and Earth chemistry. Elemental abundances are important, but in many situations it is the abundance of vital compounds that is most important, rather than the abundance of elements *per se*. A familiar example involves nitrogen. As the most abundant substance in the atmosphere, pure nitrogen (N_2) is never in short supply to land organisms, but higher plants and animals cannot assimilate the pure nitrogen that surrounds them. Before it is useful to them it must be converted to other forms (*e.g.* nitrates or organic nitrogen compounds). The only route by which atmospheric nitrogen becomes accessible to most living things is through the activity of nitrogen-fixing microorganisms. When these microbes are in short supply, higher organisms may be starved for nitrogen in the midst of plenty. Chapter 9 will pursue these considerations and related problems, such as the inadequate cycling processes which sequester nutrients in "sinks" within the biosphere.

ENERGY AND THE ATMOSPHERE

Having viewed Earth as a chemical habitat, let us see how life's demands for external energy supplies are met. Some "chemosynthetic" microorganisms capture energy from the oxidation reactions of inorganic compounds, but life's prime energy-trapping process is photosynthesis, and the trapped energy comes from the sun. Visible radiation (light) constitutes about half the solar beam and infrared radiation (heat) most of the remainder. Despite its destructive capabilities, ultraviolet radiation constitutes only a small portion of the sun's energy output.

Since all energy is convertible to heat, units of heat can be used as a common quantitative measure of any sort of energy. The most frequently used energy unit is the **calorie,** the quantity of heat required to raise the temperature of 1 gram of water 1 degree Celsius at sea level pressure. The chemical energy of foods can be expressed in calories if they are burned in closed chambers (calorimeters) and the heat produced by their combustion is measured. Relatively large quantities of energy are often expressed in units of 1000 calories (kilogram calories or kilocalories).

Solar Income

The sun varies slightly in the quantity of radiant energy it produces, and the earth varies in its distance from the sun. But if a flat surface were held perpendicular to the sun's rays at a point just outside the earth's atmosphere, its average dose of solar energy would be 1.94

cal/cm²/min. This quantity is known as the solar constant. Because the atmosphere intercepts considerable radiation, and because no point on the earth's surface remains perpendicular to the sun's rays for long, the radiation received at any location in the biosphere is almost always less than the value of the solar constant. But in some instances, particularly on high mountains, the solar constant occasionally can be exceeded; this occurs when direct insolation is supplemented by radiation reflected to an unshaded location from clouds and from invisible atmospheric particles.

In the absence of clouds, the atmosphere modifies the incoming solar beam as shown in Figure 2.6. Ozone, oxygen and nitrogen in the atmosphere absorb significant quantities of ultraviolet, and water vapor and CO_2 absorb some infrared before the radiation reaches land or oceans. This absorbed radiation serves to warm the atmosphere. Of the remaining solar input, some is scattered (reflected at various angles by atmospheric particles) and some passes directly to the earth's surface. The scattered light suffers two fates: some returns to space and some, reflected downward, reaches the earth's surface. Blue and ultraviolet wavelengths are particularly subject to scattering. In consequence the sky appears blue, because light coming from the sky is scattered sunlight, dominated by the blue end of the spectrum. Conversely, the sun appears red at sunrise and sunset because its beam slants toward the observer through a greater thickness of atmosphere than at midday, and is more completely scattered. The red end of the visible spectrum, which is least subject to scattering, predominates in the sunlight directly reaching the observer at these times.

Table 2.3 provides a balance sheet of the fate of solar radiation directed toward the earth. Sources used to compile such a table do not fully agree, and the figures obviously represent estimates. In any particular locality the degree of cloud cover causes cnsiderable departures from these average figures. Heavy clouds prevents as much as 80% of incoming radiation from ever reaching the biosphere. Although the proportional fate of incoming solar energy can only be estimated, the amount actually reaching a point on the earth's surface can be easily measured by instruments such as thermopiles. Some of this energy is already in the form of heat; the remainder, chiefly visible light, is converted to heat as soon as it is absorbed by water or land. Heat is also the eventual fate of light radiation trapped by photosynthesis.

Table 2.4 shows the latitudinal and seasonal distribution of solar input to the northern hemi-

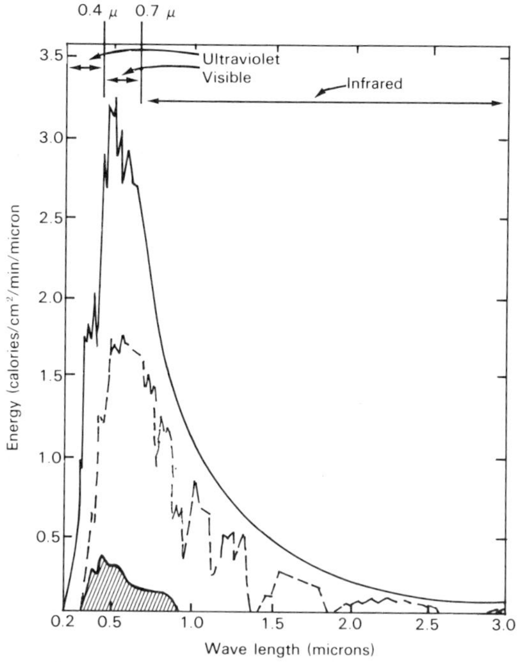

FIG. 2.6. Spectral distribution of solar energy at the outer edge of the atmosphere (*solid line*) and at sea level (*dashed line*). The *hatched area* shows the spectral distribution of scattered sky light alone. Cloudless conditions are assumed. (Adapted from D. M. Gates: *Scientific American*, September 1971, with permission from Scientific American, Inc., All rights reserved.)

TABLE 2.3

Fate of Solar Radiation Incoming to Earth

Data are given as percentages of total energy of all incoming solar wavelengths.

Absorbed by atmosphere (ultraviolet)	1–3%
Absorbed by atmosphere (infrared)	17–19%
Reflected back to space by clouds[a]	24–27%
Returned to space by atmospheric scattering	approx. 6%
Reflected back to space by land and water	approx. 2%
Absorbed by water, land and living organisms	approx. 50%

[a] Assuming a mean world cloudiness of 50%.

TABLE 2.4

Geographic Variation of Solar Radiation Received at Sea Level in the Northern Hemisphere

Latitude	Estimated Input on the 15th Day of Each Month Assuming 50% Mean Cloudiness												Estimated Yearly Total	Representative Cities
	Jan.	Feb.	Mar.	Apr.	May	June	July	Aug.	Sept.	Oct.	Nov.	Dec.		
	cal/cm²/day												*cal/cm²/yr*	
0°	388	415	429	420	405	393	397	421	445	443	410	384	150,800	Quito, Nairobi, Singapore
15°	344	388	441	465	465	459	461	459	437	393	346	324	152,000	Guatemala City, Khartoum, Bangkok
30°	290	307	390	458	500	506	494	459	397	324	252	226	138,400	New Orleans, Cairo, Shanghai
45°	123	198	297	401	471	506	490	435	336	231	146	105	113,000	Minneapolis, Milan, Vladivostok
60°	62	125	194	340	454	508	488	386	239	113	38	19	90,400	Anchorage, Stockholm
75°	0	6	69	227	417	516	482	322	126	26	0	0	66,600	
90°	0	0	0	125	382	540	494	245	22	0	0	0	55,450	

1. Values are adjusted from those of Hutchinson (1957) for 6% addition as scattered light and 25% reduction due to clouds.
2. For comparison, daily and yearly values of the solar constant (assuming a 12-hour daylight period) are 1398 cal/cm²/day and 510,000 cal/cm²/yr. At sea level, regardless of latitude, solar energy received is less than this.
3. The latitudinal irregularity in summer exists because both the sun's angle of incidence and the day length affect the amount of radiation received.

sphere, and illustrates several features of great biological importance. The equator receives relatively even amounts of solar energy throughout the year, although the input rises slightly when the sun is directly overhead at the vernal and autumnal equinoxes. Total yearly radiation, surprisingly, is higher at 15° N than at the equator, and declines relatively slowly to 30° N. Poleward from this latitude, however, the yearly solar input declines quite sharply. At high latitudes the total input is squeezed into a few months, and more energy pours into the high Arctic during June than into any other part of the hemisphere during any month. Under this influence Alaskan gardeners produce cabbages of legendary size, and the tundra explodes into growth. The northward movements of caribou and the masses of migratory birds that nest at high latitudes testify to the intensity of the short Arctic growing season. The southern hemisphere, of course, is a seasonal and latitudinal mirror image of the northern hemisphere in terms of solar input.

Before leaving the subject of solar input we should look at the absorption characteristics of green plants. Plants photosynthetically capture less than 5% of the solar energy impinging on them; this energy becomes the stored chemical energy of organic matter. A good deal more solar energy is absorbed by plants but converted immediately to heat. In a spectral analysis one finds that plants absorb considerable quantities of visible light, particularly in the blue and red absorption peaks of chlorophyll. (Hence the light *reflected* from foliage is preponderantly green.) Ultraviolet radiation is also heavily absorbed, but radiation in the far-red and infrared wavelengths is very largely reflected. This greatly reduces the heat load that must be borne by leaves exposed directly to the sun; such leaves would otherwise warm to unbearable temperatures on still days. (Reflection of infrared wavelengths makes possible the modern practice of mapping vegetation densities by aerial infrared photography.) At very long infrared wavelengths, above 2.0 μ, leaves again show a pattern of high absorptivity, but since very little incoming solar radiation is of these wavelengths there is little additional heat load on the plant. Rather, this absorption maximum in the far infrared probably *reduces* the plant's heat load by serving as an "emission window." If heat can be absorbed at certain wavelengths it can also be emitted at those wavelengths. Much of the radiation absorbed at shorter wavelengths, however, is converted to long infrared wavelengths within the leaf; so through its window in the far-infrared a plant probably radiates far more heat to its surroundings than it receives at this wavelength from the sun (Gates, 1963). Altogether, the absorption pattern of green plants seems remarkably adaptive.

Terrestrial Back-radiation and Atmospheric Circulation

The land and sea do not retain all the heat they absorb from the sun, as we know from the cooling that occurs at night. Especially on clear nights there is intense back-radiation toward the cold depths of space. A traditional ice-making technique of India makes use of this phenomenon, and illustrates its intensity. Water is placed under the stars in shallow black pans, set in sawdust to insulate them from the ground. Even in relatively warm weather there usually is a layer of ice by morning; underlain by the black radiating surface, the water has lost most of its heat to the night sky (Schmidt-Nielsen, 1964).

Unlike the diverse solar input, Earth's back-radiation is entirely long-wave thermal radiation. Much of this radiant heat never reaches outer space, but is absorbed in the lower atmosphere by water vapor and CO_2. These minor atmospheric constituents are responsible for moderating the day-night extremes of temperature on the earth's land surfaces, and for keeping the average temperature of the globe as high as it is. Water vapor and CO_2 absorb none of the incoming visible radiation, and only some of the infrared. But a large fraction of Earth's outgoing, long-wave radiation is captured by these gases, and the atmosphere thus is heated mainly from below. This is the atmosphere's **greenhouse effect**—it collects heat on the way out. Besides absorbing radiant heat, the atmosphere withdraws heat directly from the oceans by evaporation; the heat of vaporization of water (Table 2.1) is transferred to the vapor entering the atmosphere.

The yearly balance between solar input and back-radiation is tipped in favor of input in the tropics and subtropics, but at latitudes above 38° there is a net loss of heat from the earth's surface (Fig. 2.7). It would appear that low latitude temperatures should steadily increase and high latitude temperatures should gradually decrease toward those of outer space. This does not happen because the atmosphere and the oceans redistribute the excess heat of the tropics toward the poles. In this vital task the atmosphere is the chief principal.

The atmospheric circulation that distributes heat is itself caused by heat differentials A simplified scheme of circulation in the lower atmosphere (Fig. 2.8) illustrates this fact. Warm air is less dense than cool air, and much of the global circulation can be explained in terms of (*a*) rising masses of warm air (producing surface regions of low atmospheric pres-

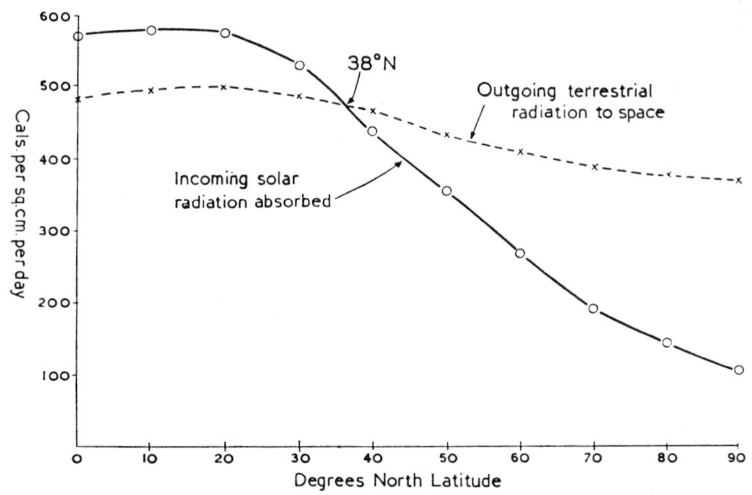

FIG. 2.7. Latitudinal differences in the average solar energy absorbed by the Earth and the average heat energy reradiated by the Earth to space. There is a net gain at lower latitudes and a net loss at higher latitudes. (From F. K. Hare: *The Restless Atmosphere*. New York: Harper Torchbooks, 1963; after Houghton (1954). Reproduced with permission.)

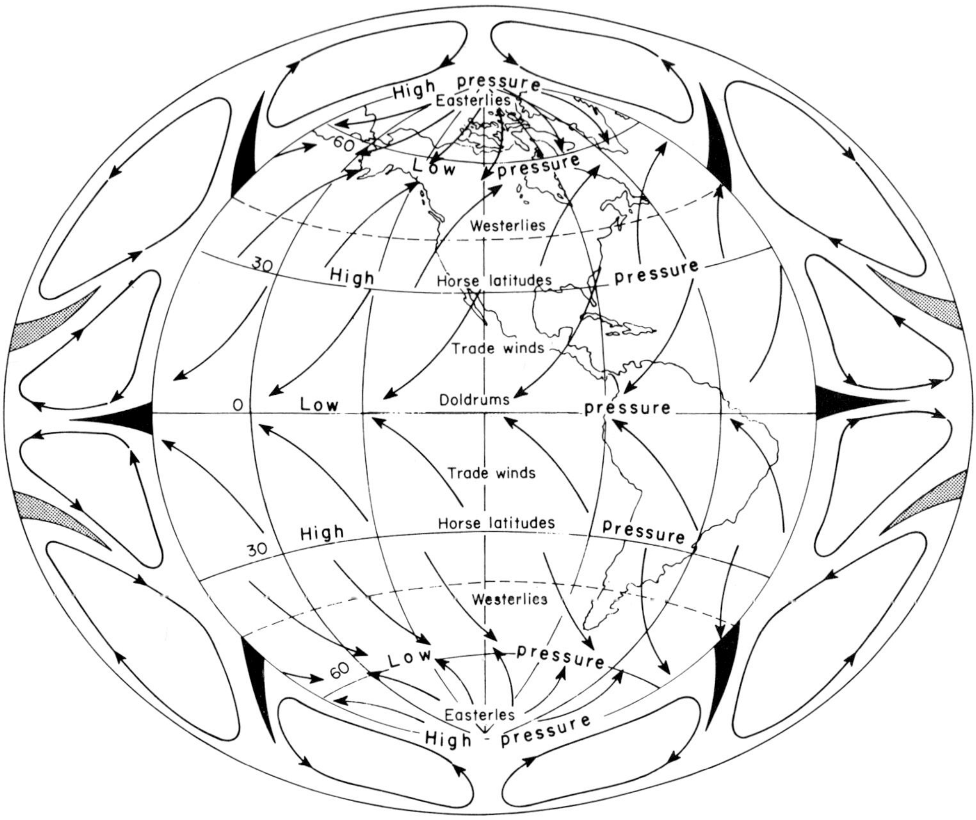

FIG. 2.8. General, idealized circulation of the atmosphere. In both northern and southern hemispheres there are tropical, mid-latitude and polar cells of circulating air associated with latitudinal belts of high and low pressure. (From G. R. Rumney: *Climatology and the World's Climates.* New York: Macmillan Co., 1968. Reproduced with permission.)

sure), (*b*) descending masses of cool air (producing regions of high pressure), and (*c*) surface air currents from the latter to the former, filling the void left by the rising air.

Figure 2.8 depicts the atmospheric circulation as a set of cells existing at different latitudes and driven primarily by the density differences of cooler and warmer air. Despite its oversimplification, this model describes many of the major climatic patterns of the earth. To explain the model let us start at the equator. Always warm, the equatorial belt is a perennial region of low pressure, *i.e.* of rising air. Air at the bottom of the atmosphere absorbs heat from sun, land and water. This air rises, cooling as it does so by adiabatic expansion. (Any gas becomes cooler when a reduction in pressure allows it to expand, and pressure on a rising air mass decreases with altitude because the rest of the atmosphere is progressively rarefield with altitude.) From high latitudes cooler air, hugging the earth's surface, flows toward the equator to replace the rising tropical air. This equatorward flow represents the trade winds, among the most constant air currents of the globe. Reaching the equatorial low pressure belt the trade winds are transformed into the warming, rising equatorial air mass. Near the equator the predominant air movements become vertical rather than horizontal, and the erstwhile trade winds are light and erratic; this is the doldrums, the region of calms dreaded in the days of sail.

Equatorial air masses do not continue to rise. Cooling and becoming denser, they do not sink in place but are shouldered aside, to north and south, by the continually rising air beneath them. This sinking equatorial air eventually reaches the earth's surface again in the region of latitudes 30° N and S (the horse latitudes). The air warms as it descends, and since warming air can hold increasing amounts of water vapor, these latitudes are belts of relatively little rainfall. Many of the world's deserts occur here, in the horse latitudes. Because they are characterized by dense, descending air, the horse latitudes are regions of predominantly high atmospheric pressure. From these latitudes air flows down pressure gradients toward the equator, completing the two equatorial circulation cells; some air also moves poleward from the horse latitudes, joining temperate circulation cells driven by low pressure belts near latitudes 60° N and S (Fig. 2.8). These temperate cells add contributions of descending air to the high pressure belt in the horse latitudes. Poleward flow within the temperate cells is represented by prevailing winds from the west, the westerlies of mid-latitudes. The existence of low pressure regions near 60° N and S is less easy to understand than is the equatorial low pressure region, but the lows at 60° are presumably due partly to the warming of cold, dense air that approaches from the poles, and partly to the higher pressures created by descending air farther down in the horse latitudes. Some of the air rising from these high latitude low-pressure belts travels northward to complete the polar circulation cells.

These cells of circulating air are creatures of the lower atmosphere, or troposphere (Fig. 2.9), which is characterized by turbulent vertical air movements as well as the horizontal winds of the earth's surface, and by the progressive cooling with altitude that is so familiar in mountain regions. Averaging 6.55° C/km of altitude (2° C/1000 feet), this vertical temperature differential is known as the lapse rate of temperature with altitude; it results primarily from the adiabatic cooling of rising (expanding) air and the adiabatic warming of descending (contracting) air. Vertical air movements cease, and consequently so does the lapse rate, at the upper boundary of the troposphere. Above this boundary (the tropopause) lies the stratosphere, an atmospheric zone characterized by relatively stable, predominantly horizontal air movements. The tropopause averages perhaps 10 km in altitude, but it is higher in the tropics than toward the poles (Rumney, 1968), as might be expected if the strongest rising air masses occur near the equator. The tropopause is not a continuous boundary but in each hemisphere is broken latitudinally into three segments, whose overlapping edges differ greatly in altitude.

These zones of tropopause overlap are regions of marked, high altitude temperature contrast; they are the location of jet streams: ribbonlike, high velocity eastward air currents that are probably the product of this thermal stress. One can think of the breaks in the tropopause as the points of contact of the tropospheric circulation cells we have been discussing. Some air is exchanged between stratosphere and troposphere, but these atmospheric layers are largely self-contained, as indicated by their maintenance of separate circulation patterns. It is the tropospheric air movements that determine the climates of the biosphere.

Questions arise even from so simple a model of tropospheric circulation. If the major pressure belts are latitudinally arranged, why is not the major pattern of air flow directly north and south, along the steepest pressure gradients? Phrased differentially, why do the trade winds approach the equator at an angle, and the westerlies blow from the west? Also, what seasonal shifts occur in these belts of high and low pressure? Certainly seasonal shifts should be expected, if temperature differences are the cause of the belts.

Dealing first with the question of wind directions, one would certainly expect winds to follow the shortest possible route from regions of high to regions of low pressure. But the tendency of air to flow directly down a pressure gradient is modified by several other forces: centrifugal force, friction with the earth, and—most significant here—the Coriolus force or deflecting force of the earth's eastward rotation. If one takes an apple and tries to draw a straight line from the top or bottom toward the middle, while at the same time turning the apple toward the right, one finds that the line curves in the same manner as do the trade winds as they approach the equator. If one now draws a line from "equator" to either "pole" of the apple while maintaining its rotation toward the right, the line curves to assume the direction of the mid-latitude westerlies. The explanation of this deflecting "force" is that points near the equator traverse a much wider arc than do points near the pole during the same period of time. Singapore, on the equator, travels 40,000 km in a day's rotation; the barber pole at 90° N, coinciding with the earth's axis of rotation, goes nowhere. Wind moving toward the equator is constantly "falling behind" the longitude at

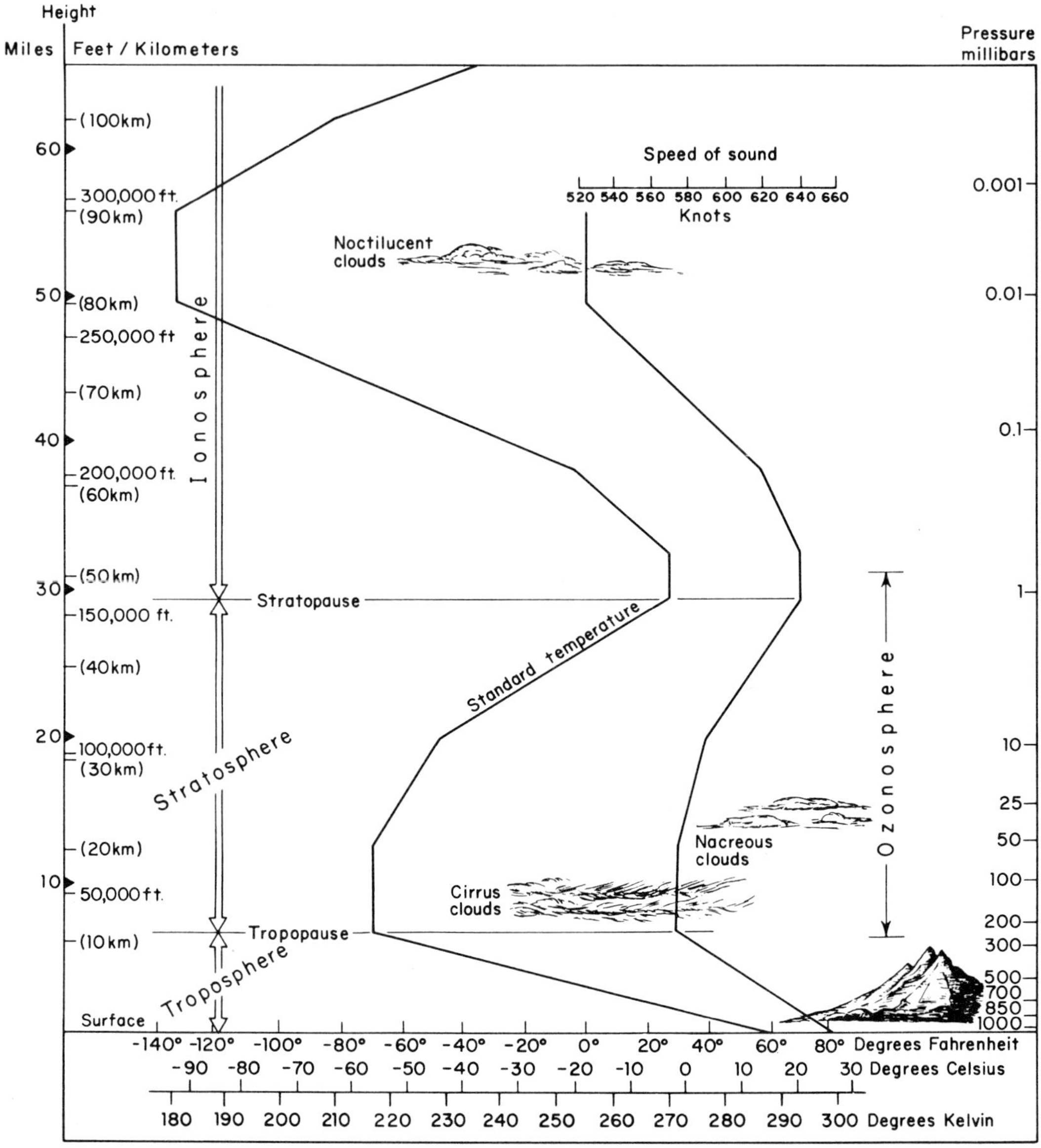

FIG. 2.9. Temperature distribution in the atmosphere, together with other features. (From G. R. Rumney: *Climatology and the World's Climates*. New York: Macmillan Co., 1968. Reproduced with permission.)

which it started—equivalent to saying that it is assuming an easterly direction. The reverse is true of air moving away from the equator. The upshot is that winds such as the westerlies blow more at right angles to the pressure gradient than along it. But the air eventually gets from high to low pressure areas nonetheless! (Descriptions of this process are much more difficult to understand than demonstrations with a turning sphere. Try the apple demonstration for yourself.)

Turning to the question of seasonal shifts in the tropospheric circulation cells, these do exist—for example, the doldrums move northward in the northern summer and southward in the southern summer—but before we delve more deeply into the matter a confession must be made. Figure 2.8 is an oversimplification, and, at least in the norther hemisphere, the latitudinal arrangement of high and low pressure belts is more ideal than reality. This is because of the strikingly different thermal properties of land and water. The southern hemisphere is mostly water, and the relatively small land masses do not seriously alter the pattern we have discussed (see Fig. 2.10). In the north, however, the large continents and the oceans interact in complex ways to affect atmospheric circulation patterns. Why? because land and ocean heat and cool at different rates and to different degrees. Water, with its great heat capacity, is much more stable thermally than is land (Table 2.1). The ocean changes temperature slowly and grudgingly; the same amount of heat gained or lost will change the temperature of water far less than that of land.

Consider the Asian monsoon, which in summer completely disrupts the northeast trade winds between New Guinea and the west coast of Africa (Fig. 2.10). The summer monsoon circulation results from the development of a giant low pressure system over Central Asia, caused by heating of this great land mass. Air is drawn to this low pressure area even from south of the equator, and from highs as far away as the North Pacific and North Atlantic. Why do these nothern oceanic highs exist? Because relative to the land the northern oceans are cool in summer, their temperature increase moderated by the great heat capacity of water. The air over these oceans is likewise cool, and heavy. In winter the monsoon circulation is reversed. Central Asia cools rapidly relative to the oceans, and becomes a high pressure region of bitterly cold dense air. The North Pacific and Indian oceans, now warmer than the land, become regions of lower pressure toward which the continental air flows. Both in winter and summer the Himalayas introduce a barrier into the airflow patterns, causing winds leaving and entering Central Asia to assume curving paths.

The Asian monsoon is a giant seasonal version of the diurnal sea and land breezes so common in summer along many seacoasts. The sea maintains a relatively stable temperature, but during the day air over the land warms appreciably. As this air rises, cooler air flows in from the sea and takes its place. The breeze dies near sundown as the land begins to cool and the air above it stops rising. Later in the evening the breeze resumes, but in the opposite direction; the air over the land, continuing to cool, sinks and flows out beneath the now warmer sea air.

Frontal Systems

Thus far we have talked about the relatively constant, predictable side of meteorology. But we know that the weather changes from day to day, not just from season to season: the red face of the weatherman frequently belies the predictability of atmospheric patterns.

Weather prediction is particularly risky in middle latitudes—in the United States, for example, or southern Europe. Despite the heat-distributing movements of the atmosphere, there exists as we know a latitudinal temperature gradient, and this gradient is particularly steep between latitudes 30° and 60°. Below 30° there is essentially a constant circulation of the same, low latitude cell of air. Above 60° the cells of polar air have a similar relative integrity. The mid-latitude cells receive contributions of air from both north and south, but the steep latitudinal temperature gradients in these regions reflect the incompleteness with which these cells capture and mix air from the equatorial and polar cells.

These steep thermal gradients are related to the meteorological instability of the mid-latitude belts. There are seasonal squeezings northward and southward of the mid-latitude circulation cells because of the seasonally changing strengths of their northern and southern neighbors. And there are short term, nonseasonal waxings and wanings in the contribution to mid-latitudes of polar and equatorial air. These contributions tend to arrive not in constant quantities but in irregularly timed packets (air masses) which vary in size and impetus. There is a constant interplay in the mid-latitudes between warm air masses pushing poleward and cold air masses pressing equatorward. The sharp boundaries between these air masses are termed fronts; they are regions of thermal contrast and atmospheric instability, and often of wind and rain. When we speak of a warm front we mean a warm air mass moving aggressively against, and displacing, a mass of colder air. A cold front is the opposite. Within the mid-latitudes meteorologists speak of the *average* seasonal location of frontal zones, *i.e.* the latitudinal boundaries above and below which

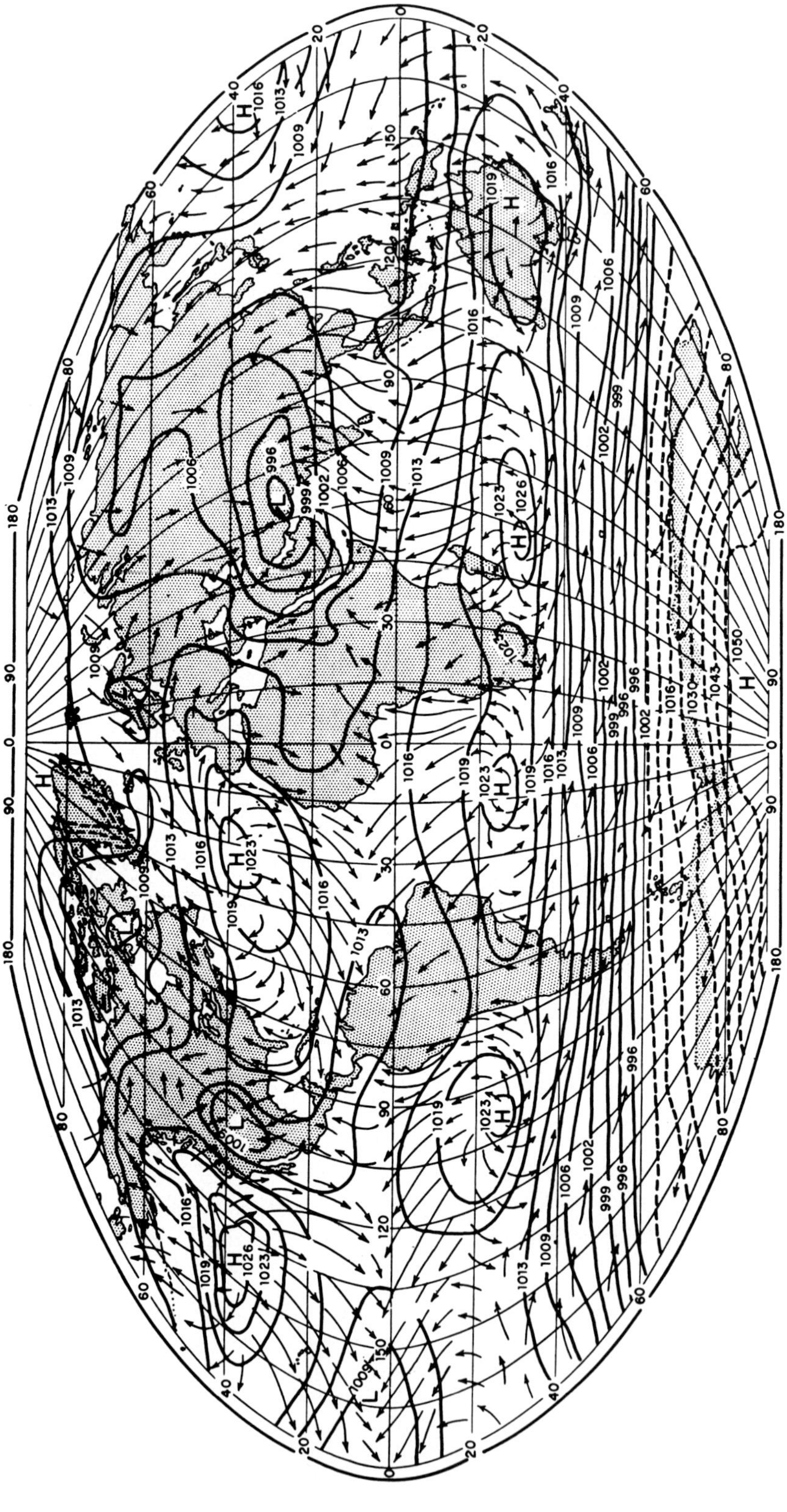

FIG. 2.10. Average sea level pressures and wind directions in July. (From V. C. Finch, G. T. Trewartha, A. H. Robinson, and E. H. Hammond: *Elements of Geography*. New York: McGraw-Hill, 1957. Reproduced with permission.)

polar air and equatorial air seasonally predominate. As we shall see, average frontal positions are important in demarcating certain major terrestrial vegetation types. But intrusions of tropical and polar air masses are constantly crossing these boundaries and are constantly creating temporary frontal disturbances.

Rainfall

Rain is primarily a consequence of cooling air, which ordinarily means it is a consequence of rising, expanding air. The capacity of air to hold water vapor diminishes logarithmically with temperature. At 30° C (86° F) a cubic meter of air can hold 31 g of water. At 15° C air is saturated when it contains 12.5 g/m^3 of water vapor, and at 0° C the saturation level is only 4.9 g/m^3. Thus air in polar regions, even when saturated, contains very little water. Precipitation is consequently low near the poles, and it is only the slow rate of evaporation and the impermeable frozen substratum that make the tundra soggy in summer. Some scientists speculate that ice ages are initiated not by cooling but by slight polar warming, with consequent increases of available moisture for precipitation at high latitudes.

What causes air to rise, and hence lose moisture? It may simply be the warming of air near the earth's surface, causing it to become less dense. As this air rises and cools it may form thunderheads, and brief but heavy local rainstroms frequently result. Such convective thunderstorms are the chief source of precipitation in some warmer parts of the world, *e.g.* Amazonia and the United States Gulf Coast in summer. Rainfall also occurs where moving air masses collide. At the point of frontal contact the warmer air mass is wedged upward by the cooler, denser air mass, becomes cooler itself, and loses some of its water. For much the same reason, mountains that obstruct a surface flow of air are centers of high rainfall. Upon reaching a mountain the air is forced to rise, cools, and cannot retain all of its moisture. Mountainous islands in the path of the trade winds, which blow constantly from the same direction, receive some of the world's heaviest rainfall on their windward slopes. Having crossed the crest of the mountains, however, the air has not only lost much of its moisture but is free to sink, and its water-holding capacity increases as it warms. For this reason islands such as New Caledonia and Madagascar are quite dry on their leeward sides, although their windward slopes support luxuriant rain forest. On the Hawaiian island of Kauai the summit of Mt. Waialeale receives an average of 451 inches (1145 cm) of rain annually. Twelve miles to the southwest, a point low on its leeward slope has annual rainfall of less than 16 inches (41 cm) (Rumney, 1968). This phenomenon, known as the rain shadow effect, occurs wherever mountains stand in the way of prevailing winds, particularly if these winds come off the oceans and carry abundant moisture. The mountain ranges of the American West create a series of moisture-intercepting barriers to the prevailing westerly winds; to the east of each range is a rain shadow area of low rainfall. The contrast between the western and eastern sides of the Sierra Nevada is a striking example (Fig. 2.11). Another rain shadow, affecting the Great Rift Valley in Kenya, has been vividly described by Isak Dinesen:

> The winds run straight against the Ngong Hills, and the slopes of the hills would be the ideal place for setting up a glider, that would be lifted up by the currents, over the mountain top. The clouds, which were travelling with the wind, struck the side of the hill and hung round it, or were caught on the summit and broke into rain. But those that took a higher course and sailed clear of the reef, dissolved to the West of it, over the burning desert of the Rift Valley. Many times I have from my house followed these mighty processions advancing, and have wondered to see their proud, floating masses, as soon as they had got over the hills, vanish in the blue air and be gone.
>
> *Out of Africa* (1937)

OCEAN CURRENTS AND THE LAND

Like the atmosphere, the hydrosphere is capable of circulation, though in slower and more predictable patterns. It serves as a second vehicle for the redistribution of heat across the globe.

The Causes of Ocean Currents

Wind and gravity are the two forces that drive the circulation of the oceans. The major surface currents are propelled by wind (with the aid of the Coriolus force), as is evident from a comparison of Figure 2.12 with Figure 2.8. The trade winds drive great tropical currents westward until they pile up against the oceans' western shores. In the Pacific, the western lands reflect

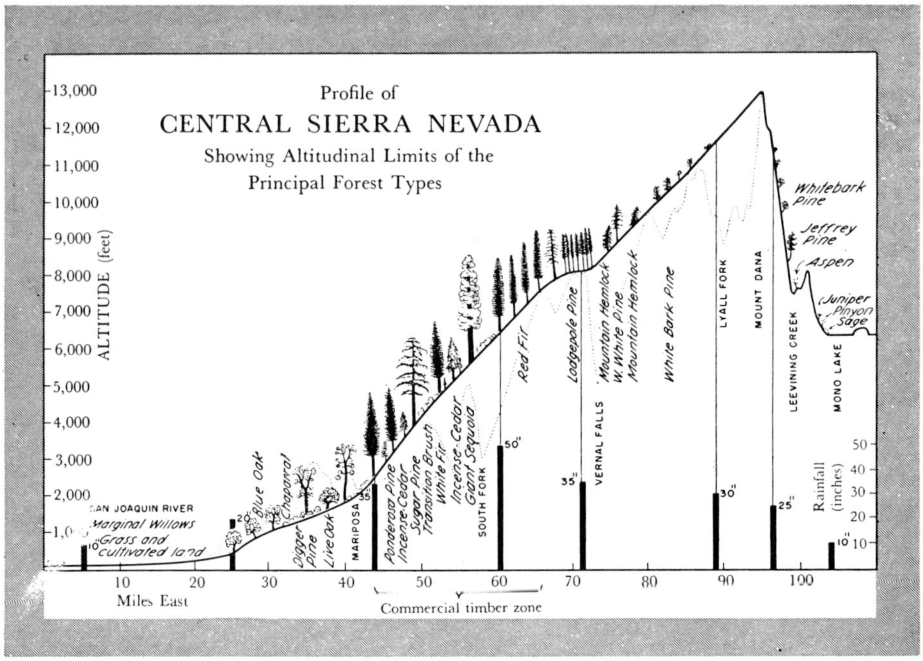

FIG. 2.11. Profile of the Sierra Nevada showing rainfall changes and associated vegetation changes. Prevailing westerlies bring moisture to the west slopes; air forced upward by the mountains loses much of its moisture even before reaching the highest elevations on the west side, and a severe rain shadow occurs on the east. (From *Trees: The Yearbook of Agriculture 1949.* Washington, D. C.: U. S. Department of Agriculture.)

some of this water straight backward as a thin, equatorial counter-current, but the majority of equatorial water in the western Pacific as well as in other oceans is deflected to the north and south. This water flows poleward, close to the coasts of the western land masses, until it falls prey to the prevailing westerlies, which drive it back toward the east. Beneath the southern continents this eastward flow is unimpeded and is known as the West Wind Drift, but nearer to the equator in both hemispheres the easterly flow eventually encounters continental obstacles, is deflected toward the equator, and ultimately begins its westward journey again. These are the main surface patterns. In addition, in the North Atlantic, a portion of the Gulf Stream escapes to the Arctic above Western Europe; in turn, surface water escapes southward from the Arctic Ocean around both sides of Greenland and through the Bering Strait.

The *deep* ocean currents are driven by gravity, operating through the density differences of water. The density of water is inversely related to temperature (at values above 4° C), and is directly related to the concentration of solutes. Water is heavier the cooler or the more saline it is. In the deep ocean circulation these two influences on density are frequently at war.

The predominant direction of deep flow is from the poles toward the equator. The process is similar at both poles, but the contribution from the south pole is much greater and we will describe it only. Frigid meltwater from the Antarctic ice cap begins the circulation. Initially, despite its low temperature, it is not too dense because it is quite dilute, having its origin in snow. Rather than sinking immediately, this meltwater mixes with the waters of the West Wind Drift, which forms a cold circumpolar current around Antarctica. Unlike other wind-driven currents, the Antarctic circumpolar current is no mere surface current but extends from surface to bottom as a mass of isothermal, freely mixing water. Cooled by the contribution of meltwater but remaining at least moderately saline, this hybrid water mass buds off dense offspring that make their way toward the equator, hugging the deepest parts of the abyssal floor. These deep currents from the south actually penetrate far beyond the equator into parts of the northern oceans. They follow the bottom because they are cold; they are actually

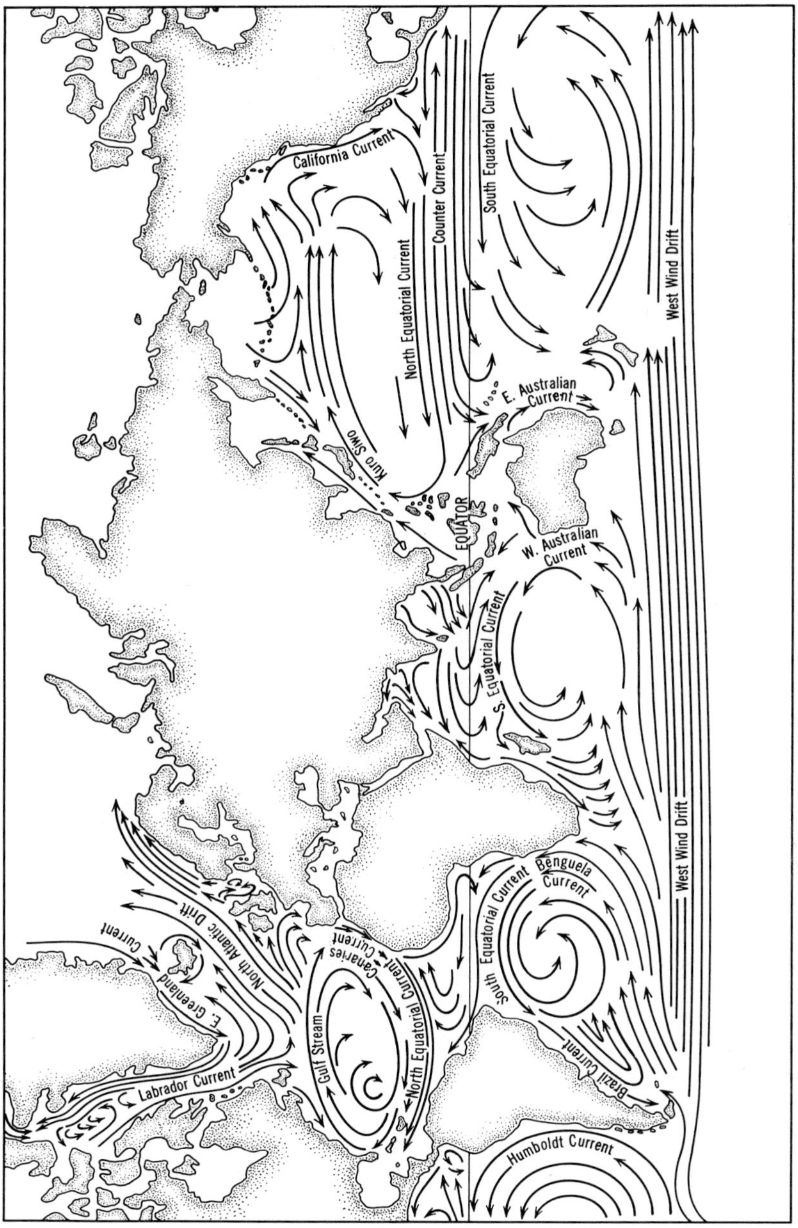

FIG. 2.12. Surface ocean currents of the world. (From R. H. MacArthur and J. H. Connell: *The Biology of Populations*. New York: John Wiley & Sons, 1966. Reproduced with permission.)

less saline than the warmer overlying waters, whose salinity is increased by evaporation. Gradually the density difference between cool, low salinity polar waters and warm, high salinity tropical waters disappears, and the polar waters mix upward and lose their identity. The deep, equatorward circulation we have described seems to be a one-way flow, but the cycle is at least partly completed through the atmosphere, which replenishes the polar ice caps with precipitation. Compared to surface currents the deep water circulation is extremely slow, but it provides a continuous supply of oxygenated water to most parts of the abyss.

Influence of Ocean Currents on the Land

It is clear from Figure 2.12 that the water bathing the eastern edges of the continents is derived from equatorial currents, while the water bathing the western edges of the continents has crossed the oceans at higher latitudes. If water temperatures have any influence on adjacent land temperatures, we might presume that at any given latitude the west coast of a continent should be cooler than its east coast. In the southern hemisphere this deduction has considerable truth, but in the northern hemisphere the poleward narrowing of the oceans and the intrusion along continental east coasts of Arctic water masses such as the Labrador Current almost reverse our expectations. The Gulf Stream and Japanese Current are still relatively warm when they reach the northwest coasts of Europe and North America, and temperatures on these coasts are consequently very mild for such high latitudes.

In Europe this mildness is not only coastal; unimpeded by north-south mountain ranges such as those of the Pacific Northwest, warm, moist air associated with the Gulf Stream moves far inland in Western Europe and moderates the climate of a large part of the continent. In the Pacific Northwest, warm oceanic air is carried against coastal mountains by the westerly winds and deposits abundant moisture on their seaward faces (and on soggy cities such as Seattle). Farther down the Pacific coast there is a curious reversal. The waters of the Japanese Current, now known as the California Current, enter latitudes where the land is warmer than the sea, either seasonally or permanently; air moving inland off the ocean is now warmed, and clings tightly to its moisture. Off California a seasonal oscillation occurs in the relative warmth of sea and land; most of the coastal rainfall is in winter, when the land is cooler than the winds off the sea. Further south the sea is always cooler, and the result is the year-round aridity of Baja California. At this latitude another factor enters the picture: the prevailing winds here blow not from the west, but down the coast toward the equator. Oceanic air thus seldom penetrates far inland—but even if it did, lt would cling tenaciously to its moisture.

The southern hemisphere provides examples of climatic situations very similar to those of California and Baja California. Cold currents bathe the west coasts of South America, southern Africa and Australia, bringing rain to the tips of these continents in winter when the land is even cooler than the oceanic air. Equatorward on these same coasts, currents constantly cooler than the land, together with prevailing winds that blow parallel to the shore, bring year-round coastal aridity to the Atacama, Namib and west Australian deserts.

More General Maritime Effects

The climatic effects of particular ocean currents are part of a more general tendency for temperatures to change more gradually and to be seasonally less extreme along coasts and on islands than in the interiors of continents. Because of the thermal stability of water the sea warm slowly in the spring and cools slowly in the fall, and these influences are felt in the climates of nearby land areas. Midsummer heat waves and midwinter cold snaps are also moderated in coastal areas by the more conservative thermal behavior of the ocean. Known as the **maritime effect,** this phenomenon is exemplified in the western United States by the year-round moderation of temperatures in San Francisco, and on the American east coast by the extension northward, along a thin coastal fringe, of sourthern forest trees. American holly, for example, grows along the east coast as far north as Long Island. Along this coast, too, the persistence of leaves on the trees well into the fall and the tardy resumption of growth in spring illustrate the maritime effect (Fig. 2.13). The influence on western Europe of the Gulf Stream, a maritime effect of unusually great extent, has resulted in genetic adaptation of the trees of that region. In the cities of northeastern America one can deduce from the time of

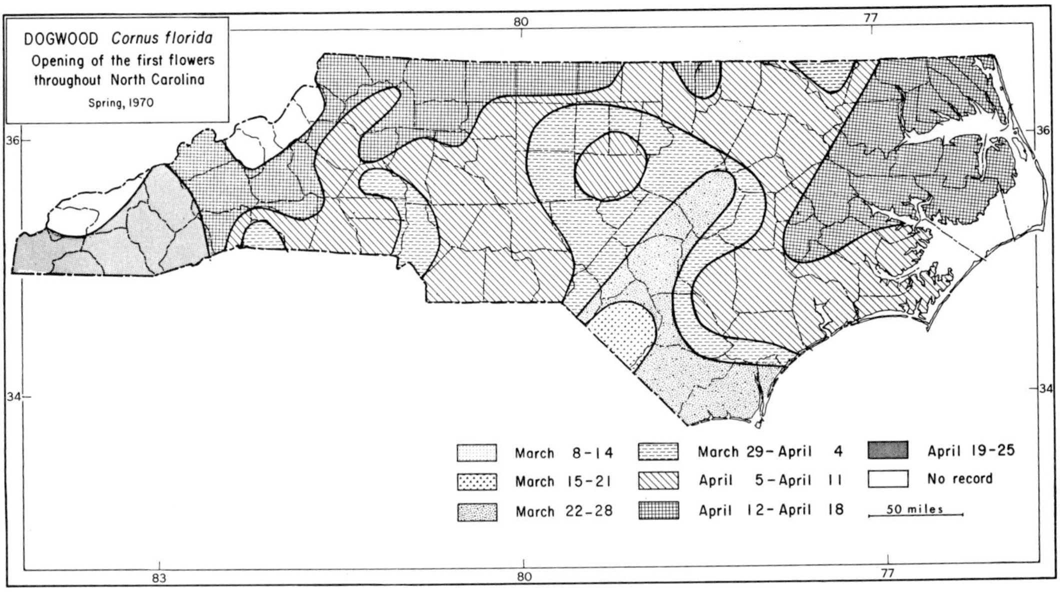

FIG. 2.13. Maritime effect is illustrated by the spring flowering dates of dogwood in North Carolina. Flowers appear later on the coast than inland at the same latitude. (In the western part of the state rising elevations also delay the season of flowering.) (From H. Lieth and J. Radford: *Bioscience 21:* 62–70, 1971.)

autumn leaf fall which species of deciduous tree are natives and which are introductions from western Europe (Table 2.5). Although their ancestral home lies much farther north than that of American deciduous trees, the European trees retain their leaves later into the fall; they are still attuned to the mild, gradually cooling autumns of western Europe. The American trees lose their leaves while the days are still relatively long, an adaptive defense against the swifter and more unpredictable onset of cold temperatures that characterizes eastern North America.

MAJOR PATTERNS OF TERRESTRIAL LIFE

In this chapter we have dwelt mainly on conditions that permit life on Earth and on how these conditions vary within the biosphere. Lest the title of the chapter be thought a betrayal, we must inquire more directly into how the patterns of life itself reflect the nonbiological patterns we have been discussing.

Biomes and Their Causation

The major types of terrestrial ecosystem are so familiar that they need hardly be listed. Many are of great extent, and each type possesses distinctive biological and environmental attributes. These ecosystem types intergrade, but terms such as tropical rain forest, temperate deciduous forest and tundra carry immediate meaning. If asked, we could easily list a series of organisms and environmental attributes descriptive of each and inappropriate to any other. These distinctive types of terrestrial ecosystem are termed biomes. Biomes are usually not spatially continuous, but are subdivided by geographic or climatic barriers. The desert biome, for example, includes all the scattered deserts of the globe.

It is no accident that most biomes are identified primarily in terms of characteristic vegetation types. Vegetation differences are conspicuous and form a visible basis for classification. More important, the vegetation is a critical determinant of the rest of the biological community in a given locality. Perhaps most significant, the vegetation of an area is an extremely sensitive reflection of the total environment of that area. Vegetation serves as an integrator of environmental conditions as does no other single measure (Billings, 1964). The vegetation of an area is the product of the climate, the soil, the grazing intensity of animals, the prevalence of fires, and other variables. When talking of the tropical rain forest biome or the tundra biome we thus are specifying not only distinctive vegetation types but also characteristic animal types and particular physicochemical conditions.

What factors primarily account for the existence of distinct biomes? Generalizations at this level may be too broad to be of much value, but most of the biomes locate themselves neatly on a two-axis grid of moisture and temperature (Fig. 2.14). For temperature one might substitute yearly solar input without seriously upsetting the pattern, but temperature seems the better choice of these two possible determinants. The existence of tundra and boreal coniferous forests on low latitude mountains as well as in extensive high latitude belts supports the primacy of temperature over solar input as a

TABLE 2.5

Dates of Leaf Fall Recorded for Different Species of Maple

The contrasting patterns of leaf fall shown below characterized maples growing on the campus of Franklin and Marshall College (Lancaster, Pa.) during the autumn of 1972. The trees were checked twice a week; each x indicates the recorded date of 100% leaf loss in a tree of the indicated species. See text for explanation of patterns. (Data were collected by Patrick Francis as part of a course project.)

Species	10/31	11/3	11/7	11/10	11/14	11/19	11/22	Native Habitat and Notes
Acer rubrum (red maple)		xxxxxx	x	x				Newfoundland to Florida; several races
Acer saccharum (sugar maple)		xxxx	xxxx	xxx	x			Nova Scotia to Maryland (to Georgia in mountains)
Acer saccharinum (silver maple)	x	x	xxxx		x			Maine to northern Florida
Acer saccharinum var. (Weir maple)				x	x			Horticultural variety of silver maple
Acer platanoides (Norway maple)				x	xxx	xxx	xx	Northwestern and western Europe; climate influenced by Gulf Stream
Acer pseudo-platanus (sycamore maple)				x	x	xx	xx	Northwestern and western Europe; climate influenced by Gulf Stream
Acer japonicum (Japanese maple)				x	x			Japan (Japanese Current should have effect similar to that of Gulf Stream)

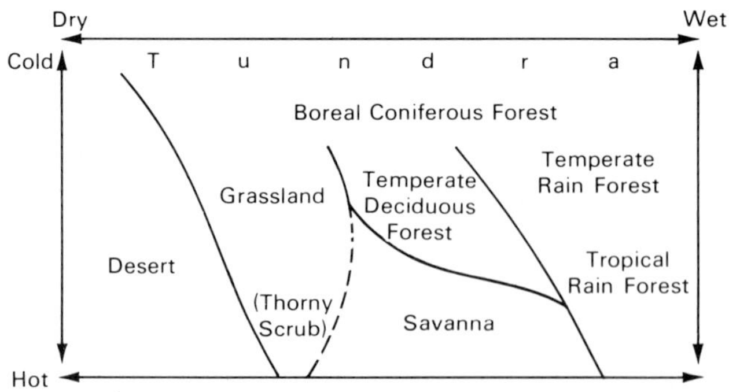

FIG. 2.14. Distribution of biomes with respect to moisture and temperature. Grassland areas in the tropics may all be the consequence of forest destruction by fire and grazing. In semi-arid tropical regions a thorny deciduous scrub may constitute a separate biome. This sometimes is called savanna, but the usual meaning of savanna is grassland with scattered trees.

biome determinant: there are thermal and biological similarities between the tops of the Alps and the Arctic tundra, but great disparities in their seasonal and yearly income of solar radiation. Temperature also seems more directly responsible for the skewing toward the right that occurs at the bottom of Figure 2.14. The skewing occurs because evaporation is increased at higher temperatures, reducing the fraction of total rainfall that is actually available to organisms.

The pattern of Figure 2.14 seems convincing, yet plant ecologists have argued for decades about whether vegetation type is controlled chiefly by climatic factors or by **edaphic** (soil) **factors.** The textural characteristics of soil bear as greatly on the amount of moisture available to plants as does the actual rainfall, and nutrients other than water also vary with soil type, so the advocates of edaphic control have room for argument. Further, they can cite examples of striking vegetation differences that are obviously due to soil, not climate (Fig. 2.15). And what do the proponents of climate say when they are told that the name of almost every biome in Figure 2.14 can be replaced by the name of a distinctive soil type? Each biome, in short, is characterized by a particular type of soil as well as by a particular climate (Table 2.6). Who is to say which is the more important determinant?

Most ecologists would probably agree to the suggestion that moisture and temperature *do* determine the major biome type, while soil acts at a secondary level, determining local variants of vegetation within particular biomes. Soil is

FIG. 2.15. Edaphic effects on vegetation in the White Mountains of eastern California. The light colored limestone soils in the background are deficient in phosphates but are relatively moist and cool. The vegetation these soils support is largely bristlecone pine (*Pinus aristata*). In the foreground is sagebrush (*Artemisia tridentata*), growing on a sandstone soil that is dryer, warmer and richer in phosphates. Soil characteristics here clearly determine the vegetation difference. (From a study by Wright and Mooney (1965). Photograph courtesy of Harold Mooney.)

TABLE 2.6
Association of Soil Types with Particular Biomes

Biome	Typical Soil
Tropical Rain Forest	Red laterites: deep, water-leached soils rich in iron oxides, with little humus
Boreal Coniferous Forest	Shallow grey podzols: acid humus overlying a grey, leached layer, the minerals of which have become concentrated in a deeper layer; silicates dominate the mineral composition
Temperate Deciduous Forest	In warmer climates, reddish or yellowish and semi-lateritic; in cooler climates a grey-brown podzol, less acid and less thoroughly leached than boreal podzols
Grassland	Chernozems: predominantly black soils (sometimes brownish or reddish), with a thick, rich humus layer and subsoils rich in calcium carbonate
Desert	Reddish or greyish soils, usually with little humus and little vertical differentiation; often shallow, sometimes saline

an important determinant, but the characteristics of a particular soil are themselves determined by climate, organisms, topographical situation and time. Soils derived from different parent materials tend to converge in their characteristics when weathering under the same climatic and topographic conditions. Thus the correspondence of particular soil types with particular biomes can be viewed as the consequence of both being determined in large measure by climate. But the local variants in vegetation produced by distinctive soil types can be very extreme, resulting even in the intrusion of one biome into the home territory of another. Such is the case with the Kentucky Barrens, a prairie of several thousand square miles that occupies distinctive limestone soils in the middle of a deciduous forest region.

The rainfall-temperature grid of Figure 2.14 accomodates most but not all of the earth's biomes. At least two will not fit: the broad sclerophyll and tropical deciduous forest biomes. The broad sclerophyll biome is named for its characteristic vegetation: broad-leaved (rather than coniferous) plants, the leaves of which are tough and leathery (from the Greek *skleros*, hard; and *phyllon*, leaf). Broad sclerophyll vegetation consists more of shrubs than of large trees. To Americans this biome is familiar as the coastal chaparral of southern California, but broad sclerophyll is also the biome of Mediterranean lands, of southwest Australia, of the southern tip of Africa—in fact, wherever one finds the combination of cool, wet winters and hot, arid summers.

Readers should not confuse the loose popular usage of "chaparral" with its much more specific ecological meaning. On the other hand, the popular image of "jungle" is too narrow. It is a mistake to think, as most people do, that the forests of tropical Asia, Africa, and South America consist only of evergreen rain forest. Wherever there is abundant but highly seasonal rainfall in the tropics, one encounters tropical decidous forests. It is apparent that if we could make Figure 2.14 three-dimensional, a third dimension that would accommodate the broad sclerophyll and tropical deciduous biomes would be some measure of rainfall seasonality.

The global location of particular biomes is intimately related to the atmospheric and oceanic circulation patterns described earlier in this chapter. Some of these relationships we have already noted. Deserts occur in either of two situations: (*a*) in the belts of high pressure near 30° N and S of the equator, or (*b*) where ocean currents are perpetually cooler than adjacent coasts. Elements of the broad sclerophyll biome often occur just poleward from coastal deserts, where the land is warmer in summer but cooler in winter than the nearby ocean. The temperate rain forests of the Pacific Northwest and of southwestern New Zealand are found where a warm current lies offshore and westerly winds carry moisture-laden air against coastal mountains. The boundary of the tundra and boreal forest biomes coincides with the average summer position of polar air fronts, while the boundary between boreal and temperate deciduous forests seems to coincide with the average *winter* location of polar fronts (Bryson, 1966). The relationship of tropical deciduous forest to monsoon circulation patterns should be easy to discern; frequently, in fact, this biome is called monsoon forest.

Indeed, the classification of climates most widely followed by climatologists, that of Köppen, is based on vegetation types. It is illuminating to know that climatologists see such a close relationship between biomes and climate; but to one wishing to understand more fully the relationship of biomes to environmental factors there is a certain circularity in defining climates in terms of biomes. Climatic classifications constructed independently of vegetation therefore have much to recommend them. One such classification, based on biologically significant parameters, is that of Thornthwaite (1948), which pinpoints striking contrasts of seasonal water shortage between different biomes (Fig. 2.16).

Biogeographic Realms

We have been speaking of ecologically determined patterns of terrestrial life. There are also historically determined patterns, for which evolutionary isolation is responsible. One of the greatest students of these patterns was Alfred Russell Wallace:

> Leaving Bileling, a pleasant sail of two days brought us to Ampanam in the island of Lombock. . . . Birds were plentiful and very interesting, and I now saw for the first time many Australian forms that are quite absent from the islands westward. Small white cockatoos were abundant, and their loud screams, conspicuous white color, and pretty yellow crests, rendered them a very important feature in the landscape. This is the most westerly point on the globe where any of the family are to be found. . . .
>
> During the few days which I stayed on the north coast of Bali on my way to Lombock, I saw several birds highly characteristic of Javan ornithology. . . . On crossing over to Lombock, separated from Bali by a strait less than twenty miles wide, I naturally expected to meet with some of these birds again; but during a stay there of three months I never saw one of them, but found a totally different set of species, most of which were utterly unknown not only in Java, but also in Borneo, Sumatra and Malacca.
>
> *The Malay Archipelago* (1869)

In this passage Wallace, the curious naturalist, relates his discovery of a major discontinuity in the geography of life: the division between a dominantly Australian fauna and a fauna linked to Asia. The change was so abrupt that Wallace was able to trace a line through the middle of the Indonesian archipelago separating these faunal realms. The contrast between Bali and Lombok is especially striking because these islands lie close together in the middle of a continuous island chain (Fig. 2.17). The climatic differences between the two islands are certainly too slight to account for the biological differences.

What is the explanation? Wallace points out that Borneo, Bali and the islands to the west all rise from a shallow sea, the floor or which is part of the Asian continental shelf. During the Pleistocene ice ages the sea receded and these islands were part of a wholly or nearly continuous Asian land mass. In the Lombok Strait, however, and northward between Borneo and the Celebes, the sea deepens abruptly to over 200 m. This border (Wallace's Line) has apparently constituted a water barrier for a very long time, restricting east-west movements of land animals. The biological differences between Bali and Lombok reflect geographical isolation, not environmental contrasts.

Just as the earth's land surfaces are characterized by ecologically distinct biomes, so they can be divided into biogeographical realms of separate evolution. The Australian realm is one of the most distinct of these. The unique Australian biota has evolved in splendid isolation from the other continents; that is why it is unique. South America (termed the Neotropical Realm) is likewise a highly distinctive biographical region. Like Australia, South American has been an isolated continent for much of its history; the Isthmus of Panama is of quite recent origin. Besides these two continents, zoologists recognize four other major biogeographical regions: the Ethiopian realm (most of Africa plus Madagascar), the Oriental realm (southern Asia and western Indonesia), and the Palearctic and Nearctic realms (northern Eurasia and North America, respectively). The biogeographical realms recognized by botanists are similar but somewhat more numerous. Isolated islands such as Hawaii and New Zealand have such distinctive biotas that, if larger, they too might be recognized as separate biogeographical realms.

Each biogeographical realm is a self-contained, internally contiguous region, as might

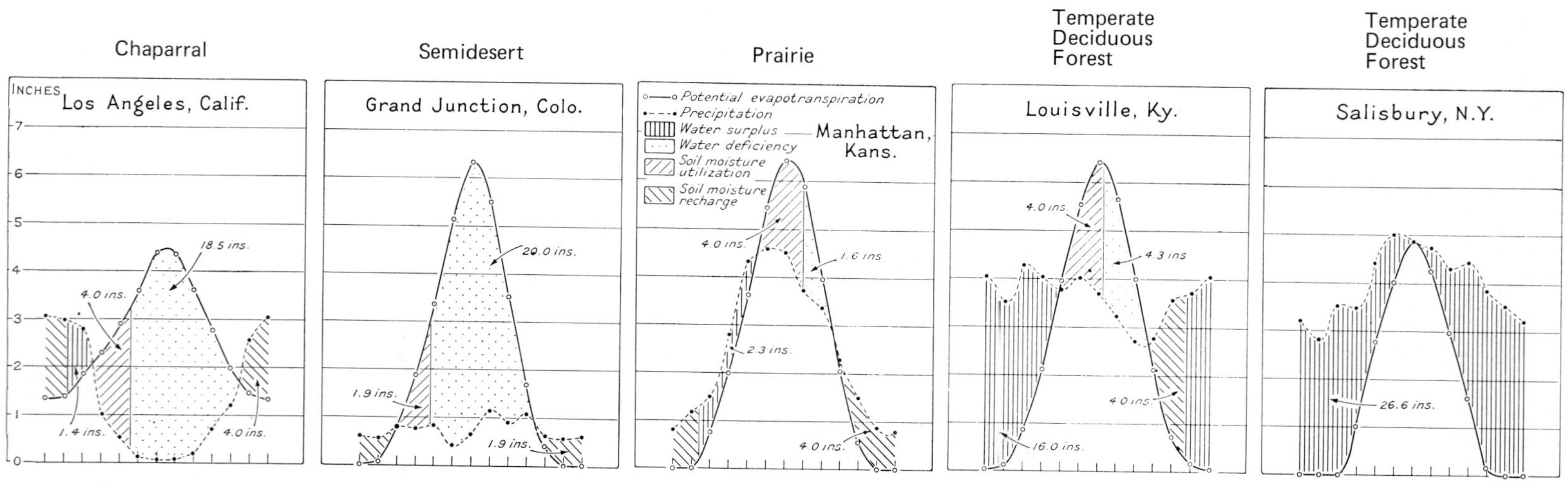

FIG. 2.16. Climatic contrasts in an east-west traverse across the United States. Potential evapotranspiration (PE) is the quantity of water that would be lost by evaporation from soil and vegetation if unlimited water reserves were available. When seasonal precipitation and potential evapotranspiration are plotted together, their interaction defines four possible periods during the year: a period of water surplus, when the ground is essentially saturated and rainfall exceeds PE; a period when PE exceeds rainfall but there is adequate residual soil moisture for plant use; a period of water deficiency when PE continues to exceed precipitation and soil moisture reserves run low; and a period of soil moisture recharge when precipitation again exceeds PE. PE is chiefly dependent on temperature and day-length; it thus varies geographically, and it also varies seasonally in the same locality. Note the differences in degree and longevity of the period of water deficiency, and also the seasonal differences in rainfall patterns. (From C. W. Thorthwaite: *Geographical Review*, *38*: 55–94, 1948. Redrawn with permission.)

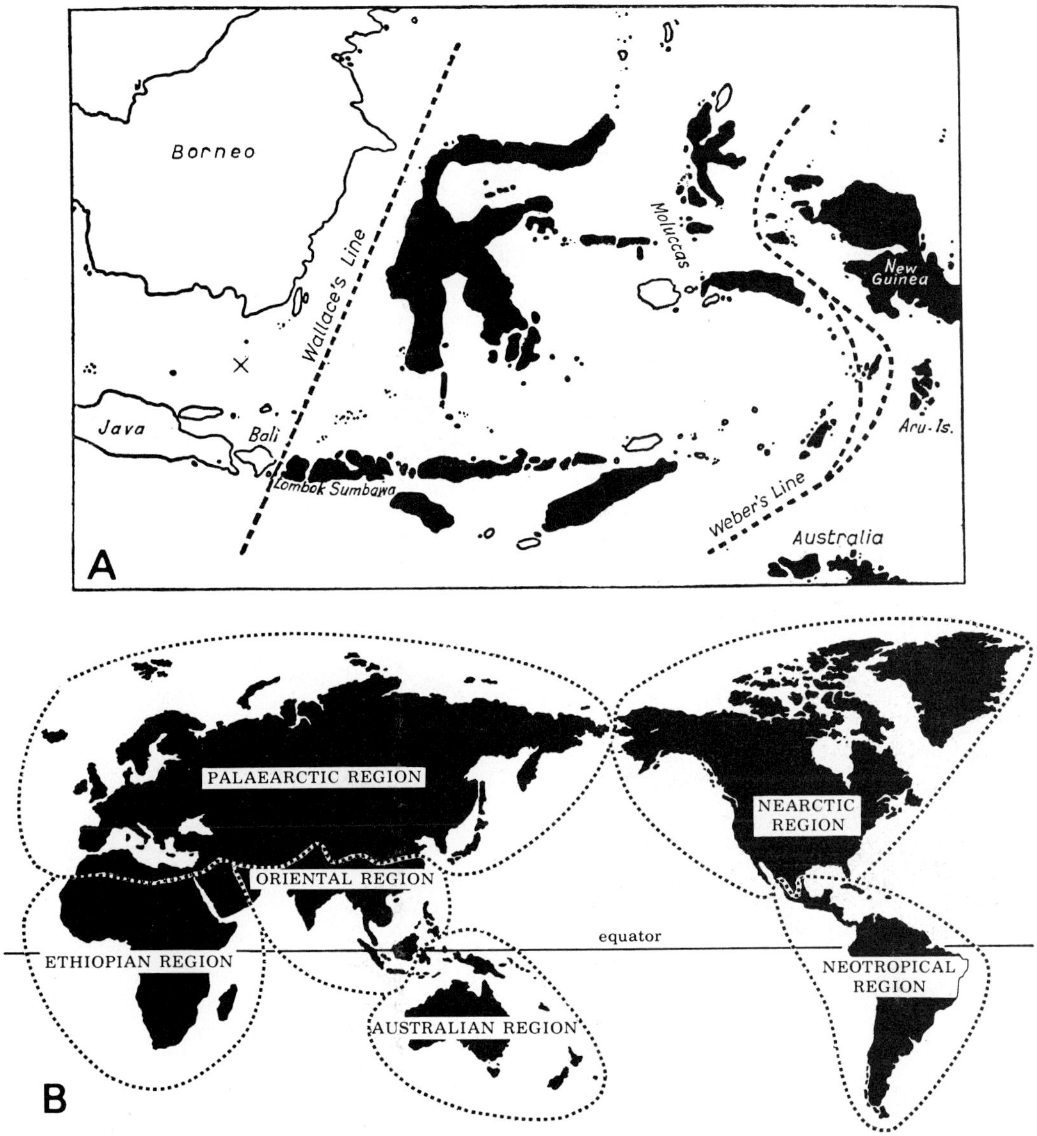

FIG. 2.17. (*A*) The distribution of cockatoos in the East Indies illustrates the sharp biogeographic discontinuity noticed by Wallace. Since Wallace's day a species of cockatoo has been discovered on one small island (denoted by an *X*) west of Wallace's line. Weber's line, farther east, is a less sharp biogeographic boundary that sets apart a somewhat distinct New Guinean element *within* the Australian realm. (After B. Rensch, from C. S. Elton: *The Ecology of Invasions*, New York: John Wiley and Sons, Inc., 1958.) (*B*) The major biogeographic regions of the world, Wallace's line is represented here, but not Weber's. (From W. T. Keeton: *Biological Science*, Ed. 2. New York: W. W. Norton & Co., Inc., 1972.)

be expected of a separate center of evolution. A single biome, as already emphasized, may be scattered all over the globe; there are deserts in five different biogeographical realms and tropical rain forests in four. What relationships exist between organisms in geographically isolated parts of the same biome? There is usually little or no relationship in terms of evolutionary ancestry; but under similar environmental influences similar communities have arisen, containing organisms that fill similar roles and that often even look remarkably alike (Fig. 2.18). This is the phenomenon of **convergent evolution.** When different organisms in different segments of a biome fill similar roles they are termed "ecological equivalents," whether or not

P.B.

they look alike. The marsupial wolf of Tasmania and the unrelated placental wolf of North America are ecological equivalents; but so, too, are the bison and certain of the New Zealand moas! Until introduced by man, grazing mammals did not exist in New Zealand. Flightless birds took their place in that land's grassland biome.

OCEANIC LIFE PATTERNS

The oceans, too, have ecological and biogeographical patterns of life. Distinct community types exist both on the ocean floor and permanently suspended in the medium. The floor and the waters above it constitute two great ecological divisions of the ocean, the benthic and pelagic regions (Fig. 2.19). Each of these regions can be subdivided according to depth into the marine equivalent of biomes. In differentiating these community types the light gradient is highly important, although other factors also are involved. The light gradient is steep near the surface, and biological communities here change swiftly with depth. Once below the zone of light, biological changes are much more gradual. As we have pointed out, the fauna of the lightless abyssal floor is remarkably similar over a vertical range of 8 or 9 km.

What of horizontal life patterns? Though less sharply defined than on land, biome-like horizontal patterns exist in surface and shallow coastal waters and seem to be determined chiefly by temperature. Coral reef ecosystems, for example, occur only in warmer seas. Besides such ecological patterns, shallow-water communities also exhibit biogeographical patterns; Elkman (1953) recognizes 16 zoogeographical realms of littoral marine animals. Similar phytogeographical realms probably exist for seaweeds.

To a greater extent than on land, the ecological and zoogeographical patterns of coastal waters coincide, suggesting that evolutionary isolation in the oceans has been brought about through the combination of geographical separation and ecological separation by temperature. Temperature extremes are less in the ocean than on land, so it is somewhat surprising that this factor serves so effectively to isolate coastal marine organisms.

The vast region of the abyss is very uniform thermally and relatively continuous geographically. Thus, it shows little of the patterning, either ecological or biogeographical, that is found in shallow and surface waters.

The patterns we have been discussing, both terrestrial and oceanic, are patterns of groups of organisms. Other sorts of patterns also characterize life on Earth—patterns of productivity, for example, or patterns of species numbers. These will be described in later chapters.

ORGANISMS AS MODIFIERS OF THE ENVIRONMENT

One should not conclude at this point that organisms are purely at the mercy of climate, soil, water chemistry and the like. In discussions of microenvironments, of ecological succession, of the chemical histories of atmosphere and ocean, we will see that organisms play important roles in determining their own physicochemical environments. Indeed, the environmental changes wrought by man pale (though not to insignificance) beside the global changes in environment accomplished by organisms in past ages. Even in the present, when we automatically consider man the world's most overpowering species, it is well to set things in perspective. Considering only other animals, is man really more important in the global scheme of things than, say, the species of coelenterates that build coral reefs, or—heretical thought—than the abyssal holothurians (sea cucumbers)? According to Heezen and Hollister,

> It is an inescapable conclusion that holothurians are the dominant large animals of the major part of the earth's surface. Holothurians feeding on bottom sediments mix and till the surface muds on an enormous scale, producing features more widespread and more visibly evident than those produced by any other animal on earth.
>
> *The Face of the Deep* (1971)

FIG. 2.18. Convergent evolution among rain forest mammals of two biogeographic realms. Those on the *left* are from tropical Africa (Ethiopian Realm), those on the *right* from South America (Neotropical Realm). *From top to bottom and left to right:* pigmy hippotamus and capybara; African chevrotain and paca; royal antelope and agouti; yellowback duiker and brocket deer; terrestrial pangolin and giant armadillo. Each pair of animals is drawn to the same scale. The pairs are virtually unrelated in terms of ancestry, despite their morphological similarity. (From F. Bourlière, in B. J. Meggers and E. S. Ayensu, eds.: *Tropical Forest Ecosystems in Africa and South America*. Washington, D.C.: Smithsonian Institution Press, 1973.)

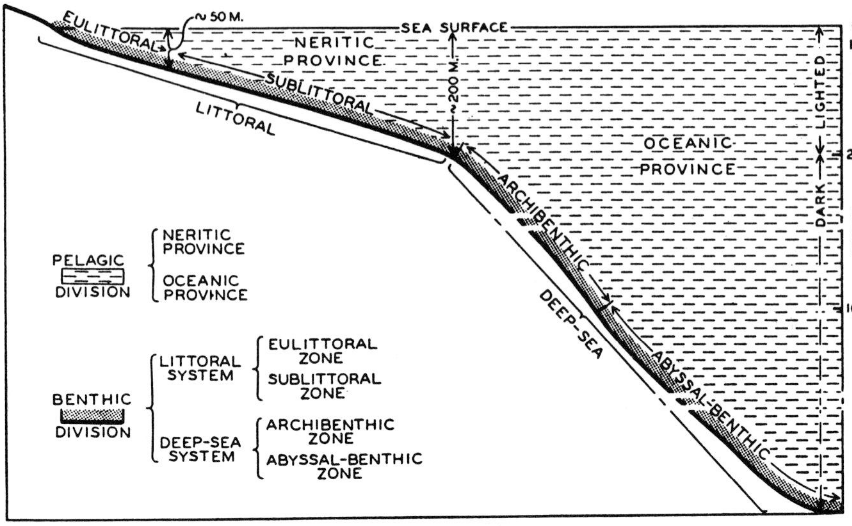

FIG. 2.19. Main habitat divisions of the marine environment. (From H. U. Sverdrup, M. W. Johnson, and R. H. Fleming: *The Oceans: Their Physics, Chemistry and General Biology*, © 1942, renewed © 1970. Reprinted by permission of Prentice Hall, Inc., Englewood Cliffs, N. J.)

Of course, sea cucumbers and coelenterates are agents of the status quo. Man's claim to preeminence is as an agent of ecological change. At this moment in Earth history he *is* unchallenged at that.

SUGGESTED READING

Aeolian zone and life at high altitudes

HOCK, R. J. The physiology of high altitude. *Scientific American* (Feb., 1970).

SWAN, L. W. Ecology of the heights. *Natural History* (Mar. 1963). See also the references to Swan at end of this book.

Abyssal floor

HEEZEN, B., and C. HOLLISTER. The Face of the Deep. New York, Oxford University Press, 1971. Marvelous photographs and text.

SANDERS, H. L., and R. R. HESSLER. Ecology of the deep-sea benthos. *Science* 163: 1419–1424 (Mar. 28, 1969).

Earth as a habitat

The Biosphere. Special issue of *Scientific American* (Sept. 1970). See especially articles by Hutchinson, Oort and Deevey.

BERRILL, N. J. Biology in Action. New York, Dodd, Mead & Co., 1966. An introductory biology text with an excellent section on "The Planet Earth: Past and Present."

GATES, D. M. Energy Exchange in the Biosphere. New York, Harper and Row, Inc., 1962.

HARE, F. K. The Restless Atmosphere. New York, Harper Torchbooks, 1963. An excellent paperback on weather and climate.

HENDERSON, L. J. The Fitness of the Environment. Gloucester, Mass., Peter Smith Co., 1970. A biochemist's classic inquiry into the suitability of Earth for life. First published in 1912.

HUTCHINSON, G. E. The biogeochemistry of aluminum and of certain related elements (concluded). *Quarterly Review of Biology* 18: 331–363 (1943). A classic discussion of the chemistry of the earth in biological perspective.

THORNTHWAITE, C. W. An approach toward a rational classification of climate. *Geographical Review* 38: 55–94 (Jan. 1948).

Biomes and biogeographical realms

BILLINGS, W. D. Plants and the Ecosystem. Belmont, Cal., Wadsworth Publishing Co., 1964. Useful descriptions of the biomes, their climates and soils.

CARLQUIST, S. Island Life: a Natural History of the Islands of the World. Garden City, N.Y., The Natural History Press, 1965. An absorbing introduction to the whole subject of biogeography.

EYRE, S. R. Vegetation and Soils: a World Picture. Chicago, Aldine Publishing Co., 1963.

RUMNEY, G. R. Climatology and the World's Climates. New York, The Macmillan Company, 1968. The biome approach to climatic classification, also with excellent sections on the atmosphere and its circulation patterns.

SCHIMPER, A. F. W. Plant Geography upon a Physiological Basis. W. R. Fisher. Trans. Oxford, Clarendon Press, 1903, and (reprinted) New York, Hafner Publishing Co., 1960. The first and still one of the most fascinating books to use the ecological approach to plant geography.

PART II

Communities and Species Adaptations

FIG. 3.1. "Grazing succession" in the grassland community of the Serengeti Plain. The sequence in the foreground goes from *right to left*, zebra being followed into a pasture by wildebeest and topi, and they by Thompson's gazelle. Each species in its grazing exposes elements utilized by the next species in the sequence. See text for further description. (Drawing by Sarah Allan.)

chapter 3

Introduction to Communities

It is interesting to contemplate a tangled bank, clothed with many plants of many kinds, with birds singing on the bushes, with various insects flitting about, and with worms crawling through the damp earth, and to reflect that these elaborately constructed forms, so different from each other, and dependent upon each other in so complex a manner, have all been produced by laws acting around us.
Charles Darwin (1859) *The Origin of Species*

TWO COMMUNITIES, TWO CONCEPTS

East African Grassland

Until recently, the grasslands and savannas of East Africa were frequented by nomadic herdsmen, big game hunters and photographers, but seldom by ecologists. This no longer is true. The rich complexity of African plains communities, the relative ease of studying animals in open habitats, and a desire to understand and aid East Africa's beleaguered wildlife before it is too late: all these have conspired to carry ecologists to the great game parks in suddenly swollen numbers. East African governments realize that if their game is to be saved, it must be better understood, and have encouraged ecological studies in areas where the large mammal fauna is still abundant. We now know far more about African grassland communities than we did even in the early 1960s, when pioneers like Lee and Mary Talbot and Desmond Vesey-Fitzgerald toiled nearly alone.

To ecologists the fascination of African grasslands lies in the complex interrelationships that seem to characterize their fauna and flora. The grassland community exemplifies beautifully the so-called **organismic concept** of the community, in which the various species are envisioned as parts of a tightly knit, interdependent community unit rather like a superorganism. A rancher seldom puts sheep and cattle together in a pasture: they compete for the same food and overgraze the land. On the Serengeti Plain, then, how can a dozen or more species of grazing mammal subsist on the same pasture? How can so many types of meat-eater—lions, leopards, cheetahs, hyenas, wild dogs, plus a host of smaller carnivores—live together in one landscape? How do the grazers influence the character and specific composition of the vegetation, and how are they influenced by it?

Partial answers are beginning to appear. Gwynne and Bell (1968) discovered significant differences in food habits among four of the Serengeti's principal grazers, the zebra and three species of antelope (Table 3.1). The smallest species, Thompson's gazelle, eats significant quantities of the small, broad-leaved herbs that grow among the taller grasses. The other three species specialize in grass, but tend to select different parts of the plant. Zebras concentrate most heavily on stem, wildebeest and topi on sheath (the portion of the leaf that clasps the stem). Leaf blade, an unimportant dry season food item for all species, is much more prominent in the wet season diet, the contrasts evident in Table 3.1 being maintained at this time. For example, leaf blade was found to comprise 61% of the food of wildebeest in the wet season, but only 37% of the zebra's diet at this time. Table 3.1 indicates remarkably similar food habits for wildebeest and topi, and

TABLE 3.1

Dry Season Foods of Serengeti Grazers

The following percentages are derived from analysis of the freshly ingested mouth, rumen and stomach contents of four gazelles and five of each of the other species. Data are from Gwynne and Bell (1968).

Species	Dicots	Grasses	Grass Leaves	Grass Sheaths	Grass Stems
			%		
Zebra	0.1	99.9	0.2	48.7	51.0
Topi	0	100	9.4	53.4	37.2
Wildebeest	0	100	17.2	52.7	30.1
Thompson's Gazelle	38.7	61.3	2.9	37.5	20.9

probably fails to pinpoint the real dietary difference between these two species. Although they feed on similar parts of a grass plant, wildebeest and topi tend to prefer different species of grass, and to predominate in areas of different grass type (Heady, 1966). In the locality studied by Gwynne and Bell, topi were not nearly as common as wildebeest; in other areas topi are common and wildebeest scarce.

Zebra, wildebeest and Thompson's gazelle stagger their use of the same pasture (Fig. 3.1) in a fashion that reflects their dietary preferences and suggests that these herbivorous species are as much cooperating as competing. Zebra move into an area first, when the grass is tall and fruiting stems extend above the leaves. Concentrating on stem, they mow the pasture back and increase the proportion of sheath and leaf in the herbage remaining. Wildebeest then enter the pasture and crop the grass even shorter, increasing the accessibility, and probably also the growth, of the low dicotyledonous herbs. Wildebeest thus condition the range for the gazelles, last in the sequence to appear. In the words of Gwynne and Bell,

> The relationships between species in such a grazing succession can thus be seen to be facilitative rather than competitive, with the result that, within limits, their numbers are positively rather than negatively interdependent.

The same conclusion was reached for a different group of herbivores by Vesey-Fitzgerald (1960 and 1965), who conducted studies on the seasonally inundated grasslands around Lake Rukwa in southern Tanzania. In these pioneer investigations of the grazing succession phenomenon, Vesey-Fitzgerald described the following sequence as the receding lake uncovered flood plain grasslands at the end of the wet season. The first terrestrial herbivores to utilize the pasture were elephants, which appeared while the tall grasses (up to 2-m height) were still partly submerged. The elephants not only fed themselves but made trails that were used by smaller mammals when the water became shallower. At the same time, hippos made equally useful trails from the lake to higher ground. As the waters continued to fall, buffalo herds followed the elephants into the flood plain, grazing and trampling the taller grasses; this stimulated vigorous sprouting of the once tall species, as well as growth of smaller grasses. The result was fine forage for zebra and three locally common species of antelope, which followed the buffalo into the floodplain pasture. Continued recession of the lake through the dry season provided continued opportunity for this sequence to occur; under steady grazing pressure the short, lush, final pasture persisted and provided ample food until the next rainy period. Although he did not study food habits in detail, Vesey-Fitzgerald observed selective choice of different grass species or grass parts by the different animal types. His descriptions emphasize the benefits accruing to the smaller grazers from the presence of the larger, and he points out that the animals actually maintain their food supply more than they deplete it, since their activity stimulates continued sprouting of fresh forage throughout the dry season. In the drier Serengeti, grazing perhaps has less of a renewing effect on the grasses, but Vesey-Fitzgerald feels that at Rukwa the pasture can be kept in good condition for grazers only by the maintenance of adequate populations of all the species involved. It is a case of "one for all and all for one."

Not only the condition of African grassland but its actual species composition changes dramatically under different grazing pressures (Heady, 1966). This may lead to another sort of succession. When the natural pasture is overgrazed, species of short grass replace the tall species favored by wildebeest, buffalo and elephant, and these animals may seek other areas. However, the short grass stage is preferred by many of the smaller antelope species, *e.g.* gazelles. Continued heavy grazing of short grass pasture may result in the replacement of grasses by broad-leaved weeds and thornbushes, a habitat more suitable to still other antelopes, such as impala. At the other extreme, very low grazing intensities usually result in the conversion of African grassland to forest, again with a concomitant change in herbivore species. This occurs if burning (which favors grassland) is also curtailed.

Among the carnivores of African grassland a similar picture is emerging. The carnivores are perhaps less interdependent than the herbivores, and their influence on the vegetation is at most indirect (through their influence on herbivore populations). But they minimize direct competition with each other through preferences for different foods and for slightly different habitats, and by hunting in different ways and at different times (Fig. 3.2). Their habits and those of the grazers are intertwined: lions, which favor zebras as prey, often hunt in teams, taking advantage of the long grass to set up ambushes into which they drive their quarry. Cheetah, feeding principally on gazelles at the rear of the grazing succession, hunt on the short grass with little concealment or teamwork, running their prey down by sheer speed.

We are left with the strong impression that African grassland communities are tightly knit, interdependent groups of plants and animals, which collectively order their own destinies. Without grazers there would usually not be grassland; without some species of herbivores, others probably would not exist. Carnivores' hunting techniques are influenced by the grazing succession, and their food habits by the food habits of other carnivores. The communities are so tightly integrated that species which one might expect to be competing fiercely for the same food may instead depend upon one another, as parts of cooperating units.

The Susquehanna Flood Plain Forest: a More "Individualistic" Community

One can stand on the banks of countless rivers in the eastern United States and see on the other shore a blue-green fringe of vegetation, distinct from the deeper green of the slopes beyond. The blue-green foliage is that of silver maple, a characteristic tree of the flood plain forest. Where hills rise abruptly behind a narrow flood plain, the blue-green fringe is a sharply demarcated strip. Where the land slopes more gently, the boundary with upland forest is less easily distinguished; here it may be misleading to speak of a boundary at all.

Several American forest types could be called flood plain forest. Some are better called swamp forest, since they remain wet the year round but are never subject to swiftly moving water. Other forests are inundated only during floods, but at such times are exposed to the full force of the current. The silver maple forest is of the latter kind, and constitutes a distinctive community type within the temperate deciduous forest biome. The silver maple forest, and similar flood plain communities dominated by trees such as river birch and black willow, are subject to these environmental peculiarities: (*a*) submergence to varying depths and for varying periods by flowing water, the force of which often results in considerable physical damage; (*b*) sudden accumulations of soil, or equally sudden erosional removal of soil that was present; trees may be half buried in mud as spring floods recede, only to be excavated below root level the following spring; (*c*) where the soils are sandy, the occurrence of "feast or famine" situations with regard to moisture; waterlogged at some seasons, the porous sands later become parched, and smaller plants wither and die if their root systems are shallow.

Animals in such a forest encounter large and unpredictable fluctuations in available food. Stranded in drying pools, fish and invertebrates are a rich but temporary food source for skunks, raccoons, and oppossums. Foliage-eating insects may ordinarily have plenty to eat, but a major flood may strip the leaves from their favorite food plants.

How are members of the community adapted to the situation? Many trees, silver maple and sycamore among them, are exceptionally fast-

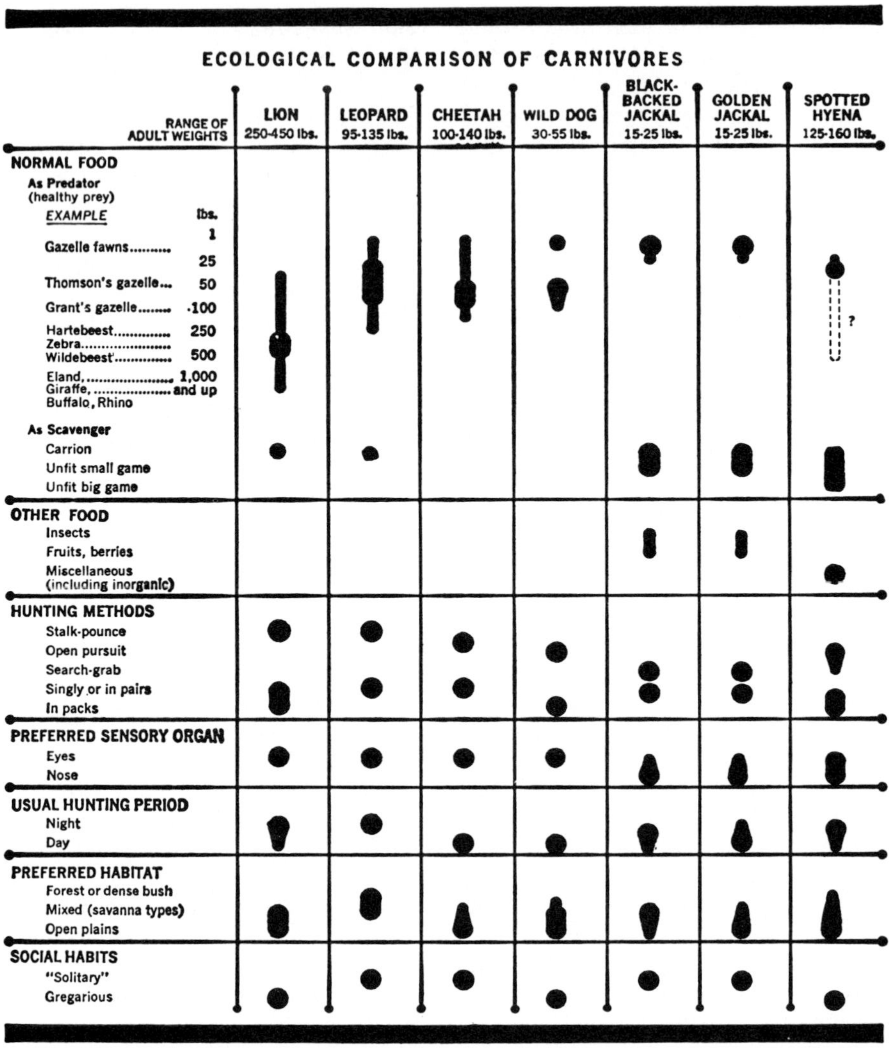

FIG. 3.2. Comparisons of Serengeti carnivores. Differences in size and behavior result in a variety of preferred prey species, allowing these predators to coexist with minimal competition. (From R. D. Estes: *Natural History Magazine*, p. 46, March 1967. Reproduced with permission from The American Museum of Natural History, New York.)

growing; this is advantageous if seedlings are to gain a foothold between floods, or win the race against drying sand. Black willow relies only partly on seeds; the roots of established willows also send up sprouts, which are better anchored than are seedlings against untimely floods. Silver maple, sycamore, and river birch sprout also, but less as a mode of reproduction than of regeneration after damage. Sprouts of different ages on a single tree often can be used to date former floods (Fig. 3.3). Along the Susquehanna, willow and river birch grow in the most exposed positions and their extensive root systems are often largely unearthed by erosion. Most of these trees remain securely anchored nonetheless, because here and there the tough roots have wedged themselves into cracks in boulders or bedrock. The supple stems of these two species also are a factor in their success; at least when young, the stalks bend before powerful floodwaters but seldom break.

These are but a few of the specific adaptations of flood plain plants; others can be discovered on a trip to the nearest river. From the standpoint of adaptive autecology, the flood plain forest is a fascinating community.

Many plants of the flood plain, though not those mentioned so far, are found in other habitats as well. What one really sees as one approaches the river is a gradient of forest composition (Fig. 3.4). The vegetation is a mixture of true flood plain species and more cosmopolitan species. Only near the river, where the floods are most frequent and severe, do the cosmopolitan species disappear.

What of flood plain animals and their adaptations? There are insects which feed specifically on certain small flood plain plants, and perhaps they have special abilities to survive submergence. So, too, may some of the soil invertebrates, and studies of these groups may be fruitful in revealing adaptations. But many larger animals, like skunks and raccoons, are transients. They divide their time between the flood plain and adjacent upland communities, and their mobility makes unnecessary the specific adaptations required of flood plain plants. There *are* vertebrate species characteristic only of the flood plain habitat, an example being the prothonotary warbler. But these animals, like the transients, can escape floods when necessary and probably have not adapted to them morphologically or physiologically.

To sum up, the flood plain forest is a distinctive community type, but it contains many species that are not unique to it, and others that are not permanent residents. Where topography permits, it intergrades with adjacent community types. In contrast to the African grasslands the flood plain forest impresses one as a loosely

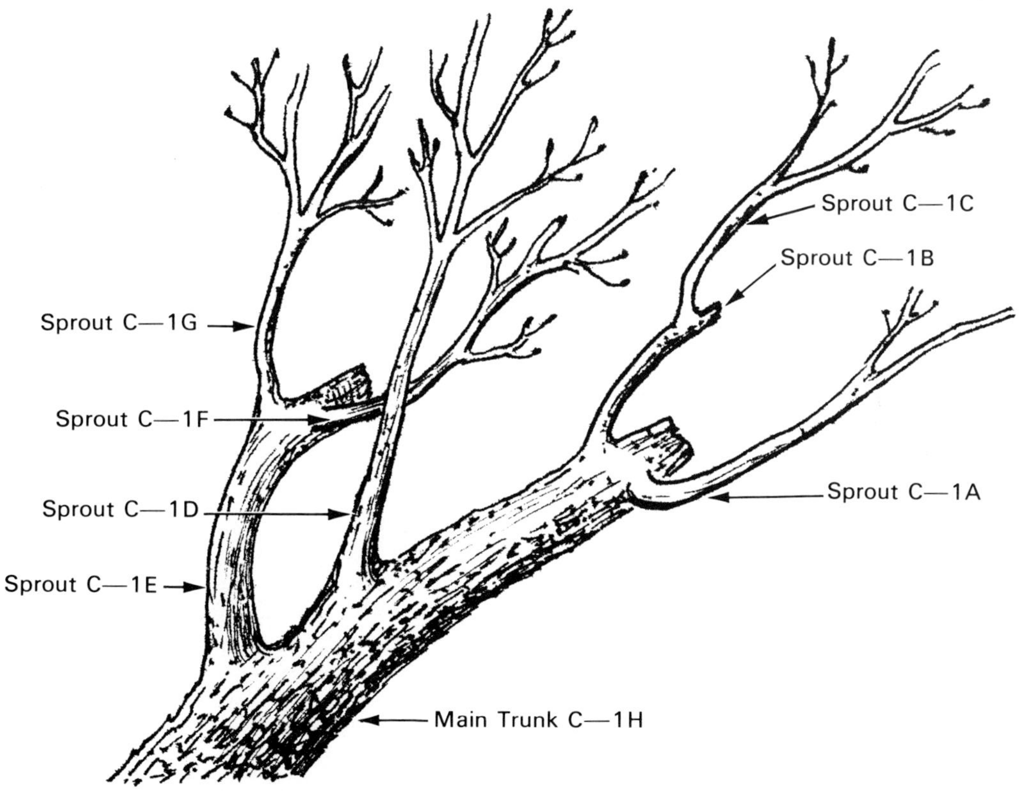

FIG. 3.3. Leaning, half-buried silver maple in the Susquehanna flood plain bears testimony to many past floods. The labeled sprouts were aged by their growth rings; initiation of every sprout can be traced to a large recorded flood, illustrating the utility of this method in botanically dating past floods where records are not available. The main trunk apparently was pushed over and damaged either in 1947 (when sprouts *C-1A* and *C-1B* were initiated) or in 1951 (when the later sprout *C-1E* was initiated). Sprout *C-1B*, in turn, was damaged by a later flood, resulting in sprout *C-1C* (1956). Likewise, damage to sprout *C-1E* can be dated to 1964, when sprout *C-1G* was initiated. Sigafoos (1964) describes similar studies. (From an undergraduate thesis by Lance Landvater.)

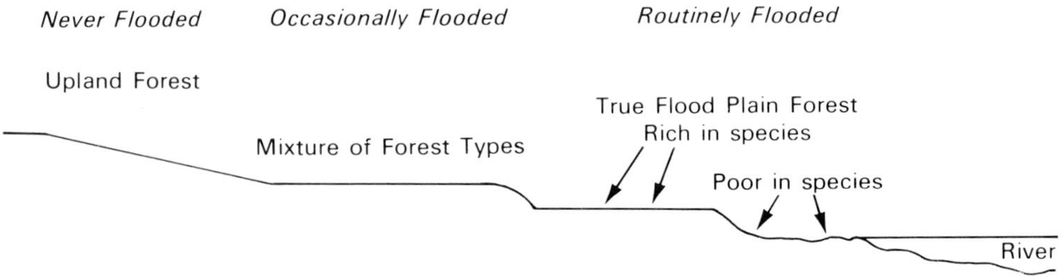

FIG. 3.4. Idealized profile of forest gradation across a flood plain. Although most flood plains consist of a series of fairly well-defined stepped terraces, one often finds a relatively continuous gradation from upland forest to true flood plain forest as the river is approached. Along the Susquehanna, the sharpest discontinuity occurs as one descends from the last terrace to the river. On this slope and the rocky, seasonally inundated flats at its foot, the number of tree species is markedly reduced compared to the richer forests on the terraces.

organized community; the species interact, but it is hard to see that they are highly adapted to one another, or collectively determine their own destiny. Those species specifically adapted for life in this environment seem adapted primarily to the unpredictable, dominating river. Each species, it seems, makes it own way.

Superorganism or Collection of Individuals?

What is the essence of a biological community? The Serengeti and Susquehanna examples point in different directions, toward the two viewpoints—**organismic** and **individualistic**—that ecologists have adopted regarding the nature of communities. The organismic concept, championed by F. E. Clements and his followers, holds that each community is a distinct entity, its species integrated so tightly that the whole is not simply the sum of its parts but requires those parts to persist. No adherent to the organismic viewpoint would argue that all parts of a community are as indispensable to its functioning as are the vital organs of a body. But we can read that implication into phrases such as Vesey-Fitzgerald's (1965), regarding the Rukwa Valley grasslands:

> Any disruption of the grazing sequence or deviation from the established pasture rotation will certainly be prejudicial to the maintenance of a correctly balanced fauna.

What kinds of evidence support the organismic concept? For one thing, its proponents point to the fact that communities are so easily recognized as distinct types. If you tell a North Carolina forest ecologist that you have an 80-year-old oak/hickory stand behind your house, he can immediately tell you 30 other species of plant and animal that you should find there. He need not visit your particular patch of forest, he can *predict* what it contains, particularly if you tell him something about its topographic location. If you tell him the forest is only 40 years old, he will predict another set of species; and he can tell you what it will contain 100 years hence, if left undisturbed.

Thus one argument for the organismic concept is simply that particular communities are so invariable as to be predictable. Clements marshalled strong support for the organismic concept by pointing to the replacement of one community by another during ecological succession. (This is a phenomenon different from grazing succession, as we shall see later in this chapter.) Clements claimed that each community in a successional sequence is distinct and predictable not only in composition but in order and time of appearance. There is much truth to this statement—in some instances.

Proponents of the organismic concept also emphasize the special relationships existing between species in a community, particularly relationships in which the species derive benefit from one another and appear interdependent. A cohesiveness is implied here, one that suggests that a community is more than just the sum of its parts. This argument takes on special force when a special relationship can be shown to involve several species, as in African grazing sequences. A marine example is provided by multi-species schools of small tropical reef fishes; these heterogeneous groupings are thought by Ehrlich and Ehrlich (1973) to provide protection from predation to each species in the association. Perhaps the most familiar examples of mixed species associations exist among certain foraging birds. If one spends much time in a forest in late summer, it soon becomes apparent that the birds move about in groups. The branches are suddenly alive with a variety of birds, all hunting busily for food. A few minutes later, the woods seem empty.

Mixed-species flocks of birds appear after the end of the breeding season and can be found through the winter when not disrupted by migration. Particular species tend to associate in these flocks (Table 3.2), and we may ask what advantages they derive from traveling together. Perhaps they uncover food for one another by their different feeding techniques or by moving in groups, or perhaps they avoid the wasted effort of picking over an area that has already been foraged by others. There is no unanimity of opinion regarding the possible advantages accruing to the members of mixed bird flocks (Moynihan, 1962; Morse, 1970; Cody, 1971), and perhaps separate answers must be sought for each case. Often the associated species are all insect eaters or all seed eaters. Here, as in the case of the Serengeti's mammalian herbivores, one would expect competition for the same food among the associated species. As in that example, competition may be subtly avoided or minimized. We may presume, at any rate, that some benefit outweighs any negative factor of competition, or natural selection would have discouraged the habit of flocking with other species. Though we do not always understand the benefits of mixed-species associations such as these bird flocks, and though they are but subsets of whole communites, their existence supports the organismic concept.

Adherents to the second school of thought, the **individualistic concept,** maintain that a community is a collection of species brought

TABLE 3.2

Typical Mixed-species Associations of American Birds

Deciduous Forest, Southeastern Pennsylvania[a]	Fields and Open Land, Southeastern Pennsylvania[a]	Open Pine Forest, Northern Arizona[b]	Mohave Desert, California[c]
Flock Type A	Flock Type A		Front of Flock
		White-breasted nuthatch	
Black-capped chickadee	Slate-colored junco	Pigmy nuthatch	Chipping sparrow
Tufted titmouse	Song sparrow	Audubon's warbler	Brewer's sparrow
Golden-crowned kinglet	Field sparrow	Pine siskin	Black-throated sparrow
Ruby-crowned kinglet	Chipping sparrow	Western bluebird	(Sage sparrow)
Downy woodpecker	Goldfinch	(House finch)[d]	
White-breasted nuthatch			
Brown creeper			
Flock Type B	Flock Type B		Center of Flock
Slate-colored junco	White-throated sparrow		White-crowned sparrow
White-throated, Tree or Song sparrow (depending on season)	Cardinal		Black-chinned sparrow
	Goldfinch		(Golden-crowned sparrow)
	Purple finch		
	Flock Type C		Rear of Flock
	Goldfinch		House finch
	Pine siskin		Oregon junco
	Purple finch		(Black-headed grosbeak)
			(Green-tailed towhee)

[a] Pennsylvania associations were recorded by Russel Smith (undergraduate honors thesis, Franklin and Marshall College, 1972).

[b] Arizona association was recorded by the author near Flagstaff, Ariz.

[c] Tripartite flock of the Mohave Desert is described by Cody (1971).

[d] Species in parentheses are less characteristic in these flocks than the others, but even the characteristic species are not always present in a particular flock type.

together by similar environmental circumstances, but operating more on their own than as essential cogs in a larger entity. This concept of the community was first forcefully espoused by H. A. Gleason in response to what he considered the overstated organismic notions of Clements. Individualists are able to turn many organismic arguments around. They would emphasize, for example, that the mixed-species flocks of Table 3.2 are actually more variable in composition than the table suggests. Any particular flock may not contain all the species that form a "typical" flock of that type; there may be additions, subtractions or substitutions, suggesting that no species is indispensable to the function of the flock. With respect to forest types, individualists acknowledge that a particualr kind of forest is likely to contain a particular group of species, but they point out that it may well not contain *all* the species one would expect—and certainly the proportional abundances of the component species are not identical in different samples of the same forest type. Even the predictable stages of ecological succession are not predictable in absolute detail; in some successional sequences, they are hardly predictable at all.

Individualists also cite paleoecological evidence for their views. Pollen studies indicate that when the glacial ice sheets of the Pleistocene receded, the communities we now know did not migrate northward as units; rather, the component species moved individually, forming temporary species associations quite different from the communities in which they exist today. If the organismic concept were correct, say individualists, one would expect to find communities maintaining their integrity, both in the past and in the present. In the present, too, the interface between adjacent communities should be sharp, and the communities on each side of the boundary should have their full complement of species if these species are essential cogs in their respective communities. Instead, real communities often merge gradually across an ill defined boundary, forming what is known as an **ecotone** or transitional community (Fig. 3.5). Can one conceive of a hybrid superorganism? This is what an ecotone represents, and strict proponents of the organismic concept should find ecotones embarassing. In Minnesota, deciduous forest species drop out one by one as we move toward the prairie, until finally there is virtually a one-species forest of bur oak, still functioning as a deciduous forest despite the disappearance of other tree species. Obvious, gradual ecotones such as the forest-prairie transition are not always found, however; where there is an extremely steep gradient of environmental factors, the sharp community interfaces that seem required by the organismic concept do appear (Fig. 3.6). But in such situations, sharp zonation patterns are equally explicable by the individualistic concept; individualists

FIG. 3.5. Prairie-forest ecotone in western Oklahoma is marked by discontinuous thickets, dominated by dwarf oaks and scattered in a sea of grass and herbs. Within the dense thickets the "trees" are protected somewhat from desiccating winds, and tend to grow taller than individuals at the exposed edges. This generates the familiar domed profile of such thickets. (Photograph by author.)

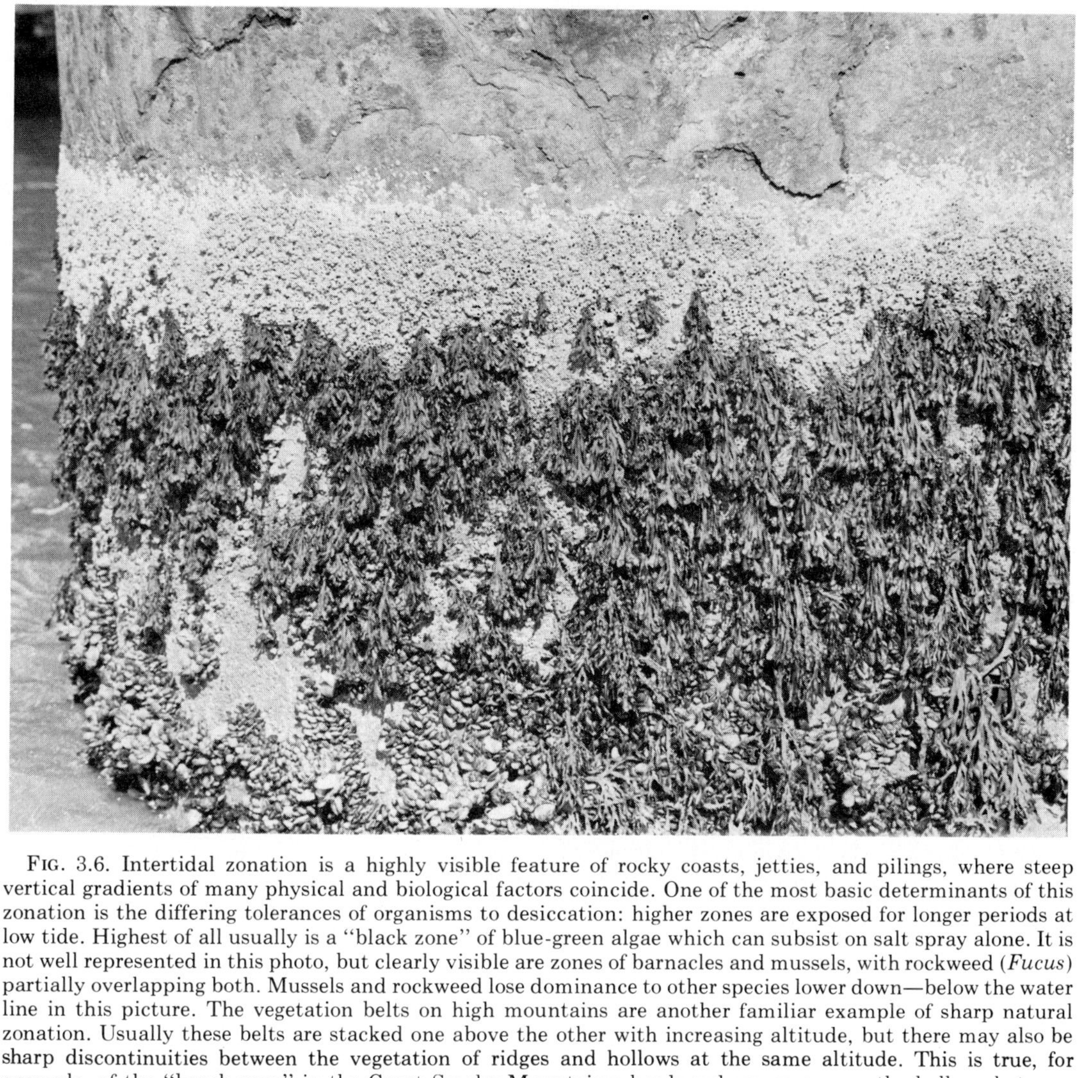

FIG. 3.6. Intertidal zonation is a highly visible feature of rocky coasts, jetties, and pilings, where steep vertical gradients of many physical and biological factors coincide. One of the most basic determinants of this zonation is the differing tolerances of organisms to desiccation: higher zones are exposed for longer periods at low tide. Highest of all usually is a "black zone" of blue-green algae which can subsist on salt spray alone. It is not well represented in this photo, but clearly visible are zones of barnacles and mussels, with rockweed (*Fucus*) partially overlapping both. Mussels and rockweed lose dominance to other species lower down—below the water line in this picture. The vegetation belts on high mountains are another familiar example of sharp natural zonation. Usually these belts are stacked one above the other with increasing altitude, but there may also be sharp discontinuities between the vegetation of ridges and hollows at the same altitude. This is true, for example, of the "beech gaps" in the Great Smoky Mountains: hardwood groves occupy the hollows between exposed, conifer-clothed ridges. (Photograph courtesy of David Miner.)

simply say that when environmental factors change very rapidly, one would expect many individual species to meet their limits within a common, narrow geographical zone.

The banner of individualism has been taken up by the modern school of continuum analysis, exemplified by ecologists such as Curtis, McIntosh and Whittaker (*cf.* Curtis and McIntosh, 1951; Curtis, 1959; Whittaker, 1967). These workers have found that even when it appears otherwise, the transition between communities is generally smooth and progressive rather than abrupt. Environmental factors such as temperature and moisture exist in nature as unbroken gradients between extremes, and a corresponding unbroken continuum of overlapping species is found along these gradients. Each species has an optimum somewhere along each gradient. The continuum may contain regions where certain species are bunched because of similar optima, and one may be deluded into believing that these species are integral parts of a distinct, organismic community. But it is a delusion, say the individualists, because one cannot really set "typical" communities apart from the intermediates that actually occur in nature.

One of the classic continuum studies of recent years was conducted in adjoining desert and montane communities of the arid Southwest, where a strong correlation between temperature, moisture and elevation causes the vegetation to be conspicuously zoned with altitude. But even here it was discovered that:

Major growth forms appear as a procession of broadly overlapping waves of dominance in the . . . continuum from subalpine forest to desert, in the sequence: [spruces and firs], pines, evergreen broadleaf trees, pigmy conifers and evergreen broadleaf shrubs, monocot shrubs, narrow-leaved shrubs, grasses, and desert forms . . . . No discontinuity of growth-forms was observed.

Whittaker and Niering (1965)

In other words, even the sharp interfaces evident in the zonation of montane vegetation become more fuzzy when examined at close range. The continuum nature of flood plain forests is illustrated in Figure 3.7.

What do we make of organismic and individualistic arguments? Must we espouse one viewpoint and discard the other? The individualistic concept presently is more widely accepted, and most ecologists feel that the claims formerly made by Clements' organismic school were both too simple and too extravagant. But surely Clements' grand vision has continuing value when balanced by the individualistic viewpoint. A community is less tightly knit than an organism, and particular species are not as indispensable as particular organs. But species do exist in groups, and the analogy of community and organism is not useless. Clearly, in this regard, the perspective of an ecologist is colored by the communities with which he is familiar. It is not difficult to imagine that the prairie environment in which Clements grew up was instrumental in his development of the organismic concept. The native herbivores of the Great Plains were less varied than those of African grasslands, but they probably had similar influences on each other and on the composition of their natural pasture:

The bison, pronghorns, jackrabbits, prairie dogs, pocket gophers and other grazing mammals, of course, did interact with one another, and together they exerted a maximum effect on the growth of vegetation. There is reason to believe that control of the grasses by bison provided more forbs for pronghorns and better visibility for prairie dogs, thus aiding in their survival.

D. F. Costello (1969), *The Prairie World*

Does one wonder that the organismic viewpoint found its champion in a man from the plains?

RELATIONSHIPS AMONG SPECIES IN COMMUNITIES

Having illustrated species interaction in specific communities, we now move to a more general categorizing of the relationships that can prevail between coexisting species. Table 3.3 lists the possibilities, with examples. The symbols indicate whether individuals of the two interacting species are benefitted (+), disadvantaged (−) or unaffected (0) by the interaction. We should emphasize that the symbols refer to individuals, not populations, of two species. This distinction sometimes makes a difference. In the case of predation, for example, one can argue that a prey *population* is benefitted by predation (which may prevent overabundance and remove diseased individuals), but one cannot claim any benefit to the *individual* victim of a predator.

Table 3.3 includes both specialized and gen-

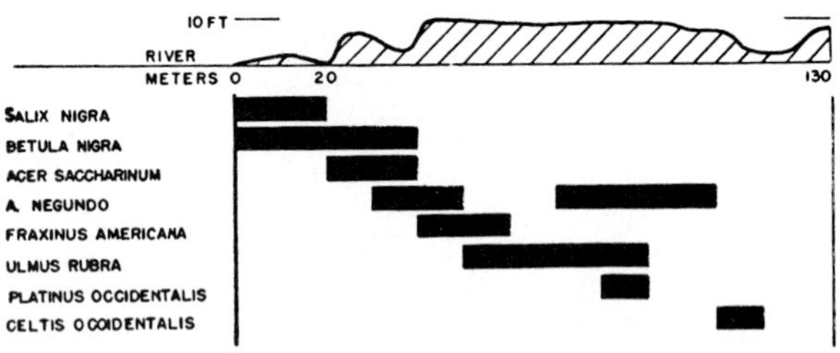

FIG. 3.7. Overlapping distributions of tree species in the flood plain of the Raritan River, New Jersey. A profile of the flood plain, vertically exaggrated approximately three times, is shown at the top. Common names of the species from *top to bottom*, are black willow, river birch, silver maple, ashleaf maple, white ash, slippery elm, sycamore, and hackberry. Lindsey et al. (1961) describe somewhat more complex continuum patterns for flood plains in Indiana. (From W. A. Wistendahl: *Ecological Monographs 28:* 129–153, 1958).

TABLE 3.3
Possible Types of Relationship between Members of Two Coexisting Species

Relationship	Species A	Species B
Neutralism	0	0
Symbiosis		
Mutualism and Protocooperation	+	+
Commensalism	+	0
Mimicry		
Batesian	+	0 (−)
Mullerian	+	+
Amensalism	0	−
Predation (and Parasitism)	+	−
Competition	−	−

eralized relationships, as well as the nonrelationship of neutralism. It can never be demonstrated that two species in a community affect each other in no way (neutralism), but this is a theoretical possibility. Quite probably, however, every species in a community affects every other, indirectly if not directly. The complex web of community interrelationships exemplifies Francis Thompson's famous line, "Thou canst not stir a flower without troubling a star."

As used in Table 3.3, the term symbiosis means any physically close special relationship that benefits at least one member of a species pair and harms neither. The last three words of this definition are sometimes deleted, thus admitting parasitism as a form of symbiosis. But parasitism can equally well be considered a special form of predation, as it is in Table 3.3. The distinction between mutualism and protocooperation is between an obligatory and a nonobligatory relationship. Cleaning symbiosis among tropical reef fishes is in most cases protocooperation: the larger fishes are rid of ectoparasites and the smaller cleaners obtain tidbits of food, but neither party *requires* the services of the other. However, the relationship between the algae and fungi composing a lichen is usually mutualistic: the partners need one another to survive.

Mimicry is functionally close to symbiosis, but lacks the close physical proximity that is one criterion of symbiosis. Instances of concealing mimicry, as when the color pattern of a moth resembles the bark of a tree, should perhaps be assigned to commensalism, since the moth only benefits from the relationship when it is resting on the bark. However, in most instances the partners need only be present in the same community for benefits to occur. The usual benefits of mimicry are protection from predators; occasionally a predator may benefit from mimicking a nonpredator. A minimum of two species is involved in mimicry, but sometimes a "model" has several mimics. The interaction of the monarch and viceroy butterflies (Fig. 3.8) is a case of **Batesian mimicry:** the tasty viceroy looks very similar to its model, the bitter-tasting monarch, and insectivorous birds avoid both of them. The monarch receives no benefit from this relationship, and could conceivably be harmed. If viceroys become too abundant, birds may learn (by trial and error) that most striped, orange butterflies are delicious, and the rarer monarch will be vulnerable to "accidents" in which they are mistaken for viceroys. Thus, Batesian mimicry is most effective when the mimic is rarer in the community than the model. In some instances, however, the mimic is protected to some degree even when the model is more rare. The poisonous coral snake is mimicked by the scarlet snake, scarlet king snake, and several species of milk snake, all nonpoisonous and together much more common ordinarily than the coral snake. Similarly, several common nonpoisonous species vibrate their tails when alarmed. They are aural rather than visual mimics: in dry grass or dead leaves the noise resembles that of a rattlesnake, and may scare off predators even when true rattlesnakes (the model) are relatively rare.

Mullerian mimicry involves the convergent evolution of two or more species, each of which is dangerous or unpalatable to predators (Fig. 3.8). Mullerian look-alikes provide a mutually reinforcing deterrent to possible predators. They benefit each other because the trial-and-error learning of predators is usually more thorough the more frequently they encounter unpalatable organisms of a particular appearance. Hence Mullerian mimics benefit from a "safety in numbers" principle. In Mullerian mimicry there is no distinction between mimic and model; each species is both.

The most general and widespread types of interspecies relationship are predation and competition. These relationships are the "glue" of communities and the impetus to many forms of evolutionary adaptation. Parasitism,

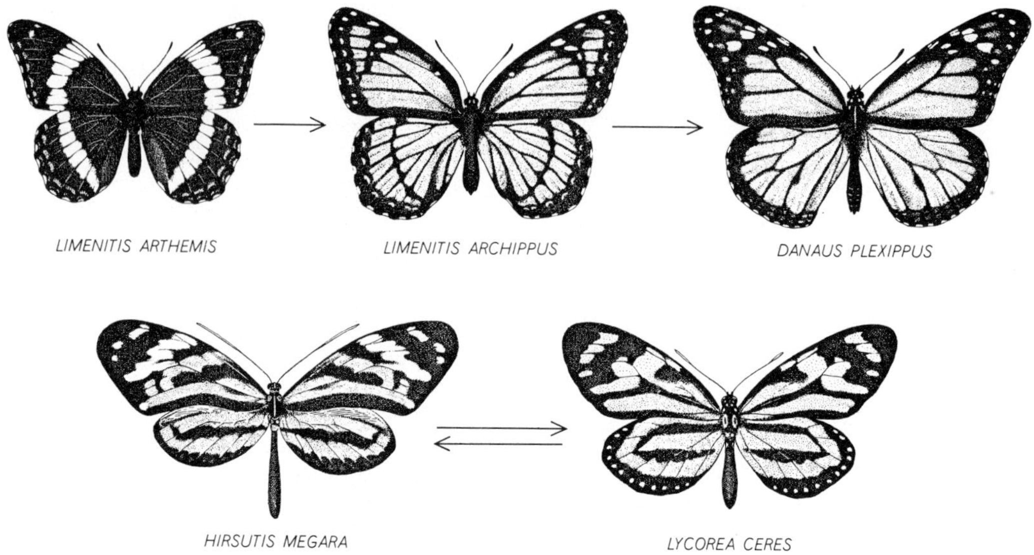

FIG. 3.8. Mimicry in butterflies. In the upper example (Batesian mimicry), the tasty viceroy (*Limenitis archippus*) has come to resemble the unpalatable, unrelated monarch (*Danaus plexippus*) far more closely than it resembles its own true relatives. The two species below, both unpalatable, are unrelated Mullerian mimics. (From L. P. Brower: *Scientific American*, February, 1969. Reproduced with permission from Scientific American, Inc. All rights reserved.)

which can be thought of as specialized predation in which the prey is not killed, is also a very common type of interaction. Amensalism once was considered relatively rare in communities, but actually it may be quite common. For example, it often is a by-product of the chemical warfare (allelopathy) that is becoming recognized as important among plants. Black walnut trees, for example, produce a compound that inhibits the growth of other plant species nearby. The chemical inhibitor does not discriminate between true potential competitors (*i.e.* other trees) and small plants, although the presence or absence of the latter makes no real difference to the walnut. The relationship between the walnut and the small plant species is amensal. In the analogous case of penicillin, many microorganisms that are inoffensive to *Penicillium* are affected by its antibiotic action. Similarly, ducks that feed on rotting aquatic vegetation may die of botulism, innocent victims of the microbial toxins that accumulate when anaerobic conditions develop. Ecologists now recognize that the classic practitioner of amensalism is man himself, in his unwitting destruction of innocuous organisms through pollution, wide spectrum pesticides and habitat destruction.

THREATS TO SURVIVAL AND ADAPTIVE RESPONSE

In communities not victimized by human amensalism, predation and competition are the chief actors in Darwin's "struggle for existence." Nonbiological factors—storms, droughts and the like—also are major threats to survival, but predation or competition accentuate nonbiological threats by driving species into marginal habitats where they are particularly vulnerable to physicochemical hardships. Of the two biological threats, competition is usually the more serious and persistent, since the intensity of predation is at least partly density-dependent. We mean by this that feeding pressure from predators usually diminishes as the prey becomes scarce: this happens either because the predator also becomes scarce (through starvation) or because it turns to other food sources. Thus, at least theoretically, predation should not cause extinction of the prey species. But competition can drive into oblivion the weaker member of a competing pair.

The Principle of Competitive Exclusion

Figure 3.9 illustrates one of the most famous of all ecological experiments. The Russian microbiologist G. F. Gause cultured, both separately and together, two similar species of *Paramecium*. In each culture flask the growth medium contained an identical inoculum of bacteria as a food source. Two aspects of the results are noteworthy: (*a*) in the mixed culture, one of the species was eliminated after a period of coexistence; and (*b*) in this culture both species, even the eventual winner of the competition, grew more slowly than they did alone.

The second point demonstrates that competition is, indeed, a −/− relationship as depicted in Table 3.3; even the stronger competitor is disadvantaged. But the first point is responsible for the experiment's fame. Gause had demonstrated experimentally a principle now thought to be basic to community organization: the principle of competitive exclusion (known also as Gause's Law). The principle states that two species cannot have identical ways of life in the same community; or more succinctly, that they cannot simultaneously occupy the same niche. The principle of competitive exclusion had been developed in the 1920s by Lotka and Volterra, in their mathematical explorations of two-species relationships; Gause's experiment vindicated their models.

Laboratory experiments such as Gause's are the only clear-cut, "live" demonstration of the principle of competitive exclusion. These experiments usually employ a vast oversimplification of natural conditions, and their significance for real communities is therefore uncertain. Perhaps the most nearly natural experiments that test this principle are those of Thomas Park and his associates at the University of Chicago (*cf.* Park, 1962). Park's experiments principally involve two species of flour beetle, *Tribolium confusum* and *Tribolium castaneum*. The natural habitat of these beetles—stores of milled grain—is so simple and homogeneous that it can easily be recreated by Park in the laboratory. The community of "real world" flour bins is likewise simple, consisting of little more than these two species. Park has demonstrated competitive exclusion in hundreds of experiments, and has been able to vary the victor by conducting the experiments under conditions of different moisture and temperature (*T. confusum* does better in cool, dry flour; *T. castaneum*

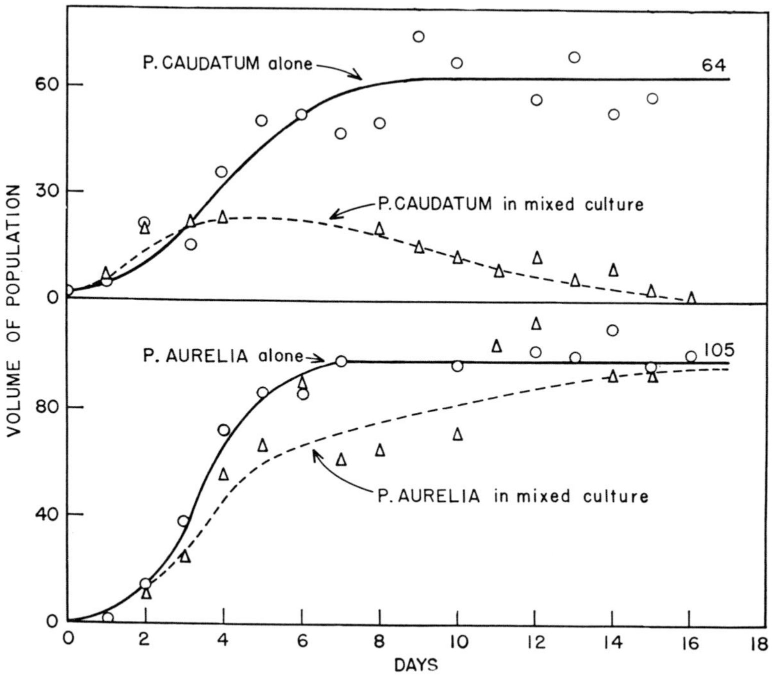

FIG. 3.9. Gause's competition experiment. The curves show the fates of two ecologically similar species of *Paramecium* when cultured alone and when cultured together. (After G. F. Gause, from W. Allee, A. Emerson, O. Park, T. Park and R. Schmidt: *Principles of Animal Ecology*. Philadelphia: W. B. Saunders Co., 1949.)

perfers it warm and moist). But the experimental climates have been kept unnaturally constant; under a fluctuating climate both species might conceivably persist through continuing reversals of their competitive advantage. In terms of climatic optima, the niches of the two species thus differ somewhat, perhaps sufficiently to permit coexistence in an uncontrolled flour bin despite the reliance of both species on the same food.

In nature, one often suspects that a species which disappears from a community has been eliminated by competition, but it is difficult to rule out other contributory factors. The principle of competitive exclusion is occasionally challenged (see the exchange between Ayala (1969 and 1970) and Gause (1970)); but it is generally accepted and enormously important in ecological thought. Ecologists in the field often use the principle in reverse: if one finds two species coexisting in a community, they *must* differ in their habits or requirements. This assumption has stimulated many careful investigations of the ecological differences between superficially similar species inhabiting the same community, and these studies have led to enlarged understanding of how communities actually function. Gwynne and Bell's study of African grazers is such an investigation. Every biologist who uncovers competition-reducing differences between ecologically similar, coexisting species adds indirect support to the principle of competitive exclusion.

Strategies for Avoiding Exclusion

Many biological characteristics of plants and animals derive from evolutionary responses to the threat of competitive exclusion. The term **evolutionary strategy** is used by many biologists as shorthand for "viable evolutionary alternative," and if we realize that it carries no purposive connotation, it is a convenient term. We can think of several evolutionary routes, or strategies, that might cope with competition. One way the threat can be met is by the acquisition of characteristics that serve a species in head-to-head competition; an example is the evolution of increased height by trees. Such a strategy results in a war of escalation among competing species, but each sooner or later reaches a limit set by structure or physiology. There seem to be physical limits, for example, to the heights that are attainable by trees in their quest for adequate light. A head-to-head strategy might not result in any clear winner; it would certainly result in many losers among species less capable of evolving into giants, if this were the only strategy employed. Obviously there are other strategies; many successful tree species achieve only modest size, yet manage to persist. Thus, although species do employ the head-to-head strategy, they seldom rely on it exclusively. There are at least three other ways by which they may cope with the threat of exclusion. They may evolve ways (*a*) of avoiding potential competitors, or (*b*) of keeping other species away, or (*c*) they may enlist the aid of other species in the competitive struggle.

a. Getting out of the way. This strategy can be followed in three dimensions: space, time or occupation. Spatial escape from competition is exemplified by weeds, which typically produce numerous seeds specialized for wide dispersal. A weedy species can rapidly invade a newly cleared habitat where initial competition is slight and it typically grows to maturity within 1 or 2 years. By then the habitat is more competitive due to the appearance of other species. The weedy pioneer is poorly equipped for head-to-head competition, and is gradually eliminated from its now crowded site. But before this population disappears, its wide-ranging offspring have discovered and successfully colonized new sites that are suitably bare. Weedy species usually persist through time by leading a gypsy existence. Egler (1971) fully realized this when he wrote of the futile irony of current programs to control roadside ragweed by application of herbicides. The resident ragweed is killed but so are other plants, creating competition-free habitats in which the most successful colonizer often is—ragweed!

Other species follow a different strategy: they remain in the same location, but escape one another in time by staggering their periods of activity. Such **temporal niches** minimize competition between coexisting species even if they have similar resource needs. In tropical rain forests growth occurs year-round, but the periods of flowering and fruiting of different tree species are staggered with respect to one another. The common need for animal pollinators and seed dispersers has been solved by the evolution of timing mechanisms that "share the wealth." In this case, the forest trees have done more than avoid competition; their cooperation assures year-round supplies of specialized foods (nectar, fruit or seeds), permitting the existence of mammals and birds that in turn are specialized for nectar-, fruit-, or seed-eating. The evolution of temporal reproductive niches by rain forest plants and of specialized feeding habits by rain forest animals must have gone hand in hand; it is one of many examples of coordinate evolution **(co-evolution)** among species in a community. Co-evolution is a mutually reinforcing process involving feedback between two or more species; a change in the first stimulates a change in the second, which pushes the first toward further change (often, as in the above example, further in the same direction). This then stimulates the second, and so, back and forth, to the evolution of relationships that may be highly specialized.

The nectar-, fruit- and seed-feeders of the preceding example exemplify a third way by which species escape from competition—escape into different "occupations." The caterpillar of the monarch butterfly feeds on milkweed, a plant that is avoided by other herbivores because the "milk" contains bitter alkaloids. The parrot fish of tropical reefs crushes coral with its curious jaws, and has this abundant food resource nearly to itself. The bristlecone pine of the Sierra Nevada has evolved a tolerance to the harsh timberline environment where competition from other trees is minimal. This last example represents a form of escape in space, but one very different from that of species such as weeds, that continually must colonize new sites. Since individual bristlecone pines may live 4000 years or longer on one site (certainly long enough for potential competitors to find the site!), they seem more analogous to monarch caterpillars and parrot fish, *i.e.* occupational specialists, than to weeds. Probably all species are ecological specialists in some sense, though often less obviously than in these examples.

b. Making others get out of the way. How do organisms "persuade" potential competitors not to compete? We have already mentioned that black walnut trees produce a chemical which inhibits the growth of other plant species. This is an example of **allelopathy,** a phenomenon whose prevalence is becoming increasingly recognized. Whittaker (1970) has suggested that one of the major directions of evolution in higher plants has been toward the acquisition of biochemical defenses against potential competitors. He believes that allelopathy may be an important determinant of the plant diversity of modern communities, and may strongly influence rates of successional replacement by plant species on a site. Of course, plants have evolved chemical defenses not only against competitors but also against animal predators; only the former represent allelopathy.

When extreme, allelopathy prevents even the appearance of competitors. It then must be classed as a special form of competition, **interference competition,** a phenomenon analogous to preventive medicine. Among animals, territoriality provides an imperfect parallel to allelopathy. Having established a territory, an animal is recognized as owner of the property and thereafter may seldom have to defend it; he has exclusive use of its resources. Unlike allelopathy, territoriality usually functions to forestall competition between members of the *same* species rather than between members of different species. But both allelopathy and territoriality are spacing mechanisms that remove organisms from direct competition with one another.

c. Obtaining competitive assistance from other species. Symbiotic relationships often give their possessors competitive advantages. The lianas and epiphytes of rain forests are relatively small, but they use trees as ladders or platforms and thereby can compete successfully for light. Remoras hitch rides on sharks, assuring themselves of food by being present at the kill. These are commensal relationships benefiting only the smaller species, but "competitive mutualism" is possible as well; in this regard, think of the benefits shared by primitive human hunters and their dogs. Janzen (1967a and 1967b) has described analogous associations between ants and several species of swollen-thorn acacia in Central America. The acacias are fast-growing, shade-intolerant trees; in their enlarged thorns they provide homes for symbiotic ants, which they also feed with sugars from specialized nectaries, and with oils ad proteins from nutritive organs on the leaves. Why this tender loving care? It is merely tit for tat. The ants kill vines and other plants that come in contact with the acacia, and also kill herbivorous insects that feed on its foliage. When Janzen destroyed the ants in some acacias with insecticide, the unprotected trees were largely defoliated by herbivorous insects; their growth rates were consequently slowed, and they were shaded out by the growth of other trees. In this complex example the acacias use their symbiotic ants for defense against both competitors and predators; in return, the ants avoid competition by being occupational specialists with regard to food.

The cooperative arrangements between two species in a community often seem unrelated to direct competition, but they may confer secondary competitive advantages. Thus the mutualism of nitrogen-fixing bacteria and leguminous trees such as acacias may allow the latter to grow more rapidly than their competitors: of course, we usually think of the bacteria simply as providing the legume with a needed nutrient, not as helping it to compete. Similarly, in the example of the remora one thinks initially of the shark as providing food, not helping the remora to best its competitors. Of course the shark does both; other animals would soon consume the scraps from the shark's meal if the hitch-hiking remora did not get there first.

Two Success Stories

The preceding examples illustrate the molding influence exerted by competition on species characteristics. The threat of predation is equally influential. Together, the two are responsible for much that we recognize as distinctive in a species. To illustrate this statement, let us look closely at two familiar plants, the pasture thistle and the dandelion. Anyone familiar with thistledown and dandelion "puffs" (see cover) will recognize that, reproductively, these species are typical weeds. Producing numerous, highly mobile seeds, they readily colonize new and browner (not greener) pastures, thereby sidestepping potential competitors. Growth to maturity takes but one or two seasons, and almost certainly some of the wide-ranging seeds of each generation will find cleared areas suitable for growth.

Yet life in open habitats carries predatory if not competitive hazards. Each isolated plant is likely to be noticed by a passing herbivore. For a pioneering weed the strategy of hiding from herbivores seems especially difficult, but this strategy—exemplified by small size and inconspicuous features—certainly is a possible one. If the plant is insect-pollinated, however, it cannot *afford* to be inconspicuous to pollinators. How does it attract pollinators while at the same time avoiding herbivores?

Pasture thistles (though not all thistles) make no pretense of hiding. They bear large, fragrant, butterfly-luring blossoms on tall stalks, and their bristly armor substitutes effectively for the hiding strategy as a deterrent of herbivores. Indeed, if one counted thistles in a cow pasture and on adjacent ungrazed land one probably would find many more thistles occurring in the pasture. We may infer that the thistles not only are relatively immune to the cattle but actually are using them in the competitive struggle against other plants (Fig. 3.10). One is used to thinking of the relationship between herbivores and plants as one of simple offense and defense; here something more subtle than simple warfare is involved.

Turning to the dandelion, we might introduce it as the plant perfectly adapted to lawnmowers. Sweating and satisfied, the weary suburba-

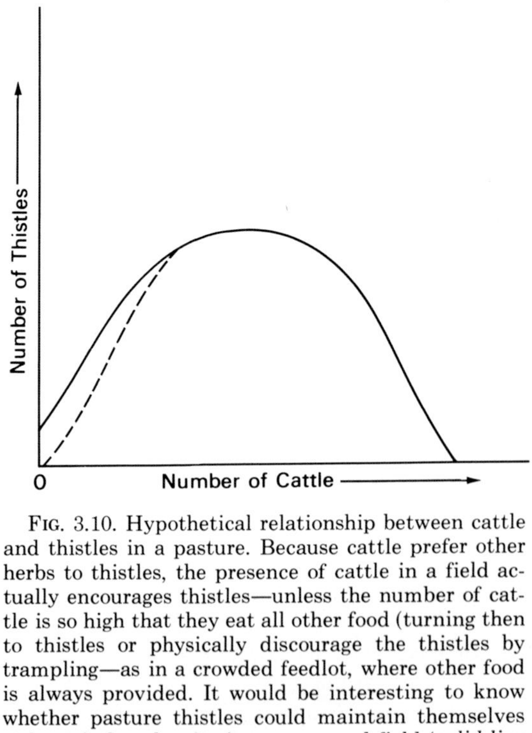

FIG. 3.10. Hypothetical relationship between cattle and thistles in a pasture. Because cattle prefer other herbs to thistles, the presence of cattle in a field actually encourages thistles—unless the number of cattle is so high that they eat all other food (turning then to thistles or physically discourage the thistles by trampling—as in a crowded feedlot, where other food is always provided. It would be interesting to know whether pasture thistles could maintain themselves at least in low density in an ungrazed field (*solid line* descending to ordinate) or whether they would be competitively excluded by other plants (*dashed line*). If the latter is the case, we could call thistles "obligate bovophiles."

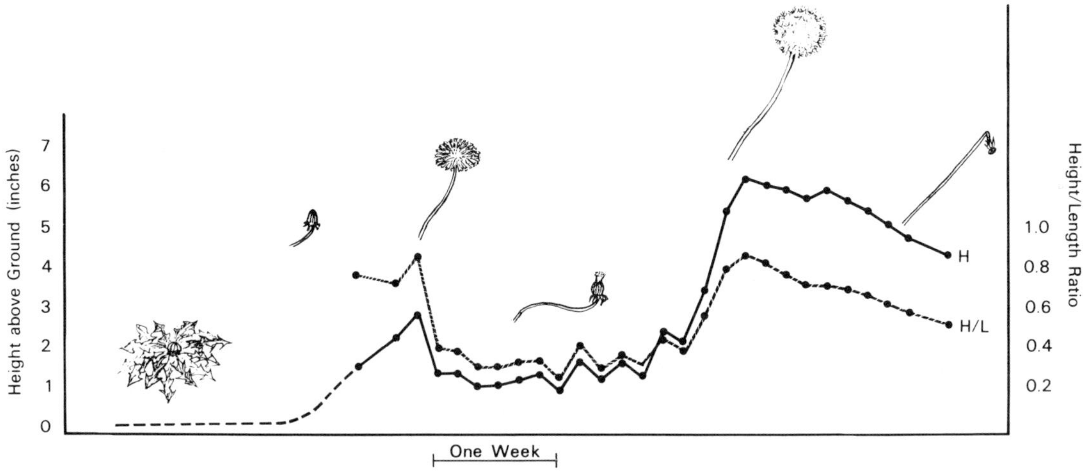

FIG. 3.11. Growth pattern of the flowering and fruiting stalk of a dandelion growing in the author's back yard. The time interval between points is 1 day. The two curves show the actual height above ground and the height: length ratio of the stalk which is a measure of the degree of erectness. See text for further description.

nite puts his mower away. Next day, the treat awaiting his unappreciative eye is a myriad impudent yellow heads and slender-stalked powder puffs, nodding smugly above the manicured lawn. How have they escaped the blade? By growth habits so self-serving as to make one almost suspect evolutionary forethought. The flower stalk of a dandelion does not grow at a constant rate (Fig. 3.11). When a bud appears, it hugs the ground for days; only when the flower is fully developed does the stalk suddenly elongate. Then, raised above the surrounding leaves, the flower advertises boldly to passing insects. After only a day or two it closes, and while the seeds mature the stalk usually droops and the flower head lies supine. Then the stalk stiffens again and there is another spurt of growth. Raised even higher than before, the head becomes an easy target for the wind (or small children) to spread the seeds abroad. For a plant that must enlist the aid of insects and wind but must also guard against lawnmowers, one cannot imagine a more perfect growth strategy than that of the dandelion.*

One must not be so captivated by these reproductive habits as to overlook the dandelion's vegetative strategy. The plant's only stalk is that of the flower head; the leaves typically hug the ground in a rosette, escaping all but the closest blade. This growth pattern probably also serves the dandelion competitively, by discouraging the establishment of other plants in its immediate vicinity. Although their supine position is helpful, the leaves of dandelions frequently do suffer damage; but beneath is a carrot-like tap root containing enough stored energy to produce a new crown of leaves even when the plant is completely scalped.

What of the toothed shape of the dandelion leaf? The teeth of plants such as holly are known to deter caterpillars from feeding on them, but surely the soft teeth of dandelion leaves could not be very effective in this regard. But thistles often grow in the same habitat as dandelions, and the leaves of thistles are deeply lobed. Might dandelions be mimics of thistles? The hypothesis is at least plausible (Fig. 3.12). Certainly it does not hurt a dandelion to resemble a thistle; and if an herbivore is not fooled by this resemblance, it yet may be deterred by the acrid juice of the dandelion stalk and the bitter flavor of the older leaves.

In fact, we can understand much of the ingenuity of dandelions in terms of adaptation to herbivorous predators. Dandelions really seem preadapted to their modern mechanical predator; they possessed the features comprising "the lawn mower strategy" long before the first Toro blade started whirring. Dandelions and thistles, in fact, are similar in many ways. They both illustrate that species having a

*In fact, this strategy no longer is a perfect one for the dandelion. Probably relatively recently, its flowers have evolved the ability to produce viable seed without fertilization. Both the growth pattern of the flower stalk and the gaudy flowers themselves, however, indicate that dandelions formerly were dependent on insect pollinators.

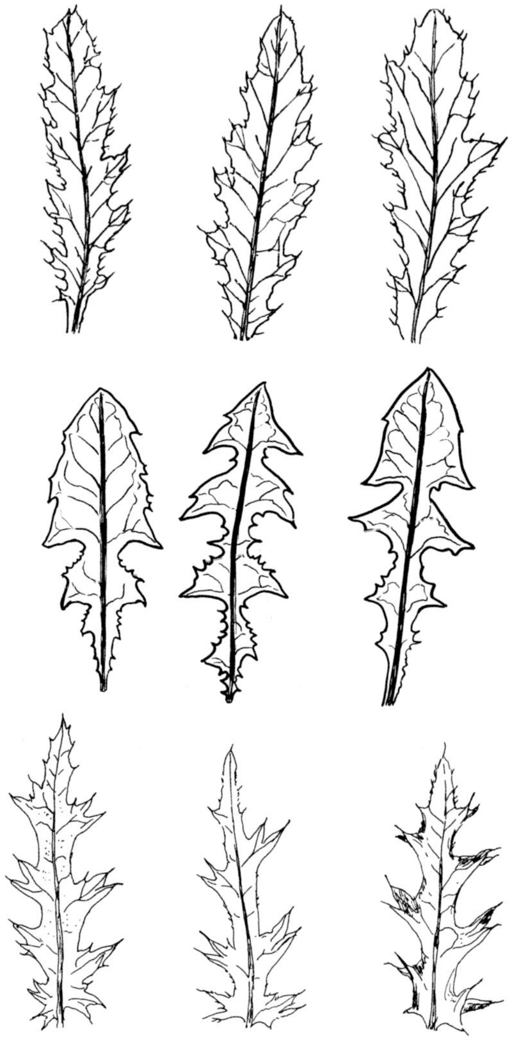

FIG. 3.12. Leaves of Canadian thistle (*above*), dandelion (*center*), and bull thistle (*below*). The shapes of dandelion leaves vary, depending on the age of a plant and whether it is growing in the shade or in the open (Sanchez, 1971). In shade, red light is preferentially filtered out by overhanging foliage, and the resulting preponderance of far-red striking the dandelion inhibits its lobing of the leaf. When growing in the open (*i.e.*, in sites where it is more conspicuous to grazers), the dandelion receives more even proportions of red and far-red light; here its leaves become deeply lobed and more closely resemble those of thistles. Whether there has been selection for protective mimicry in the dandelion is not certain; many plants growing in direct sunlight have deeply incised leaves, and there may be other reasons for the evolution of this growth form. For example, lobed leaves are less subject to overheating by the sun. (Drawing by Sarah Allan.)

primarily escapist response to competition may need ways of withstanding unusually heavy threats of predation. Once these defenses have been acquired, such species may even turn their predators into allies against competitors, and thus depart from a strategy of pure escapism.

THE ECOLOGICAL NICHE

We have already illustrated by example the essential meaning of the term "ecological niche." A species' niche is its position and role in a community. This includes how and when if feeds as well as on what it feeds (if an animal); what feeds on it; where it lives; its preferred combination of physical and chemical conditions, and the range of these conditions that it can tolerate. Sometimes the term **niche** is used interchangeably with **habitat,** but this is misleading. Odum (1963) has suggested that habitat whould be considered an organism's "address" and niche its "profession." By this distinction, species that are ecological equivalents in different communities occupy the same niche but not the same habitat. However, the term niche is also used to mean *both* the profession and address of species. By this definition, ecological equivalents occupy similar but not identical niches.

In either case, niche is a more inclusive concept than habitat. G. E. Hutchinson (1957 and 1965) suggests that a species' niche can be visualized as a multidimensional space, the shape and volume of which can be defined by that species' environmental requirements and tolerances. The many dimensions of this "hypervolume" represent environmental parameters, on which one mentally plots the species' limits. One cannot actually draw a volumetric niche of more than three dimensions; it must be imagined. But one can graphically approach Hutchinson's concept by substituting a two-dimensional area for his multidimensional hypervolume. This is done in Figure 3.13, where each axis represents the possible range of some environmental factor. The axes are made to intersect at the point on each gradient which is optimal for the species in question, and a perimeter line defines its maximum and minimum tolerable limits. The two-dimensional area that results is a representation of the species' niche. Further aspects of Figure 3.13 will be explained shortly.

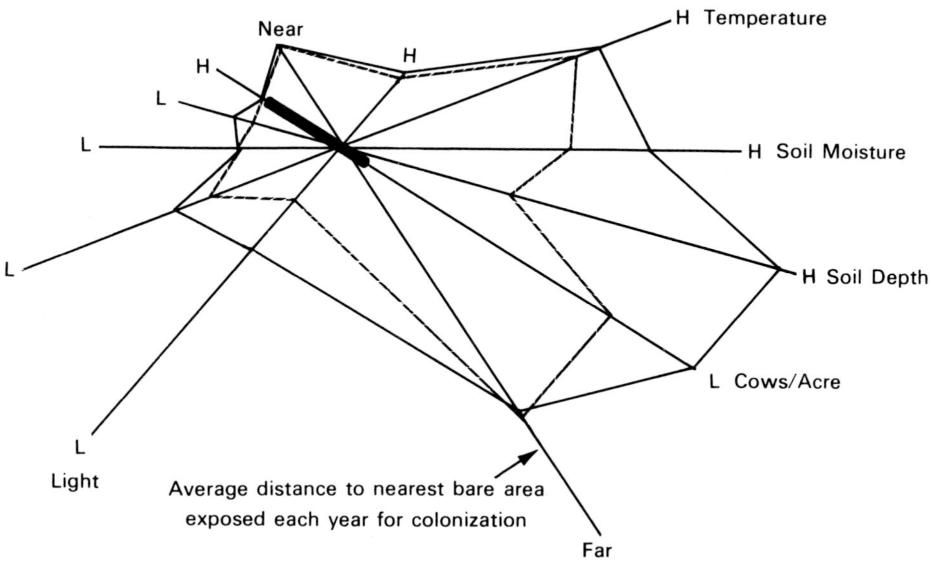

FIG. 3.13. Two-dimensional niche diagram for a hypothetical species of thistle. The fundamental and realized niches (see text) are represented by the outer and inner perimeter lines, respectively. A part of the axis representing cattle density is thickened to indicate that the thistle species is essentially free of competition at relatively high levels of cattle stocking. No other axis contains a segment that in itself confers freedom from competition.

A niche diagram of this sort is helpful, but for two reasons it is quite artificial. In the first place, a species' tolerance range for a particular environmental factor is not absolute but relative; the limits depend on whether the species is under stress from other factors. For example, a crab's tolerable salinity range is narrower when the temperature is very high or very low than when the temperature is optimal. Conversely, its tolerable temperature range is reduced at salinities that are not optimal. This phenomenon, termed **factor interaction,** makes it unrealistic for our niche diagram to indicate absolute ranges of tolerance for individual factors.

Moreover, a diagram based on tolerances tends to show the space a species *could* occupy in theory, not the space it *does* occupy in actuality. The latter is considerably smaller, primarily because of competition with other species. Competition is not included in the niche diagram as a specific axis; rather, it is considered a factor that compresses a species' niche from many sides. Thus one can draw an inner perimeter line on our diagram, enclosing the space that actually is occupied in nature by the species. This space is called its **realized niche.** The outer perimeter line, set by tolerance limits and abilities in the absence of competition, is the species' **fundamental niche.** In practice, field observations can help ascertain the limits of its realized niche, and ideas regarding the limits of its fundamental niche can be obtained in laboratory experiments. The difference in size of a species' realized niche and its fundamental niche theoretically constitutes a measure of the intensity of competitive pressure to which the species is subject.

An example may be helpful here. It is the zone *outside* the realized niche but *inside* the fundamental niche that is exploited by man when he successfully cultivates species of shade tree or crop under climatic conditions more extreme than those in which the species naturally occurs. Man accomplishes this largely by reducing the competition, thus allowing his chosen species to more nearly approach the limits of its fundamental niche.

The concept of the niche occupies a prominent place in modern theoretical ecology. Terms such as "food niche" or "temporal niche," of course, simply are a convenient shorthand for particular dimensions of a complete, or ecological, niche.

In addition to defining the niches of real species, one can talk of niches that *could* exist in a community, *i.e.* hypothetical ecological slots for hypothetical species. Having defined a hypothetically valid slot (*e.g.* with a mathematical model), one can then look at real communities and seek species that actually fill that slot.

By considering ecological niches in the abstract, rather than always associated with real species, one can also visualize a community as a collection of filled and vacant niches. As an illustration, the American deciduous forest now has several hypothetical niches vacant for higher carnivores. These niches are not completely hypothetical; we believe they exist because wolves and mountain lions once roamed these forests, and their former food is still abundant. But we cannot be sure that some other niche requirement of large carnivores is not now lacking in these forests. Perhaps a viable vacant niche does still exist for the mountain lion, since there are signs of it making a comeback (United States Bureau of Sport Fisheries and Wildlife, 1973). A real niche for the wolf perhaps does *not* still exist in the Eastern United States, but the role it vacated perhaps represents a viable niche for a carnivore with slightly different requirements.

In other communities, there are apparently vacant niches which have never been occupied. The streams and lakes of New Zealand, for example, formerly contained eels but no other large fishes. Trout were stocked in these waters; their great success and growth to mammoth size indicates that before their introduction unoccupied niches were available. In America, the rapid range extension of newcomers like the cattle egret (Fig. 3.14) probably has a similar explanation.

ROLES SPECIES PLAY IN COMMUNITIES

The position of any species in a community is a collection of characteristics: what it feeds on, how abundant it is, whether it is persistent or transient, and so forth. If we are to understand at all how a community operates, how its species composition is related to its functioning, or how a particular species fits into the total picture, we must pin down these contrasting characteristics and be able to recognize them. The aim here is not to create artificially rigid categories into which real species must be forced, but simply to define the *possible* roles a species may play in several different "theaters" within the community. In most cases we will define these roles in terms of alternative extremes, not because we always can classify a species as either this extreme or that but because we will better be able, once we know these extremes, to evaluate the characteristics we actually see in a species and thereby define certain biological aspects of its niche.

Food Chain Roles

The food relationships within a community define many broad roles which are reflected in community food chains: **producers** (plants) to **consumers** (animals that depend on living or freshly killed organic matter), and within the consumer category herbivores and several classes of carnivore. Primary carnivores feed on herbivores, secondary carnivores on primary carnivores, and so on. These various links in the chain are called **trophic levels,** and a food chain composed of the categories just cited is termed a **grazing food chain.**

Decomposers comprise a second sort of food chain. Like consumers, decomposers are classed as primary, secondary, or tertiary depending on their position in the food chain. A secondary decomposer, for example, subsists on organic residues already processed by another set of decomposers.

Most organisms have varied food habits, and in consequence they may straddle two (or more) trophic categories. Omnivorous man is an example. Nevertheless, the practice in many ecological studies is to assign such species to a single trophic level. When trophic complexity is simplified (for convenience) in this manner, species are assigned to the level at which they *primarily* operate.

Specialists and Generalists

These terms hardly need definition, but do describe clearly distinguishable roles a species may play with regard to any dimension of its niche. It may be narrowly and specifically adapted, *i.e.* specialized, or it may be broadly tolerant or flexible, *i.e.* adaptable. We may ask why these alternative modes of coping with the environment exist. Are they equally suitable

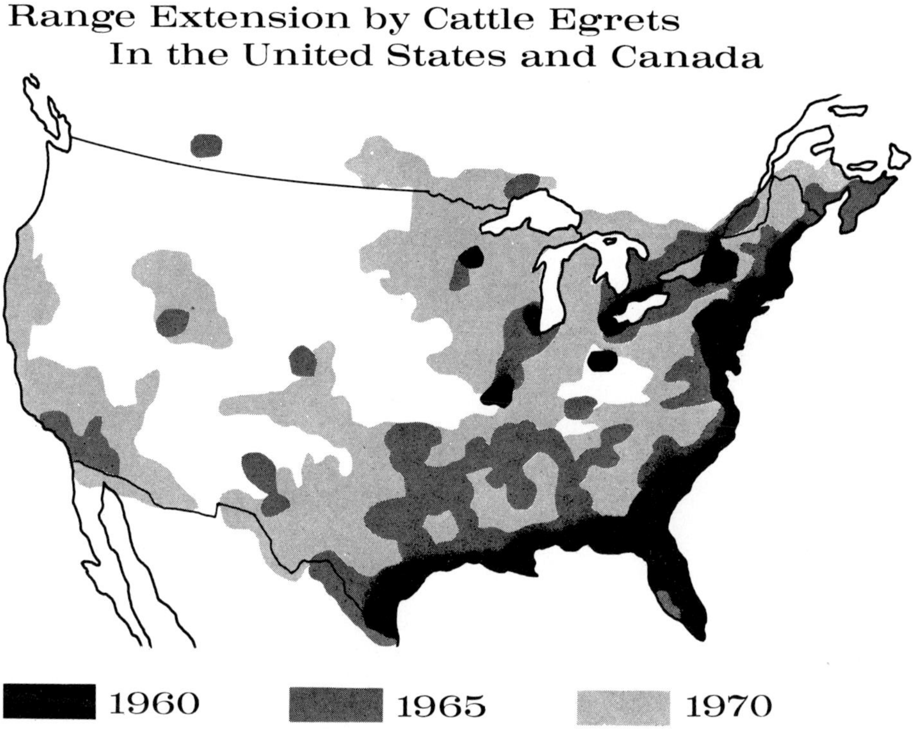

FIG. 3.14. Range extension by cattle egrets in the United States and Canada. An immigrant to South America from Africa in the 19th century, the cattle egret has spread northward, reaching Florida in the 1940's and continuing to expand its range since then. Cattle egrets follow livestock, feeding on the insects stirred up as the animals graze. These birds also have learned to follow plows and grain combines. It is suggested that this specialized feeding niche is incompletely exploited by other American birds, accounting for the cattle egret's success. (From W. J. Weber: *Natural History Magazine*, p. 61, February 1972. Reproduced with permission from the American Museum of Natural History, New York.)

evolutionary strategies? There are dangers to becoming rigidly specialized, but if the alternative, enhanced adaptability, were always the best strategy, highly specialized plants and animals should not be nearly as prominent as they are in many natural communities. Actually, specialists often have distinct advantages over generalists. By becoming a specialist, a species minimizes competitive interference from other species. We have seen how zebra, wildebeest, and Thompson's gazelle benefit by specializing in their food habits. Each is assured of a relatively exclusive resource. The alternative policy of each species eating everything in sight would probably result in one species acquiring the whole resource and the others being eliminated from the community. This policy represents gambling for high stakes: there are big winners and big losers; specialization is a conservative policy by which species get out of each other's way.

Specialization has the further advantage of producing experts. The zebra's teeth and digestive system are specialized for a fibrous, high cellulose diet and allow it to exploit stem more efficiently than can antelopes (Gwynne and Bell, 1968). The gazelle has become very efficient at finding and digesting small herbs. Similarly, specialization has allowed the bristlecone pine to perfect the characteristics that allow it to survive at timberline: slow respiratory rate, efficient moisture retention, sturdy wood. If it were also trying to live farther down the mountain, natural selection for rapid growth among competitors might push the species in an opposite direction, toward high photosynthetic and respiratory rates, extravagant water use, and more porous, rapidly produced wood. It would then be a less successful timberline species.

We thus see that the evolutionary route of adaptability has its drawbacks: a generalist

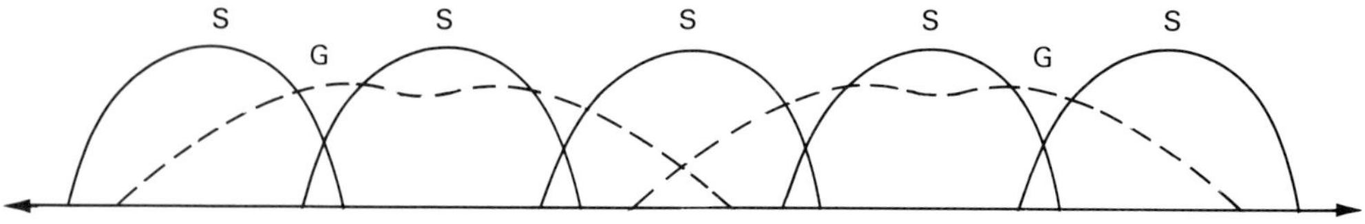

FIG. 3.15. With regard to any varying factor in the environment, for example current in a stream, specialists may divide up the gradient of conditions in the manner shown, each species having a different optimum and a rather narrow range of tolerance. Competition with other specialists thus is reduced. Generalists, with a broader range of tolerance, may span the optima of several specialists. If the specialists' optima are well separated, generalists may find greatest success in the ranges of conditions that lie between the specialists' optima. Probably more important to the success of generalists, however, are environmental fluctuations. Specialists cannot cope with these as well as generalists (see text).

cannot become so finely tuned to a particular resource, cannot live in such extreme environments, is not so insulated against competition. The *advantage* of being an adaptable generalist, of course, lies in being better able to cope with change. The Serengeti has its generalist herbivores, though we have not emphasized them. The elephant eats grass (all of it), digs up roots, browses on tree foliage, strips and eats bark. It competes with grazing herbivores, burrowing herbivores, browsing herbivores, and is probably less efficient than any of these in their specialty. But when one resource is gone, the elephant turns easily to others.

Figure 3.15 illustrates diagrammatically the relationships of coexisting specialists and generalists in a community. It can be seen that theoretically there is niche space for both. And in a natural community, we usually find both. But one measure of the environmental stability of a community is the proportion of specialist to generalist species that it contains. Extreme specialists are less able to cope with erratically fluctuating environments than are generalists, but may well be favored in constant (or even predictably fluctuating) environments. The most adaptable of all species, man, rose to dominance and spread across the globe relatively recently, during a period of extreme climatic fluctuation. This climatic instability probably selected strongly for adaptability in man, and helps to explain the rapid evolution of man's potential in this direction. But man is less well suited to extreme environments, such as deserts and high mountains, than are the specialist plants and animals of those regions. Specialists seem favored by evolution both in physically or chemically extreme environments and in nonextreme, constant environments; generalists seem favored in nonextreme, unpredictable environments.

Equilibrium Species and Opportunistic Species*

Certain species in a community are more numerically constant and more persistent through time than others. Constancy and persistence are the defining traits of equilibrium species. A community consisting largely of such species is an equilibrium or near-equilibrium community, slow to change and relatively stable except when disrupted by catastrophic forces.

Other species are opportunists, possessing the characteristics we already have noted for weeds. They seldom persist long in one place, and communities of opportunists are typically short-lived. These species are capable of rapid dispersal and rapid rates of increase under favorable circumstances, but the circumstances in any one place are usually only temporarily favorable to them. Hutchinson (1951) has termed such species **fugitive species** because they are always on the move; another name for them is **colonizing species.** The term fugitive species is sometimes reserved for the most extreme of opportunists, species which are so transient that they seldom are part of a recognizable, many-species community.

These are categories of species that have adopted different strategies of coping with competition. Opportunists avoid it; they get out of the way in both space and time. Equilibrium species live with competition but minimize its effects, often by adaptively accommodating their needs to those of other species. They carve

*Also called K adapted and r adapted species. See Chapter 14.

out occupational or temporal niches that are sufficiently distinct to permit their long term persistence in the community. The temporal niche of an equilibrium species should be considered distinct from the escape in time that is part of the opportunist strategy; the former is a specialization that allows an equilibrium species to remain in a community, the latter simply an aspect of the colonizing, "get there first" strategy of opportunists.

Purely opportunistic and purely equilibrium species both are specialists with regard to competition. However, many species lie somewhere between; they perhaps combine the dispersion tactics of opportunists with tactics that retard their competitive elimination from the community. The pasture thistle is such a species; it can persist indefinitely in a pasture, with the help of cattle, but it is also a very good colonizer. The characteristics of a pure opportunist are: (*a*) great reproductive potential (fecundity), with consequent ability to mount a population explosion when conditions are suitable; (*b*) rapid growth and early maturation of individuals; (*c*) inability to compete on a sustained basis with most other species; and, as a consequence, (*d*) restriction to transient habitats that are low in competitors. The characteristics of a pure equilibrium species are nearly the opposite: (*a*) low reproductive potential, the evolutionary emphasis being less on large numbers of offspring than on reduced mortality of offspring; (*b*) slow growth of individuals, hence often slow maturation, the direction of selection being toward an ability to endure temporarily unfavorable conditions (particularly those competitive hardships caused by the presence of other individuals of the same or different species); (*c*) a consequent inability to mount population explosions or to rapidly colonize new habitats; and therefore, (*d*) localization in long-lived, relatively stable habitats.

Now for some examples. We have already mentioned weeds as typical opportunists. Many weed species produce great numbers of seeds (Table 3.4) that are light and specialized for long distance dispersal by wind (Fig. 3.16). Other weed species vary the pure opportunist strategy; they produce fewer seeds but ensure their dispersal by enclosing them in burs, sticktights, or fruits that are distributed willingly or unwillingly by animals. The seeds of many weeds can remain dormant for considerable periods, but once they begin growth they must be relatively free of competition if they are to survive; so ordinarily they grow and mature quickly. In a long term study of an Arizona field, for example, McDougall (1958 and 1966) recorded a sudden and overwhelming explosion of tumbling mustard, the seed-producing champion of Table 3.4; this species achieved complete dominance of the field for one summer, then declined to unimportance the following year. This is typical behavior for a pure opportunist. McDougall notes, however, that during its moment in the sun the tumbling mustard *was* able to outcompete other species in the field; massive germination and rapid growth of tumbling mustard were triggered by spring rains, and it gained such a head start on the other species that later in the summer it was able to monopolize space and available moisture.

TABLE 3.4

Seed Production by Some Common Weeds

The following numbers of seeds have been recorded from single plants in a single season (Data from Encyclopaedia Britannica, 1952 Edition, Vol. 23, p. 474. Chicago, Encyclopaedia Britannica, Inc.)

Tumbling mustard (*Sisymbrium altissimum*)	511,208
Horseweed (*Erigeron canadensis*)	243,375
Green amaranth (*Amaranthus retroflexus*)	196,405
Tumbleweed (*Amaranthus graecizans*)	180,220
Common nightshade (*Solanum nigrum*)	178,000
Black mustard (*Brassica nigra*)	58,363

Harper *et al.* (1970) have listed the expected outcome if selective pressures push a plant in the opposite direction from opportunism, *i.e.* toward the equilibrium strategy. Such a species would tend to evolve: (*a*) larger and fewer seeds, each having greater viability because it has few sibling competitors and contains large energy reserves; (*b*) low dispersibility, "if only as a consequence of having larger seeds;" (*c*) lower reproductive effort, *i.e.* less energy channeled into seed production and more into longer, stronger stems that increase the species' ability to compete for light; and (*d*) increased vegetative reproduction (*e.g.* from roots or rhizomes) as opposed to seed reproduction. Vegetative reproduction allows continued nourishment of the shoot by the parent. These characteristics may

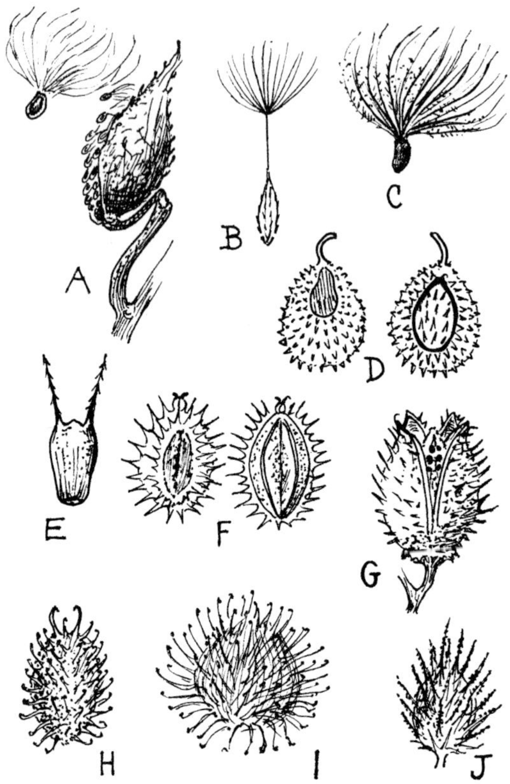

FIG. 3.16. Seeds of common weeds, specialized for wide dispersal by wind or animals. *A*, common milkweed; B, dandelion; *C*, Canadian thistle; *D*, hounds-tongue; *E*, beggar-ticks; *F*, wild carrot; *G*, jimson-weeds; *H*, cocklebur; *I*, burdock; *J*, sandbur. (Redrawn from W. C. L. Muenscher: *Weeds*. New York: The Macmillan Co., 1935.)

be compared to the list of general traits we have already given for equilibrium species. A species evolving a mixed strategy might show selection both for the ability to reproduce vegetatively and for the production of many light seeds. Tree species that combine these characteristics include quaking aspen, cottonwood, and black willow.

Opportunistic and equilibrium animal species display similar contrasts. For example, Hurley (1973) identifies a subtidal species of California barnacle (*Balanus pacificus*) as a fugitive species (extreme opportunist), basing her designation on the following characteristics. (*a*) It typically is found on small rocks and other objects partially buried in sand. These small substrates are isolated, transient habitats, being subject to complete burial or to erosion and movement by currents. On these rocks, *B. pacificus* is usually the only species to be found. By contrast, other species of barnacle are most common where *B. pacificus* is rare: on large, permanent substrates such as the cliffs and boulders of rocky coastlines, where more complex and permanent communities are the rule. (*b*) Newly matured *B. pacificus* produce an estimated 420,000 young per year. These young are planktonic, resulting in wide dispersal, and maturation is often completed in 3 to 4 months. Other less opportunistic barnacles described by Hurley produce far fewer young (7000 to 100,000 annually) and take longer to mature; their larvae, however, also are planktonic. (*c*) Many barnacles breed only seasonally, but *B. pacificus* breeds year-round; there is thus an ever-present supply of *B. pacificus* larvae ready to colonize ephemeral habitats whenever they become available.

In another marine example, Turner (1973) claims discovery of the first pair of opportunistic species known from the floor of the deep sea. This is a habitat whose constancy appears to afford little scope for the opportunistic strategy. The species described by Turner both are boring bivalve molluscs; they colonized experimental blocks of wood that he had partially embedded in bottom sediments at a depth of 1830 m. When the blocks were recovered after 104 days they were so riddled with burrows that they resembled fragile honeycombs. Many of the molluscs recovered from these burrows already contained large numbers of eggs, although they presumably had colonized the wood blocks as young, free-swimming larvae only 3½ months before. (Adults are incapable of moving to a new piece of wood, and die when their temporary bonanza is exhausted.) Rapid maturation and high fecundity, plus their proven ability to colonize new and scattered habitats, makes these molluscs ideal examples of opportunistic species. We may contrast them with the ooze-feeding species of the abyssal community, most of which must be typical equilibrium species. Subsisting on their regular but generally sparse food supply, the species of the ooze no doubt grow more slowly, using less of their hard-won energy harvest to produce new offspring and more to maintain themselves; doubtless they also possess specialized food habits, enabling permanent sharing of their common resource. The wood-boring molluscs are food specialists as well, but their opportunistic habits provide them with a nearly private resource, not one that must be widely shared.

Dominant, Rare, and Facilitator Species

Dominance in a community can be defined in various ways, but the term has an intuitive meaning that is easy to grasp. Relative dominance among plant species may be measured by the proportion each contributes to the total photosynthetic productivity of the community; by the percentage of the total plant cover that each species occupies; in forestry, by summed cross-sectional areas (basal area) of the stems of each species; or by the frequency with which each species is encountered in the community. Among animals, one usually considers dominance separately for each trophic level, and often measures it in terms of the relative numbers of individuals in each species or relative species biomasses (aggregate living weights).

Communities differ sharply in the degree to which they are dominated by their most common species. Does one species of plant or herbivore overshadow all the rest, or is there a more equal sharing of the available niche space? Intercommunity contrasts in this regard will be discussed in later chapters; here we wish chiefly to stress the interesting fact that certain species seem often to be dominant in the communities that harbor them, whereas certain other species seem never to be dominant.

Why should this be so? According to the continuum concept of species distribution, each species should be rare toward the fringes of its natural range and more common in habitats where its environmental optima coincide. But even where conditions favor them most, some species remain rare.

Can we characterize the kinds of species that are commonly dominant, and those that are never dominant? Certain generalizations are possible, the first being that there are several roads to dominance. We can envision a dominant species as being any of the following.

1. A pure or nearly pure opportunist which can saturate a new habitat before other species arrive in significant numbers. The dominance of such a species is temporary but sometimes nearly absolute, as in the example we have given of tumbling mustard.

2. A partial opportunist with staying power, *i.e.* a species which can overrun a habitat quickly and, once established, can suppress species which in other circumstances could out-compete it. An example is the dense thickets of laurel and rhododendron that occur in parts of the Appalachians. Once established, laurel and rhododendron cast year-round shade, monopolize water and nutrients, and make survival difficult for seedlings of other species, although if a tree species could persist long enough to overtop the thickets, it might well turn the tide of competition.

3. An equilibrium species having long term competitive abilities and a fairly wide fundamental niche. Imagine a species of tree which can grow tall enough to become part of the canopy, and which can flourish in a fairly broad range of envionmental conditions. It can become a dominant in the community by virtue of its generalist's tolerance range plus its ability to reach the light. However, it is unlikely that a species of this type will ever be heavily dominant, because communities usually contain several species having these characteristics. A species with a narrower fundamental niche may dominate *parts* of a community more heavily, but it will be a patchy dominance centered on the species' rather narrow zone of environmental optima.

4. A specialist living in an extreme environment, *e.g.* bristlecone pine in the Sierra Nevada. This is almost the only tree species able to survive the physical conditions of the timberline environment in which it lives; hence, it dominates the timberline vegetation.

5. A specialist which feeds on a dominant. Koalas feed only on the leaves of certain species of *Eucalyptus*, but if those eucalypts should predominate in a community, koalas would have the potential to be the dominant herbivore.

A species which fits none of these descriptions may well never be very abundant in any community. Characteristically rare species have, almost by definition, a small realized niche; but this alone is not sufficient to explain their rarity. In addition, they usually are specialists in some uncommon resource, and usually are weak competitors for at least some resources they must share with other species. A forest floor orchid, for example, may be pollinated only by one or a few species of insect, with which it has evolved a specialized, mutualistic relationship (Fig. 4.11); it must grow in the shade of larger plants, and is probably further limited in its distribution by competition with other small plants. One can see why it would ordinarily—perhaps always—be rare.

Dominant species play large functional roles in communities, and rare species usually play minor roles. Topi are the least common of the four Serengeti species described by Gwynne and Bell; their importance to the three species that dominate the grazing sequence, and their role in controlling the composition of the pasture, is no doubt relatively minor in the habitat sudied by these men. But a separate category seems useful for species that fail to dominate a community by ordinary criteria, but that play a crucial role in its functioning. We can call these species **facilitator species;** they have a hidden importance that is out of proportion to their biomass or share of resources in the community.

Facilitator species are best defined by example. In a study of Georgia salt marshes the ribbed mussel, *Modiolus demissus*, was found to be a common animal, but in biomass (11.5 g/m^2) and amount of energy consumed (56 Kcal/m^2/year) it was not particularly impressive (Kuenzler, 1961a and 1961b). Nevertheless, the ability of this species to filter phosphate from the water and make it available to the rest of the community proved to be crucial in maintaining the fertility of these marshes. In any salt marsh, the incoming tide contains considerable phosphorus of three different types: inorganic dissolved phosphate, dissolved organic compounds containing phosphorus, and phosphorus-rich organic particles (*e.g.* living and dead planktonic organisms). Without the mussels, most of this nutrient supply would be wasted; it would leave the marsh when the tide receded. But as they feed, mussels draw extraordinary amounts of water through their gills, and in this filtering process they extract a very significant fraction of the particulate phosphate from the water. Most of this particulate material is not even ingested by the mussel but is trapped in slime, ejected, and deposited on the bottom as "pseudofeces." Of the particulate phosphate actually ingested, more than half passes through the mussel and emerges in the feces. Neither of the forms of dissolved phosphate is trapped by the mussel's filtering process, but its efficiency in trapping particulate phosphate resulted, in Kuenzler's study, in the retention of more than 13% of the total waterborne phosphorus within the marsh as pseudofeces and feces. The mussel population trapped far more phosphorus than was captured by any other process in the marsh, leading Kuenzler to conclude that mussels are a major agent in the extraordinary fertility of salt marsh communities.

Mound-building termites furnish a terrestrial example of facilitator species. In the tropics termites are not restricted to the woodwork, but build conspicuous mounds of varying size in forests and grasslands. Large mounds may be 3 to 4 m high and 15 to 20 m across. In terms of their soil properties, these mounds are distinct microenvironments. Watson (1967) discovered that the soil of a large termite mound in Rhodesia differed from surrounding soils in the following ways: it was finer-textured, considerably more alkaline, richer in available base minerals (*e.g.* calcium and magnesium) and richer in phosphate, organic carbon and mineral nitrogen. Not surprisingly, old or inactive termite mounds support vegetation quite different from that around them. Often this distinctive vegetation takes the form of dense thickets; in effect, each mound supports a microcommunity made possible by the termites. But their facilitator role extends further in the case of "termite savannas" such as those of western Uganda (Fig. 3.17). The human population of this region, while primarily agricultural, hunts wild animals on the savanna and uses fire to concentrate the game and destroy its cover. But this technique is only partly successful: the grass is burned but the termite thickets remain mostly unburnt (Fig. 3.18). Upon reflection, it is apparent that termites make possible the continued coexistence of man and game animals in this area and are the major facilitator of the whole savanna community, not merely of the thicket microcommunities. Without termites there would be no cover; antelopes and other game would have been hunted out long ago; seasonal burning by man might have been abondoned; and the area would probably support woodland rather than grassy savanna. Even the way of life of the human population would be significantly affected.

Other examples of facilitator species spring to mind when one thinks of man. We will mention but one, and leave others for the reader to discover. In the coniferous forest of eastern Canada the deciduous paper birch is a conspicuous *but relatively minor* element. Its bark, however, is peculiarly suitable for constructing

FIG. 3.17. Termite savanna in western Uganda: aspect before burning. (Photograph by author.)

canoes and teepees. More than any other species, it made the area habitable for aboriginal Indians and shaped their semi-amphibious way of life in this lake-dotted landscape. And by facilitating the presence of man, the paper birch obviously exerted major indirect influence on the rest of the forest community.

ECOLOGICAL SUCCESSION

Depending on its modifying adjective, "succession" has different meanings in ecology. We have spoken of grazing succession; two much more general phenomena are **ecological succession** and **seasonal succession.**

The term ecological succession refers to a developmental or maturation process at the community level, typically noncyclic and generated by the organisms of the community itself. Ecological succession involves the appearance on the same site of a series of community types, each usually more long-lived than its predecessor.

The key phrases in this definition are "noncyclic" and "community-generated." These features distinguish ecological succession from seasonal succession in which the changes one observes are cyclic and are imposed on the community by cyclic, nonbiotic events *e.g.* seasonal weather patterns). True ecological succession can be set back to earlier stages, sometimes repeatedly, by outside agents such as storms, fire, or human disturbance. In such instances ecological succession may appear to be cyclic, but the changes in the community that occur between setbacks are wrought by the organisms themselves, and in the absence of interference would not be repeated.

In some communities ecological succession is an important and obvious process; in others it is insignificant. If this point were emphasized more often, arguments about succession might be fewer among ecologists. Ecological succession characterizes those situations where organisms really are capable of modifying the physicochemical conditions within their community. Where they cannot do this in any community-wide sense, ecological succession scarcely occurs. As we shall see in the next chapter, the

FIG. 3.18. Termite savanna after burning. Thickets remain, though browned at edges, providing cover for game. (Photograph by author).

classic examples of ecological succession come from forested regions. This is no accident: more than most other communities, a developing forest modifies its own environment in important and progressive ways, particularly with regard to soil composition and light. These changes often diminish the abilities of resident species to survive, and encourage other species whose needs are more suited to the new conditions. The new species, in turn, may alter *their* environment in self-defeating ways but make it suitable for still other species. In forests the plants are the major causal agent of succession but the animals change too, largely as the result of vegetation changes. In other communities, of which grassland may be an example, animals may play a greater causal role in the successional process.

In contrast to a forest, consider now a sandy beach exposed to waves and tides. An ocean beach harbors a well defined community; on the Atlantic shore it might include the surf clam (*Donax*), burrowing crustaceans such as the mole crab (*Emerita*), several characteristic species of polychaete annelid, and the sandpipers and larger birds that feed on these. The organisms are beautifully adapted to their environment but almost completely incapable of changing it. Tides come and go, surf pounds the beach, and the sand is constantly shifting. Ecological succession is an impossibility.

This example is extreme, but in many communities the organisms cannot modify their collective environment to the degree necessary to initiate any real pattern of ecological succession. Oceanic communities are largely of this nature. In *The Forest and the Sea* (1960), Marston Bates develops a wonderful analogy between the two habitats of his title. In each, life is distributed in horizontal layers: the photosynthesizers and herbivores are massed at the top, the decomposers on the floor. Daily fluctuations in climate become muted with depth in each habitat. There are vertical gradients in light, in temperature, and in turbulence; each decreases as one moves downward. Bates likens the huge, flimsy leaves of plants growing on the quiet floor of a rain forest to "the delicate and fantastic growths in the quiet depths of a tropical lagoon." Even the oceanic patterns of diurnal vertical migration, exemplified by zooplankton which move toward the surface at night and descend by day, have their rain forest equivalent: mosquito species migrate daily betwen the treetops and the forest floor.

But in analyzing these similarities, Bates makes a significant point: the gradients in the ocean result primarily from properties of the watery medium: its thermal stability, viscosity, and light-absorbing characteristics. The forest gradients largely represent the insulating and filtering influences of the vegetation—here organisms of the community are determining its environment. In the oceans, the "vegetation" consists of tiny phytoplankton and seaweeds which also are usually relatively small; these plants seldom are significant influences on the physical environment. In fact, when one thinks of ecological succession in the sea, one thinks easily of only one example: the progressive changes induced by developing coral reefs. Elsewhere in the ocean, the mudflows that occur on the continental slope may be repopulated by a series of successional communities, but this statement is conjectural. Likewise, in the intertidal zone, a rock that has been cleared of its attached community may be recolonized in a fairly predictable sequence, but this sequence usually has more to do with the dispersal abilities of the colonizers than with their progressive modification of the physicochemical environment. This is true of the earliest stages of forest succession as well, but is much less true of the later stages. In the oceans as a whole, and in physically extreme environments on land, ecological succession plays a relatively minor role.

To understand the literature of ecological succession one must know the meaning of several terms. A particular sequence of successional communities, *i.e.* the sequence that normally occurs in a particular context, is termed a **sere.** Each community type in the sequence is called a seral stage. The "low altitude New England landslide sere" would mean the typical series of communities that repopulate a landslide-denuded slope in that part of America. Analysis of successive seral stages would reveal a trend from early communities dominated by opportunistic species to later communities composed more and more completely of equilibrium species. The final stage of a sere, termed the **climax,** theoretically is a community at equilibrium with itself, which will not change appreciably unless disturbed by a climatic shift or other outside influence.

Unless otherwise specified, the climax community of a sere is assumed to be a *climatic* climax, *i.e.* a community which, like the biome to which it belongs, is largely determined by climate. However, one may specify a fire climax, edaphic (soil-determined) climax, grazing climax, topographical climax, etc., if one of these influences actually overrides climate in determining the terminal community in a sere. The two distinct community types of Figure 2.15 are determined by soil differences; they are the end points of different edaphic seres, and one of them is quite different from the typical climatic climax of the region.

One also speaks of **primary succession** and **secondary succession,** depending on whether succession is occurring on a virgin site not previously occupied by similar communities, or whether it is occurring again on a site from which the former community was removed. The colonization of rock by lichens and the rock's gradual conversion to a soil supporting other forms of life is an example of primary succession; the return of forest after lumbering is an example of secondary succession. Because secondary succession usually starts on a substratum preconditioned by its former inhabitants (*e.g.* on an existing soil that already contains organic matter), it generally runs its course much more rapidly than does primary succession. However, rates of both primary and secondary succession vary greatly, depending partly on physicochemical conditions (Figs. 3.19 and 3.20) and partly on the organisms involved. Succession of blue-green algae on a rock thrown into a hot spring may be completed in a few weeks. By contrast, the redwood forests of coastal California are successional communities in a sere that culminates in forests of fir and western cedar; but the redwood stage alone lasts as long as 2000 years! (It may last even longer if fires occur relatively frequently.)

Having introduced the terminology of ecological succession we now should note that some ecologists are very uncomfortble with it. Niering and Goodwin (1974) make clear the reasons

FIG. 3.19. Primary succession is proceeding very slowly on this 900-year-old lava flow at Sunset Crater, Arizona. The lava still is largely bare of vegetation, except for occasional lichens (light patches in foreground) and stunted shrubs or trees springing from crevices where organic matter—probably largely wind-introduced—has collected. This lava apparently is quite resistant to breakdown, and the relatively cool semi-arid climate does not foster rapid decomposition. The volcano itself, in the background, consists of small pumice fragments which can be colonized more easily, despite problems of erosion. On the southern slope, however (right side in photograph), evaporative moisture stress is more severe and retards plant growth. (Photograph by author.)

for this discomfort: terms like succession, sere and seral stage imply an inflexible, lock-step pattern of community replacement on a particular site, marching with precise, measured tread toward climax. Succession, however, really is a much more fluid and varied process than this. Seral stages often merge into one another and cannot be sharply defined. Supposed climax species sometimes appear embarrassingly early in the process. At other times they may not appear at all: for example, a forest sequence may be arrested nearly indefinitely in a dense thicket stage like the tangles of laurel and rhododendron in the Appalachians that we mentioned earlier. Niering and Goodwin recommend, in fact, that terms such as "community development" and "relative stability" be used instead of "succession" and "climax," in an effort to erase the lock-step mental image that the latter terms invite. Their suggestion is not without merit. In this book we will retain the original terminology because it is so widely used, but will also endeavor to keep before the reader the flexibility of successional processes.

THE BALANCE OF NATURE—MYTH OR REALITY?

Communities during succession obviously are not balanced in the sense of being self-perpetuating, but the concept of the climax community implies a self-perpetuating equilibrium. Is this equilibrium more conceptual than real? Is the climax community only an idea?

It seems heretical to question the balance of nature. This idea is one of the eternal verities, learned on our parents' knees as we listened to the tales of Old Mother West Wind. Yet in a rare moment of harmony two famous biologists (Norman Borlaug, the DDT-defending developer of miracle wheat, and Paul Ehrlich, the environmental firebrand) have agreed that the concept of community balance is a myth. Several years ago, in an interview for *Farm Profit* magazine, Borlaug described his early experiences as a forest ecologist in Idaho:

> That area . . . was essentially untouched by man, but still I found things weren't in balance with nature. We had big bark beetle epidemics, and in the summer, any number of fires would be started by lightning. . . . This balance of nature talk just makes me sick.

Borlaug is sickened chiefly by the argument of environmentalists that DDT should be banned because it upsets the balance of nature. Ehrlich, though not at all fond of DDT, treats with similar scorn the idea that communities show any real constancy through time (*cf.* Ehrlich and Birch, 1967).

Should we be equally scornful? The balance of nature can be questioned at two levels. First, one can ask whether mature communities really are structured so as to be self-perpetuating. Second, one can question whether the physicochemical environment really allows them to be self-perpetuating, even if they have that propensity.

Both questions are difficult to answer with certainty. But with regard to the first, we can suggest that in terms of their constituent species, mature communities do indeed seem geared toward self-perpetuation. Equilibrium species predominate in such communities, and as we have seen, the characteristics of such species are the product of an evolutionary strategy that emphasizes persistence of the species in the community. A community is the sum of its constituent species; hence it is reasonable that the final stage of succession should be an equilibrium community.

But can this propensity toward equilibrium be brought to fruition? What about periodic catastrophes—storms, droughts, and similar

FIG. 3.20. Primary succession occurs very rapidly on the young lavas of Hawaii. They may be softer than those at Sunset Crater, but the warm, wet climate doubtless is a major factor also in speeding the successional process. These plants, *Sedleria cyatheoides* (amaumau), have colonized a lava flow less than 3 years old. (Photograph by Glen Kaye, Hawaii Volcanoes National Park.)

FIG. 3.21. Dead *Eucalyptus* trees on the Berkeley hills, killed *en masse* by an unusual spell of freezing weather, are removed as potential fire hazard. (Photograph courtesy of *The Oakland Tribune*.)

natural events? Are they so severe and so frequent that what we imagine to be mature, stable communities are, in reality, communities always struggling back from the last catastrophe?

Consider an example. The winter of 1972–73 was unusually cold in the San Francisco area. On the hills above Berkeley, forests of Australian *Eucalyptus*—large, well established trees which had flourished in their new environment for 60 years or more—were completely devastated (Fig. 3.21). These forests seemed secure and relatively stable; they were not.

Another example: in June 1972, Hurricane Agnes nearly drowned parts of eastern America. The Susquehanna River poured an enormous overdose of fresh water into Chesapeake Bay, reducing the salinity in much of the Bay to levels lethal to shellfish and many other estuarine organisms. Marine biologists estimate that it may take 60 to 100 years for these communities to recover. But will they ever reach stability? Figure 3.22 shows the recurrence interval of floods of various magnitude on the Susque-

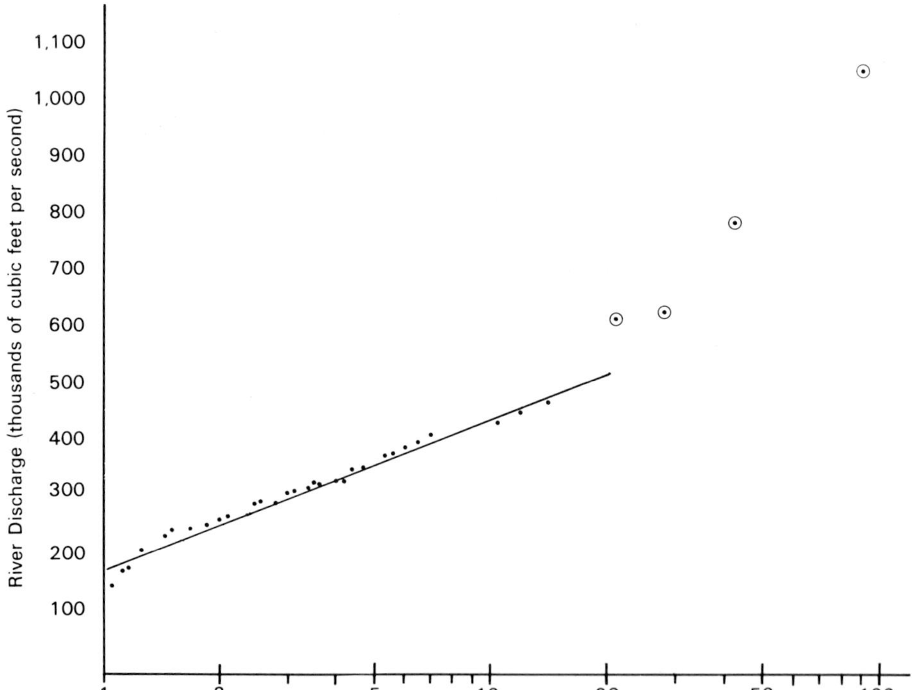

FIG. 3.22. Recurrence interval of Susquehanna River floods of given magnitudes since 1889. The graph shows, for example, that every 10 years a flood of at least 440,000 cubic feet per second is to be expected. (U. S. Geological survey data, compiled by Lance Landvater.)

hanna. The graph is interesting because it suggests the existence of two distinct classes of flood. Smaller floods (below 500,000 cubic feet per second) decrease in frequency in a predictable pattern as their magnitude increases. Larger floods, though relatively rare, occur much more frequently than is predicted by the recurrence pattern of smaller floods. Extrapolation of the small flood pattern suggests that a flood the size of Agnes should occur, on the average, every 1500 years. But if one draws a separate line of best fit for recurrence of floods larger than 500,000 cubic feet per second, one would expect another Agnes in only 90 years. One may speculate that the marine communities of upper Chesapeake Bay are *always* in a state of recovery. If they appeared stable before Agnes, it perhaps was only because of the shortness of man's memory and his failure to recognize the slow but progressive changes that were occurring in the Bay communities.

What of the flood plain forests along the Susquehanna itself? One might argue that they would develop to a "flood climax" state: those species accustomed and adapted to floods might develop an equilibrium community geared to the regular recurrence of floods of a given size. But it is difficult to see how the community could be equilibrated both with the usual small flood and the occasional catastrophic flood. The damage caused by the latter would occur sufficiently frequently to keep the community constantly out of balance. Probably we must recognize flood plain forests as communities perpetually in succession.

Are *any* communities truly in equilibrium? The reply depends on what one means by equilibrium. If one means the evenly weighted sides of a laboratory pan balance, the answer is no. If one will admit that children on a seesaw are balanced, even when tipping up and down, the answer is yes. Nature does achieve balance, but it is dynamic, not static. In an equilibrium community there are bound to be oscillations in numbers, but there should not be significant oscillations in kinds. The famous fluctuations of lynx and hare populations in Canadian forests (Fig. 3.23) do not mean that this community lacks equilibrium, but merely that it is regulated at a level of coarse rather than fine adjustment. The species of a mature tropical rain forest also fluctuate in numbers, but less dramatically; this system is also in balance, but is more finely tuned. We can speak in analogous terms of the physiological homeostasis of different organisms. The body temperature of man fluctuates, but only within narrow limits; it represents a finely tuned equilibrium. The body temperature of the camel is more coarsely tuned, but is no less an equilibrium (see Fig. 6.15).

To really deserve the term "balance," community equilibrium should have a significant time dimension. It is difficult to do more than assume this time dimension in the case of many mature communities. In the sediments of an African lake, however, I found a marvellous stratigraphic record of the persistence of a single plankton community for more than 3500 years. Five or six algal species were dominant in the lake's plankton throughout this period and though they fluctuated in relative abundance

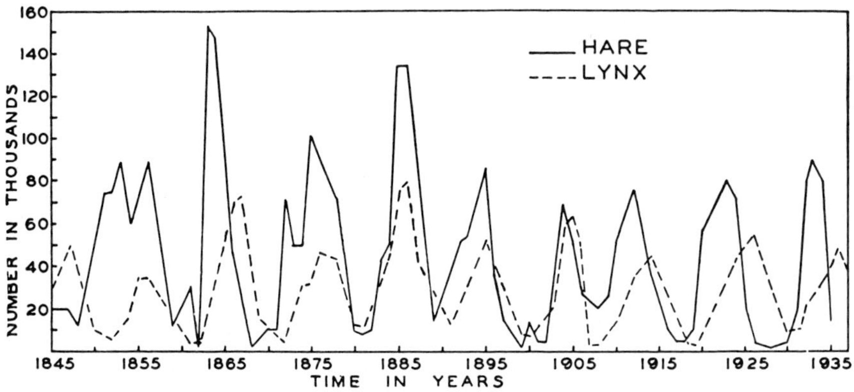

FIG. 3.23. Changing abundances of lynx and snowshoe hare in the Hudson Bay watershed of Canada, as indicated by long term trapping recoveries of the Hudson Bay Company. (From E. P. Odum (1971) after D. A. MacLulich: University of Toronto Biological Series, No. 43, 1937. Reproduced with permission.)

they remained important, continuing members of a distinct species association. For these species the period of dynamic equilibrium represents some 100,000 generations, certainly sufficient to justify the term "balance." The balance ended 5000 year ago, when a change of climate produced catastrophic changes in the lake's physical and chemical properties, which had been relatively stable during the period of community equilibrium. In its later history, the lake has continued to suffer physicochemical fluctuations which have prevented the establishment of any highly persistent communities.

Any community is subject to some degree of environmental unpredictability, but we need not believe that all communities are continually recovering from catastrophes rather than existing in true equilibrium. It seems probable that the species common in most mature communities are those able to endure not only the usual range of environmental conditions but the periodic extreme events as well. At least this should be true of the longer-lived species, which must endure such events if they are to persist in the community. An undergraduate study of forest succession in Pennsylvania provides pertinent evidence: in progressively more mature forest stands growing on the same slope, Stackhouse (1973) found species of mesic trees (those with relatively high moisture demands) being replaced by more xeric (dry climate) species. He inferred, quite logically, that the end point of succession in this region is a forest that can withstand periodic droughts, and that these events weed out less resistant species during the long drawn out later stages of succession. This example suggests that the process of maturation supplies a mature community with a buffer against environmental extremes of all but the most catastrophic sort.

One can conclude that mature communities are indeed structured in a manner that facilitates self-perpetuation. The concepts of equilibrium community and equilibrium species are not simply abstractions. The fluctuations of species populations within mature communities do not refute the idea of balance; they may indeed help to *assure* equilibrium at the community level. In a perceptive inquiry into the meaning of community stability, Preston (1969) illustrates this last statement with the example of Alaskan salmon. Upriver breeding runs vary greatly in size from year to year, but these population fluctuations can be viewed at the species level as a strategy of self-perpetuation. Bears feed on the migrating salmon, and trout on salmon eggs, but the fluctuations in their food supply make it impossible for these predators to build up large populations dependent only on salmon. The large salmon runs suffer less predation because of the small runs that precede them, and the several salmon species probably are more secure (and perhaps maintain higher average populations) than if all the runs were of equal size.

Even mature, well buffered communities are not immortal. True disruptions occur, often, alas, at the hand of man, but also by other means. For such disruptions, Preston (1969) provides a succinct analogy:

> A brick is lying on its face; we tilt it up slightly and let it go. It promptly flops back into its original position. It is stable against *small* displacements. We now tilt it up considerably, and suddenly it leaves us and flops down on another face. It is not stable against *large* displacements, or rather it has another equilibrium position into which it will then move. If we try to balance it on one edge . . . it can theoretically be done, but it is now unstable against even the smallest displacement. Any displacement will grow of itself.

Thus not only a sudden catastrophe, but also an accumulation of more subtle stresses can destroy the equilibrium of a natural community. One insult, piled on many similar ones, may break its back as surely as an avalanche. For illustration, one can look at the San Bernardino National Forest, near Los Angeles. Dominated by ponderosa pine in an apparent fire climax, sections of this forest are now dying. Many species of pine are susceptible to air pollutants (Dochinger *et al.*, 1970), but for a time the forest withstood the gradual increase in smog levels. No one realized that the brick was about to flop.

SUGGESTED READING

Bates, M. The Forest and the Sea. New York, Random House, 1960. (Also published as a Mentor paperback.) A nontechnical but delightfully personal and perceptive look at many communities and habitats.

Colwell, R. K., and E. R. Fuentes. Experimental studies of the niche. *Annual Review of Ecology and Systematics* 6: 281–310 (1975). A major review article.

Connell, J. H., D. B. Mertz, and W. W. Murdoch, eds. Readings in Ecology and Ecological Genetics. New York, Harper and Row, 1970. This and the next title are good paperback anthologies of original papers.

HAZEN, W. E., ed. Readings in Population and Community Ecology. Philadelphia, W. B. Saunders Co., 1975 (3rd edition). Contains the papers by Hutchinson and Schoener, below, plus many others.

HUTCHINSON, G. E. Homage to Santa Rosalia *or* Why are there so many kinds of animals? *American Naturalist* 93: 145–159 (1959). A whimsical but very influential inquiry into some aspects of community organization.

HUTCHINSON, G. E. The Ecological Theater and the Evolutionary Play. New Haven, Yale University Press, 1965. An extremely lucid discussion of the niche, including Hutchinson's famous "hypervolume" idea, is included in this delightful volume.

ODUM, E. P. The strategy of ecosystem development. *Science* 164: 260–270 (1969). A long-time student of ecological succession summarizes a great deal of information in this paper.

SCHOENER, T. W. Resource partitioning in ecological communities. *Science* 185: 27–39 (1974). A recent theoretical and experimental review of niche investigations, at a rather advanced level. Perhaps Chapter 7 of this book should be read first.

climates. Here, too, flowering and photosynthetic activity are staggered, and the blossoms are more conspicuous through the absence of leaves. It is certainly conceivable that this function of deciduousness is the oldest, evolving in some angiosperms at the same time as insect pollinators were evolving. Insects or drought—no one can really say what was the initial stimulus to the evolution of deciduous forests.

DIFFERENTIATION OF TEMPERATE DECIDUOUS FORESTS

Like most biome types, the temperate deciduous biome is discontinuous. Its three main divisions (in eastern North America, East Asia, and western Europe) are separated by geographic and climatic barriers. In the southern hemisphere few areas are climatically suitable for such forests, and the biome is poorly represented there. Climates favoring the temperate deciduous biome are roughly the inverse of those favoring the broad sclerophyll biome. The latter, also a temperate biome, occurs (as we have noted) under conditions of summer drought and winter rain—therefore, implicitly, in regions of mild winter temperatures. The temperate deciduous biome occurs in cooler regions which experience considerable spring and summer rain, followed by the "physiological drought" of a cold winter.

North America contains the most extensive temperate deciduous forests in the world. More or less continuous before European settlement, these forests even now extend through a considerable climatic range of both temperature and moisture. Consequently, these North American forests exhibit climatically determined regional differences in their climax plant and animal associations. The major regional subtypes are identified in Figure 4.1.

Not shown in this figure, however, are subdivisions of these regional forest types that result from nonclimatic agencies such as peculiarities of soil, drainage or incidence of fire. Many of these nonclimatic forest subtypes are limited in extent, but some, such as America's southeastern pine forests, are very widespread. Some southeastern pine forests are immature successional forests on abandoned farmland; others are maintained by recurrent fires and may be called a "fire climax." On a scale still smaller than these nonclimatic subtypes, we find very local forest differences such as those occurring on the two sides of a hill or gorge (Fig. 4.2). Here climatic contrasts again are often responsible, but on a very local scale.

The overall picture of the American decidous forest resembles a mosaic: large regional pieces fitted together and containing smaller pieces, each smaller piece in turn composed of very local variants. The analogy is apt, but we must realize that the boundaries between forest pieces are less sharp than between the pieces of a mosaic. All biomes probably have a mosaic structure, this being particularly true of biomes which, like temperate deciduous forest, contain many species.

What are the implications of mosaic structure? simply, that the organisms making up the community differ in their ecologic requirements and preferences, sometimes in quite subtle ways; and, moreover, that the local environment is sufficiently varied to accommodate each species in a suitable microhabitat. In the pages ahead we will consider both the microenvironmental aspects of a forest and the niche differences among some of its species.

THE INTERNAL CLIMATE OF A FOREST

Climate is the basic determinant of terrestrial biome types; and as we have just seen, when a biome segment is as extensive as the American deciduous forest, it may be regionally differentiated by climate. Here climate modifies the community. But a forest community also modifies climate, to a degree that is not only noticeable but biologically important.

Light

Part of the light striking the top of any forest is reflected; a great deal more is absorbed by foliage. The amount reaching the forest floor is variable but is often surprisingly low—less than 5%, for example, in a dense beech forest in Austria (Fig. 4.3). Because of the heavy shade cast by the canopy layer, beech forests have only sparse growth of smaller trees, shrubs and herbs. In more open forest types these lower layers of foliage are well developed; here the attenuation of light is less smooth than in Figure 4.3 and follows a somewhat stepped pattern. But in forests generally, the canopy intercepts the lion's share of light. In grassland, on the contrary, attenuation of light by the vegetation may be equally complete but it occurs more evenly (Fig. 4.3). Dense tropical rain forests, with their nearly closed canopy and many lower layers of foliage, probably inter-

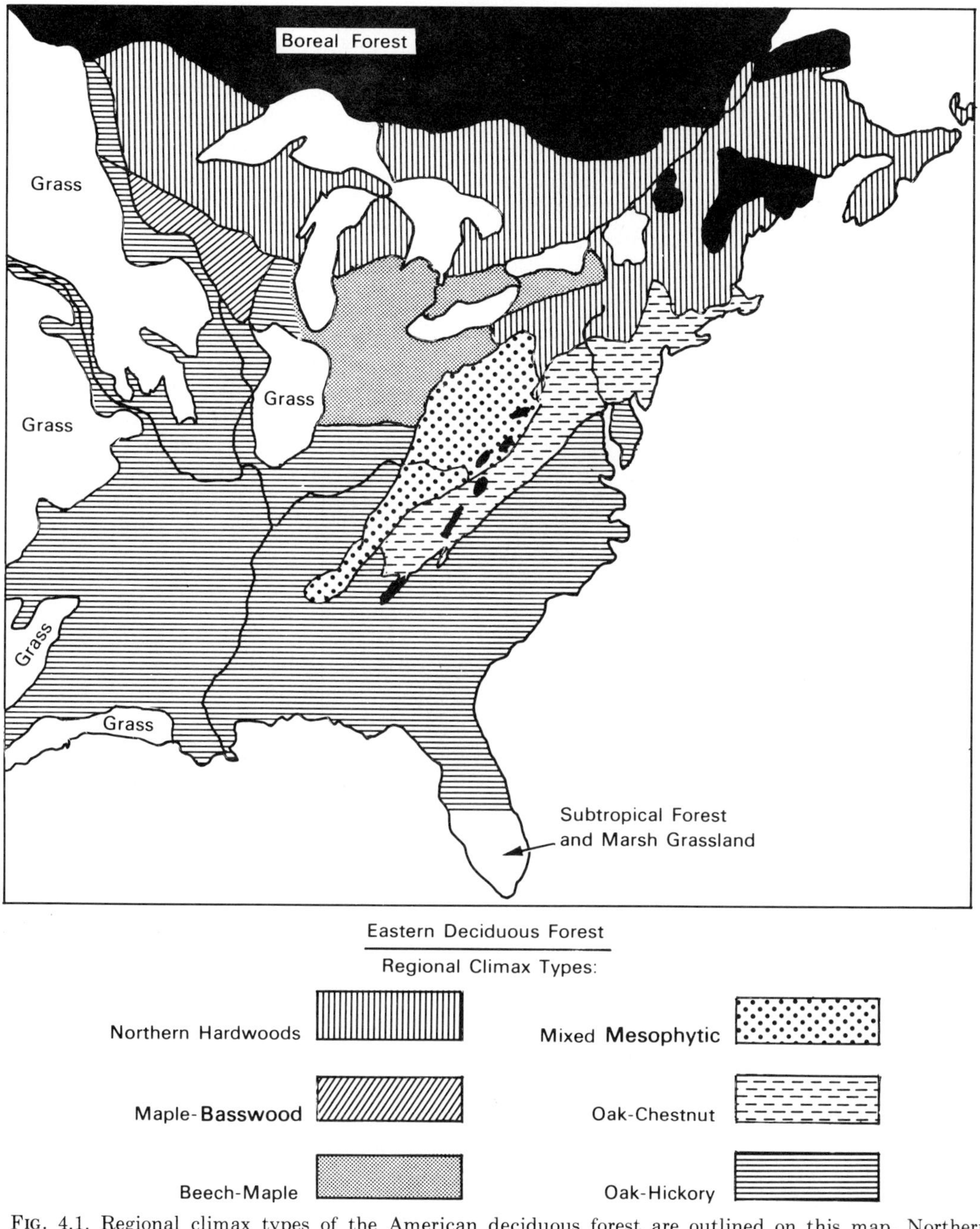

FIG. 4.1. Regional climax types of the American deciduous forest are outlined on this map. Northern hardwood forests actually are a mixture of conifers (hemlock, white pine, red pine) and hardwoods (sugar maple, red maple, yellow birch); this forest type represents a broad ecotone between the deciduous forest proper and the boreal coniferous forest. Within the deciduous region, true boreal forest caps the high Appalachians in discontinuous islands as far south as northern Georgia. On its southern borders the deciduous region grades into subtropical evergreen forests in Florida and eastern Mexico. Westward it merges via oak savanna and thickets into prairie (Fig. 3.5), but along rivers more diverse valley forests push far into the grassland. The extensive oak-hickory climax type, dominated by white oak, has been replaced in many localities by successional pine forests; if allowed to mature these forests would return to oak-hickory. The oak-chestnut forest type has been effectively converted to a mixed-oak type by the blight which decimated chestnut earlier in this century. The richest regional subtype is the mixed mesophytic forest, centered just to the west of the southern Appalachians. It is "mixed" because many different tree species share importance, and "mesophytic" because it is relatively moist. More than any other, mixed mesophytic forests are believed to typify the richness of pre-Ice Age deciduous forests in America. (Much modified from Dansereau, 1957.)

FIG. 4.2. Gorges and ridges, particularly if they run east-west, often support conspicuously different vegetation on their two slopes because of differing exposure to sunlight and evaporation. In this photograph of Walnut Canyon, Arizona, note the tall forest of ponderosa pine on the cool snowy north-facing slope and the scrubby forest of pinyon pine and juniper on the warmer south-facing slope, where snow disappears quickly. In the bottom of the gorge, where moisture is highest and solar heating least, one finds tall douglas firs. On nearby mountains douglas fir occupies cool, moist upper slopes; lower down is a zone of ponderosa pine and still lower a semiarid zone of pinyon and juniper. All three zones are represented at Walnut Canyon, reflecting the sharp climatic contrasts that exist very locally within the gorge. In the deciduous region topographic differences create similar contrasts in local climate and vegetation (*cf.* Cantlon, 1953; Shanks and Norris, 1950), but these contrasts are not as visually striking as in the simpler forests of the American West. (Photograph by author.)

cept more light than any other vegetation type. Geiger (1959), a leading authority on microclimatology, suggests that light reaching the ground in some rain forests may be no more than 0.1% of that striking the canopy. Of course, the figures we have mentioned are averages for particular forests. Within any forest, the canopy and lower plant layers do not constitute a homogeneous screen; dappled light patterns are so characteristic of the forest floor that they are mimicked by fawns and jaguars.

In deciduous forests, the prominence of smaller vegetation is controlled not simply by the *amount* of shade cast by the canopy layer, but also by the *duration* of the canopy shade phase. This varies considerably in forests dominated by different deciduous trees (Fig. 4.4) and one can expect undergrowth to be more dense in situations where the shade phase is relatively short-lived.

Up to now we have spoken only of light and shade, not of spectral changes. The light passing downward through a forest is not only reduced in intensity but changed somewhat spectrally. While relatively unimportant ecologically, foliage-induced spectral changes do influence some organisms, as we noted in Chapter 3 with respect to the leaf shape of dandelions.

Temperature

The forest canopy intercepts both incoming solar radiation and nightly heat radiation from the forest floor; consequently the daily range of temperature beneath a dense deciduous canopy is less than outside the forest—sometimes as much as 5° C less (Geiger, 1959). The forest canopy itself, however, may heat to tempera-

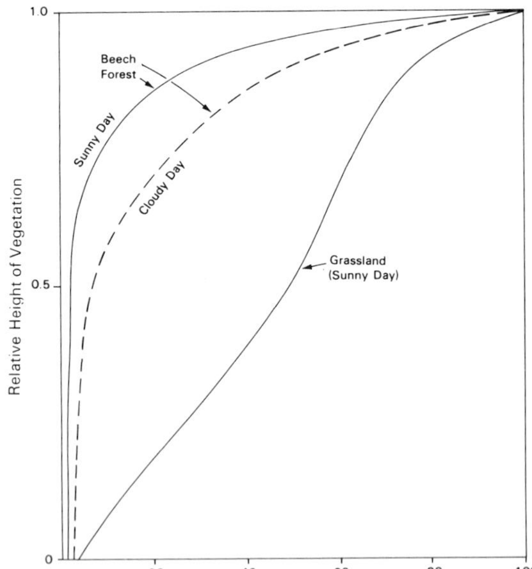

FIG. 4.3. Light attenuation patterns in an Austrian beech forest on sunny and cloudy days, and in a tall grass meadow or prairie. (Modified from R. Geiger: *The Climate Near the Ground*. Cambridge, Mass.: Harvard University Press, 1965.)

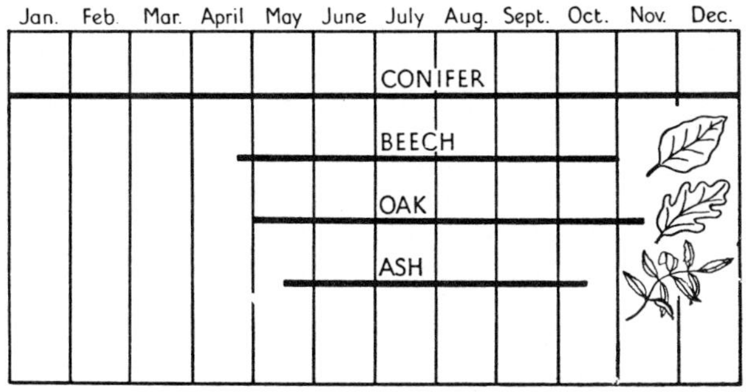

FIG. 4.4. Duration of the shade phase in English woodlands having different canopy species. (From E. Neal: *Woodland Ecology*, Cambridge, Mass.: Harvard University Press, 1965.)

tures well above those outside the forest, and in sunlit forest glades summer temperatures near the ground may also be very warm because of impeded air circulation. In winter, of course, the climate within a deciduous forest differs less from that outside. This statement, unexpectedly, is true also for temperate evergreen forests because thermal contrasts between the inside and outside of any forest are primarily a function of the intensity of incoming radiation. In summer the intense radiation warms the canopy but is prevented from warming the air and ground beneath it. In winter the lesser intensity of incoming radiation results in smaller thermal contrasts even if the canopy retains leaves.

These effects of the forest on its own thermal environment are distinct from temperature contrasts that may exist in a forest because of small scale topographic differences. We already have noted the importance of topography as a local influence on climate.

Wind and Atmospheric Gases

Winds within a forest are greatly reduced, and the floor is almost permanently calm. Even in winter a deciduous forest maintains a gradient of decreasing turbulence beneath the forest canopy (Table 4.1). During the growing season, forest air remains more moist than that outside, due to the evaporative contribution of water vapor from leaves. Forests tend, in fact, to be relatively constant environments in terms of both air movement and humidity.

Due to biological activity, however, gases such as oxygen and carbon dioxide may be more variable within a forest than outside, particularly during calm weather. There tends to be a gradient of decreasing CO_2 upward from the forest floor, where it is produced by decomposition, toward the canopy, where during the day it is removed by photosynthesis. At night, CO_2 released by respiring leaves supplements the CO_2 released by decomposers, so that there is also a distinct diurnal cycles of CO_2 concentration in the forest atmosphere. Obviously the amplitude of this cycle changes seasonally, depending on the degree of biological activity. Daily oxygen gradients and cycles, while probably less discernible, can be expected to be the reverse of those characterizing CO_2.

Rainfall

One might expect rainfall over forests and other dense vegetation to be augmented by the quantities of moisture transferred to the air by the plants. Even extensive forests, however, provide no clear-cut evidence that the vegetation significantly affects the amount of rainfall

TABLE 4.1

Vertical Gradients of Wind in a European Oak Forest

These measurements were made by R. Geiger and H. Amman on 2 days when the wind velocity above the canopy was approximately the same. The first measurements were made before the leaves had opened, the second when the leaves were fully expanded. Adapted from a graph in Geiger (1965). Canopy had an understory of beech saplings.

Height above Forest Floor	Wind Velocity	
	Trees leafless	Trees in leaf
m	*m/sec*	
28 (above canopy)	3.4	3.5
24 (in canopy top)	2.5	1.5
20 (middle of canopy)	1.9	0.9
13 (lower part of canopy)	1.5	0.6
4 (beneath canopy)	1.2	0.8

a locality receives. What forest vegetation *does* affect is the amount and microdistribution of rain water reaching the ground. When precipitation is slight, virtually all the water may be intercepted by the canopy. This is especially true in coniferous forests, where each needle collects an independent drop; by contrast the drops accumulating on the broad leaf of an angiosperm tend to coalesce into rivulets which are not retained by the leaf (Geiger, 1959). The harder or longer the rain, the greater the fraction of total precipitation that passes through the canopy.

In addition to this general canopy effect, each tree individually influences the amounts of water reaching the ground beneath it. Some rain flows down branches to the trunk and thence to the ground, causing the soil immediately adjacent to the trunk to be relatively moist. Direct penetration through the canopy, however, increases with increasing distance from the trunk. Thus common sense is vindicated by measurements: you really do stay drier by standing under a tree in a storm, and the nearer to the trunk you stand the drier you will be (except for your feet, which will probably be in a puddle, and your back, if you make the mistake of leaning against the trunk).

Overall, the *internal* rainfall climate of a forest is largely reflective of the *external* rainfall climate; and the latter, plus soil and topographical characteristics, is the main determinant of the moisture available to a particular patch of forest. But the moisture available to any particular *plant* is also determined by the prevalence of other plants competing for water. This fact is dramatically illustrated by forest trenching experiments such as those of Korstian and Coile (1938). When a small plot of forest floor is isolated by trenching, and the roots of trees leading into that plot are cut, it quickly develops a luxuriant growth of seedlings and herbs (Fig. 4.5). One may infer that the trees previously had tapped most of the available soil moisture in the plot and had thereby prevented the growth of small plants. These results confirm the climatic studies of Thornthwaite (1948) (Fig. 2.16), which clearly show that moisture stress can be expected late in the growing season in at least some deciduous forest regions.

Forest Microclimates

The foregoing discussion makes clear that sites separated by only a few meters or even centimeters in a forest may differ climatically in ecologically important ways. An herb whose foliage is 10 cm above the ground is exposed to different levels of CO_2 than the meter-high shrub above it. The quantity of light received by the herb is also quite different than that received by the shrub—or than the light received by a nearby herb under another shrub. The temperature range and humidity within the burrow of a shrew contrast sharply with those on the ground surface above. These very local climates (microclimates) may be more relevant to survival and distribution, particularly of small organisms, than is the "average" climate of the habitat or region.

Many microclimatic effects are easy to recognize. On your next visit to a forest, examine the distribution of moss on and around a single tree. The flow of rainwater down the trunk is not uniform, but is channeled by the location of branches and by the uneven contours of the stem. One side of the tree may remain moist longer after rain because it is less exposed to sunlight. The distribution of moss (and of green algae on the trunk) will reflect these microclimatic differences. A patch of moss will be found high in the tree, in a crotch between branch and trunk where water is trapped longer than elsewhere. Near ground level, the moss may grow on some root bases but not others, and on the ground itself, more lushly between two diverging roots than between others. Different species of moss may dominate these adjacent areas. The microclimatic basis of such differences is usually quite evident, but to actually measure these microclimates is often difficult. Special needs may require special (but not necessarily elaborate) micrometeorological instruments, the invention of which test the ingenuity of the investigator.

To avoid giving a false impression with this example, we should add that the distribution of mosses in a forest is not only a matter of moisture. Some mosses thrive in quite dry conditions. Desiccated rock mosses in the American Southwest spring miraculously from gray "deadness" to green activity within an hour after rain. Forest mosses often are restricted to rocks and tree bases not because these are particularly moist locations—sometimes the opposite is true—but because these areas are relatively free from smothering leaf litter and from the competition of taller, more deeply rooted plants. Thus, for a moss, as for

FIG. 4.5. Before-and-after aspect of a forest-floor quadrat freed from influence of the roots of surrounding trees. Three years after the perimeter of this quadrat was trenched and all tree roots were cut, the plot contained many thriving saplings and small shrubs, while areas outside the quadrat remained bare. (Photographs by C. F. Korstian, courtesy of W. D. Billings.)

any species, microclimate is only part of a total microenvironment that also includes the influence of other organisms (Fig. 4.6).

FOREST SOILS

The soils of a forest are as varied as the microclimates. They differ in depth and texture, and in chemical characteristics such as humus content, acidity, and mineral nutrients. Several of these variables, in turn, affect the moisture-holding capacity of the soil and the degree to which this moisture is available to plants. Soil characteristics directly influence not only plants but also the soil fauna of decomposers and burrowing animals; and through the plants and soil fauna, soils also

FIG. 4.6. Moss advancing across a rock slab creates a striking microenvironmental boundary on the floor of a Connecticut forest. Moisture and soil conditions within the moss mat are far less harsh than on the lichen-covered rock just beyond, and allow grasses and saplings to gain a foothold. (Photograph by author.)

affect the above-ground animals of the community.

Soil Profiles

As it matures over time, a soil develops a series of distinct layers, or horizons. These layers are the product of (*a*) downward leaching by rain water; (*b*) the movement of water drawn upward by evaporation; (*c*) biological contributions of organic matter from above; and (*d*) mechanical and chemical weathering of the parent rock beneath. Where the influence of water and organisms is slight, or where soils have been removed or deposited in bulk, the typical soil horizons may not be present. Horizons thus are frequently absent in the soils of deserts, steep slopes and flood plains.

The major horizons of a well differentiated soil profile (an exposed vertical section) are as follows (see Fig. 4.7):

A horizon. This is the upper, organic soil, site of most decomposition and most soil fauna. Organic matter usually is concentrated in the upper part of the horizon, where separate layers may be recognizable (A_{00}, loose litter; A_0, partly decomposed leaf mold; A_1, largely decomposed humus). The lower part of the A horizon (A_2) is lighter in color, lower in organic matter and relatively coarse in texture; it is the zone of maximum downward leaching by water.

B horizon. This horizon usually is low in organic matter (though humus sometimes leaches into its upper portion), but relatively high in soluble minerals; this is the soil zone in which mineral matter and finer soil particles (clays) become concentrated after being leached down from above. Internal layering may be evident, but distinct boundaries with the upper A horizon and lower C horizon sometimes are lacking.

C horizon. This is lightly weathered parent material consisting of stones and soil that are little altered chemically from the underlying rock. Soils that are thin or young may possess A

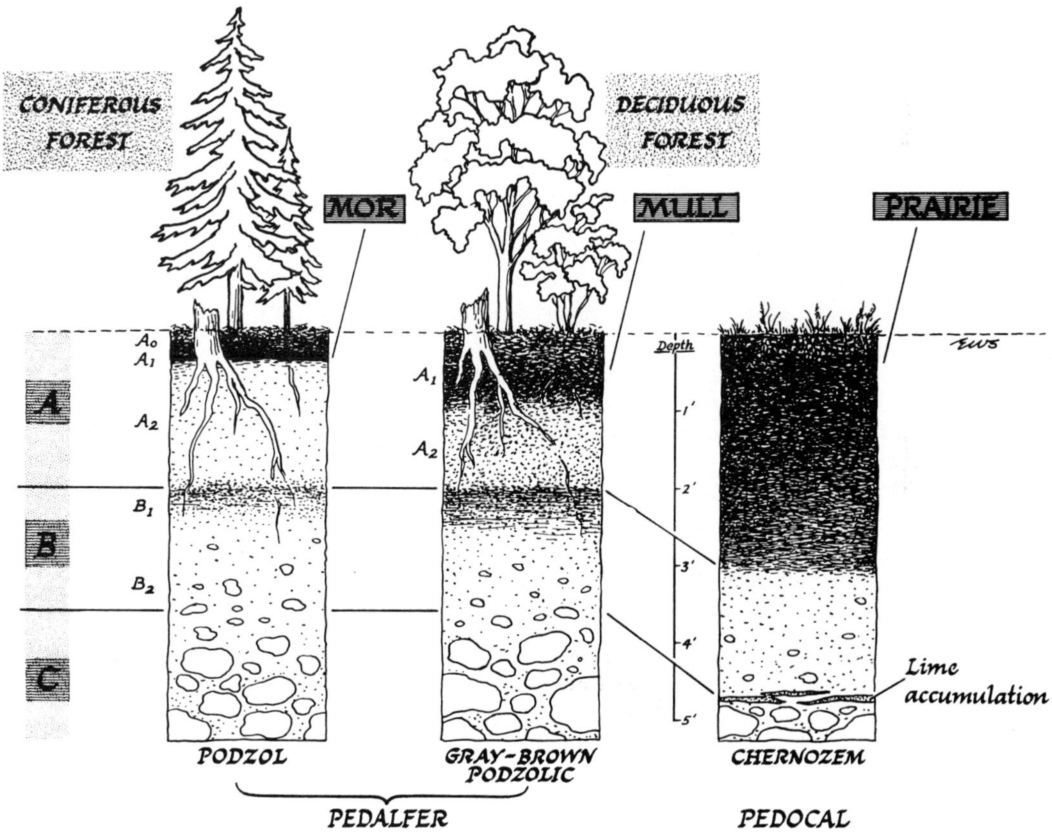

FIG. 4.7. Mature soils of most biome types exhibit horizontal layering (horizons), as described in the text, but the thickness and physicochemical character of the layers differ depending on vegetation cover and rainfall. Mor humus is quite acid, compact, and slow decaying, with fungal decomposers predominant; mull humus is less acid and better aerated, with bacterial decomposers predominant and a rich invertebrate fauna. Calcium-rich prairie soils are termed pedocals, chernozem being the typical pedocal of moderately moist grassland regions. The varied forest soils of temperate and north temperate regions are grouped under the term pedalfer. (From G. B. Moment: *General Zoology*, Ed. 2. Boston: Houghton Mifflin Co., 1967. Reproduced with permission.)

and C horizons but no separately recognizable B horizon.

Despite local variability, distinct regional soil types exist within the American deciduous forest. It is instructive to compare these with each other and with soils of adjacent biome types (Fig. 4.7). In the center of the deciduous forest region the soils are gray-brown podzols: moderately leached, calcium-poor, usually no more than a meter deep, and possessing a rich, organic surface layer. To the north, these soils grade into the true gray podzol that typifies the boreal forest. Boreal podzols are often shallow, and are highly altered in their upper portions by the acid leachate of decaying conifer needles. The organic layer is thin, and the leached A_2 horizon is so depleted of mineral salts and of iron and aluminum that it has a distinctive grayish white color resulting from the predominance of quartz and other silicates. This ash-gray layer gives the soil its name: "podzol" is derived from the Russian word for gray rock salt. True podzols are deficient in calcium and in nitrates; because of the acidity there are few animal decomposers, though fungi often are abundant.

In southern decidous forests the soils tend to be much deeper and older than the young postglacial soils of the north. Southern forest soils are related to the red lateritic soils of tropical rain forest. In regions of relatively high rainfall the texture of these weathered southern soils is fine and clayey, and they have a yellowish or reddish color. Persistent leaching over long periods has removed compounds of calcium, magnesium and potassium, and also the silicates, leaving a soil richly colored by iron and aluminum oxides, which are the dominant remaining minerals. Warmth and moisture promote rapid decay, so soils in the American southeast and in the wet tropics accumulate little organic matter in the A horizon.

In states bordering the upper Mississippi, deciduous forest soils approach the condition of prairie soils. The latter are made highly distinctive by their vegetation as well as by their climate. Typical prairie soils are deep, often having developed from thick deposits of fine, wind-blown sediment. Because prairie rainfall is relatively low, leaching and decomposition are slower than in deciduous forest soils. Calcium and other soluble salts are leached weakly from the A horizon and accumulate in the B horizon, often forming a limy hardpan at the depth where downward leaching of these minerals is balanced by their evaporative precipitation during drier seasons. The A_1 horizon of prairie soils is dark, highly organic, and remarkably thick, resulting from the nearly complete, above-ground death of the vegetation each year and its incomplete decomposition. Dead grass is initially degraded more rapidly than are the leaves of many deciduous trees, so little fresh litter accumulates on prairie soils; but the large yearly contributions of dead vegetation and the slowness of later decomposition lead to the accumulation of humus. Prairie soils thus are rich in both organic nutrients and mineral salts, and are among the most fertile in the world.

Table 4.2 illustrates some of the chemical contrasts we have described for different soil types. Unrepresented in this table is a soil of the southern deciduous type, which would typically be relatively low in the indicated nutrients and high in water-retaining capacity.

Soil Moisture

The amount of moisture in a forest soil depends partly on the topographical position of the site, partly on recent rainfall history, and partly on the intensity of evaporation and transpiration. In addition, however, characteristics of the soil itself affect how much rainfall it can absorb and how tenaciously it clings to its moisture. As a rule, the higher its organic content or the finer its mineral texture, the more water a soil can hold. Humus-rich soils promote growth not only because they are rich in organic nutrients, but also because they absorb large quantities of moisture during rains; moreover, the humus retards subsequent evaporative loss of this water from the soil, but does not withhold it from plants. Fine-textured clay soils also have a large water-holding capacity, but much of the water adheres so tenaciously to the soil particles that it cannot be utilized by vegetation. Any soil, as it dries, loses its least tightly bound water first, but in fine clay soils so much of the water is bound that plants begin to wilt when the water content is still as high as 15% of soil weight. In coarse soils, plants are able to extract nearly all the accessible water, and even "tender" plants do not wilt until soil moisture levels fall to about 1% (Daubenmire, 1959, p. 119). But because coarse-textured soils can hold so little water in the first place (see

TABLE 4.2

Characteristics of A_1 Horizon Soils from Different Vegetation Types in Wisconsin

These data represent averages for several soil samples from each vegetation type. All data are from Curtis (1959).

	Available Nutrients				pH	Water-retaining Capacity[a]
	Ca	K	P	NH_4		
		ppm				%
Boreal conifer forest	2785	150	35	18	5.1	223
Moist northern hardwoods (maple/hemlock)	4215	93	?	?	5.6	247
Moist deciduous forest (maple/basswood)	3655	96	34	9	6.9	75
Dry deciduous forest (oaks)	2210	84	25	12	6.2	73
Oak-prairie transition	5220	124	22	12	6.5	83
Dry prairie	5425	94	85	7	7.8	62
Sand barrens (grasses and herbs on deep sand)	50	18	30	5	6.2	29

[a] Water-retaining capacity is the amount of water a soil can hold, expressed as a percentage of its weight when dry.

Table 4.2, sand barrens example), they tend to be dry environments even in regions where rainfall is high. From the standpoint of water usable by plants, optimal soils are therefore soils rich in humus and having medium texture.

INTERRELATIONSHIPS OF FOREST PLANTS

Deciduous forest vegetation beautifully exemplifies both individualistic and organismic aspects of community organization. Individualistic species patterns are implied by the continuous regional and local variation of forest composition; the organismic aspect is illustrated by the remarkable integration of forest plants through mutual adaptation. The many species may not absolutely depend on one another, in the strictest Clementsian sense of a superorganism, but the ways of life they have evolved are clearly related to the habits of coexisting plant species. Adaptations with regard to light and reproduction illustrate this statement particularly well.

Photosynthetic Patterns and Growth

In temperate deciduous forests the waning of winter is first signalled by the flowering herbs of the forest floor. Some even thrust up through snow, and long before the trees put out leaves the floor of the forest is green. This **prevernal** (early spring) habit of growth is nearly unique to deciduous forests, but it has evolved again and again among small forest herbs. Why do these plants brave the late frosts and sluggish temperatures of early spring? apparently because this is when light is most available to them. Trees control the light climate within a deciduous forest; and for trees the time most conducive to growth is early summer, a time of warm temperatures, overhead sun, and still-abundant soil moisture. The trees are **summergreen** plants. Because summer brings limiting light intensities to herbs of the now shadowed forest floor, their spurt of activity in early spring is a beautiful timing adaptation to the "facts of light" beneath a deciduous forest canopy.

The prevernal photosynthesis of many herbs and the summergreen pattern of the trees are just two of the five activity patterns characterizing deciduous forest plants. E. J. Salisbury (1925), one of the great early students of deciduous forests, recognized two other patterns, **wintergreen** and **evergreen** in his studies of English woodlands. These types, plus a fifth, more recently recognized pattern (the **winter annual** pattern) are contrasted in Figure 4.8. Like prevernals, wintergreen and winter annual plants are typically small and herbaceous; unlike prevernals, both have life patterns geared in part to the *autumn* period of high light, after the leaves have fallen. Wintergreens actually are year-round green perennials, differ-

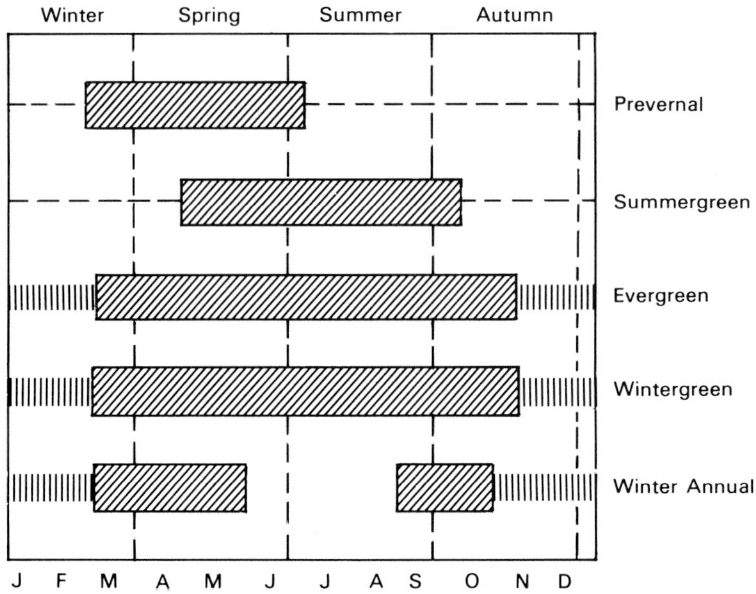

FIG. 4.8. Classification of deciduous forest plants according to their periods of photosynthetic activity and growth. For explanation, see text. (Modified from E. Neal: *Woodland Ecology*, Cambridge, Mass.: Harvard University Press, 1965.)

ing from typical evergreens only in that they add new leaves in autumn that persist through the winter; Christmas fern is an example. Winter annuals, of which bedstraw (*Galium aparine*) is an example, germinate from seed in the fall, producing a low rosette of leaves that grow slowly in the winter and more rapidly in early spring. The life cycle of winter annuals is complete by late spring or early summer; by then they have produced seeds and died. The evergreen class in deciduous forests contains plants of all sizes: trees (*e.g.* hemlock), shrubs (mountain laurel) and herbs (teaberry). The evergreens grow slowly if at all in the winter, but they can take advantage of both spring and fall light maxima within the forest.

Let us return to the prevernals, the category most unique to deciduous forests, and examine how profoundly their shade-avoiding photosynthetic strategy has affected their total biology. In nonforest habitats the majority of small, herbaceous plants are summergreen annuals. The typical deciduous forest herb, however, is not only prevernal but also perennial. Carefully examine a trout lily, spring beauty, or other prevernal plant and you will find a corm or tuber at its base. Energy stored in this perennial root is what enables the plant to spring to life so rapidly and so early. When the leaves of a trout lily ("dogtooth violet") appear, their photosynthesis captures energy less for immediate use than for storage and use the following spring. By early summer the leaves of a prevernal die and the plant goes into underground dormancy.

Risser and Cottam (1968) studied the carbohydrate cycle of four prevernal species and found the following pattern in each. With the appearance of leaves in spring, the bulb rapidly begins accumulating carbohydrate reserves. This stored energy initially is starch rather than sugar, and conservation of energy thereby is facilitated during the summer months. Even though dormant, the bulb of a prevernal is alive in summer and continues to respire; if its stored carbohydrates were sugars, which are more easily metabolized than starch, the high temperatures might cause greater respiratory loss of energy reserves.

In fall or winter the bulb breaks dormancy and begins to produce roots and an underground shoot. Now the stored starch is progressively converted to soluble sugar, providing energy for growth and conferring a secondary advantage: the sugar acts as an antifreeze, probably playing a particularly important role in the sensitive young shoot and roots. By late winter the plant is well fueled, already differentiated, and ready to burst into photosynthetic activity.

Not all small forest plants follow one of the strategies (prevernal, wintergreen, winter annual or evergreen) that avoids summer shade problems. Some herbs and many shrubs, such as blueberry and viburnum, possess the summergreen pattern of the deciduous trees above them. How do they manage? by having extremely low light requirements for growth. The photosynthesis/light intensity relationship of green plants has two critical points (Fig. 4.9). The lower point is the compensation intensity—the intensity of light at which sufficient photosynthesis occurs to balance respiration. At intensities below the compensation point the plant respires away more energy than it captures, and when stored reserves are gone it dies; but when available light exceeds the compensation intensity, photosynthetic production exceeds respiratory consumption and the plant can grow or store energy (assuming that other needs, such as moisture, are adequately satisfied). But beyond an upper critical light intensity, the saturation level, the photosynthetic rate fails to increase with increasing light; the plant's photosynthetic apparatus is "saturated," and additional light is wasted. Plant species differ in their compensation intensities and saturation levels, as well as in the sensitivity of their photosynthetic response to light differences below the saturation level. (Differences in slope of the curves in Figure 4.9 illustrate such differences in sensitivity.) All these physiological contrasts relate to a species' normal habitat. Species that typically grow in unshaded habitats have high compensation intensities but also high saturation levels, enabling them to use more of the light energy that bathes them. On the other hand, the summergreen herbs of

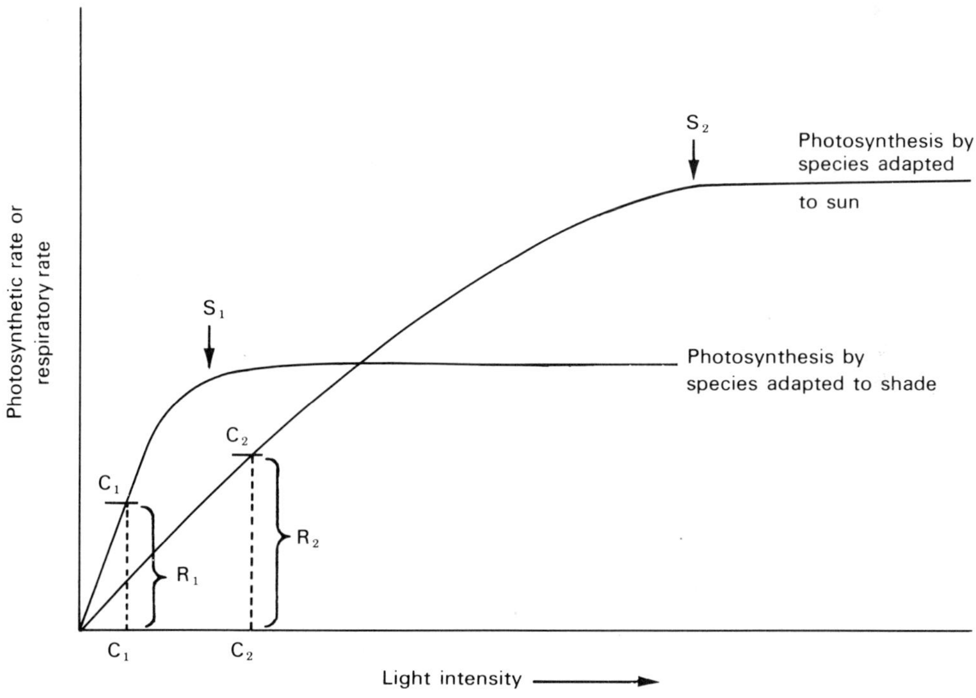

FIG. 4.9. Photosynthetic and respiratory contrasts between a shade-adapted plant (S_1, R_1, C_1) and a sun-adapted plant (S_2, R_2, C_2). Shade-adapted species are relatively efficient in their use of low light intensities, as seen in the fact that slight increases of light result in large increases of photosynthetic rate. The compensation intensity (C_1) is low, but so is the light saturation level (S_1). Sun-adapted species are less efficient in using low light intensities, and have higher compensation intensities (C_2), but they also have higher saturation levels (S_2), and at high light intensities can exceed the maximum photosynthetic rate of shade-adapted species. Sun-adapted species also typically have higher respiratory rates (R_2) than shade-adapted species (R_1). (For simplicity, the diagram depicts respiratory rates as independent of light intensity, although this is not strictly true.)

deciduous forests have extraordinarily low compensation intensities, and can therefore grow in deep shade (Table 4.3). They also have low saturation levels, but this is no liability because light beneath the summer canopy seldom exceeds their saturation levels. The strategy of small, summergreen forest plants is not one of timing, but of scaling down their need for light. To a lesser extent, this strategy also is followed by small plants that do use timing strategies; all the compensation intensities of Table 4.3 are strikingly low, but those of the summergreen herbs are lowest.

Given their access to summer light, the summergreen activity pattern of forest trees is not surprising. Understory trees like cherry, however, tend to leaf out earlier than do the canopy species (*cf.* Salisbury, 1925). Even canopy species, if they are to persist in a forest, must be able to survive low light intensities as seedlings. The abilities of tree species in this regard are very unequal (Table 4.4): those with low compensation intensities tend to be equilibrium species, capable of persisting in a mature forest, while those with high compensation intensities often are successional species. In Table 4.4 sugar maple, beech and hemlock are equilibrium species, typical of old forests. Hemlock persists in such forests, despite its relatively high compensation intensity, perhaps because its evergreen habit allows young trees to utilize spring and fall light for growth. White pine does not reproduce well in mature forests, and though long-lived, it usually is considered a successional species. Northern white cedar and larch characterize boggy areas of the northern hardwoods region, where together with black spruce they constitute islands of boreal coniferous forest in a deciduous landscape. It is often assumed that these species flourish in bogs south of the coniferous biome because the bogs are cool, moist habitats climatically akin to the boreal forests of the north. But the data in Table 4.4 suggest an alternate explanation: possibly larch and white cedar favor bogs in the sourthern parts of their range because bogs are relatively open habitats where the high light requirements of these species are easily satisfied. The coolness and moisture of the bogs may be conditions that these trees tolerate but do not require. These suggestions are supported by additional observations: in the deciduous region white cedar grows not only in bogs but on limestone uplands, which tend to be dry habitats with relatively sparse competing vegetation; and larch flourishes on bare, hot strip mine spoils in parts of Pennsylvania. Both these unusual habitats provide plenty of light, but lack the cool, moist conditions of northern forests.

In summary, we cannot doubt that competition for light has been a major evolutionary force molding the character of deciduous forest

TABLE 4.3

Compensation and Saturation Intensities of Herbs from a Deciduous Forest in Southern Ontario

Experiments were run at a temperature of 25° C. Data are from Sparling (1967).

	Compensation Intensity	Saturation Intensity
	ft-candles	
Summergreen herbs		
Maianthemum canadense (Canada mayflower)	12–20	250
Polygonatum pubescens (hairy Solomon's seal)	13	200–300
Oxalis montana (common wood-sorrel)	18	250
Actaea pachypoda (white baneberry)	19–26	150–250
Wintergreen herbs		
Polystichum acrostichoides (Christmas fern)	22	250
Prevernal herbs		
Dicentra canadensis (squirrel corn)	150–220	ca. 3000+
Claytonia caroliniana (spring beauty)	180	ca. 1500+
Caulophyllum thalictroides (blue cohosh)	70–130	ca. 3000+
Erythronium americanum (trout lily)	45–84	ca. 3000+

TABLE 4.4

Compensation Intensities of Tree Seedlings

The values are percentages of full winter sunlight in Maine, and are derived from 3-hour experiments. Data are from Burns (1923).

White cedar (*Thuja occidentalis*)	18.6
Larch (*Larix laricina*)	17.6
White pine (*Pinus strobus*)	10.4
American hemlock (*Tsuga canadensis*)	8.4
American beech (*Fagus grandifolia*)	7.5
Sugar maple (*Acer saccharum*)	3.4

plants, nor that it remains a primary form of interaction in today's forests. But we should recall, too, the evidence that competition for moisture and perhaps other nutrients inhibits seedling growth in many forests (Fig. 4.5). The staggered activity periods of the plants in deciduous forests serve to reduce competition for moisture as well as for light, particularly between large and small species. Moreover, contrasts such as that in Figure 4.10 clearly demonstrate that deciduous forest plants have adapted to one another in more ways than simply with regard to light.

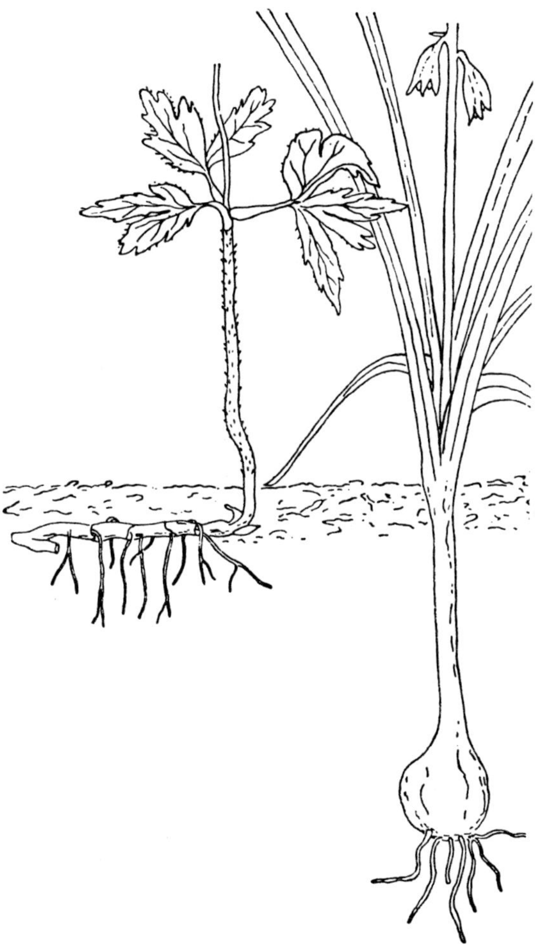

FIG. 4.10. Root systems of wood anemone and bluebell. These prevernal species of English forests utilize different soil layers, thus minimizing competition for soil nutrients. (From E. Neal: *Woodland Ecology*, Cambridge, Mass.: Harvard University Press, 1965.)

Reproductive Patterns

Actually it is flowers, not foliage, that first make the prevernal herbs conspicuous in the spring. The flowers often appear even before the leaves, and it is worth inquiring into the timing of their appearance. The showy prevernal flowers imply the use of animal pollinators; wind-pollinated flowers inspire little poetry because there has been no selective pressure to make them conspicuous. Are insects abroad in late winter or early spring? Indeed many are, and early blooming flowers profit from the lack of concealing foliage and the relative scarcity of competitors for the insects' attention. Neal (1965) found the flowering times of insect-pollinated plants, both large and small, to be considerably staggered in English forests, further minimizing competition for pollinators. He found, however, that almost all plants in English forests, even those that are wind-pollinated, bloom before midsummer. Foliage, by its wind-breaking effect, as surely decreases the effectiveness of wind pollination as it decreases by concealment the effectiveness of insect pollination.

One interesting exception to the early flowering forest pattern is witch hazel. I remember the astonishment with which I first noticed this small tree blooming in Pennsylvania forests in late October, and I relive that astonishment in my students each fall. In a nearly forgotten paper written in 1906 Sigmund Graenicher recognized the flowering strategy of witch hazel for what it is: the successful mirror image of the flowering habit of most other plants in the deciduous forest. By being a nonconformist, witch hazel avoids competing for pollinators against such showy potential adversaries as dogwood, its frequent neighbor. Insects are still abroad in late fall: Graenicher recorded 44 species visiting the flowers of witch hazel, mostly after its early-withering leaves had fallen.

Successful pollination is one side of the coin; successful dispersal and establishment of seeds is another. In deciduous forests, seed dispersal mechanisms tend to be related partly to species size (Harper *et al.*, 1970). Herbs and low shrubs often bear hooked or hairy seeds that cling to the coats of passing animals. Taller shrubs and small trees (*e.g.* blueberry, dogwood, elderberry and cherry) typically *attract* animals by means of fleshy fruits. Large trees, if they are equilibrium species, often produce heavy seeds (*e.g.* nuts) that also are attractive to animals, though not necessarily designed to be. Opportunistic tree species more often rely on light, numerous, wind-dispersed seeds.

Although they rely on animals both for pollination and seed dispersal, perennial forest herbs possess the insurance of vegetative propagation, which in many cases seems far more effective than propagation from seeds. A notable exception is forest orchids, which have evolved more specialized relationships with insect pollinators than have most other plants (Fig. 4.11). Orchids do not reproduce vegetatively but produce quantities of tiny seeds well suited for dispersal by the weak breezes of the forest floor.

In general, however, seedlings of forest herbs face many hardships, and it is not surprising to find vegetative propagation so common. Seedlings must survive the difficulties posed by leaf litter, as well as the shade of midsummer or the coolness of other growing seasons. Propagation by vegetative means minimizes these difficulties because a nutrient lifeline usually persists between the established parent and the struggling offspring. Perhaps one reason why winter annuals are not more prominent in forest communities is that they rely solely on reproduction by seed. The seeds of a winter annual appear in spring but typically lie dormant during the summer and germinate in autumn. Autumn in a deciduous forest, however, is the time of leaf fall, and new seedlings are likely to be buried in new litter. The reproductive strategy of winter annuals seems, in fact, to be an evolutionary response not to forest pressures but rather to the pressures of open habitats—rock outcrops, pastures, etc.—that experience summer drought or disturbance (Baskin and Baskin, 1972). Winter annuals are much more characteristic of such habitats than of forests, and the relatively few winter annual species of forests frequently occur in open habitats as well.

The adaptive peculiarities of smaller plants in the temperate deciduous forest are easily remembered if we compare their characteristics with those of typical meadow plants. Deciduous forest herbs are most typically perennial, prevernal in both flowering and photosynthesis, and vegetative in their most common mode of propagation. The herbs of meadows are typically summergreen, and summer- or fall-flowering; in much greater proportion they are annuals or biennials, and they reproduce largely by seed. We can understand these differences largely in terms of the environmental contrasts created by the presence of a tree canopy in the forest and its absence in the meadow. Let us now consider the animal component of deciduous communities, before returning to the trees for a discussion of forest succession.

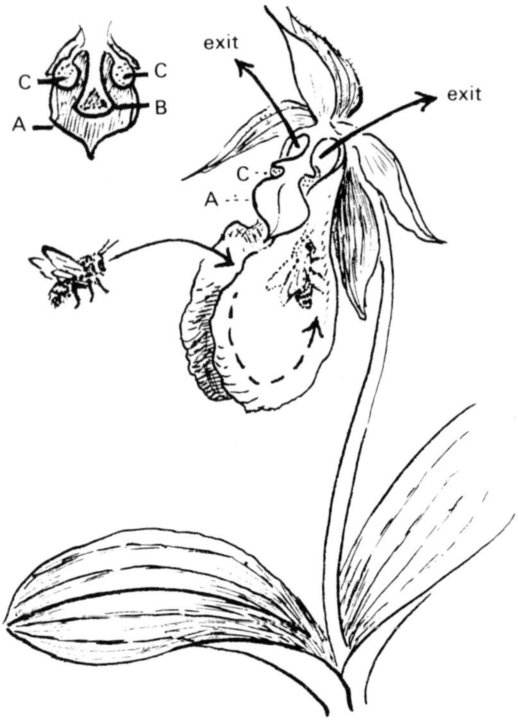

FIG. 4.11. Charles Darwin was one of many to be fascinated by the elaborate associations of insects and orchids. He describes the pollination mode of the lady's slipper (*Cypripedium acaule*) in his book *On the Fertilization of Orchids*, published in 1862. To reach the nectar of this orchid the small flies which typically pollinate it must force their way between the inturned lips of a highly modified petal that is the flower's most conspicuous feature. Unable to escape the way it entered, the insect crawls out one of two small openings at the base of the petal. In so doing it brushes unavoidably against the overhanging reproductive structures of the flower, first contacting the stigma, then one of the anthers. Pollen already adhering to the insect (from a previous flower) is removed by the sticky stigma, and a new load is picked up from the anther as the insect makes its exit. The inset shows the positioning of fertile anthers (*C*) and stigma (*B*) on the underside of the sterile, shieldlike anther (*A*) that guards the escape route. (Drawing by Sarah Allan.)

FOREST CONSUMERS

Many characteristics of the fauna of deciduous forests relate to the structure and periodicity of the vegetation. The food source of most forest animals lies either in the canopy layer or on the forest floor. The space between is relatively sterile, although tree trunks provide food for some organisms.

Most large forest herbivores are either arboreal or semi-arboreal. A few are restricted to the forest floor, for example deer and rabbits in the forest proper and moose, muskrat and beaver in marshes or at the margins of streams or lakes. These mammals all are most common where the forest is young or open, providing an abundance of ground-level vegetation. (Beavers, of course, also have ways of reducing tall foliage to their own level!) More typical of forest habitats, however, are mammals that can climb to the abundant food of the canopy: species of squirrels and mice, porcupines, opossums and raccoons. Many of these animals are seed and fruit eaters, and a number of them are partly carnivorous (especially insectivorous) as well. Varied food habits are useful to animals in a habitat where important food resources are scattered and only periodically available. Long-lived animals of temperate forests have less opportunity to become fruit or seed specialists than do the animals of tropical rain forests, where year-round stocks of these foods are available.

In deciduous forest, moreover, even the seasonal abundance of specific plant foods is unreliable. Graybill (1971) has reported as much as 5-fold year-to-year variations of fruit and nut production in western Pennsylvania forests. Beech, white and black oak, and sugar maple varied more than 10-fold in their seed production during a 3-year period on Graybill's study plots, but yearly fruit production of black cherry was relatively constant. It would be interesting to know whether the variability in seed production by the first four species is due primarily to yearly variations in climate, available moisture, etc., or whether it is at least partly an evolutionary strategy (analogous to that of the salmon in Chapter 3) that forestalls the development of too large populations of seed predators. Nut producers rely on animals such as squirrels for seed dispersal, but successful seedling reproduction depends on some seeds not being eaten. To black cherry, however, there is no strategic value to evolving a variable annual production. The more of its fruits that are eaten the better, since the seed within the fruits are dispersed but not destroyed by animals.

The dominant pure herbivores in deciduous forests are insects which feed either on foliage or wood. Unlike mammalian herbivores, many insects are able to be foliage specialists because their life cycles can be completed during the period of foliage availability; wood-eating insects, of course, have food available year-round, and many remain active during the winter.

Foods available on the forest floor include not only fallen leaves and limbs, but the animal life that subsists on this dead vegetation. Many of the tree-climbing mammals descend to feed on the larger decomposers (*e.g.* earthworms and insect larvae) of the litter, rotting wood and humus. Joining in this activity are pure carnivores of the forest floor such as moles, shrews, salamanders and predatory arthropods, which in turn sustain larger forest floor carnivores such as weasels, skunks and foxes.

Deciduous forests, particularly in America, boast unusually rich and varied bird faunas. No doubt this is partly because birds can exploit both canopy and floor food sources more effectively than most other animal groups. Many insectivorous species of bird feed in the canopy; some pick their food off leaves and twigs (*e.g.* most warblers and vireos), others probe the bark (nuthatches and woodpeckers) and still others capture insects on the wing (*e.g.* flycatchers). A separate group of insectivorous species (thrushes, brown thrasher, and towhee) does almost all its feeding on the ground. Still other forest birds are primarily seed and fruit eaters or are omnivores; again there is some division into canopy species (chickadee, grosbeaks, cedar waxwing) and forest floor species (ruffed grouse, wild turkey). Still other birds, such as the blue jay, may be generalists. But more than other vertebrates, the birds of deciduous forests have the opportunity to be food specialists. In part, this opportunity is a consequence of their great mobility; when insect or plant foods become seasonally unavailable, the birds that depend heavily on them migrate to warmer climates to obtain similar foods. It is not surprising that the most characteristically migratory bird groups of deciduous forests are those dependent on highly seasonal, foliage-feeding insects; the woodpeckers and other bark-probing birds tend to remain during the winter, because their insect food also remains.

The distinctiveness of the deciduous forest fauna is best emphasized by comparing it with the faunas of other biome types. Compared to grassland or savanna communities, for example, deciduous forests contain remarkably few grazing and browsing mammals, and also few seed-eating birds. Instead deciduous forests are unusually rich in insectivorous birds and omnivorous, semi-arboreal mammals. This description serves for most other forest types as well; in the tropical rain forest, where permanent arboreal niches are possible for mammals, Bourliere (1963) mentions an instance where the biomass of arboreal primates was more than 10 times that of the forest floor ungulates—and he states that arboreal rodents are even more abundant than the primates! He reports in a later paper, however, that biomasses of arboreal and forest floor mammals are more nearly equal than this in most rainforests (Bourliere, 1973).

In forests of all types, the dispersion patterns of vertebrates also contrast strikingly with those of grassland. Forest vertebrates rarely form large herds or flocks, as do many ungulates and seed-eating birds of grassland. Instead, forest vertebrates live in relatively small groups or singly. We will explore these contrasts further in Chapter 13, but may point out here that the food supply of most deciduous forest vertebrates is relatively sparse; they make meager direct use of vegetation for food, and hence their population densities are relaively low (Table 4.5). If only for this reason, we should not expect the large aggregations of herbivorous vertebrates, so common in grasslands, to be duplicated in deciduous forests. In America there is one glaring historical exception, the passenger pigeon, to this pattern of sparse, spread-out populations of forest vertebrates. We will consider this exception later.

FOREST DECOMPOSERS

Where do the unutilized plant residues of a deciduous forest go? to the decomposers. Pearson (1948) has estimated that 125,000 earthworms inhabit an average acre of undisturbed Pennsylvania forest. The weight of this single group of decomposers, Pearson calculates, is more than 10 times that of the resident mammals. The weight of microscopic decomposers is perhaps 100 times greater than that of the earthworms. Bray (1964) has found that in temperate deciduous forests herbivores consume only 6 to 11% of the leaves before they fall. The annual contribution to the floor of these forests averages about 4 metric tons per hectare (3568 pounds per acre), about 70% of this debris being dead leaves, the remainder twigs and stems (Bray and Gorham, 1964). But there is little or no cumulative increase of litter or humus from year to year, and we may conclude that decomposers consume virtually the whole yearly crop of litter.

Soil organisms and their ecology can only be touched on here. The soil subcommunity poses formidable taxonomic problems to the beginner, but studies are possible that largely skirt the need for detailed identification. One can, for example, determine the vertical distribution of invertebrates within the soil, by extracting the organisms from different layers of a profile (Table 4.6). This extraction is easily done by modified "Berlese funnels," *i.e.* ordinary glass or metal funnels topped by a screen that retains the soil or litter, but is of sufficiently large mesh to allow the passage of small organisms. Soil is placed on the screen and illuminated from

TABLE 4.5

Typical Densities of Mammal Species in a Deciduous Forest in Pennsylvania

The data are based on trapping and hunting records and on early records on Indian populations. From Pearson (1948).

Species	Number per Acre	Total Weight per Acre (g)
Deer mouse (*Peromyscus leucopus*)	5.0	100
Short-tailed shrew (*Blarina brevicauda*)	4.0	72
Red-backed mouse (*Clethrionomys gapperi*)	0.5	11
Woodland jumping mouse (*Napaeozapus insignis*)	0.5	12
Long-tailed shrew (*Sorex cinereus*)	0.2	0.7
Flying squirrel (*Glaucomys volans*)	0.1	7
White-tailed deer (*Odocoileus virginianus*)	0.04	1700
American Indian, circa 1600 A.D. (*Homo sapiens*)	0.0005	25

TABLE 4.6

Depth Distribution of Soil Invertebrates in a Coniferous Podzol Soil of New York State

The data are expressed in numbers per square foot. From Lutz and Chandler (1946).

Depth	Soil Description	Mites (Oribatidae)	Springtails (Collembola)	All Others
inches				
0–2	Fresh, largely unaltered litter	15,000	1980	140
2–6	A_0 and upper A_1 horizons	3800	1420	20
6–10	Lower A_1 horizon	2400	500	20
10–13	A_2 horizon	150	0	0
13–16	B_1 horizon	120	30	0
16–23	B_2 horizon	70	70	0

above; as it heats and dries, the organisms migrate downward, pass through the screen, and fall into a bottle of preservative. The Berlese extraction technique can also be used to compare faunas in the same horizon (or all horizons) of different soil profiles (*cf.* MacFadyen, 1963). Other simple methods in soil ecology include the "buried slide technique" for sampling microorganisms, and the "flushing technique" for sampling earthworm populations. If a soil is drenched with dilute formalin or potassium permanganate solution, earthworms will be driven to the surface where they can be collected or counted.

One can also study the role of different groups of organisms in contributing to the breakdown of litter. One simple procedure is to compare decomposition rates of dead leaves that have been buried in nylon bags having different mesh sizes. Large meshes will permit the entrance of earthworms and other large invertberates, while smaller meshes will selectively exclude them (Edwards and Heath, 1963). Species succession in decomposing litter can also be studied with the aid of mesh bags (Crossley and Hogland, 1962); in this case the bags would be of the same mesh size and be filled with the same type of litter, but would be buried for different lengths of time. Or, again making use of mesh bags, one can compare the rates of decomposition of different types of litter. The leaves of different plant species vary in chemical content, acidity and palatability to decomposers, and these differences can be identified by comparative studies of decomposition rates. References useful for methods or taxonomy, or as introductions to the literature of soil ecology, include Jackson and Raw (1966), Kevan (1962), Lutz and Chandler (1946), Wilde (1958), Doekson and Van der Drift (1963) and Reichle (1970).

Decomposers are linked in intricate ways to the rest of the forest community. Large decomposers such as earthworms and grubs are eaten by many forest animals, and thus belong both to grazing food chains and to decomposer food chains. Between soil organisms and plants, too, there exist many reciprocal relationships which derive partly from differences in the type of litter produced by different plant species. These differences have pronounced effects on the nature and richness of the soil biota. In turn, the unequal competence of soil organisms to quickly decompose different leaf types affects the plant community, by determining the depth of leaf litter throughout the year. Oak leaves, for example, are decomposed slowly; thus, in oak forests, herbs and seedlings must contend with a thick year-round bed of litter that may prevent them from germinating or growing successfully.

Plants and decomposers also interact through having similar needs. Both utilize soil moisture, and they can be thought of as very unequal competitors when the supply becomes low; at such times, the activity of decomposer becomes much reduced, as can be demonstrated by measuring the rate of release of respiratory CO_2 from the forest floor (Reiners, 1968; Froment, 1972). Finally, at the most basic level of community integration, decomposers convert organic matter to simple nutrient compounds that can be used again by the plants, and through this activity the decomposers facilitate continuation of the whole community.

Among decomposers themselves, as might be expected from the term "decomposer food chain," there are important relationships. The soil invertebrates process materials into forms more readily used by bacteria and fungi, so that where larger soil animals are numerous, so also are soil microorganisms. Beyond this broad statement we will not proceed; interested readers can consult the review of Edwards *et al.* (1970), which considers both general and specific relationships among soil organisms.

NICHE SEPARATION OF CLOSELY RELATED SPECIES

In any community, the coexistence of species that are closely related taxonomically creates particularly interesting questions as to how they differ ecologically. Let us examine several groups of deciduous forest species with this problem in mind.

Maples of Mt. Monadnock

The gaunt upper slopes of Mt. Monadnock, in southern New Hampshire, beckon from afar to hikers. But to reach these upper flanks one must pass through rich northern hardwood forests lower down, and here one may find coexisting in the same community four of the seven species of maple native to eastern North America. One occasionally is rewarded for being out of condition: several years ago while trekking rather slowly up the mountain I noticed this profusion of maples and asked the obvious question. How do the four species manage to coexist? How do their niches differ?

Part of the answer was easy: only two of the species, red maple and sugar maple (*Acer rubrum* and *Acer saccharum*), vie for a place in the canopy. The others, moosewood and mountain maple (*Acer pennsylvanium* and *Acer spicatum*), are small understory species. In addition, however, each of the four species has a recognizably different altitudinal range on Monadnock, and to some degree a different preferred habitat at elevations where all exist together. Along the Marlboro Trail, which climbs the southwest ridge, mountain maple has by far the narrowest distribution. It is concentrated along a small stream near the bottom of the slope, and occurs slightly higher in a patch of young second growth forest. Its restriction to these relatively open habitats suggests that mountain maple may be rather light-demanding. Ecologically it is perhaps a small version of the Eastern sycamore described in Chapter 1. Sycamore does not occur in this cool, northern hardwood forest, and mountain maple is by no means a perfect equivalent; but this maple probably needs more light than most understory species receive, and hence occurs only where the canopy is broken.

Moosewood, the other small maple of Monadnock, is found much more generally through the forests of the lower slopes, and it also extends to quite high elevations. Both its distribution and its remarkably large leaves suggest that it is a species that can grow in deep shade. The distinctly green trunk and branches of this species contain chlorophyll year-round, and perhaps contribute some photosynthetic activity during periods before and after the canopy trees are in leaf.

Red maple is the most widely distributed species on the mountain; one finds it on both gentle and steep slopes, in marshland at the mountain's base, and also among spruces in the boreal conifer zone higher on the mountain. The largest red maples are found on a steep, rocky but organic slope just below the first spruces; here they are beyond the upper limit of sugar maple, the fourth species of maple on the mountain. Sugar maple is generally distributed on the lower slopes, but seems to prefer broad hollows where it often dominates the canopy layer, sometimes in company with beech. Unlike red maple, it is not found in the marshes at the base of the mountain. Since all but bare rock habitats on Monadnock are relatively moist, one can say little about the relative drought tolerance of any of these maples, but red maple seems most tolerant of waterlogged soils.

The distribution of these maples on Mt. Monadnock is quite an accurate index of their general geographic range. Red maple, the most wide-ranging and perhaps most versatile species on Monadnock, is found from Newfoundland to southern Florida and west to Minnesota and eastern Texas, growing in both uplands and lowlands. It is somewhat less shade-tolerant than sugar maple, a fact we might have suspected from its prominence in relatively well lit sites on Monadnock (steep, side-lit slopes and open marshes). Sugar maple is one of the most important trees in mature northern deciduous forests, but it extends neither as far north nor as far south as red maple. Mountain maple and moosewood both are restricted to cool deciduous forests, and both have quite similar geographic ranges, extending from eastern Canada down the Appalachians to Georgia, and west to Michigan and Minnesota. It is probably safe to say, however, that within this range moosewood occupies a wider variety of local habitats, as it does on Monadnock, than does mountain maple.

Further study certainly is called for, but the maples seem to provide a relatively easy problem of niche differentiation; separating them ecologically takes some initial scrutiny, but they are fairly distinct in their preferences and the number of species is not too great. Only three other species of maple are prominent in eastern United States. Black maple (*Acer nigrum*) is visually similar to sugar maple, and possibly occurs with it on Mt. Monadnock. These two species present perhaps the most interesting niche problem among American maples, if indeed they are separate species. Some authorities think, however, that black maple is but a geographical variant of sugar maple. Silver maple (*Acer saccharinum*) is primarily a flood plain species, as is ash-leaf maple (*Acer negundo*). Careful studies, however, reveal distributional differences on flood plains where these two species coexist (see Fig. 3.7). Ash-leaf maple has one of the widest geographical ranges of any American tree; it is as much at home in Oak Creek Canyon, Arizona as it is along the Potomac.

Coexisting species of oak present a more difficult problem of niche differentiation than do the maples. Numerous species of oak may occur in the same habitat, and they are often difficult to tell apart. Several, like white oak, are quite variable all by themselves. Besides taxonomic problems, the oaks present relatively subtle niche differences, which have intrigued a number of ecologists. Although considerable success has been achieved in sorting out these differences, often by continuum analysis (see Curtis, 1959; Racine, 1971; Keever, 1973), many answers remain elusive. Of course, questions of niche differentiation can be asked of the plants in any type of forest; good examples from the western United States are the studies of Pharis (1966) and of Hodges and Scott (1968).

Vireos in Maryland

Interest in niche separation among animals received enormous impetus from two careful field studies published about 15 years ago—that of Robert MacArthur (1958) dealing with warblers in the coniferous forests of northern New England, and that of Joseph Connell (1961a) dealing with the zonation of species of barnacle on rocky coasts. MacArthur's work demonstrated the quantity of new information that could be derived by the simplest of ornithological methods—patient observation with binoculars—when one starts with specific questions and uses a comparative approach. Connell demonstrated the value of field experiments—alteration of habitat and removal of suspected competitors—in delineating niches.

Both studies have stimulated considerable work in both the theory and actuality of animal niches. In deciduous forests, studies of birds by Root (1967) and Williamson (1971) are close spiritual relatives of MacArthur's work. Penelope Williamson, whose investigation we will use as an example, attempted to delineate niche differences among three species of vireo and two other bird species, all of whom belong to what Root (1967) has termed the "foliage-gleaning guild" of eastern deciduous forests. Root coined the term "guild" to denote groups of coexisting species that "exploit the same class of environmental resources in a similar way," hence have presumably overlapping niches. The vireos of Williamson's study, for example, all feed primarily on caterpillars inhabiting the leaves of deciduous trees. Since these birds feed on similar food, Williamson theorized that they might minimize competition by different foraging behaviors; and, armed with binoculars, a stopwatch and a portable tape recorder for recording her observations, she set out to test her thesis. Williamson also captured and color-banded vireos of each species so that she could recognize variations of behavior among individuals. The foci of her study were the red-eyed vireo (*Vireo olivaceus*, perhaps the most common of deciduous forest birds), and the rarer white-eyed and yellow-throated vireos (*Vireo griseus* and *Vireo flavifrons*).

Williamson observed that all three species of vireo feed primarily from small branches or twigs, collecting their food by reaching out from their perches to nearby leaves. This type of behavior is only one of several possible ways of collecting insects; it helps to reduce vireos' competition with other insectivorous birds which rely more on hovering (which permits feeding on outer leaves inaccessible to perching birds), hawking (capturing insects in flight, as is done by most flycatchers) or extracting insects hidden in bark. But the vireos' similarity in feeding mode only accentuates the question of how the three species minimize competition among themselves. Williamson did find that red-eyed vireos use the hovering tactic more than the other two species; but this minor difference in *how* the three species feed seems to be considerably less important than differences in *where* they feed (Fig. 4.12). At least during the breeding season, the white-eyed vireo feeds in young patches of second growth or at the edges of older forests, while the other two vireos typically feed within mature forest stands. In these older stands, yellow-throated vireos feed primarily in the tops of tall trees, occupying a relatively narrow layer of the upper canopy. Red-eyed vireos usually forage beneath the yellow-throated vireos; in addition, male and female red-eyed forage at different heights (Fig. 4.13), helping to reduce the strong *intraspecific* competition one might expect in a species as common as the red-eyed vireo. These forms of spatial separation seem to be the chief niche-separating features among the three vireos; their breeding seasons (which represent periods of maximum food demand) nearly coincide, but the nests of the three species occur at quite different heights (Fig. 4.12). In addition, of course, the white-eyed vireo nests in a different type of forest than the other two. During the breeding season, widely spaced pairs of yellow-

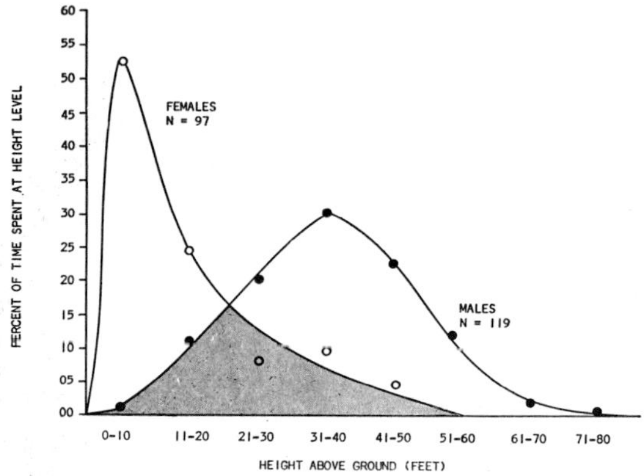

FIG. 4.13. Separation by height of foraging male and female red-eyed vireos during the breeding season. Females forage in the vicinity of the nest, males in the vicinity of their higher breeding perches. *N* is the number of observations. (From P. Williamson: *Ecological Monographs* 41: 129–152, 1971.)

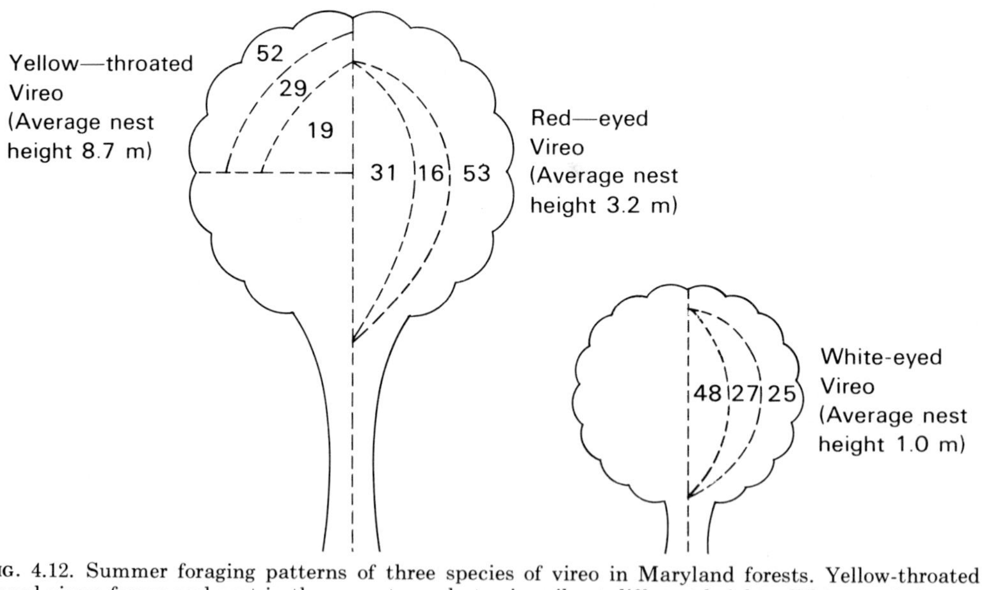

FIG. 4.12. Summer foraging patterns of three species of vireo in Maryland forests. Yellow-throated and red-eyed vireos forage and nest in the same trees, but primarily at different heights. White-eyed vireos prefer the small trees and shrubs of clearings with young, regenerating forest. The numbers represent percentages of foraging time spent in outer, middle, and inner portions of trees. (Adapted from P. Williamson: *Ecological Monographs* 41: 129–152, 1971.)

throated vireos occupy pancake-shaped, treetop territories, averaging 10 acres in area but only 40 feet thick; they feed in these territories and defend them against other members of the same species. The territories of red-eyed vireos are much more compact in area (averaging 1.5 acres), but are also much much deeper (about 80 feet), extending from tree top to forest floor. Williamson makes no mention of the two species defending their territories against each other, and presumably they do not; this testifies to the effective separation of their feeding niches.

Williamson recorded many other aspects of the feeding behavior of the three vireo species, as well as of the two associated insectivorous birds (American redstart and Acadian flycatcher) that she studied during more than 1700 hours of observation. The patterns described here are niche separation patterns during the breeding season; Williamson found marked differences in the species' behavior at other times of the year. In autumn, for example, red-eyed vireos moved to the low, young forest habitats of white-eyed vireos, joining them and other species (*e.g.* warblers and Carolina wrens) in mixed foraging flocks, and feeding heavily on berries as well as insects. Whether red-eyed and white-eyed vireos were primarily competitors or cooperators in these flocks is uncertain. (If competitors, presumably they *both* benefited from the activities of the other species in the flock.)

Salamanders in Indiana

Our final example of a forest niche investigation is an unpublished study by Kelly Williams (1973). Salamanders of the genus *Ambystoma* excite brief attention each year because of their mass migrations to forest ponds to breed. Then they seemingly vanish. Actually, during the remainder of the year they lead secretive lives underground. Very little was known of the adult, nonbreeding habits of these salamanders before the 4-year study by Williams, whose interest in *Ambystoma* was sparked by the discovery that four species of *Ambystoma* (*A. maculatum*, *A. jeffersonianum*, *A. opacum* and *A. texanum*) were using the same pond in southern Indiana for breeding.

Using trapping techniques that allowed him to census the salamanders as they came to the pond to breed, Williams found a beautifully staggered pattern of reproduction. *A. opacum* (the marbled salamander) breeds in autumn, laying its eggs not in the pond but in depressions around the edges. The young, however, make their way to the pond after hatching. The other three species migrate to the pond successively between late December and March. *A. jeffersonianum*, the species constituting the earliest wave, was found burrowing through soft ice at the edge of the pond in even slight January thaws; it actually breeds beneath the ice. The breeding period of *A. texanum* overlapped that of *A. jeffersonianum* but usually began about 2 weeks later. *A. maculatum* did not arrive at the pond until the beginning of March.

To some extent these sequential breeding times probably minimize competition for food among the larvae. By late spring the young of all four species are living in the pond together, but because they are of different sizes, they utilize somewhat different foods. (Young larvae of the fall-hatching species, *A. opacum*, actually feed on eggs of the others, so competition may not be the *only* selective force leading to the staggered breeding pattern).

To study the population dynamics of each species from year to year, Williams used traditional toe-clipping methods to mark individuals he captured, then set them free. He also employed a less traditional marking method—implanting a radioactive wire in the tail—to study the habits of each species during the nonbreeding season. Using a radiation detector, he was able to trace these individuals to their underground lairs, and again he found remarkable, species-specific patterns. *A. jeffersonianum* lived only in relatively deep burrows made by small mammals. *A. maculatum* (the spotted salamander) sometimes was found in small mammal burrows, but also inhabited dead-end burrows which it may have made itself. *A. texanum* remained near the pond, and lived only in the burrows of crayfish. *A. opacum* sometimes inhabited shallow small mammal burrows, but frequently lived on top of the soil beneath moist litter. Thus, through habitat specificity during most of the year and temporal separation during the breeding season, these four species of *Ambystoma* effectively partitioned their common environment into recognizable distinct niches which allowed all four to coexist.

ECOLOGICAL SUCCESSION

In parts of the American deciduous region primary succession is still occurring on boulders and rock slopes scoured clear by the last glaciation, or by more recent landslides. Most of the eastern United States, however, has undergone forest clearance either once or repeatedly, and is now in various stages of secondary succession. Our discussion will center on this process.

Examples of Secondary Seres

Although we can suggest typical seres, innumerable local variants result from topographic, edaphic and other physical influences, as well as from differences in the species pool available to take part in the sequence. Ecological succession has been intensively studied in many areas, but there remains ample room for interesting and original detective work in elucidating local successional patterns.

Succession on abandoned fields. "Old field succession" is the term used to describe the progressive recolonization by wild vegetation of land completely cleared for agriculture. The classic studies of old field succession were carried out in the rolling Piedmont lands of the American Southeast. Here the soil is old, rather leached, and relatively poor in nutrients. Tobacco, a major crop, heavily depletes those nutrients that do occur, and without fertilization the soil soon becomes marginal for agriculture. In consequence, former farming practices in the Southeast often involved shifting agriculture—tilling a plot until it became unproductive and then abandoning it. This practice is still common in the tropics where soils often are even poorer than those of the Piedmont. Most southeastern Piedmont landscapes are in some stage of old field succession; the sere takes as much as 200 years in this region.

Table 4.7 depicts two typical old field seres of the American Southeast, as well as analogous seres in other deciduous regions and (for contrast) a prairie sere in Oklahoma. Some typical old field landscapes of varying ages in the deciduous region are shown in Figure 4.14.

We may identify certain common features of these old field seres. All begin with a weed stage characterized primarily by annual, opportunistic plants. On all sites, even the prairie, there is a tendency for perennial plants to become predominant in succeeding stages. Following the weeds, grasses often assume importance; many old fields in deciduous regions temporarily look like prairie, and contain either true prairie grasses or related grass species. Conifers frequently come next in the old field sequence—loblolly and shortleaf pines in North Carolina, red cedar in many other areas, white pine in New England. These generalizations are not to be taken as commandments, even as the seres in Table 4.7 are not to be considered archetypes for whole regions. Further studies in New Jersey, for example, have revealed significant variations, even on the same soils, to the pattern in Table 4.7 described by Bard (1952) (Buell *et al.*, 1971); and in many parts of the East a grass (yellow foxtail) rather than broad-leaved weeds, may dominate the initial old field

FIG. 4.14. Secondary succession in the deciduous region. (*A*) Fields abandoned less than a year in the mid-Atlantic states sometimes boast nearly pure stands of *Setaria glauca*, whose nodding, furry head is responsible for the common name—foxtail grass. (*B*) After 5–8 years the field may be dominated by little bluestem grass (*Andropogon scoparius*), with numerous saplings of red cedar (*Juniperus virginiana*) heralding the demise of the grass stage. (Grazing, if heavy, can maintain such a field in the bluestem-cedar stage for a very long time.) (*C*) Much later, the cedars themselves give way to more shade-tolerant hardwoods. (*D*) In the southeast a similar fate overtakes loblolly and short-leaf pines (*Pinus taeda* and *P. echinata*) which "substitute" for red cedar on the clay soils of the Carolina Piedmont. Because they grow so large, these pines lengthen the conifer stage considerably in this region. (Photographs by author.)

B
C
D

Table 4.7

Representative old-Field Seres

In North Carolina and Oklahoma distinct stages are more recognizable than in New Jersey and Illinois, where young trees appear early and gradually replace the smaller plants. Where percentages are indicated, they are values for per cent cover of the study plots by particular species. The climax community is attained in as little as 45 years in the prairie of Oklahoma, in approximately 200 years in North Carolina, and in an unknown but probably still longer period in New Jersey and southern Illinois. Note the logarithmic time scale.

Years after Abandonment		North Carolina Piedmont (Oosting 1942)		New Jersey Piedmont (Bard 1952)	Shawnee Hills, Southern Illinois (Bazzaz 1968)	Central Oklahoma Prairie (Perrino & Risser 1972)	
150	Hardwood Stage	White oak, post oak hickory, dogwood, etc.	White oak, red oak, hickory, dogwood, sourwood, etc.	White oak, dogwood, hickory, red oak, etc.	Black oak, white oak, hickory spp., etc.		Tall Grass Prairie Climax
100						Big bluestem, little bluestem, switchgrass	
60	Conifer Stage	Shortleaf pine, hardwood understory	Loblolly pine, hardwood understory	Red cedar, poison ivy, sassafras, black cherry, red maple, red oak	Broomsedge (5%), goldenrod (38%), sassafras and persimmon thickets, young trees		
40							
30				Red cedar, sassafras, black cherry, dewberry, little bluestem, poison ivy, hawkweek, aster		Little bluestem	Perennial Bunch Grass Stage
20					Broomsedge (22%), cinquefoil, sassafras and persimmon thickets, saplings		
		Shortleaf pine ↖ (drier sites)	Loblolly pine ↗ (moister sites)	Red cedar, hardwood saplings, goldenrods, aster, dewberry	Broomsedge (16.5%), tall goldenrod, shrubs		
10		Young pines, Broomsedge			Broomsedge (32%), cinquefoil, persimmon, sumac, sassafras, winged elm		
5	Grass Stage	Broomsedge, pine seedlings		Heath aster, gray goldenrod, saplings of red cedar, black cherry, red maple, dogwood	Broomsedge (69%), gray goldenrod, saplings of persimmon, sassafras, sumac, red cedar	Triple awn grass, heath aster, western ragweed, silver beardgrass	Annual Grass Stage
3		Broomsedge					
2	Weed Stage	White aster, ragweed		Gray goldenrod	Daisy fleabane, white aster, gray goldenrod		
1		Crabgrass, horseweed		Ragweed, evening primrose, sheep sorrel	Ragweed, crabgrass, horsenettle, wild garlic	Western ragweed, Johnson grass, cheatgrass, sorrel sp.	Weed Stage

stage. It is worth noting, too, that quite different seres characterize abandoned grazing land and abandoned cropland.

The southeastern old field sere provides perhaps the most striking and consistent successional pattern of any in the world, but even here there are variants, and the species composition of early stages, in particular, is quite unpredictable. Weeds and grasses dominate croplands during the first few years after abandonment, but by the 15th year there usually is a closed forest of young pines that have virtually eliminated the sun-demanding weeds and grasses. The pines, too, regenerate poorly in their own dense shade, and are eventually replaced by more shade-tolerant hardwoods. Farther north, in the New Jersey Piedmont, red cedar is by far the most prominent tree during at least the first 60 years of old field succession, but even after 60 years there is not yet a closed forest; Bard describes this 60-year New Jersey stage as an incredible tangle of emergent cedars, poison ivy, shrubs, young hardwoods and weeds. By this time North Carolina old fields support a forest of tall pines with a distinct understory of hardwoods that will eventually predominate in the climax stand. It is not known how long it takes for the 60-year New Jersey "forest" to transform into the quite dissimilar hardwood climax of the region.

The successional pattern described for southern Illinois also is much slower than than in North Carolina, and is notable for a rarity of conifers; the more advanced stages of this sere, between the 40-year stage and the presumed oak-hickory climax, are not accurately known.

Succession after lumbering and fire. Lumbering seres are highly variable, depending on the completeness of cutting. If some seed-bearing trees and considerable shade remain, there may be little in the way of stepwise species change during recovery. If a large segment of forest has been heavily lumbered, however, the sere may be quite similar to later old field stages. Bracken fern and brambles may appear first, later to be replaced by pines and/or fast-growing deciduous trees such as pin cherry, poplar and black birch. Spurr (1956) notes few differences between the trees of lumbering and of old field seres in Massachusetts.

Table 4.8 shows the vegetational changes and associated changes in the bird fauna that occur after lumbering in upland oak-chestnut forests in North Carolina. It is particularly interesting to note the specificity that many bird species show for particular successional stages; this is equally true of other forest vertebrates and invertebrates.

Succession after fires is also highly variable, depending again on the magnitude of disturbance. In an area not subject to frequent burning, a light fire may destroy litter and small plants but leave humus and large trees nearly intact, and a seral pattern will not really be initiated. A severe fire in such an area may destroy the humus and larger vegetation as well, and successional recovery will be a long term process with several seral stages, similar but probably not identical to a clear-cut lumbering sere. Areas subject to frequent fires typically have little humus remaining in the soil and support a vegetation adapted to the presence of fire; in such habitats there is little real succession between one fire and the next (Fig. 4.15). Burning frees the minerals contained in litter and humus for reuse by plants, but much of the carbon and nitrogen are lost (Spurr, 1964, p. 118). These soil changes may encourage some forms of vegetation while discouraging others; among herbaceous plants, for example, fire increases the abundance of leguminous species, whose symbiotic root bacteria can fix the needed nitrogen (Lemon, 1967).

Most studies of fire ecology describe the nonsuccessional effect of *repeated* fires on a community, rather than the successional changes that occur after an isolated fire. The papers by Vogl (1967) and Brown (1960) are among the few that provide hints of successional sequences after a fire in deciduous forest. Edwards (1954) followed the long term changes in mammalian populations resulting from a single fire; and his description, while actually specific to a coniferous region of British Columbia, is probably quite analogous to the recovery process that occurs after infrequent burns in moist northern hardwood forests of the deciduous region. The fire studied by Edwards occurred in 1926 and destroyed 200 square miles of mature valley forest.

Before this fire, mountain caribou were common winter inhabitants of the forest, where they fed on abundant tree lichens. In summer, they left the forest for high mountain meadows. Wolverines and grizzly bears followed the same seasonal pattern, while marten were common in

TABLE 4.8

Changes in the Bird Fauna of a North Carolina Lumbering Sere

Plot 4 (xeric shrubland) represents a stage of succession that appears soon after lumbering of oak-chestnut forests in the vicinity of Highlands, N.C. It contained dense thickets of brambles and black locust, sumac and cherry saplings, interspersed with grassy openings. Plot 5 was a rather uniform stand of deciduous trees, mostly 10 to 25 cm in stem diameter and dominated by oaks, hickories and maples. Plot 6 was a mature forest but with some thickets in openings created by blight mortality of chestnuts. From Odum (1950).

	Xeric shrubland (plot 4)	Intermediate deciduous forest (plot 5)	Mature climax (plot 6)
		pairs/100 acres	
Chestnut-sided warbler (*Dendroica pensylvanica*)	108		8
Towhee (*Pipilo erythrophthalmus*)	108		6
Catbird (*Dumatella carolinensis*)	68		
Field sparrow (*Spizella pusilla*)	45		
Golden-winged warbler (*Vermivora chrysoptera*)	18		
Brown thrasher (*Toxostoma rufum*)	18		
Indigo bunting (*Passerina cyanea*)	18		
Goldfinch (*Spinus tristis*)	9		
Flicker (*Colaptes auratus*)	4		
Ovenbird (*Seiurus aurocapillus*)		50	38
Red-eyed vireo (*Vireo olivaceus*)		45	12
Blue-headed vireo (*Vireo solitarius*)		38	22
Black-throated blue warbler (*Dendroica caerulescens*)		23	24
Black-and-white warbler (*Mniotilta varia*)		15	4
Wood pewee (*Contropus virens*)		15	2
Scarlet tanager (*Piranga olivaceus*)		15	
Wood thrush (*Hylocichla mustelina*)		8	8
Robin (*Turdus migratorius*)		8	2
Downy woodpecker (*Dendrocopus pubescens*)		2	4
Hairy woodpecker (*Dendrocopus villosus*)		2	4
Carolina chickadee (*Parus carolinensis*)		1	2
Veery (*Hylocichla fuscescens*)			12
White-breasted nuthatch (*Sitta carolinensis*)			4
Ruby-throated hummingbird (*Archilochus colubris*)			4
Tufted titmouse (*Parus bicolor*)			2
Ruffed grouse (*Bonasa umbellus*)			2
Totals	396	222	160

the forest year-round and beaver occurred permanently along the streams. The fire virtually eliminated martens from the affected area, caused a serious long term decline in caribou and reduced the populations of grizzlies and wolverines. As revegetation began after the fire it was not these animal species, except for beaver, that returned; instead, other species invaded from outside and from small, relatively open patches and meadows existing before the fire. Fireweed and willow appeared early, then abundant growths of young birch and aspen; these created a low, rich, but intermittent vegetation cover that was highly favorable to mule deer, ground squirrels, deer mice, beaver and black bear. These animals increased in abundance, some spectacularly, as did carnivores such as cougars and coyotes. In the 1930s moose colonized the valley and multiplied rapidly; their abundance apparently stimulated an increase in wolf populations. By the early 1950s young conifers were beginning to repopulate the burned area, but their return was slowed by the continuing abundance of squirrels and

FIG. 4.15. Forests subject to chronic, recurrent fires are characterized by a permanent *arrested succession* known as a **fire climax.** Sequoia forests of California, ponderosa pine forests of Arizona, long-leaf pine forests of the Carolinas, and the Pine Barrens of New Jersey all are examples (Cooper, 1960, 1961, McPhee, 1968). The Pine Barrens, shown here, receive ample rainfall but the soils are so sandy and porous that they dry rapidly. A fire-adapted vegetation has dominated this area for at least 10,000 years (Florer, 1972). (*A*) Soil profile showing the sandy, infertile, highly leached soils typical of well-drained Pine Barrens sites. Humus has been burned away, and *B* horizon is stained darkly with iron oxides. (*B*) Pitch pine, the dominant tree, possesses thick, fire-resistant bark that protects dormant buds beneath it. Fire may kill exposed foliage but stimulates these buds to grow and produce new needles, as on the small tree in the foreground. Note fire scar near base of larger tree. Fire also stimulates closed cones to open, and seeds fall onto a fertile bed of fresh ash. If the whole aboveground portion of a tree is killed, sprouts spring from the roots and sometimes bear new seed within 3 years. (*C*) Like pitch pine, scrub oak sprouts prolifically

FIG. 4.15 (*cont'd.*)
from uninjured roots after fire, and the sprouts quickly mature to seed-bearing age. Its bushy growth habit and slow-decaying leaves enable scrub oak effectively to trap its own leaves and recycle their nutrients—a great advantage in the infertile, easily leached soils of the Pine Barrens. (*D*) In the region of most severe and frequent fires, the forest is reduced to a stunted tangle of pitch pine and scrub oak. Elsewhere the forest is taller and more diverse. (*E*) Low-lying, permanently boggy areas serve as refuges for fire-prone species like red maple and Atlantic white cedar. (Photographs by author.)

mice, which eagerly sought out their seeds. Deer, together with coyotes and cougars, had declined somewhat as the deciduous shrubbery grew taller, but moose were still extremely abundant and prefire inhabitants such as caribou and marten were still rare. Twenty-eight years after the fire, succession was still far from completion.

Probing the Successional Process

Examples of particular successional sequences raise questions about the successional process in general. Why is the species composition of seral stages predictable? Why is predictability not absolute? Why do the seral stages of secondary succession replace one another at a progressively decreasing rate (Table 4.7)? Why does the same type of sere, *e.g.* the old field sere, take longer to run its course in some places than in others? Do animals influence the forest successional process? To some of these questions we already have partial answers, but further exploration is fruitful.

The predictability of seral communities. Although patterns of species replacement vary to some degree even on very similar sites, successional sequences often can be predicted with surprising accuracy. Predictable species changes are a consequence of predictable environmental changes. In any region there is a pool of available species, and if we know their relative colonizing ability, their competitive strategies, and their specific environmental requirements we can predict in which stages of a sere they will be most common, and how long they will be able to persist.

Autecological contrasts among specific forest trees serve well to illuminate this argument. Once we know the "biographies" of these species, it is not difficult to predict their successional behavior.

Sycamore and black cherry. Consider first a species like sycamore. We know from Chapters 1 and 3 that sycamore is primarily a flood plain species, but sometimes it appears on abandoned fields. The myriad lightweight seeds produced by sycamore (Table 4.9 and Fig. 4.16) make this species a swift and opportunistic colonizer, but the germination requirement of bare ground means that colonization is limited to the earliest phase of succession. Seedlings established at this time, however, are likely to persist for many years; sycamore grows at extraordinary rates and reaches extreme heights so that established individuals can satisfy their high light needs by keeping ahead of their competitors. But eventually the old trees die and sycamore disappears from the maturing forest.

Now consider a different tree: black cherry. This species produces abundant but relatively large seeds (Table 4.9) which can germinate in a wider variety of soil conditions—including litter-covered forest soils—than can sycamore. Distant dispersal of these seeds is assured by the attractiveness of the fruit to birds. Black cherry is frequently one of the first trees to appear in old field succession, but it also successfully colonizes relatively small forest

TABLE 4.9

Reproductive Contrasts among Deciduous Trees

Data are from Fowells (1965), Curtis (1959) and Downs & McQuilkin (1944).

	Approximate Number Seeds per Pound	Dispersal Mechanism	Notes
Sycamore	200,000	Wind	Millions of seeds produced per tree.
Sugar maple	6,100	Wind	Records of up to 8.5 million seeds per acre in a pure stand, but 1.5 million/acre is probably average in such a stand.
Black cherry	4,800 (when cleaned of fruit)	Birds and Local mammals	Trees are prolific seed producers, beginning as early as 10th year.
White oak	300–400	Local mammals	78,000 seeds per acre recorded as an average in a stand of mixed white and black oaks. A single tree may produce an average of 2,000 ripe seeds per year in its best age period.

gaps such as those created by the death or windthrow of individual large trees. Black cherry has relatively high light requirements, but rapid growth and early maturation in such gaps makes possible the perpetuation of the species even in mature forests, though usually in small numbers (Fig. 4.17). Auclair and Cottam (1971), who have studied this species in the oak forests of southern Wisconsin, consider black cherry "an ecological opportunist par excellence" primarily because of its habitat versatility and high reproductive and dispersal potential; but for a pure opportunist it has remarkable staying power, and some of its characteristics (such as the ability of individuals to grow slowly in moderate shade and to shoot up when the shade is removed) are more typical of equilibrium species. Knowing its mixed strategy, we can predict black cherry's successional behavior; it should be most important rather early in succession, then decline to reduced but relatively stable numbers.

White oak and sugar maple. These are species typical of late and final stages of succession, and they differ from sycamore and black cherry in growing more slowly and being quite tolerant of shade (Table 4.4). Both reach large size and enter the forest canopy, but neither reach for light in the dramatic way that successional giants like sycamore, white pine and tulip tree do. Thus far, then, white oak and sugar maple seem similar, but reference to Table 4.9 reveals striking differences in their reproductive strategies. Reproductively, sugar maple somewhat resembles an opportunist. It produces large quantities of moderately light seeds having papery wings which assist in wind dispersal. The seeds are carried as much as 100 m from the parent tree, and a pure stand of sugar maples produces several million seeds per acre in a good seed year (Fowells, 1965). Because they are small and contain limited food reserves, many seedlings perish, but sugar maple's "saturation strategy" makes it likely that some seeds will land on ground where the litter is not too deep to prevent seedling establishment. The leaves of sugar maple decay quite rapidly after falling, so in this sense parent trees make it easy for their offspring to become established. The sum of several characteristics —low compensation intensity, abundant production of small seeds and rapid leaf decay—allows sugar maple to persist in a mature forest; but limited distance of seed dispersal (in comparison, say, to black cherry or sycamore) and slow growth rate make it unlikely that this species will be prominent in early succession.

The reproductive strategy of white oak (and other oaks) is the antithesis of that of sugar maple. The seeds are large and heavy, and relatively few are produced per tree. But oak leaves decay slowly, and the litter in oak-dominated forests is thick; large seeds are needed if the roots of seedlings are to penetrate the litter and reach the soil before food reserves in the seed are exhausted. Thus, again, we see a reproductive strategy related to the species' leaf decay characteristics. But this is partly digres-

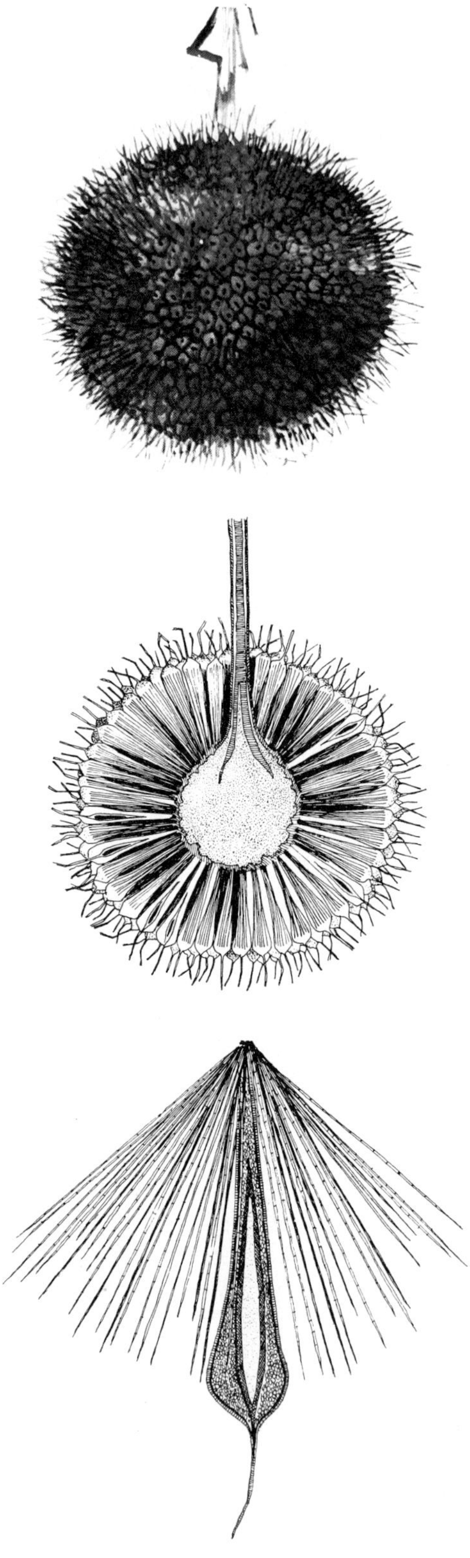

sion: the point regarding succession is that the seed characteristics of white oak are such that we would expect its acorns to reach a successional site more slowly and in smaller numbers than the seeds of other local trees but to germinate quite successfully whenever they arrive. Moreover white oak is relatively shade-tolerant so should persist once it makes its appearance.

From these examples we can see that autecological differences in such characteristics as dispersal ability, growth rate, longevity and shade tolerance allow considerable predictability regarding the species composition of successional forests of a given age. A species' shade tolerance is not always directly synonymous with its compensation intensity (page 103) though the two are closely related. For example, studies of loblolly pine indicate that its successional demise in the old field sere of North Carolina does not occur simply because parental shade exceeds the compensation intensity of its seedlings. Among shaded loblolly seedlings, photosynthesis usually exceeds respiration, but not sufficiently to permit development of a deep and extensive root system. In the late summer droughts that are quite frequent in this area, shaded loblolly seedlings die of thirst, not directly from low light levels (Billings, 1938).

We have been speaking of successional predictability, which is related to intrinsic characteristics of the species involved. Our examples have dealt with familiar trees; Keever (1950) and Bormann (1953) provide similar autecological explanations for the earlier stages of succession, suggesting why, for example, broomsedge replaces broad-leaved weeds, and why conifers (rather than hardwoods) replace broomsedge in southeastern old fields. The *unpredictability* of succession—the failure, perhaps, of two nearby fields to follow exactly the same pattern—is usually due to site differences.

FIG. 4.16. "Fruits" of sycamore trees contain innumerable tightly packed seeds, each with a surrounding bundle of lightweight hairs. These open like an umbrella when the ball is ripe, helping it to disintegrate and release the seeds. The seed hairs then act as parachutes and the tiny propagules are scattered widely by the wind. If they land in water they float, often being deposited on mudflats or sand bars where the bare, moist soil provides ideal germination conditions. (Reprinted by permission of the artist from W. K. Martin: *Natural History Magazine*, p. 46, June–July 1973.)

FIG. 4.17. Black cherry plays a perpetual game of hide and seek with tent caterpillars, whose voraciousness may so stunt its growth in old fields that other early successional trees overtop it. Even in the forest, one of the surest identification marks of black cherry in the spring is a nest of tent caterpillars. But cherry trees scattered among small forest gaps probably suffer far less damage from these herbivores than do the exposed, often clumped trees of old fields. The evolutionary strategy of black cherry may conceivably be partly a strategy of escaping tent caterpillars; at any rate, the special relationship between the two species is worthy of study. (Drawing by Sarah Allan.)

The earliest stages of old field succession, for example, vary depending on when during the year a field is abandoned, how much and what kind of fertilizers were used before abandonment, and what kind of cultivating practices were employed, as well as on moisture, soil type and other "natural" variables (*cf.* Overlease, 1969). A very important element in site differences is the availability of particular types of seed. I have seen pine seedlings well advanced in southeastern old fields that had probably been abandoned less than 3 years (judging from the abundance of horseweed and other ephemeral species of early succession). These fields, as one might expect, had mature pine forests right beside them; an abandoned field surrounded by mature hardwood forest, on the other hand, might conceivably skip the pine stage entirely and experience an accelerated introduction of oak and other hardwoods.

Why does the rate of succession slow down? As noted in Table 4.7, the time scale of secondary succession is logarithmic. Early seral stages are short-lived, but each succeeding stage persists longer—culminating in the permanence of the final equilibrium stage. This trend is explicable by the following facts.

1. The earliest plants in succession tend to be short-lived annuals, and there is a trend toward greater longevity of species as succession progresses. A short-lived species must be capable of continual successful reproduction if it is to persist in the community for any length of time. A long-lived species may remain a long time even if it doesn't replace itself successfully.

2. Environmental conditions change much

more abruptly and severely in early succession than in late succession. The complete revegetation of a bare field typically occurs rapidly, and represents a drastic change in light conditions, competitors for moisture, and other parameters, even though the vegetation that now covers the field is only grass. A similarly drastic change in the environment results from the first closed stand of trees to appear. Thereafter, the most important environmental changes in succession are more subtle and occur more gradually.

3. Succession sees a progressive shift from highly opportunistic species which are poor competitors to equilibrium species which are good competitors. The poor competitors of early succession are eliminated relatively rapidly; the good competitors of later succession are better able to persist, though eventually some of them, too, may disappear from a mature forest. Furthermore, we have seen that the colonizing abilities of equilibrium species are limited; they invade a suitable habitat more slowly than opportunistic species, and once there they typically grow more slowly, thus producing change more slowly. Grime (1966) has marshalled evidence that the equilibrium plant species of forests actually have lower metabolic rates than the opportunistic species of early succession. The former's great tolerance of shade, deriving from their low compensation intensities, is a consequence of low respiration rates; their maximum photosynthetic rates are also lower than the maxima of the opportunists (Fig. 4.9). One evolutionary strategy of equilibrium forest species seems to be to gear down the whole metabolism to levels consistent with the limited amounts of solar energy available beneath the forest canopy. Plant opportunists, on the other hand, typically have high metabolic rates, fostering the alternative strategy of high photosynthesis and rapid growth that characterizes weeds and opportunistic trees such as sycamore. But with this high metabolic rate comes a relatively high respiratory rate—hence an intolerance of shade. Thus at the most basic physiological levels we see ecological strategies at work, strategies that in the present context help to determine changing rates of succession.

A computer experiment: the effects of fertilization on forest succession. A simulation study by Botkin *et al.* (1973) illustrates in novel fashion the autecological differences we have been discussing. Botkin and his co-workers have constructed a model of forest growth in the Hubbard Brook experimental forest of New Hampshire, a site of much detailed ecosystem research in recent years. Using data on growth, fecundity, longevity, relative shade tolerance and other biological attributes of 13 native tree species, they have reproduced on the computer the natural successional sequence in this forest after lumbering. In one computer experiment using this model the investigators asked: what would be the effect on succession of fertilizing the forest so that the growth rate of all species was increased by a given percentage of their present growth rate? Such fertilization might conceivably occur in the future, they speculated, through an increase in atmospheric carbon dioxide; CO_2-enriched air is known to stimulate growth in most plants, and atmospheric CO_2 levels are rising steadily as a by-product of the burning of fossil fuels such as coal and oil.

The computer experiment was run under the assumption that growth rates of all species would be increased with no change in their shade tolerance (this is perhaps unrealistic, since as we have seen, specific compensation intensities may be coupled to specific growth rates; but at least, *relative* shade tolerances among the species would remain the same). The consequence to three species of speeding up all growth rates by 100% appears in Figure 4.18. Fertilization would increase the total standing crop of trees, but would not increase equally the final standing crop of each species in the mature forest; it would favor especially the long-lived equilibrium species. Also, the period required for succession to progress to climax would be reduced considerably, and early successional species such as paper birch would benefit only briefly from fertilization, then actually become less important than in unfertilized forests of the same age. Balsam fir, a relatively short-lived, shade-tolerant, equilibrium species, would become much more important early in succession, but after death of the first generation of firs its abundance in the mature forest would drop somewhat relative to long-lived species like sugar maple and spruce. In sum, the computer experiment suggests that patterns of forest succession are very much related to the growth rates of individual species, and that alteration of these growth rates would alter succession in ways that are not always obvious.

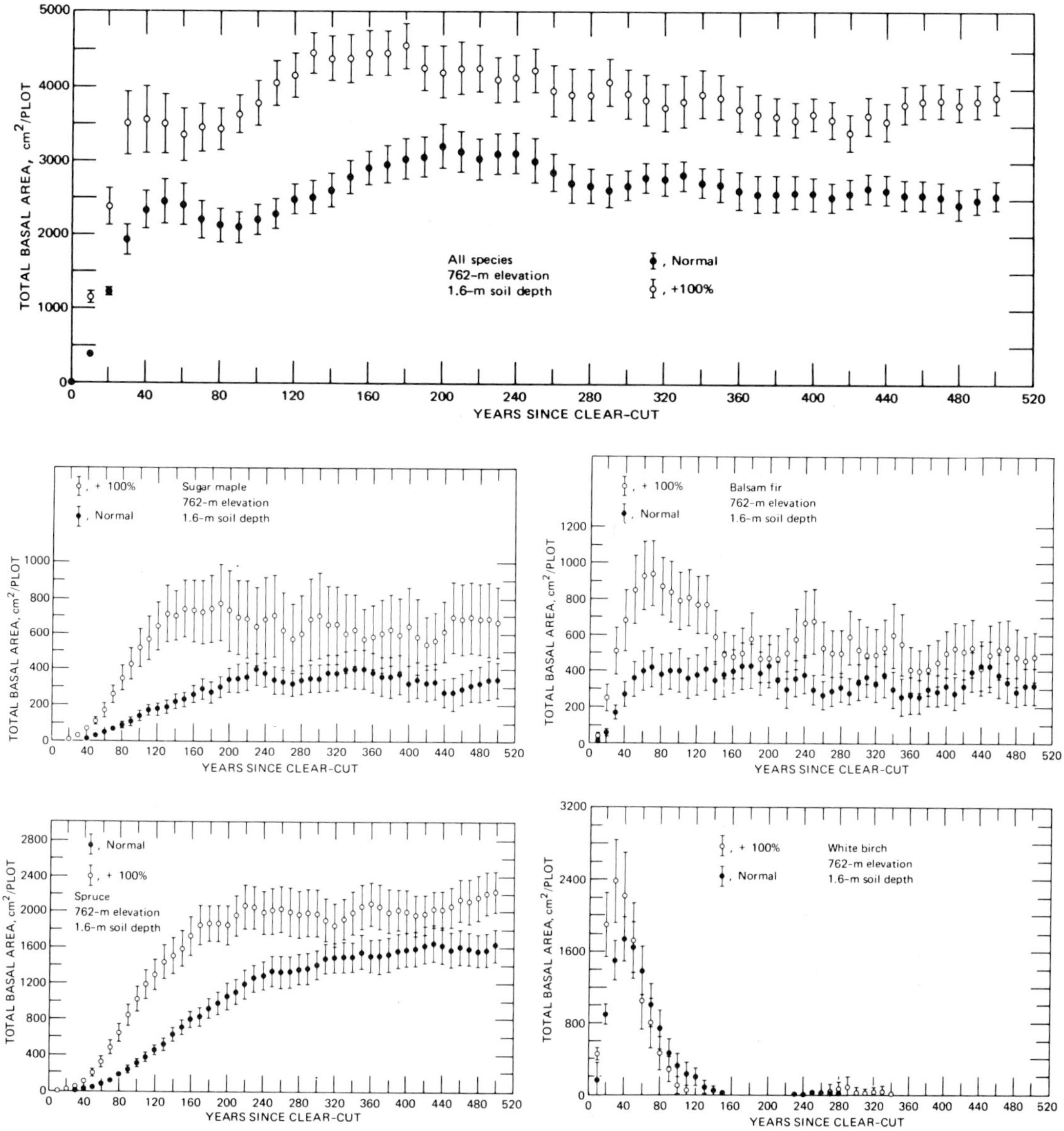

FIG. 4.18. Simulation experiment involving fertilization of a clear-cut forest plot. If the growth rates of all species were doubled, how would this affect the successional sequence and the final equilibrium forest? According to the computer, an increased total basal area (summed cross-sectional areas of all trees) would characterize the final equilibrium community (*top graph*), *i.e.*, there would be more timber. But not all species would share this increase equally. Particularly interesting are the species' varied responses in the first 200 years (the successional period); these are discussed in the text. *Circles* are mean values of 100 replicate determinations; *vertical lines through circles* indicate 95% confidence intervals. (From D. B. Botkin, J. F. Janak, and J. R. Wallis: *Brookhaven Symposia in Biology, 24:* 328–344, 1973. Reproduced with permission.)

Species diversity in successional communities. In all biomes, the early communities of both primary and secondary succession tend to contain few species. Frequently one species of plant is so overwhelmingly dominant that an early seral stage can be mistaken for an agricultural crop (Fig. 4.14A). The animals of these communities likewise are lacking in variety. Communities with such characteristics, whether successional or not, are said to be low in species diversity.

Why should early seral stages be undiverse?

The chief reason is that the ability of opportunist colonizing plants to saturate a suitable habitat is so great that the species getting a head start (because, for example, of an abundant source of nearby seeds) is able nearly to exclude other opportunists, whose weak competitive abilities preclude their playing a game of "catch-up." There are two additional considerations: (*a*) habitats at the start of succession often are relatively homogeneous, and (*b*) they often present rather extreme physical or chemical conditions that exceed many species' tolerance limits. (A bare, sun-baked field exemplifies both these characteristics.) Where initial conditions are less homogeneous, a pioneer community may be quite patchy, but each patch tends to be heavily dominated by one or two species in each trophic category.

The secondary and tertiary communities in a sere may be even less diverse than the pioneer stage; they are characterized by species which are quite rapid colonizers but which are able to dominate less by chance (*i.e.* a head start) than by a competitive ability which often gives them nearly complete, though temporary, dominance of the habitat. But these are opportunistic species too, and gradually they are replaced by species employing a variety of equilibrium strategies. Usually none of these dominates the community so completely; rather, the species are often quite obviously adapted to each other in ways that foster coexistence, as are, for example prevernal herbs to the trees in a mature forest. Late successional but preclimax communities are usually the most diverse; they contain most or all of the equilibrium species plus a residue of successional species that have not yet been completely eliminated.

Native and non-native species in ecological succession. A great many species of plant and animal have been introduced by man, accidentally or intentionally, to other parts of the world. Many of these have become established outside cultivation, but in this respect opportunistic species have been much more successful than equilibrium species. Many "typical American" weeds, and many animal pests of crops, are actually introduced species. In her New Jersey study (Table 4.7) Bard found that during the first 2 years of succession 29 to 32% of the plant species she recorded were alien, whereas in the mature forest only 5% were. This is undoubtedly a general trend: exotic species decline in importance as succession proceeds.

Why should this be? In part, the success of exotic species in early succession probably results from the fact that such species are typically short-lived, and can endure the rigorous periods of non-native climates in the resistant seed phase (or egg phase, usually, in the case of short-lived animals). They do not really adapt to their new habitat in the sense of changing their tolerance levels; but at some time during the year conditions are right for the growth and reproduction phases of their life cycle. Of course, some species of equilibrium communities also are short-lived; most insects are, for example, and (significantly) a number of introduced insect species have become established in mature forests in the United States. But in large part, late successional and climax species are perennial, and must be able to endure the extremes of climate in their habitat. We have seen (Table 2.5) how introduced perennial species may remain genetically geared to the climate of their native land and how this may be a disadvantage in a new climate. But many introduced perennials *can* endure the new climate, as the survival of exotic tree species in our towns demonstrates. Why have they not entered our native forests?

Poor dispersal ability is probably part of the answer. Conversely, exotic opportunists may be common in early seral stages partly because of their great dispersal abilities. Several alien tree species (*e.g. Ailanthus* and *Paulonia*) are successful and widespread members of early successional forests in America, despite their perennial growth habit. This argument is strengthened by the fact that native early successional species, perennials as well as annuals, tend to have wider geographical distributions than do native species of late succession (Fig. 4.19). All this suggests that introduced equilibrium species have difficulty invading late successional communities simply because of their poor dispersal abilities. But this answer is not always adequate to explain why our mature native forests should contain so few introduced species. For example, introduced trees are planted widely in New Jersey: Why should so few become established even in immediately neighboring forests? A third reason probably is involved: opportunistic species do not become closely adapted to coexisting competitors, but follow a strategy of escaping them. The species of late successional and mature communities should theoretically be more

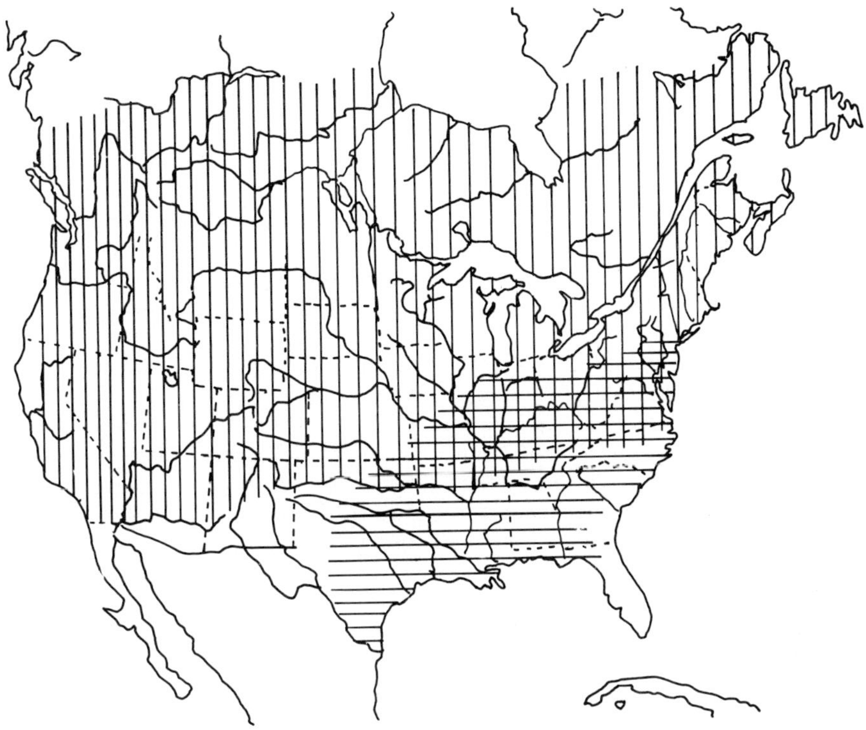

FIG. 4.19. Ranges of choke cherry (*vertical lines*) and of post oak (*horizontal lines*) exemplify the tendency for species of early succession (here, choke cherry) to be distributed more widely than those of late succession (here, post oak). (After R. J. Preston, Jr.: *North American Trees*. Ames, Iowa: Iowa State University Press, 1948 (Ed. 3, 1969). Reproduced with permission.)

closely integrated than those of early successional communities, because they must continue to live with one another. By this reasoning, alien species should have greater difficulty fitting into communities in which they did not evolve if these communities are mature. In sum, it can be argued that the differential success of alien species in early and late seral stages is related to their typically contrasting life spans (*a*) and dispersal abilities (*b*), but is also the result of (*c*) early successional communities being looser and more individualistic associations of species, and late successional communities being more organismic, tightly knit associations.

Animal influences on succession. During forest succession animals do not simply change passively as the vegetation changes, but to some degree they influence the patterns of vegetation change. Plant succession is a sequence of invasions and replacements, and both these processes can be affected by underabundance or overabundance of particular animal species. Invasion of many plant species is mediated by animal agents. The degree of dominance of red cedar in many successional sequences is undoubtedly related not only to the number of nearby seed sources, but also to the abundance of birds and small mammals that favor cedar "berries" as food, and hence spread the seed. If these animals are rare, successional trees with wind-dispersed seeds (birch, red maple or elm) may replace cedar as the dominant tree; or if songbirds that feed on sassafras fruits or on cherries are especially abundant, these trees may thrive disproportionately. Persistence after invasion is also highly influenced by animals. We have mentioned the affinity of tent caterpillars for black cherry leaves; herbivorous mammals, too, have preferences, and may channel succession in particular directions by selective feeding. For example red cedar, which has relatively unpalatable foliage, may be favored over competitors in heavily grazed sites. In his doctoral thesis for the Yale School of Forestry Douglas Sprugel (1974) described a similar phenomenon: a successional pattern in the balsam fir forest of the Adirondacks that is highly modified by snowshoe hares. When the death of mature firs creates openings, paper birch appears as a successional invader, but it almost never reaches tree size because in winter the hares devour any birch twigs projecting above the snow. As a result, the forest returns to fir more rapidly than it would otherwise, the "birch overstory stage" having been eliminated. Such speeding up of succession by animals is probably less common than its slowing down: in successional forests, deer and other browsers often prevent the establishment of shade-tolerant understory trees, or—especially on the western fringes of the deciduous region—they may delay the establishment of any trees at all. Such delay can be permanent, succession being arrested in a "grazing climax" at the grass stage or in a "browsing climax" at the shrub stage.

HISTORY OF THE AMERICAN DECIDUOUS FOREST

America's present deciduous forest communities have been molded not only by recent pressures, but by climatic and biological events hundreds, thousands, even millions of years ago. How these forests have changed, through remote and recent time, is a story that gives perspective to the present and a basis for speculation about the future.

Evolution and Extent of Arcto-Tertiary Forests

The history of broad-leaved forests begins perhaps 100 million years ago, with the rapid rise to evolutionary and ecological prominence of the angiosperms (flowering plants). This is the botanical group to which almost all broad-leaved and deciduous species belong. Angiosperm forests appeared at a time of remarkable flux in the history of terrestrial life; separation and reorientation of the earth's land masses was occurring through the processes of continental drift, and the animal world was changing nearly as rapidly as the plants. By the end of the Cretaceous period, some 60 million years ago, not only had angiosperms replaced conifers and ferns as dominant land plants but dinosaurs had retreated to near-extinction, mammals and birds were increasing rapidly in importance, and—most important for the angiosperms—insects were diversifying rapidly. It seems quite likely that the simultaneous explosion of angiosperms and insects was not coincidental, but a case of massive co-evolution resulting primarily from the mutual specialization of flowering plants and of many insect groups with regard to pollination. Ehrlich and Raven (1967) believe that an equally important aspect of this co-evolution was a continuing adaptive interplay of defense and offense: the acquisition by many angiosperms of chemical defenses against herbivores (particularly insect herbivores), the compensatory acquisition by herbivores of specialized means to circumvent the defenses of their favorite food plants, the evolution of further chemical deterrents by specific plants, the evolutionary reaction of insects, and so forth. Most of the families of angiosperms appeared quite early, and after this evolutionary spurt an apparent stagnation set in: few additional major plant groups have appeared in the last 40 million years. According to some authorities, however, this evolutionary stagnation is more apparent than real; at the more subtle levels of biochemical co-evolution and niche specialization, evolution has been continuing apace (Whittaker, 1969 and 1970).

Knowing how profound is the influence of vegetation on animal life, we can be sure that the changing forests of the late Cretaceous period were linked not only to the evolutionary spurt of the insects but to the rapidly changing fortunes at this time of reptiles, birds and mammals. This biological upheaval continued into the early Cenozoic, the geologic era spanning the last 60 million years of Earth's history. Nonbiological influences were also at work during this time: long before the Pleistocene ice ages began some 2 million years ago, climates already had started to cool and to become more differentiated by latitude. Today's tropical and temperate forests differ markedly in typical leaf morphology: tropical leaves tend to be large, and are usually untoothed and unlobed; temperate leaves are smaller and often have toothed or lobed margins or both (Dorf, 1969). Why such differences exist is unclear, but paleobotanists have used these and other clues to infer past climates from the leaves of fossil forest assemblages. They have discovered a history of temperate zone cooling that began near the end of the Cretaceous and extended up to the Pleistocene (Dorf, 1969; Axelrod and Bailey, 1969). The simultaneous, separate evolution of homoiothermy (warm-bloodedness) by both birds and mammals is further evidence of a change in climate near the beginning of the Cenozoic. The ability of birds and mammals to regulate their body temperatures physiologically is best interpreted as an adaptive defense against external temperature extremes.

It is uncertain exactly when birds and mammals actually became homoiothermic, and when some angiosperms acquired the deciduous habit; but by Miocene times (approximately 25 million years ago) a forest closely related to present temperate deciduous forests existed in an almost unbroken belt across the northern continents (Billings, 1964). This was the Arcto-Tertiary forest, a vegetation type in which broad-leaved trees predominated but which contained coniferous elements as well. The present redwood and sequoia forests of the Pacific coast, which contain many deciduous species, are remnants of the Arcto-Tertiary forest, as are the more fully deciduous eastern forests (Billings, 1964). North American topography was more uniform during the period of the great Arcto-Tertiary forest than it is now, and the western mountain ranges, so important today as determinants of temperature and rainfall contrasts, were lower and less influential at that time. The rise of these ranges, and the continued global cooling that culminated in Pleistocene ice ages, eventually fragmented the Arcto-Tertiary forest belt; we are left with biogeographical discontinuities: species of *Magnolia* and of sweet gum (*Liquidambar*) in the American southeast and in temperate east Asia but nowhere else; maples on the Pacific coast and in the eastern deciduous forest; related salamanders in all three regions. These anomalous distributions are clues to the former extent of Arcto-Tertiary deciduous forests.

The Pleistocene Glaciations

Within the past 2 million years the cooling trend of previous eons culminated in several major glacial advances over the continents of the northern hemisphere, and more limited expansions of glaciers in the southern continents. These pulses of glaciation, alternating with intervening periods of glacial retreat, may be analogous to earlier but less spectacular climatic oscillations that occurred during the gradual postCretaceous cooling trend (Dorf, 1969). The most recent glacial pulse, the Wisconsin ice sheet of North America (equivalent to the Würm glaciation of Europe) reached its greatest extent less than 30,000 years ago. Although this ice sheet obliterated much evidence of forest patterns during previous glacial stages, there is some evidence that eastern forests during the "interglacial" preceding the Wisconsin advance were quite similar to today's (Wright, 1971). Our richest paleoecological information, however, is for the last 25,000 years, the time span in which the Wisconsin ice sheet peaked and receded and today's forests took their present form.

Figure 4.20 shows the extent of the Wisconsin ice sheet in North America. Lobing of its southern fringe was caused primarily by topographic differences, but in the Northeast the ice was so thick that it rode over major obstacles such as the Adirondacks, Catskills and White Mountains (Flint, 1971). Isolated mountain glaciers possibly occurred as far down the Appalachians as North Carolina (Berkland and Raymond, 1973), although this suggestion has been challenged. South of the ice sheet lay boreal forest, extending from the Atlantic to eastern or even southwestern Kansas during the peak of glaciation; tundra may have existed in the higher Appalachians and on unglaciated plains of northeast Kansas (Wright, 1971). The vegetation of coastal North Carolina and northern Georgia contained spruce and jack pine (a northern species) together with some deciduous trees, and resembled the present vegetation of northern New England, 1000 km or more to the north (Watts, 1971). White pine and perhaps some deciduous species persisted along with spruce and fir in the Shenandoah Valley and west of the Appalachians, but the vegetation of this region (the region of present mixed mesophytic forests) certainly was disrupted much more drastically by ice age climates than had once been thought (Wright, 1971). Very little is known about the location of most deciduous species during this time of maximum glaciation, but presumably they were concentrated south of the boreal forest belt, *i.e.* chiefly along the Gulf coast.

Return of the Deciduous Forest

The Wisconsin ice sheet began to retreat rapidly some 14,000 to 12,000 years ago, and forest changes were equally abrupt. In some regions that had been ice-covered, the initial revegetation was treeless tundra (Fig. 4.21). Elsewhere, a forest dominated by boreal coni-

FIG. 4.22. Forest islands in the prairie; prairie islands in the forest. (*A*) Juniper and oak woodland clothe hills rising from grassy plains in the Wichita Mountains Wildlife Refuge in western Oklahoma. (*B*) Forests of ponderosa pine cover much of the northern Arizona plateau near Flagstaff, with other conifers and aspen mantling the lower slopes of the San Francisco Peaks. But near the center of this photograph there is a clearly defined tongue of grassland (Hart Prairie) occupying deeper soils in an outwash area from the "bowl" between the peaks. Further downslope this isolated grassland becomes mixed with light-toned stands of aspen. (Photographs by the author.)

comparison of grassland and deciduous forest climates. When one re-examines Figure 2.16, the moisture regime of Manhattan, Kansas (a prairie locality) appears to be no more rigorous than that of Louisville (a deciduous locality). But the graph suggests that this statement might be true only in an average or relatively moist year. Of the two localities, Manhattan appears more drought-prone; a small drop in precipitation during spring or summer would change the moisture picture there much more drastically than it would in Louisville, and it is possible that this difference tips the balance toward grassland.

No one doubts that relatively dry climates are at least part of the reason for the existence of American grasslands. An eastward extension of grassland, the "Prairie Peninsula," antedated the cornfields of Illinois and part of Indiana, and Prairie Peninsula outliers occurred even in Ohio and western Pennsylvania. This grassland extension coincides with a wedge of "western climate," defined to the north by the winter Arctic front and to the south by a tropical front extending northeastward from the Gulf of Mexico. Within the wedge itself, the predominant air flow in winter, and sometimes in summer, is dry westerly winds moving in from the plains (Borchert, 1950; Bryson, 1966; Wright, 1968). But in the Prairie Peninsula, as in the Great Plains, the question is not whether the region has a distinctive climate, but whether the vegetation would be grassland in the absence of fire and grazing. The retreat of the Wisconsin ice sheet coincided with the appearance in the Great Plains of man, a selective big game predator who probably set fires to aid in driving game. The disappearance of many species of large mammal at this time—mastodon, mammoth, horse, ground sloth and camel—might seem to have reduced grazing pressure on American grasslands, and according to a recent theory the hunters that probably decimated them quickly declined in numbers also (Martin, 1973). But some humans remained, adding their fires to those created by nature; and one large herbivore that escaped extinction—the bison—expanded to enormous numbers on the prairie. There is no doubt that intensive grazing by bison encouraged grassland in America, just as heavy grazing by herbivores now encourages grassland in Africa. But no one knows for certain whether extensive American grasslands would have existed in the absence of bison and Indians or whether some form of deciduous forest would have extended from the Atlantic to the Rockies.

The Deciduous Forests Encountered by European Settlers

The American forests have one very interesting quality, that of being entirely free of under or brush-wood. This owing to the extraordinary height, and spreading tops, of the trees; which thus prevent the sun from penetrating to the ground, and nourishing inferior articles of vegetation. In consequence of the above circumstance, one can walk in them with much pleasure, and see an enemy from a considerable distance.

Thomas Ashe, traveler in Pennsylvania in 1806, quoted in Bakeless (1950), *The Eyes of Discovery*

Ashe's description fulfills the image of mighty, unbroken, virgin forest that springs to our minds when we think of eastern America before the white man. These are the legendary forests out of which Boonesboro was hacked; this is the forest primeval, in which Hiawatha stalked the wary deer. These are the forests where Thomas Morton found "millions of Turtledoves (passenger pigeons) on the greene boughs . . . whose fruitfull load did cause the armes to bend" (in *New English Canaan*, a description of New England published in 1632).

Were precolonial American deciduous forests really of this character? only partly. Forests *did* cover most of the land east of the Mississippi, and they *were* often tall, dark, and relatively free of underbrush. Bakeless (1950) suggests that the predominance of big trees was due to the fact that the Indians used mostly fallen branches for firewood, and bark and saplings for dwellings. They seldom bothered to fell larger trees, except occasionally to make dugout canoes. Indians also burned the underbrush:

It is the custom of Indians to bourne the wood in November when the grass is withered and leaves dryed, it consumes all the underwood and rubbish, which otherwise would overgrow the country, making it impassable, and spoil their much affected hunting; so that by this means in those places where the Indians inhabit there is scarcely a bush or bramble, or any combersome underwood to be seene.

William Wood (1634), *New England Prospect*

But the eastern landscape was more varied than these descriptions suggest; not all of it was covered with forest, and not all of the forests were stately and mature. Passenger pigeons *were* abundant—in 1736 they sold at six for a penny in Boston—and wolves also were common, judging from the penny-a-head bounty decreed by the Massachusetts Bay Colony in 1630 (Matthiessen, 1959). These two, and others, are now extinct in their former haunts. But some animals, like the deer, probably are more abundant now than they were at the time of settlement.

How unbroken were the former forests? We may banish at least one pioneer image: Disney and Fess Parker notwithstanding, Boonesboro was not hewn out of a towering forest. Instead, Boone and his companions built their cabins in a "Great Meadow, . . . in the shade of a great elm surrounded by a turf of fine white clover, forming a green to its very stock" (Bakeless, 1950, quoting a member of Boone's party).

How widespread were forest openings such as that of Boonesboro? We have mentioned the Prairie Peninsula extending eastward through Illinois and Indiana. Christopher Gist, an Ohio frontiersman, described that land in 1750 as a mixture of forest and "large meadows, fine clover bottoms and spacious plains covered with wild rye." Bakeless (1950), the source of this quotation, describes grasslands much farther east such as the Hempstead Plain of Long Island, a grassy tract of nearly 50 square miles. This plain was dominated by little bluestem, a grass of the prairie that also thrives in the East in frequently burned fields (Niering *et al.*, 1970).

Other evidence of forest openings comes from former animal distributions. Bison were found as far east as Florida and perhaps the Carolina coast. From the Susquehanna Valley of central Pennsylvania there is a somewhat disputed report of a herd of 400 being exterminated in 1799. One would not expect these grazers to occur in mature forest habitats. River flood plains provided the low growth bison needed, and between floods many valleys may have been kept in grass and low browse by grazing pressure. On smaller streams the activities of beavers created cleared areas and, as their ponds silted up, temporary meadows. How abundant were these meadows? We can rely on fragmentary trapping records and the accounts of pioneers, but a clue also comes from the habits of a typical beaver meadow inhabitant, the meadow mouse (*Microtus pennsylvanicus*). The social and reproductive characteristics of this species seem geared to the appearance of temporary meadows, scattered like islands within the deciduous forest (Christian, 1970). True forest rodents like the deer mouse maintain relatively stable populations, but meadow mice typically build up large populations, the members of which are highly aggressive. These reproductive and social attributes virtually force emigration from crowded areas, and can be viewed as the strategy of an opportunistic species. Beaver meadows return to forest, and meadow mice must constantly have had to discover new meadows if the species was to persist. The emigration strategy that they evolved implies that beaver meadows were an abundant and recurrent feature of the deciduous forest.

Were upland forests stable and mature? Prairie patches, flood plains and beaver meadows provided grassy islands in the midst of forest. What of the forests themselves? Were they all equilibrium forests dominated by equilibrium species? If that were true, our only native successional species would be those suited to flood plains and wet meadows. How would we account for the existence of species like Kirtland's warbler, which now nests only in young, successional pine stands in central Michigan? It almost certainly evolved this habitat requirement long before the arrival of Europeans. The only reasonable explanation is that young stands of pine have existed in the Great Lakes region for many centuries. What could produce successional upland forests? Fires, for one thing; fires probably have been common in the lake states and prairie border states since the retreat of the ice sheet; along many parts of the Atlantic and Gulf coasts they probably have been common far longer. And even in the wet Northeast, the great stands of white pine found by the settlers suggest extensive earlier forest disturbance, perhaps largely by fire during severe drought years. Probably the most detailed fire history in the New World has been compiled by M. R. Heinselman (1973) for the forests of the Boundary Waters Canoe Area in northeast Minnesota. Figure 4.23 illustrates the frequency and extent to which those forests have been burnt over during the past 300 years. Almost all of the "virgin forest" in this area had its origin in a recorded fire, and 82% of the forest dates from 1863 or later.

Other disruptions in the primeval forest came from extensive windthrow during hurricanes (*cf.* Spurr, 1956). Moreover, even in forests which perhaps were less subject to extensive disturbance, we must remember that the mature forest equilibrium was dynamic, not static. In a pollen study of the Hubbard Brook forest in New Hampshire, Margaret Davis has recorded a catastrophic plunge in hemlock abundance 5000 years ago, followed by compensatory increases of birch, beech and sugar maple. Eventually—after some 400 years—there was a return of hemlock. Only hemlocks suffered; other species with similar requirements continued to thrive. The explanation Davis suggests is that the hemlocks were destroyed by a species-specific pathogen. We see similar selective destruction in our forests today.

On a much smaller scale, there must certainly have been fluctuations in food supply for forest animals. We have mentioned the year-to-year variations in seed production of many tree species. Sedentary seed eaters must live or die with these vicissitudes, and their populations probably always have been in states of considerable flux. The passenger pigeon, a mobile species which fed heavily on beech nuts and the nuts of other trees, probably maintained its huge numbers by constant searches for localities where, in a particular year, the food supply was high. Thus the year-to-year population of pigeons in any one place may have been highly variable. We could continue with similar examples, but the basic point has been made—the primeval forest was not a uniform whole but a dynamic collection of communities, even as are today's forests.

Deciduous Forests in the Wake of Settlement

We tend to think of today's deciduous forests as but a pitiful remnant of America's presettlement wilderness. The Forest Service estimates that in the United States as a whole, more than 95% of originally forested land has been cut or cleared at least once; in the eastern deciduous forests, the figure probably exceeds 99%. But surprisingly, America's temperate deciduous forests have reclaimed territory during the present century, and even today are far more diverse and extensive than those anywhere else in the world.

Land clearance for farming.

> The fall of this forest, in the East, was extensive and very rapid, for what could not be put to use was easily exported, and farm clearing was indiscriminate and wasteful. . . . The clearing, indeed, was feverish, for the settlers dreaded the dark monotone of trees, the wild beasts and savages they concealed, the wind-borne, whispered reminder of a wilderness unconquered.
>
> Peter Matthiessen (1959), *Wildlife in America*

The initial mayhem may have been severe, but agriculture has had a checkered history in much of the eastern United States. Excluding mountainous portions, most of New England once lay in croplands, hayfields and pasture. We still associate potatoes with northern Maine, dairying with Vermont, and tobacco with the Connecticut Valley; but over much of this land, stone walls that once bordered fields now thread peacefully through hardwood forests. Long ago there was an agricultural exodus from New England as the richer, more easily worked lands of the Midwest opened for settlement. In a now forested 1000-acre tract in central Massachusetts, careful detective work has indicated that 16% of the land had previously been cropland, and another 75% had been cleared pastureland (Spurr, 1964). A similar but more recent history–of clearance for agriculture, abandonment, and return to forest–characterizes many parts of the mid-Atlantic states, and is especially widespread in the south.

In highly fertile regions, however, land cleared for agriculture has remained in cultivation. For many areas of the Middle West, for instance, we have little but the writings of early settlers to tell us what the native forest types were. The same is true of fertile eastern regions, of which Lancaster County in Pennsylvania serves as an interesting example. Much of Lancaster County is a flat limestone plain, settled in the 1700s by Mennonite and Amish farmers who still work the land intensively. Only hints remain as to the nature of the former forest on this plain: here a stand of old white oaks in a country churchyard, there a small grove of ancient beeches on the slope above a stream. Lancaster's agricultural plain, however, is interrupted in places by low ridges which are thin of soil, underlain by less fertile metamorphic rocks, and more difficult to till. These ridges still support forest, including several very old stands. In Lancaster County and

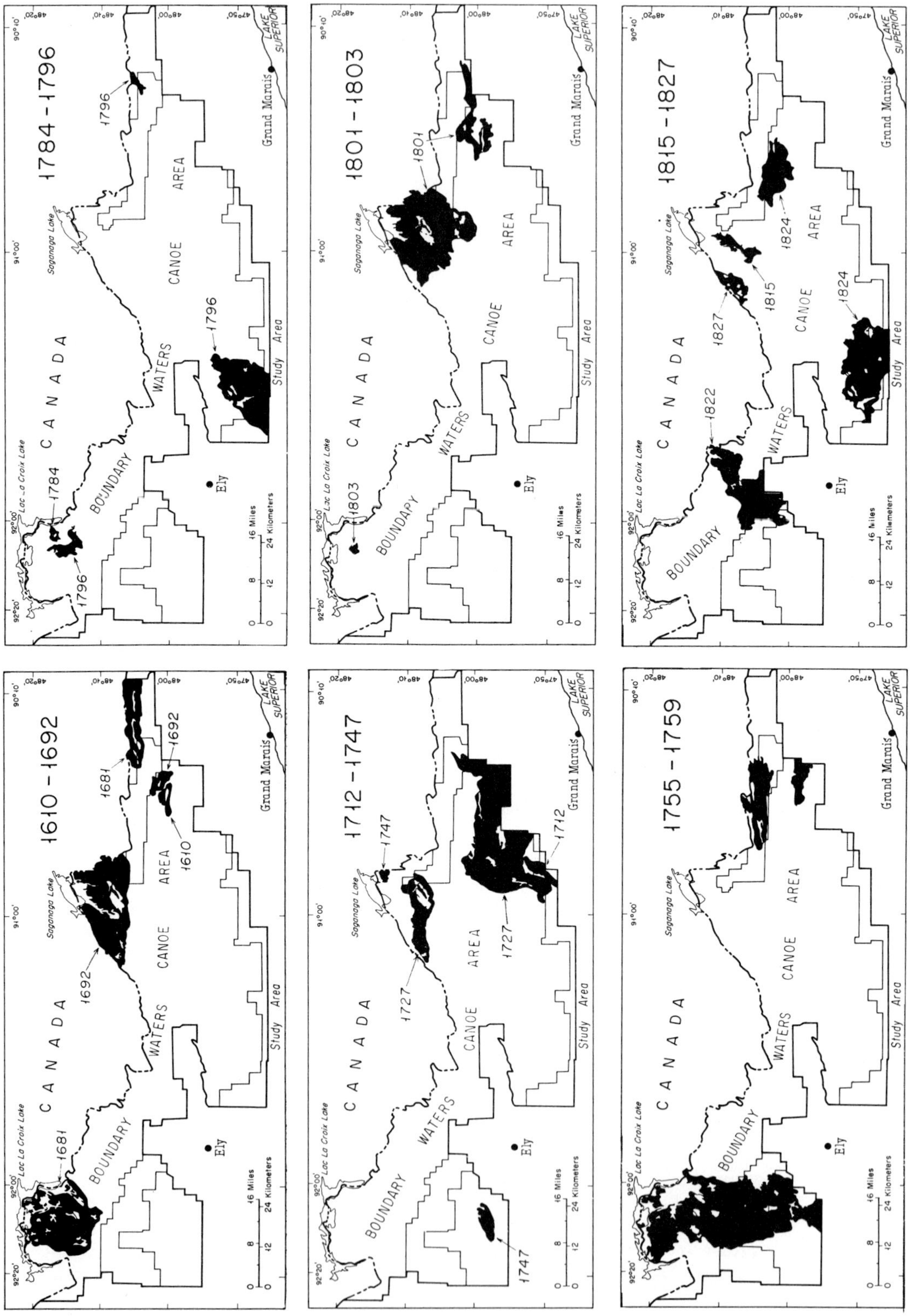

FIG. 4.23 (legend on page 138)

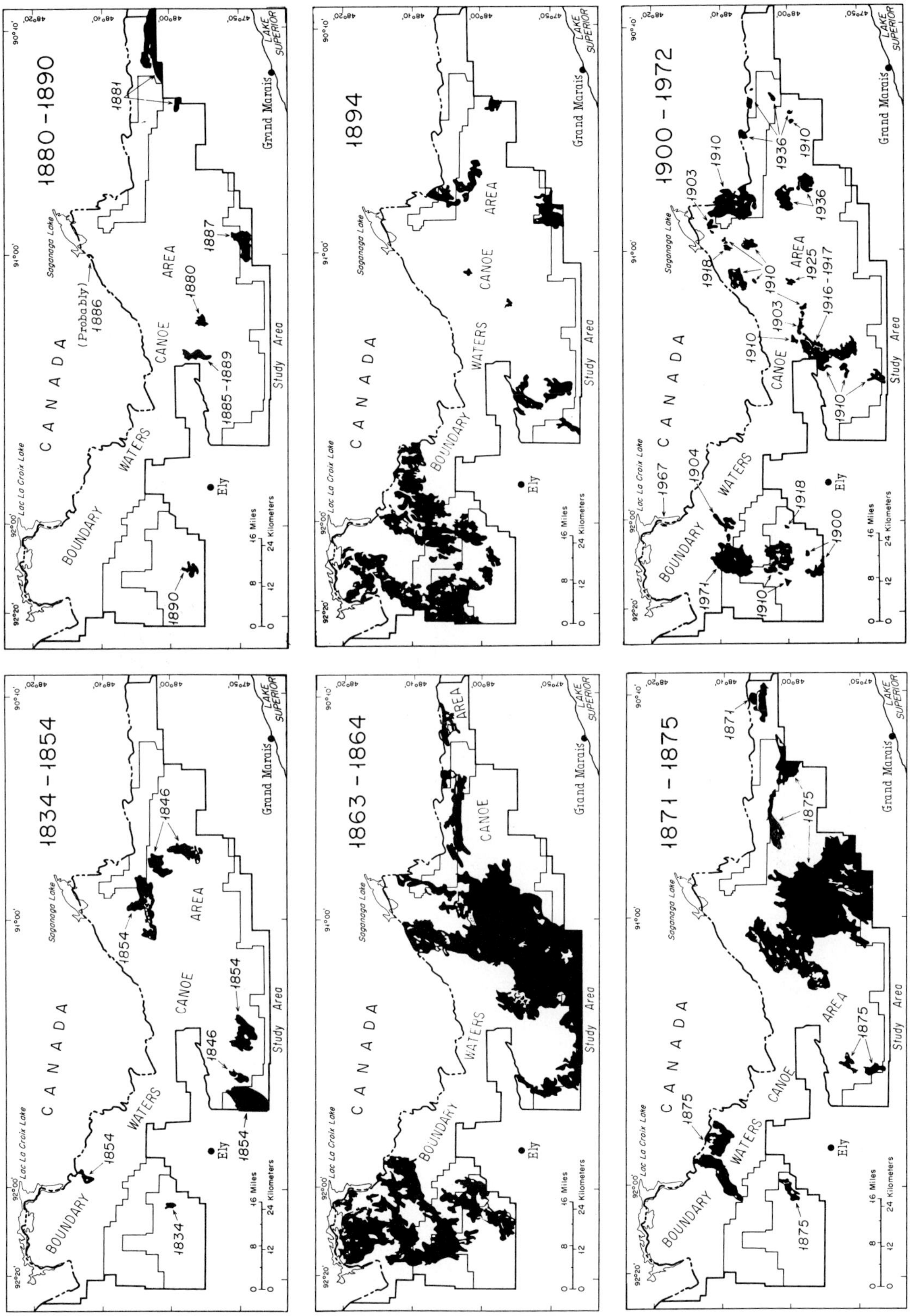

FIG. 4.23 (*cont'd.*)

elsewhere, the forest subtypes on fertile soil have been nearly eliminated but the subtypes on land unprofitable to farm have remained. Thus agriculture has selectively removed pieces from the original forest mosaic.

Lumbering.

Come all ye sons of freedom throughout old Michigan,
Come all ye gallant lumbermen, come list to a shantyman
From the banks of the Muskegon where the rapid waters flow
We'll range the wildwoods o'er while a-lumbering we go.

The music of our burnished axe shall make the woods resound,
And many a lofty ancient pine will tumble to the ground.
At night around our shanty fire we'll sing while rude winds glow—
Oh, we'll range the wildwoods o'er while a-lumbering we go.

Oh, the forests so brown at our stroke go down,
And cities spring up where they fell,
While logs well run and work well done
Is the story the shanty-boys tell.

Edith Fowke (1970), *Lumbering Songs of the Northern Woods*

The impact of lumbering, like that of agriculture, swept across the deciduous forest from east to west, but usually later in time. The early settlers found uses for many forest timbers, but even in colonial years white pine was the most sought-after tree. From pine were made houses, boat planking and masts; the hardwoods usually were put to more specialized uses:

> . . . gun stocks of black walnut, boat ribs of oak, ax handles, singletrees, and wagon hubs of hickory and black gum, birch canoes, briar pipes fashioned of laurel and rhododendron, black ash fish baskets and chair bottoms, flexible spade handles and oars of the white ash.
>
> Matthiessen (1959), *Wildlife in America*

The great age of commercial lumbering began in the early 19th century, and white pine was the prime quarry. Bangor, Albany, and Williamsport became successive capitals of the lumber industry as the wave of cutting moved westward. White pine was less abundant in the forests of western Pennsylvania and Ohio, and these forests were to some degree spared; but pine reappeared in the upper Lake States, to which the lumbermen turned their interests in the years after the Civil War. Logging in the Lake States brought with it devastating forest fires, fueled by the debris left lying on the ground. In 1871 a fire burned over a million acres near Peshtigo, Wisconsin, destroying numerous settlements and killing 1500 people. Major fires in 1894 and 1918 killed 1100 more in Wisconsin and Minnesota and destroyed a town of 12,000 people (Dasmann, 1968).

Interestingly, the last forests exploited primarily for white pine were back east, in New England. The abandonment of farmlands in this region in the early 1800s led to the appearance of widespread old field stands of white pine, and between 1895 and 1925 some 15 billion board feet were cut in New England (Graham, 1969). But the trees that repopulated these late lumbered tracts were not white pine, in the main, but less valuable, slower growing hardwoods—perhaps because they had already started to develop as a successional understory to the pines, and were not fully destroyed by the lumbering operation. Maples and oaks now grow where the New England pine forest fell in the early 1900s.

White pine is still abundant in its former range, but it no longer supports lumbering industries by itself. In Maine, the establishment of pulp mills and the refinement of processing methods saw pine replaced in the lumberman's interest by spruce, to which was later added balsam fir and, quite recently, many hardwood species. In the mid-Atlantic states, particularly

FIG. 4.23. Fire history of the Boundary Waters Canoe Area in northeastern Minnesota, a 1-million acre "virgin forest" that never has been cut or cleared by man. Fire is a chronic factor in this conifer-dominated forest, which is transitional between true boreal and deciduous types. The history was compiled by aging major stands by tree-ring counts and dating individual fires from scars on still living trees, as well as by using available historical records. Note that the time intervals on the maps vary in length: most existing stands date from five major fire years (1863, 1864, 1875, 1894, 1910) marked by summer droughts. Since 1910 fire control has been practiced by the Forest Service, resulting in less frequent burns and a dearth of normal young forest habitats. The author recommends that controlled burning be instituted to restore the natural forest community of the area: a mosaic of fire-initiated stands of differing ages. (From M. L. Heinselman: *Quaternary Research, 3:* 329–382, 1973. Reproduced with permission.)

before the development of coal resources, hardwood stands repeatedly were clearcut for fuel; as a result shade-tolerant, fire-susceptible species like hemlock and beech are now restricted in many areas to steep ravines which escaped cutting and burning (Spurr, 1964). But with the advent of mining, younger forests returned to the coal regions.

The lowland forests of the South have a lumbering history similar to New England's, except that lumbering remains a widespread industry there today. White pine was not found in these forests, and the original hardwood forests were largely cleared for farming, not lumber. But a lumbering industry sprang up at the turn of the 20th century, focusing on the successional stands of southern pines that had appeared on abandoned fields. In the South, unlike New England, the pines typically reappear after lumbering, and the timber industry of the South has developed techniques of managing large tracts of forest specifically for pine production.

Consequences of settlement for forest animals. In American deciduous forests, the impact of settlement on the vegetation can be summarized as follows. (*a*) The original forests have become greatly restricted and divided, though in some areas they are becoming more extensive again. (*b*) The vegetation has suffered selectively; lumbering interests have concentrated on particular species, sometimes to their detriment (white pine), sometimes to their advantage (southern pines). Agriculture has virtually destroyed the perhaps distinctive forest subtypes that formerly occupied particularly fertile land areas. (*c*) Mature forests are less common now than in presettlement days, and successional forests are more common; in many areas, however, mature forests are gradually returning as lands no longer farmed or lumbered are left undisturbed.

In terms of habitat alteration, the effect of these changes on forest animals has been enormous. Selective hunting and trapping have further modified the forest fauna. It sometimes is difficult to decide which influence has been more important. In 1621 Virginia boasted a thriving trade in the pelts of mink, marten, otter, wildcat and beaver (Matthiessen, 1959). Most of these species are utterly gone from the region now. Is it because they were hunted to extinction or because, like the marten of British Columbia (page 118) they require mature forest? The ivory-billed woodpecker, one of the rarest birds in the world, always has been found in a specialized habitat—the mature lowland swamp forests of the southern states—and lumbering for cypress and swamp hardwoods has much reduced this favored habitat. But Matthiessen notes that among Indians as far north as Canada there was formerly a lively trade for the bills of these birds, which make handsome ornaments; so even before settlement by Europeans the ivory-billed woodpecker may have been rare.

In general, the adverse effects of settlement were felt more by the large animals than by the small, more by the carnivores than by the herbivores, more by the species characterizing mature forest communities than by the species of successional communities. These contrasts are illustrated by the examples which follow.

Bison, woodchick and other herbivores. Bison and woodchuck both prefer grassland or meadow to forest. But the bison is gone from the East despite forest clearance; the durable groundhog thrives, to the dismay of farmers, and is almost certainly far more abundant now than before European settlement. Bison were vulnerable to hunting, and hunting pressure was great because they were a threat to crops and provided valuable meat and pelts. The woodchuck is less vulnerable to hunting both because it lives underground and because it reproduces more rapidly than the bison; furthermore, woodchuck stew (though a favorite of some) has never been a widespread dietary staple. Besides the bison, large herbivores which have nearly disappeared from the eastern United States include the elk, woodland caribou and moose; the last two, rather surprisingly, were still found as far south as Connecticut and Long Island at the time of the Civil War (Matthiessen, 1959). All these large species suffered from hunting; most of them also suffered (more than small mammals) from the fragmentation of suitable habitats. Caribou need more than a few widely scattered acres of mature forest if they are to persist, but small, relict stands might well remain viable habitats for species of small mammal that require mature forest.

Cougar and bobcat, wolf and fox, bald eagle and red-tailed hawk. Perhaps even more than herbivores, large carnivores need relatively large contiguous areas of suitable habitat. Each of the above species pairs contains a larger and a smaller member; in each case, the smaller has fared better. The large carnivores suffered more severly both from forest fragmentation and from hunting, being considered major threats to both life and livestock by the early settlers. The cougar was eliminated from the eastern United States along with the wolf, but reportedly it now is re-establishing itself from Canada as America's deciduous forests re-expand.

White-tailed deer. The great modern success story among large eastern mammals has been written by white-tailed deer. But it is a success story largely restricted to the present century. The impact of early settlement was chiefly negative; as early as 1696 Massachusetts declared a closed hunting season for deer because of reduced numbers, and by the time of the Revolution every one of the colonies but Georgia had followed suit (Matthiessen, 1959). This evident desire to preserve deer stands in sharp contrast to exploitative policies with regard to many other species, particularly predators, and protection undoubtedly helped deer to persist in many areas. But their populations continued to decline through the 19th century. In southern Vermont, deer were extinct by 1878 as a result of hunting, logging and clearing for cultivation. Even in "wild" northern Vermont they were very rare (Leonard Foote, quoted in Graham, 1947). In the Midwest, deer numbers had been reduced to less than 2000 by the early 20th century, and they had been entirely eliminated from Ohio, Indiana, Illinois, Iowa and southern Michigan (Nixon, 1970).

But since 1900, white-tailed deer have increased all over the East and Midwest, due partly to restocking and protective hunting regulations, but probably more importantly to the regrowth of young forest on abandoned farms and cut-over timberlands. It is this sort of habitat—early successional shrubs and the low browse found at the borders of fields and forests—that is optimal for the species. Graham and Nixon report that the deer population of Vermont had grown to 50,000 by 1947, and that of the Midwest to 845,000 by 1968 (the latter representing an increase of over 4000% over the numbers in 1940). A similar trend is evident for Pennsylvania (Fig. 4.24), where deer have assumed pest proportions in many areas. So far, increase in deer populations in the 20th century has far exceeded the increase in human populations in the deciduous forest region. If deer populations decline somehat in the years ahead (a prospect that seems likely), maturing forests may be a more important cause than human pressures.

Passenger pigeon and mourning dove. The opposite fortunes of these related species are perhaps the most interesting of our examples. The mourning dove, like the white-tailed deer, is thriving in the company of modern man, even though it is a sought-after game bird in more than half the United States. Its preferred habitat is fields and forest edges, its preferred food the seeds of crops and early successional plants such as ragweed and foxtail (Lehner, 1964). It is a moderately gregarious species, but the flocks it forms are relatively small. In many eastern states, the mourning dove has become remarkably abundant.

At the time of settlement mourning doves probably were not uncommon, but it was the hordes of passenger pigeons which soon became legendary. Because they formed huge flocks, passenger pigeons may have seemed even more abundant than they were, but unquestionably they were the most numerous bird species in most of the deciduous forest region. In 1810 the ornithologist Alexander Wilson estimated the size of a single Kentucky flock at over 2 billion and calculated that if each bird were to eat half a pint of food per day, the flock's daily consumption would be 17 million bushels of nuts and berries. Audubon provides a vivid description of mayhem at a nightly roost in Kentucky; the arriving pigeons created air currents "like a gale at sea," perched in "solid masses, as large as hogsheads" on every tree, and weighted down branches until they collapsed, killing hundreds of birds at a time. The mayhem was multiplied by human predators, who fired unceasingly into the roosting flock. Most of the killing was for human consumption, but two farmers drove several hundred hogs a distance of over a hundred miles to fatten up on the slaughter. (Audubon's account is reproduced in Peterson (1967).)

The flocking habit of the passenger pigeon certainly was instrumental in its demise. It is one of only two or three species of the American deciduous forest to have been hunted commercially for food (the now extinct heath hen was, briefly, another). Besides preying on the roosts, pigeon hunters took advantage of the birds'

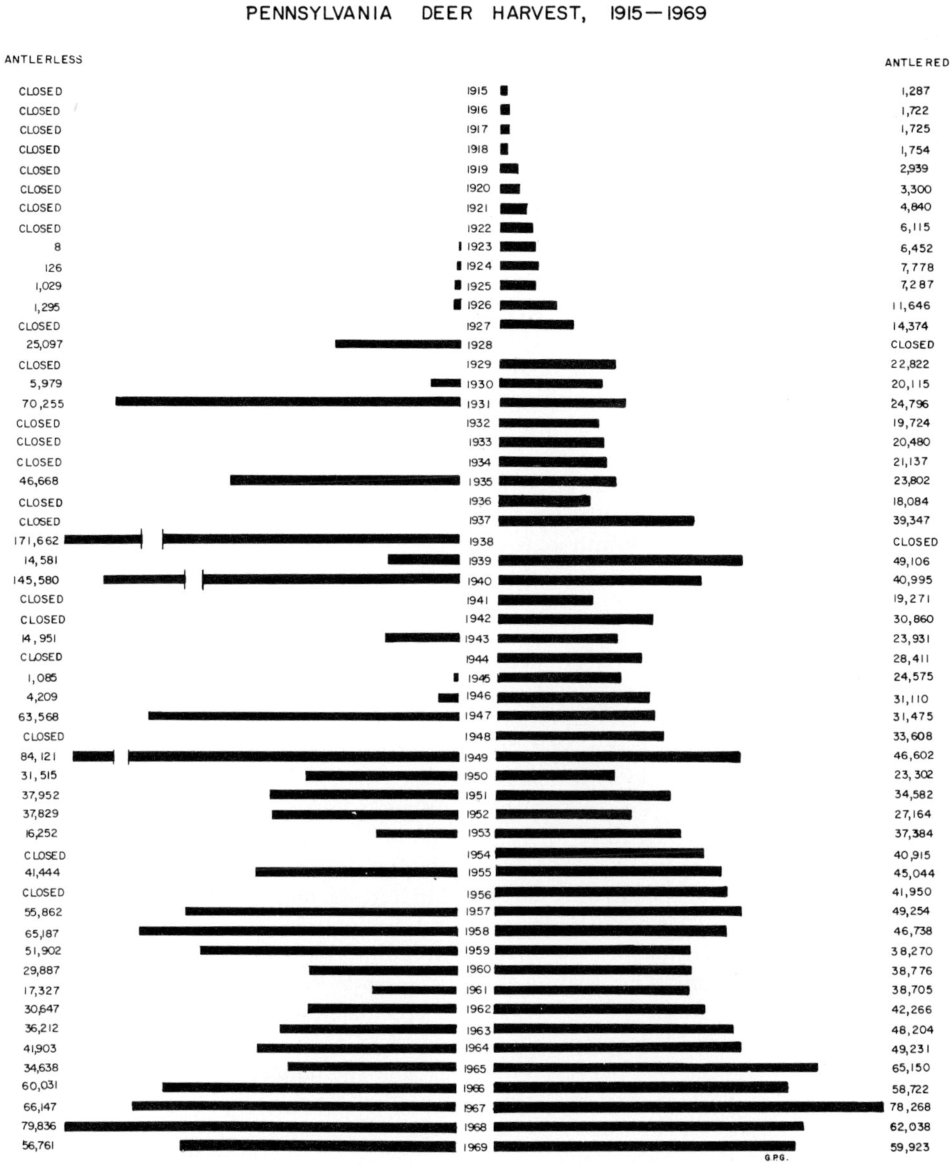

FIG. 4.24. Hunting harvest of deer in Pennsylvania has grown some 50-fold since the early years of this century, and there now is a regular season on does as well as bucks. In recent years cars have aided in the "harvest," annually killing more than 25,000 deer on Pennsylvania roads. (From S. Forbes, L. Lang, S. Liscinsky and H. Roberts: *Pennsylvania Game Commission Research Bulletin* 170, 1971.)

sociability in other ways. One trick was to capture a pigeon alive, blind it, and stake it out in a clearing as a "stool pigeon"; a curious flock of compatriots would alight beside it, and be trapped by spring nets.

Thousands of barrels of dead birds were shipped in a season to the markets of the large cities, while live birds were sent to gun clubs for use in trapshooting. The professional "pigeoners," as they were called, of which there were as many as five thousand at one time, followed the birds around the country, summer and winter, persecuting them unremittingly at all seasons.

W. E. C. Todd, (1940), *Birds of Western Pennsylvania*

Books have been written about the relatively abrupt extinction of the passenger pigeon (*cf.* Schorger, 1973), yet authorities do not fully agree why it disappeared. Intensive hunting was certainly an important factor, but also important was the cutting of mature beech and oak forests upon which the species primarily depended. In Chapter 15 we will comment further on the causes of its disappearance.

Introduced species in the deciduous forest. Few non-native plants or vertebrates have become established in American deciduous forests except in early successional communities; but introduced microorganisms and insects have had significant impact even on our mature forests. Early in this century a fungus blight introduced from the Orient rapidly eliminated American chestnut, which had been a major canopy species in much of the deciduous region. The root systems of vanished chestnut trees still sprout prolifically, but the sprouts nearly always succumb to the fungus within a few years. Where chestnut formerly grew, species of oak have generally increased in abundance, filling the vacant niche.

American elm presently is being decimated by another fungus which is carried by bark beetles. Dutch elm disease has spread throughout American elm's range during the past 40 years, and although elm may never be completely eliminated, this successional tree is rapidly declining in importance. A third fungus disease, blister rust, was introduced from Germany and has affected several species of American pine, from the east coast to the Rockies. According to Dasmann (1968), blister rust has prevented the re-establishment of white pine as a major species in eastern forests.

Of introduced insects, the gypsy moth has had perhaps the greatest effect on eastern forests (Figs. 4.25 and 4.26). Gypsy moths were introduced to Massachusetts in 1868, by a Frenchman who thought they might have silk-producing potential. Specimens escaped from his experimental containers, and in the next 60 years spread over much of New England. The powers of self-dispersal of this species are limited (the females are too laden with eggs to fly), and for many years the Hudson River served as an effective western barrier. But since 1960 gypsy moths have become solidly established west of the Hudson, and they are creating enormous damage as they spread southward. Although they feed on many species of tree, they are particularly partial to oaks. Oaks are more important in mid-Atlantic and southern forests than in New England, and gypsy moth depredation is thus likely to be especially severe as they spread southward. All manner of control measures are being tried, but the march has not been halted. Selective feeding and heavy defoliation by gypsy moths (allowing more light to penetrate the canopy) probably will affect significantly the character of infested forests in the years ahead. Thick growths of shrubs and saplings may spring up beneath the larger trees, and species not favored by the caterpillars (particularly ash, dogwood, tulip tree and wild grape) may increase in abundance. Although regretted by many, the changes will be fascinating to study.

CONCLUSIONS

Man is not primarily a forest species but a species of open country. Not only temperate forests, but also boreal and tropical forests have fallen to the axe in many parts of the world. Some forests are resilient, and return rapidly after human disturbance ceases. Others may recover slowly, if at all. In the latter category, many believe, are most tropical rain forests, which often occupy old, nutrient-poor, lateritic soils; when cleared, these soils lose additional nutrients and often harden into rocklike laterite that is difficult for forest seedlings to recolonize. Gloomy warnings abound at present as the great rain forests in the Amazon and Congo basins and in Southeast Asia undergo clearance for human settlement (see Gomez-Pompa *et al.*, 1972; Short *et al.*, 1973).

The deciduous forest of temperate regions seems to have considerable capacity to return after disturbance, but this capacity can be exceeded. The plants of eastern North America suffered almost no extinctions during the last major glaciation of the Pleistocene (Dorf, 1969); many large American mammals did become extinct, but as already mentioned, human hunters may have been instrumental in their demise. The complex deciduous forest communities encountered in America by the first European settlers were communities that in many cases had been completely reconstituted during the postglacial period, on landscapes denuded by ice sheets.

In Europe, unlike North America, many species of forest plant disappeared during the glaciations of the Pleistocene. These extinctions may have occurred because in Europe east-west

FIG. 4.25. Female gypsy moth with egg mass. Deposition of egg masses on cars, trucks, and campers is a major means of dispersal. One such mass, carried to an uninfected area, may be enough to launch a snowballing outbreak. (Photograph by author.)

mountain chains constituted a barrier to the migration of plant species as the climate changed; in eastern North America, migration was not similarly impeded. Thus, postglacial forest communities in Europe were simpler than those of America even before forest clearance was initiated in the Old World by man some 5000 years ago. Since then, the hand of European man has been heavy and selective; Europe's oak, ash and beech woodlands are beautiful but most are semi-natural at best, and they lack the rich diversity of the deciduous communities of eastern America. The deciduous forests of China and Japan have been devastated even more; in China, clearance has been so extensive that deciduous forests would return slowly even if given the opportunity, because natural seed sources frequently are so distant.

Having survived the ice sheets and escaped agriculture until recently, North America's diverse deciduous forests are today unique in the world. Their immediate future seems bright in some respects, despite chestnut blight, gypsy moths, and increasing human populations. We

FIG. 4.26. Selective defoliation by gypsy moth larvae on the Pocono plateau in Pennsylvania. The hickory in the center has hardly been touched, while oaks on each side are virtually leafless. Hickory may be accepted when preferred foods are gone, but selective feeding by gypsy moths may well encourage this species at the expense of oaks. (Photograph by author.)

have seen that forests are returning in many parts of the East, as agriculture and lumbering wane. But the modern world poses new threats to forest communities: chemical pesticides are already blamed for the extinction in eastern North America of the peregrine falcon, and for the decline of a number of other carnivorous birds; and air pollution, according to a recent study (USDA, 1973), has "reached a point in the Northeast where some form of injury to plants is believed to be occurring throughout the entire area." "Northeast" in this study includes the states from Missouri and Minnesota to Maryland and Maine. As in the case of introduced pests, the effects of air pollution are selective (Table 4.10), and can be expected to reduce the diversity that makes American deciduous communities such beautiful casebooks of adaptive integration.

The American deciduous region is a panorama of distinct community types: climatic, edaphic, and fire-adapted variants spread across its great geographic range, and whole series of successional communities are native to a single area. Perhaps the greatest threat to the remaining flora and fauna of eastern America is not industry or agriculture but a narrowness of outlook among those who would preserve forests. It is clear that native plants and animals of the deciduous region require a very great variety of habitats, not just the mature forests whose rescue alone seems to concern preservationists, or the monocultures often favored by the timber industry. The English conserve their remaining flora and fauna by jealously guarding not only their woodlands but also their fields and hedge rows. In America, not just mature forests but meadows, flood plain tangles, fire climaxes, and many types of successional habitat must be maintained if we wish to keep the species we now enjoy. If habitat diversity disappears, so will whole communities, and with them much of the biological richness we have inherited.

SUGGESTED READING

General references

BAKER, F. S. A revised tolerance table. *Journal of Forestry* 47: 179–181 (1949). A useful summary of shade tolerance information for many trees.

BRAUN, E. L. Deciduous Forests of Eastern North America. New York, Hafner Publishing Co., 1967. The most exhaustive descriptive account of America's temperate deciduous forests.

COOPER, C. F. The ecology of fire. *Scientific American* 204: 150–160 (1960).

CURTIS, J. T. The Vegetation of Wisconsin. Madison, Wisconsin University Press, 1959. A major work, useful well beyond the boundaries of Wisconsin, by an eminent plant ecologist and founder of continuum analysis.

DAUBENMIRE, R. F. Plant Communities. New York, Harper and Row, 1968. Useful text.

FOWELLS, H. A., ed. Silvics of Forest Trees of the United States. United States Department of Agriculture Handbook No. 271, 1965. A mine of information regarding individual tree species.

TABLE 4.10

Susceptibility to Air Pollutants of Trees Common to Eastern Forests

Ozone and sulfur dioxide are major byproducts of combustion in transportation and industry, respectively. These are the two most serious air pollutants in terms of widespread forest damage. Fluorides are produced in steel, glass and aluminum manufacturing. They and many other pollutants from specialized industries are serious in localized situations. Ozone and sulfur dioxide sometimes act synergistically. Eastern white pine, for example, is damaged by lower concentrations of both together than of either by itself. With the exception of red pine, most pines seem to be susceptible to ozone, while spruces and firs are generally resistant.

Data are from USDA (1973), after sources quoted therein.

	Ozone	Sulfur Dioxide	Fluorides
Deciduous Species			
Tulip tree	Very susceptible		
Sycamore	Very susceptible		
Honey locust	Very susceptible		
Green ash	Susceptible		Susceptible
Sweetgum	Susceptible		Resistant
White oak	Susceptible		Susceptible (?)
Pin oak	Susceptible		Susceptible
Red oak	Susceptible		Susceptible
American elm		Very susceptible	
Sugar maple	Resistant	Resistant	
Ashleaf maple		Resistant	Very susceptible
Coniferous Species			
Eastern white pine	Susceptible	Very susceptible	Very susceptible
Red pine	Resistant		
Eastern hemlock	Susceptible		
Northern white cedar	Resistant	Resistant	Susceptible

GEIGER, R. The Climate Near the Ground. Cambridge, Harvard University Press, 1965 (4th edition). The most widely used book on microclimate.

HEINSELMAN, M. L., AND H. E. WRIGHT, JR., eds. The ecological role of fire in natural conifer forests of western and northern North America. *Quaternary Research* 3 (Oct., 1973). Proceedings of a valuable symposium.

KOZLOWSKI, T. T., AND C. E. AHLGREN, eds. Fire and Ecosystems. New York, Academic Press, 1974. An authoritative, wide-ranging volume.

MATTHIESSEN, P. Wildlife in America. New York, Viking Compass paperback, 1964. An eloquent history of American wildlife since European settlement.

OOSTING, H. J. The Study of Plant Communities. San Francisco, Freeman, 1956 (2nd edition). A good text in terrestrial plant ecology.

REICHLE, D. E., ed. Analysis of Temperate Forest Ecosystems. New York, Springer-Verlag, 1970. Covers many topics in forest ecology.

SHELFORD, V. E. The Ecology of North America. Urbana, University of Illinois Press, 1963. A geographically organized descriptive account of all the native terrestrial community types of North America.

SPURR, S. H. Forest Ecology. New York, Ronald Press, 1964. A very useful text.

STERN, K., AND L. ROCHE. Genetics of Forest Ecosystems. New York, Springer-Verlag, 1974. The title is misleading; this is an investigation of forest communities from an adaptive, evolutionary aspect.

WILDE, S. A. Forest Soils. New York, Ronald Press, 1958.

Methods of analysis

BENTON, A. H., AND W. E. WERNER, JR. Manual of Field Biology and Ecology. Minneapolis, Burgess Publishing Co., 1965.

FRIEND, D. T. C. A simple method of measuring integrated light values in the field. *Ecology* 42: 577–580 (1961).

GRIME, J. P. Comparative experiments as a key to the ecology of flowering plants. *Ecology* 46: 513–515 (1965). Brief paper with useful approach.

PHILLIPS, E. A. Field Ecology (A BSCS Laboratory Block). Boston, D. C. Heath & Co., 1968. A very helpful manual of basic methods.

PRAMER, D. Life in the Soil (A BSCS Laboratory Block). Boston, D. C. Heath & Co., 1965.

chapter 5

Fresh Water Habitats and Communities

Man's life is but vain, and is subject to pain
And sorrow, and short as a bubble.
'Tis a hodge-podge of business and money and care,
And care, and money and trouble.

But we'll take no care if the weather prove fair,
Nor will we vex now though it rain;
We will banish all sorrow and sing till tomorrow
And angle, and angle again!

Isaak Walton (1593–1683), *The Angler's Song*

INTRODUCTION

We were led into Chapter 4 by Thoreau's serious resolve to discover what the forest could teach him. By contrast, Walton's blithe escapism seems frivolous. Yet Walton learned from his stream no less than Thoreau from his woods, and "compleat anglers" like Walton always have been aquatic experts. My own graduate studies were in limnology (freshwater ecology), so presumably I am a professional in this area; but I have a friend, a drama professor and ardent angler, who shames me in parts of my own specialty. He truly knows streams and their denizens—not only can he tell me where his beloved trout lurk, but he knows their seasonal food habits and the life cycles of their prey. He ties his flies to resemble the various nymphal or adult stages of particular species of stonefly or mayfly, and knows at which season each species and stage actually occurs in local streams. Walton would beam on this "recreational limnologist!"

Aquatic habitats are of many types: shallow and deep, salt and fresh, swift and stagnant. Some are nearly constant (*e.g.* a hot spring), others notably inconstant (a drying pool). Like the last chapter, this one is consciously selective. It provides a reasonably thorough treatment of lakes and streams in the hope that this will serve as an adequate introduction to all aquatic habitats and communities. To some degree this is impossible: ephemeral ponds, springs, estuaries, tide pools and oceans all are unique habitats which ideally should be treated separately. In the sense that all are aquatic, however, they all share common features which we can address first of all.

WATER AS AN EXTERNAL MEDIUM

What does it mean to live in water? This comprehensive question can be dealt with point by point if we refer back to Table 2.1. Compared to air, water is a thermally conservative medium: it may be hot or cold, but it changes temperature grudgingly and aquatic organisms seldom if ever experience the quick thermal changes that many terrestrial creatures encounter. Moreover, though active fishes maintain temperatures well above that of cold water, liquid water never is so cold that poikilothermic (cold-blooded) life is impossible. We see a reason, then, why homoiothermy (warm-bloodedness) should have evolved on land, not in water. Homoiothermy allows an organism to maintain relatively constant internal temperatures despite changing or inhospitable external temperatures. There is a second reason too, however: homoiothermy requires high metabolic activity, hence a rich oxygen supply. The solubility of oxygen in water is limited, and even in the best aerated aquatic habitats the available oxygen probably is insufficient to sustain a homoiotherm. Whales, seals and penguins are an obvious exception to this statement, but they obtain their oxygen from the air, not from the water, and they inherit their homoiothermy from terrestrial ancestors.

Other characteristics of water—its high surface tension and dissolving power and its relatively high density—combine to make water very much a three-dimensional habitat. The **neuston** subcommunity, for example, could not exist if water had a low surface tension. Neuston include such creatures as whirligig beetles, water striders, mosquito larvae and duckweed: small organisms which live either on top of the surface film or suspended beneath it.

Actually the whole water column of an aquatic habitat, from surface to bottom, is a region of life. This vertical third dimension is made habitable by water's buoyancy and high dissolving power, the latter providing a bath of nutrients at any depth. Nutrients can be exhausted at some levels, of course, if demand by organisms exceeds supply, but if we think of the atmospheric medium as opposed to the hydrospheric one, the contrast is very clear. Life on land is much more a two-dimensional phenomenon, bound to the substrate by the inability of air alone to provide the necessary support or nutrients.

Table 2.1 will have pertinence throughout this chapter, and we will not catalogue all its lessons now. However, the biological implications and consequences of water's buoyancy are so basic that they demand immediate further attention. For example, why are the dominant plants on land not just macroscopic but often huge, while those of water are mostly microscopic? Once pointed out, this contrast is not difficult to explain. Plants on land defy gravity only by possessing sturdy stems or trunks, and where climate allows, the dominant land plants vie for light by accentuating size. Sequoias and eucalyptus attain weights exceeding 2000 tons.

No tree, however, soars as far above the substrate as do the tiny suspended algae **(phytoplankton)** of deep lakes and oceans. To reach the light these organisms use water's buoyancy, and the smaller they are the better they can use it. The sinking rate of small bodies denser than their suspending medium is inversely proportional to their size. Since most phytoplankton *are* somewhat more dense than water they tend to sink out of the zone of light, but if reduced size means reduced sinking rate there is obvious gain in being small.* A small organism, moreover, is more easily returned to the surface by turbulent water movements.

The "reductionism" of phytoplankton evidently reflects complete reversal of the selective pressure for large size that has been so influential in the evolution of dominant land plants. These opposite solutions to the quest for light have produced, in turn, fundamental contrasts in the length of terrestrial and aquatic food chains. On land, the prevalence of large plants makes possible the existence of large herbivores which in turn can be fed on only by large carnivores. As a result, food chains on land are often relatively short, the top carnivore level being reached in three or four steps. Large herbivores exist in water, too—manatees and green turtles are examples—but they are the exception rather than the rule. Most herbivores

---

* Stokes' Law is a modification for very small bodies of Newton's general law of free-falling bodies. Stokes' Law defines all the parameters which determine the sinking rate of small organisms, and makes obvious the value of various types of planktonic adaptation, including the size reduction we have described. According to Stokes' Law,

$$s = [k \cdot d^2 (b - m)]/v$$

where s = sinking rate; d = radius (or, for a nonspherical body, a linear dimension determined by the position it typically assumes as it falls); k = a constant determined by the body's shape; b = density of the body; m = density of the medium; and v = viscosity of the medium. We see from this equation that the size, shape and density of an organism all influence its sinking rate, as do two properties of the medium, density and viscosity. A planktonic organism's size is the parameter that most sensitively affects its sinking rate. However, other features of planktonic species also may be viewed as adaptations that retard sinking. For example, some species secrete gelatinous envelopes of lower density than their bodies, or produce spines that entrap an envelope of water. These extraneous, low density substances act as if they were part of the organism and reduce its sinking rate.

in lakes and oceans feed on tiny phytoplankton, and most are little larger than their food. Tiny herbivores can be fed on in turn by minute carnivores, and indeed a typical assemblage of **zooplankton** (animal plankton) contains both. Small fishes may feed on the carnivorous zooplankton (or indiscriminately on both types of zooplankton), and several more trophic steps may intervene before the food chain culminates in the largest marine or freshwater carnivores (*e.g.* tuna, sharks, pike). Thus aquatic food chains typically are longer than those of land, and this fact, as we shall see in Chapter 10, affects importantly the processing pathways of energy in aquatic and terrestrial communities. All this derives from the buoyancy differences of water and air.

Aquatic macrophytes, of course, do exist in addition to phytoplankton. These larger plants are, in salt water, primarily the seaweeds. In fresh water they are aquatic descendents of terrestrial angiosperms. Foregoing the benefits of small size, most aquatic macrophytes are restricted to the shallow fringes of lakes and oceans, where they can live on the bottom yet obtain sufficient light for photosynthesis. Anatomically these macrophytes possess adaptations to the buoyancy characteristics of water that are as remarkable as those of the phytoplankton. These anatomical peculiarities bear eloquent witness to how different are the conditions of existence in water from those on land (Table 5.1).

LAKES AND PONDS

To a geographer, a lake may be merely a wide place in a stream, where the downhill movement of water to the sea becomes less hurried but does not fully stop. Certain lakes, however, do not fit this definition because they are end points, not way-stations. Streams flow into them, but no stream flows out. Lacking outlets, such "closed lakes" lose water only by evaporation and typically are salty or alkaline because the solutes which are continually being brought in by streams are left behind by the evaporating water. In a sense the ocean is a gigantic closed lake, and its high concentration of solutes is

TABLE 5.1

Some Common Features of Angiosperms That Have Become Secondarily Adapted to Life in Water

Like whales and penguins, aquatic flowering plants such as water lilies, eelgrass and *Elodea* evolved from land ancestors. Some common modifications exhibited by such plants are listed below, with suggestions as to their functional significance. Many aquatic angiosperms, especially those that live fully submerged, have come to resemble seaweeds by evolving small, dissected leaves and supple stems.

Modification	Examples of Plants Possessing Trait	Suggested Cause
Loss of lignin and other components of rigidity in the stem	Water lily, *Myriophyllum*	Stem, being relatively supple, does not break during movements of water, which is much more viscous than air. (Since water is also more dense than air, stems do not require great rigidity for plants to remain erect.)
Concentration of remaining vascular tissues near center of stem	Most fully submerged species	Confers additional suppleness to stem for the most rigid tissues to be centrally located.
Air spaces in stem (and in leaves if they float on surface)	Most aquatic angiosperms	Flotation serves as a device to hold plant up (this is possible in water but not in air). Also, the spaces provide rapid diffusion pathways for respiratory gasses; roots in anaerobic sediment still can obtain O_2 and rid themselves of CO_2.
Dissection of submerged leaves into small filaments, and reduction of epidermal cuticle	*Myriophyllum*, *Cabomba*	Permits more intimate exchange of dissolved gasses and nutrients between plant and medium. On land, this growth form would subject plant to too much desication.
Reduction of size	Duckweed, *Wolffia*	Plant can make use of surface tension for support; also facilitates nutrient exchange with medium.

explained by this fact. Most "open" lakes are fresh, however; the solutes brought into them do not accumulate, but are continually being flushed away through the outlet. Except in arid regions lakes and ponds are primarily of this type, so the geographer's definition carries a good deal of truth. Biologically, however, his definition is false; lakes and streams are strikingly different environments, and the species they contain also differ sharply, as any good fisherman can testify. The environments and communities of lakes and streams must be described separately.

Seasonal Environments in a Temperate Lake

In passing from the surface to the bottom of a stratified lake in summer, we can easily traverse in 10 to 20 m a range of physical and chemical conditions as great or greater than would be encountered in climbing up a hundred times that vertical range on a mountain.
G. E. Hutchinson, (1964) *The Lacustrine Microcosm Reconsidered*

Lakes and ponds typically stratify at certain seasons into horizontal layers that differ dramatically in their physical, and often chemical, conditions. Many species of organism become compartmentalized into one or another of these strata, and the lake becomes a series of partially distinct, vertically stacked communities that occupy sharply contrasting environments. What is responsible for this phenomenon, which has no real parallel either on land or in running water?

Vertical thermal contrasts and their cause. Infrared, which is the main heat component of solar radiation, is absorbed by water very rapidly. Therefore incoming radiation of these wavelengths penetrates only a short distance into a lake, and warms only the surface waters; but here, because infrared absorption is so rapid and complete, warming may be considerable despite the great heat capacity of water. Even visible light, to which water is considerably more transparent, is absorbed largely near the surface (Fig. 5.1), and the heat it liberates upon absorption is therefore concentrated in the surface region also. This localization of direct solar heating within a narrow surface layer is one of two reasons why lakes stratify thermally. The other reason is, simply, that the density of water is dependent on its temperature; differential heating of the water column therefore brings about a density stratification, and the lake so heated tends to become separated into phases just as do the vinegar and oil of a salad dressing.

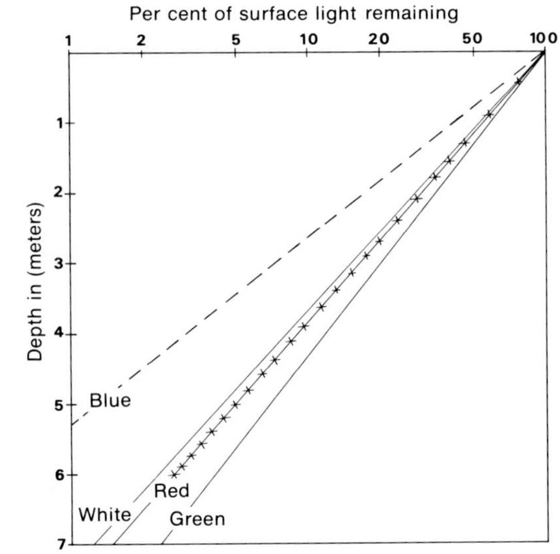

FIG. 5.1. Subsurface light attenuation in Lake Lacawac, Pennsylvania. Note the logarighmic scale, and the different rates of absorption of different wavelengths. (White light is all wavelengths combined.) In this lake of moderate transparency about 99% of entering light is absorbed within the top 7 m of the water column, 50% within the first meter. These data and data for Lake Lacawac in later figures were obtained by students in the author's ecology course during class field trips. Light readings were taken by underwater light meter equipped with colored filters.

Figure 5.2 illustrates the interdependence of temperature and density in the case of water. It is well known that pure water has its maximum density at 4° C and that ice, being colder, therefore floats. It also is well known that above 4° C water becomes progressively less dense as the temperature rises. But Figure 5.2 makes clear an additional fact: the density *difference* per degree of temperature difference also increases as the temperature rises. This is evidenced by the steepening of the curve above 5° C. This subtlety has important implications, as we shall see; but the most basic reasons for thermal stratification are the radiation-absorbing properties of water and the fact that it becomes progressively less dense as it warms above 4° C.

The thermal stratification of temperate lakes is not permanent, but occurs in winter and especially in summer. Ice-covered lakes typically exhibit "reverse" stratification: they are coolest at the top and warmest at the bottom. The ice cools water immediately beneath it to near freezing, and slightly warmer water of maximum density (4° C) becomes concentrated at the bottom by sinking. The temperature difference between lower ice margin and lake bottom is relatively small, however, and so is the thermally produced density difference. When the ice melts in spring, the density layering is destroyed, partly by solar warming of the cold surface water but partly also by wind. As long as vertical density differences are small, winds across an ice-free lake can generate enough turbulence to mix the waters into a single, thermally uniform water mass. In temperate lakes such a condition, the **spring turnover,** occurs shortly after ice breakup. If winds are strong this freely mixing phase may last a number of weeks, but sooner or later sufficiently calm weather occurs for a warm surface layer to develop under the bright spring sun. Once such a thermal discontinuity has developed, the density difference between upper and lower layers resists the mixing force by the wind, and the warm, upper layer **(epilimnion)** and cool, deeper layer **(hypolimnion)** become separately circulating water masses (Fig. 5.3). Wind continues to mix the epilimnion with itself, and it thus remains a relative homogeneous subcompartment of the whole water body, with only minor internal temperature differences. But the hypolimnion, shielded from direct solar heating and no longer able to obtain heat by downward mixing of surface water, remains cool throughout the summer. Between epilimnion and hypolimnion there develops a region of rapid temperature (and density) change, the **thermocline.** Swimmers know this region well; in early summer it may be so near the surface that, when they tread water, their toes freeze while their shoulders remain comfortably warm. The thermocline is the region of shear between the wind-circulated epilimnion and the cold, relatively quiet hypolimnion. The only heat transferred from epilimnion to hypolimnion is by conduction across the thermocline, and the conduction of heat in liquids is such a slow process as to have a negligible effect on the temperature of the hypolimnion. During the course of the summer heat transferred to the top of the thermocline gradually deepens the epilimnion, but within the hypolimnion the temperature during the summer remains the same as it was when spring turnover ended and stratification began. Although it is never directly exposed to surface winds, the hypolimnion is not unaffected by them. Strong winds indirectly set up a circulation within the hypolimnion by blowing epilimnetic waters to the windward end of the lake, where their weight forces the thermocline down and squeezes hypolimnetic waters the other way. When the winds die, the thermocline returns to a stable, horizontal position by a series of rocking, gradually dying oscillations. This rocking causes further water movements in the hypolimnion (Fig. 5.4).

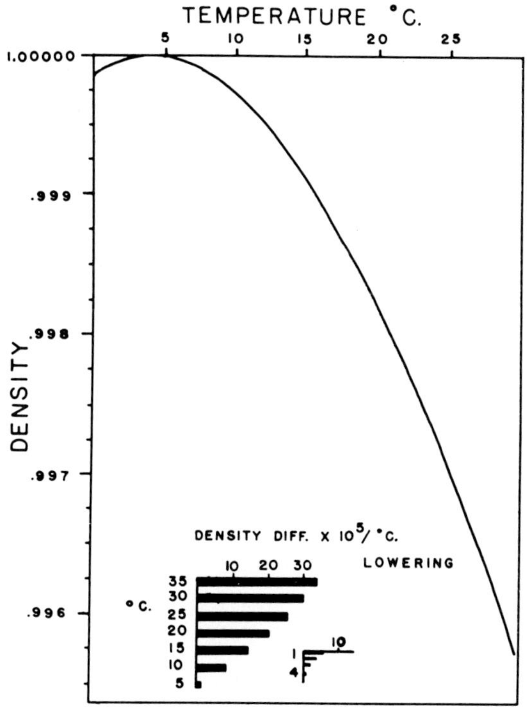

FIG. 5.2. Density of liquid distilled water as a function of temperature. The insets show, for different starting temperatures, the density difference per degree of temperature change. The smaller inset shows this phenomenon on an expanded scale between 1° and 4° C. When it freezes, water expands and its density decreases markedly (to 0.918 at 0° C); hence ice floats in even the warmest water. (From J. R. Vallentyne: *American Scientist*, Journal of Sigma Xi, *45:* 1957. Reproduced with permission from The Scientific Research Society of North America, New Haven, Conn.)

As autumn approaches the intensity of incoming radiation dwindles; eventually, more heat is lost from the surface by evaporation and nightly back-radiation than is absorbed during the day, and the epilimnion begins to cool. Now the thermocline appears to sink even deeper, but actually the epilimnion is simply growing downwards as it cools to the temperatures of the upper thermocline; the position of the lower thermocline does not change. Ultimately the

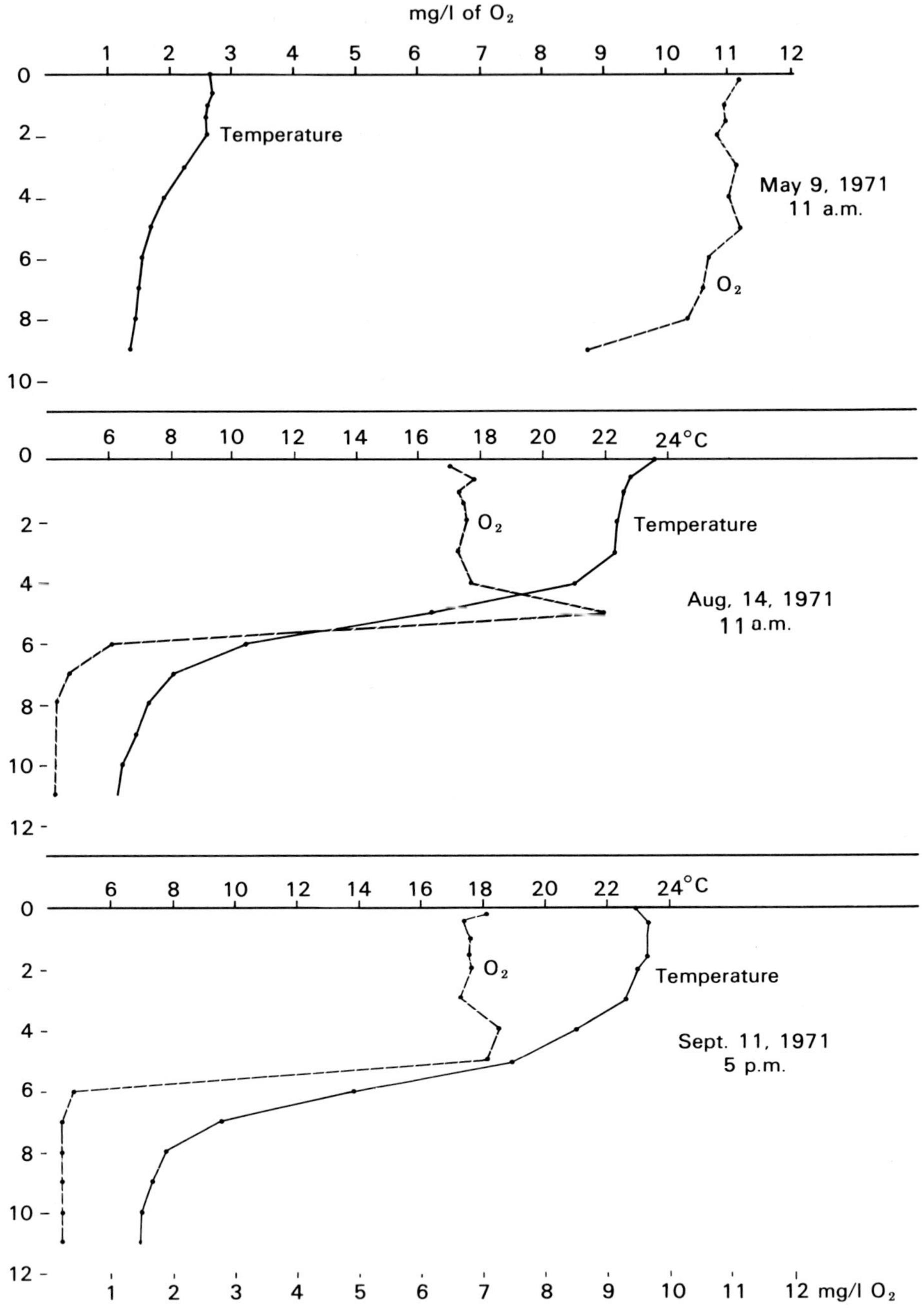

FIG. 5.3. Progression of thermal and oxygen stratification in Lake Lacawac during the spring and summer of 1971. On May 9 the lake was essentially unstratified. The small temperature gradient represents only minor density differences, and an afternoon wind could easily have mixed the lake from top to bottom. Dissolved oxygen was abundant at all depths, being slightly less near the oxygen-demanding bottom sediments. By August the situation was far different: a major thermal difference now existed between surface and bottom waters, and a pronounced thermocline marked the interface between epilimnion and hypolimnion. Oxygen levels now differed sharply between the upper and lower waters. The oxygen maximum in the upper thermocline probably represents photosynthetic oxygen production by a population of algae "trapped" within the sharp density gradient at a level determined by their buoyancy characteristics. On September 11 stratification phenomena were quite similar to those of August; autumn cooling perhaps was just beginning.

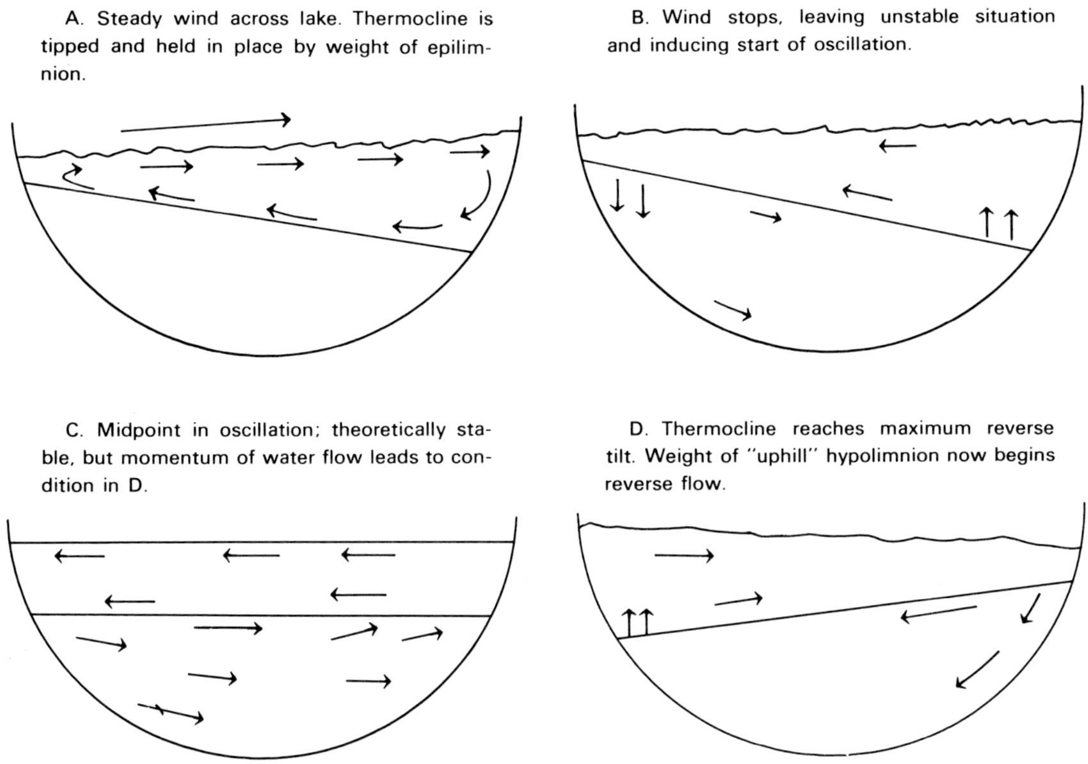

FIG. 5.4. Wind-induced tipping of the thermocline results in circulation within both the epilimnion and the hypolimnion. But the two water masses remain separate; there is no mixing across the thermocline. "Uphill" flow is opposed by gravity and downhill flow must gather momentum, so hypolimnetic water movements are greatest after the wind stops, during phases in the rocking cycle when the thermocline is virtually level. *Vertical arrows* indicate direction of thermocline movement, *other arrows* direction of water flow. (Based on a diagram in Mortimer, 1952).

epilimnion becomes as cool as the hypolimnion, the thermocline vanishes, and the density barrier to complete mixing disappears. This is the period of **fall turnover**; winds again stir shallow and deep waters together into a homogeneous whole; and if the winds fail, mixing still occurs simply by sinking of surface waters as they are cooled below the temperature of water deeper down. Fall turnover continues until ice formation eliminates the action of wind and generates the inverse temperature gradient of winter stratification.

This pattern of seasonal stratification and destratification occurs in all but the shallowest and most exposed of temperate lakes and ponds. In these, wind-caused turbulence may prevent extended stratification during the summer. One such lake, Devils Lake in North Dakota, has a surface area of nearly 50 km^2 but a mean depth of only 2.6 m; it stratifies in summer only for a few days at a time, during brief calm spells, but when winds resume the lake again becomes mixed to the bottom. In warmer portions of the temperate region, of course, lakes may not become ice-covered in the winter; in this case winter stratification does not develop, and mixing occurs throughout the cold months of the year.

In the tropics, many relatively deep lakes stratify thermally despite relatively minor seasonal changes in temperature and insolation. In these warm lakes seasonal heating of the surface waters by only a couple of degrees creates a sufficient density difference between them and underlying waters to prevent vertical mixing. (Remember from Figure 5.2 the steepness of the density-temperature relationship when water is relatively warm.)

To some degree, thermal conditions near the bottom of lakes that stratify resemble thermal conditions in the oceanic abyss. Temperatures in both are nearly constant; the winter hypolimnion experiences temperatures less cold than those near the surface, and the summer hypolimnion remains as cold as it was when spring turnover ended. In a typical temperate lake the bottom temperature may vary by 4° C between summer and winter while surface temperatures are varying as much as 25°C.

Chemical consequences of thermal stratification. During summer stratification the epilimnion is sealed off from the source of most of its mineral and organic nutrients, namely the bottom sediments and deep waters where dead organic matter accumulates. The hypolimnion is similarly cut off from its sources of dissolved oxygen: the atmosphere and the oxygen-producing photosynthesizers concentrated in the epilimnion. The chemical differences that develop between epilimnion and hypolimnion are often profound, and are brought about in large part by organisms.

In an infertile **(oligotrophic)** lake, organisms are relatively few, and the chemical changes that result from thermal stratification therefore may be hardly noticeable. But in a fertile **(eutrophic)** lake, organisms are sufficiently abundant to produce, by their metabolic activities, a chemical stratification (complete with a "chemocline" at the level of the thermocline) that is as dramatic as the thermal stratification (Figs. 5.3 and 5.5). As they grow and multiply, organisms of the epilimnion incorporate phosphates, sulfates, nitrates and other nutrients; when they die, these organisms sink through the thermocline to the bottom, taking their incorporated nutrients with them. As summer proceeds, therefore, the epilimnion is progressively depleted of nutrient substances, and these accumulate in the hypolimnion. Because epilimnetic and hypolimnetic waters do not physically mix, the only way such nutrients are returned to the epilimnion is by diffusion through the thermocline; and like heat conduction, diffusion in water is an extremely slow process. In fact, except where it extends to the shallow bottom near shore, the epilimnion is completely cut off from renewing supplies of mineral nutrients.

The hypolimnion suffers a similar fate with respect to oxygen. O_2 diffusing from the air into the surface waters of a lake is mixed by circulation throughout the epilimnion, but never reaches the hypolimnion. In a eutrophic lake the hypolimnion contains many oxygen-consuming animals and bacteria, but is too dark to support many photosynthesizers. As summer proceeds, therefore, the hypolimnion is progressively depleted of oxygen; if consumption is high enough, it may even become anoxic. On the other hand, the CO_2 produced by respiration may build to surprisingly high levels in the hypolimnion, as may other waste products of metabolism.

The terms epilimnion and hypolimnion seldom are used in discussions of winter stratification, probably because the thermal and chemical contrasts that develop are usually much less dramatic than those of summer. Metabolic activities (and sometimes populations) of organisms beneath the ice are so reduced relative to summer that chemical gradients from surface to bottom are much less sharp. When a lake is fertile but shallow, however, respiratory activity may consume virtually all the oxygen in the limited volume of water beneath winter ice, resulting in the well known "winter kills" of fishes.

Anoxic conditions, whether in winter or in summer, have important chemical consequences. Oxidized substances become reduced, often with resulting changes in their solubility (Table 5.2). The release into solution of phosphate and silicon during periods of anoxia is of great biological importance, since these nutrients often are in short supply to organisms which need them. Of course, release into an anoxic hypolimnion does not make these nutrients immediately available to epilimnetic populations, but when spring or fall turnover occurs they are carried into the surface waters and stimulate the spurt of growth by phytoplankton and other organisms that occurs at these times. A period of stratification that results in hypolimnetic anoxia therefore results in more profuse growth during subsequent turnover than does turnover occurring more frequently without anoxia.

Meromixis. The chemical stratification associated with thermal stratification obviously disappears at times of turnover, when the mixing of water from epilimnion and hypolimnion destroys both their thermal and their chemical distinctiveness. But occasionally a lake may be density layered through agencies other than temperature. Saline springs may produce a layer of dense, solute-laden bottom water, or suspended clay may create a similar density

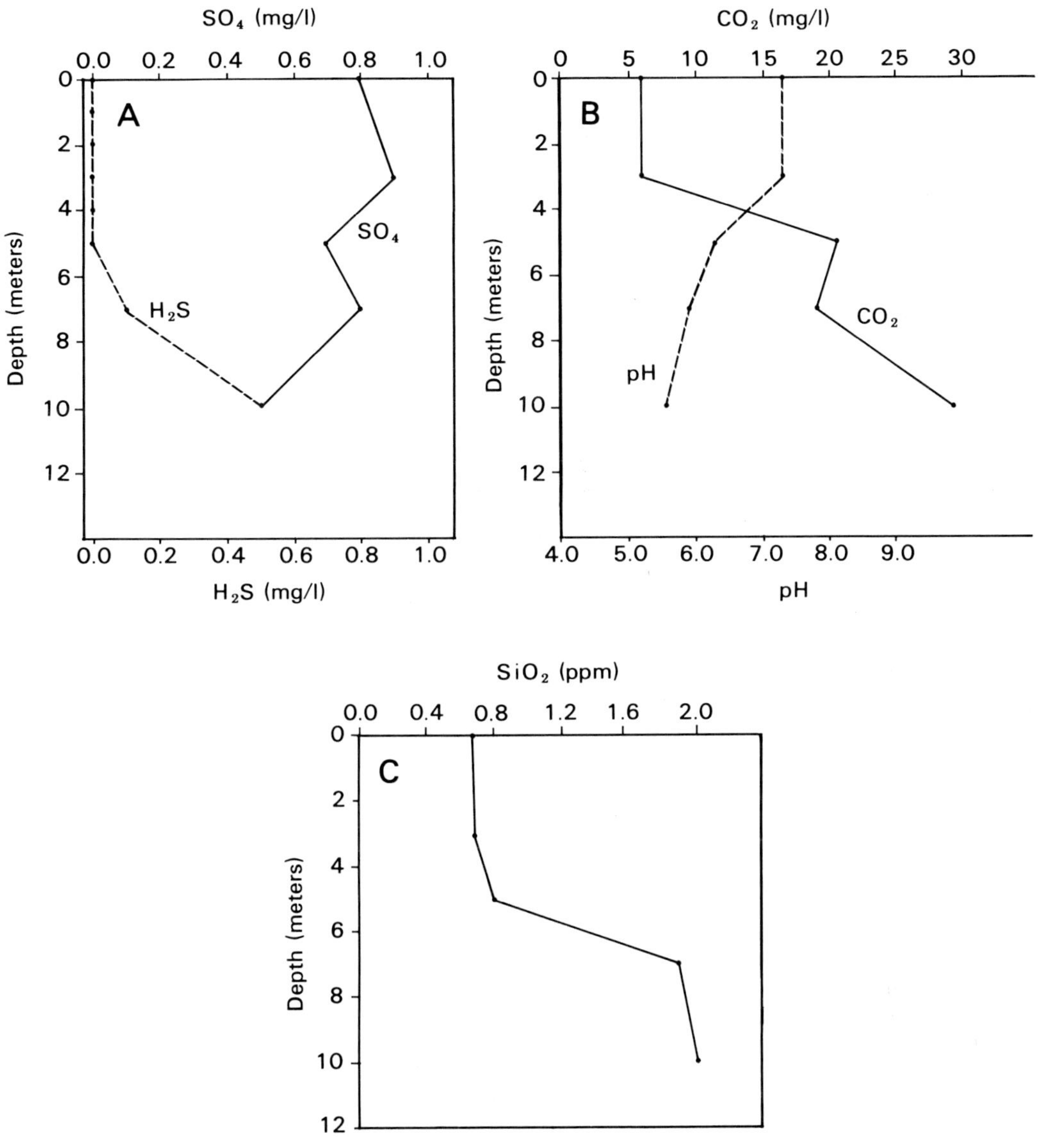

FIG. 5.5. Chemical stratification of Lake Lacawac on September 16, 1972. The thermocline occupied about the same position at this time as in September 1971 (see Fig. 5.3). Note the reciprocal distributions of oxidized and reduced forms of sulfur, and of CO_2 and pH. H_2S can exist only under anoxic conditions, hence is restricted to the hypolimnion. pH is lower in the hypolimnion because CO_2 accumulates there as a product of respiration and acts as an acid. SiO_2 is used as a nutrient by light-demanding planktonic diatoms, which largely are restricted to the epilimnion. The amounts *in solution* in the epilimnion thus decline as summer proceeds.

difference between deep and overlying waters. Dense bottom layers such as these may never lose their individuality, and the lake may never mix to the bottom.

Such lakes are termed **meromictic,** and the persistent deep layer—known as the **monimolimnion**—may become a rich nutrient sink. Meromictic lakes may stratify thermally in their upper parts, but when turnover of epilimnion and hypolimnion occurs the monimolimnion remains distinct. Meromictic lakes are often deep relative to their surface areas, since lakes with such proportions are less subject to thorough mixing by wind. One of the few

TABLE 5.2

Chemical Changes in Lake Waters as They Become Anoxic

From Vallentyne (1957) and Ruttner (1963). The transformations shown below are some of those that occur when oxygen levels fall below 2 mg/liter. In water that is becoming progressively more anoxic, these reactions occur more or less in the order in which they are listed.

1. $NO_3 \longrightarrow \longrightarrow NO_2 \longrightarrow \longrightarrow NH_3$
   (nitrate) (nitrite) (ammonia)
2. Fe^{+++} (precipitated on bottom) → Fe^{++} (dissolved)
3. $SO_4^{--} \longrightarrow \longrightarrow S^{--} \longrightarrow \longrightarrow H_2S$
   (sulfate) (sulfide) (hydrogen sulfide)

Additional changes consequent on iron behavior:

1. PO_4^{---} (as ferric phosphate precipitate) → → → → → → → → → → → → → → → PO_4^{---} (dissolved)
2. Si (precipitated in a ferric-humic-silicic acid complex) → → → → → → Si (in dissolved compounds)

meromictic lakes in America is Green Lake, in Fayetteville County, New York. This small lake occupies the deep, "plunge-pool" basin of a now vanished waterfall (Brunskill and Ludlam, 1969, *et seq.*). Many of the deep, Rift Valley lakes of Africa likewise are meromictic. Lake Tanganyika, for example, contains a monimolimnion which occupies the greater part of the lake's volume, above which seasonal epilimnia and hypolimnia occur as a result of thermal stratification (Beauchamp, 1964). The waters of the monimolimnion are deoxygenated; occasionally storms cause limited upwelling from this deep layer, bringing asphyxiation to surface-dwelling fishes while simultaneously stimulating phytoplankton growth by an influx of nutrients (Coulter, 1963).

Seasonality among small organisms. Seasonal biological patterns reflect the seasonal changes of physiochemical features that have been described. The plankton, which is composed of short-lived organisms whose populations can respond quickly and often dramatically to change, experiences particularly pronounced seasonal changes.

There are three general types of seasonal pattern among the plankton: (*a*) quantitative changes of overall plankton abundance; (*b*) successional sequences of particular species; and (*c*) within particular species, seasonal changes of form and size.

The typical cycle of overall plankton abundance in a temperate lake or pond is related closely to the cycle of stratification (Fig. 5.6). Coincident with spring turnover there is a dramatic increase in the density of phytoplankton, probably resulting from the mixing of nutrients into surface waters. As the phytoplankton increase in abundance, so do the herbivorous and carnivorous zooplankton which directly or indirectly depend on phytoplankton for food. The onset of summer stratification severs the supply of mineral nutrients to the surface, and as these nutrients are depleted in the epilimnion the overall abundance of phytoplankton—hence also of zooplankton—dwindles. Despite light and temperature conditions conducive to rapid growth, summer therefore is a time of low plankton populations in the epilimnion. In the thermocline and hypolimnion the summer decline in plankton may be less pronounced, partly because nutrient depletion is less severe and partly because these regions are usually too dark to have experienced the burst of phytoplankton growth that occurred near the surface during spring turnover.

The fall period of turnover often is accompanied by a pulse of plankton abundance similar to that in the spring but less marked, perhaps because in fall temperature and light are declining rather than increasing and serve as a brake on growth rates even though nutrients are abundant. Sometimes a fall plankton pulse fails completely; one can imagine this happening in years when stratification persists well into the fall, or when autumn weather is unusually calm and thorough wind mixing fails to occur. During the period of winter stratification plankton abundance drops to late summer levels or below; growth is retarded by the low temperature and, if ice is thick, by dim light even near the surface. Nutrients may also decline in the upper part of the water column, though not to levels as low as in late summer.

The seasonal pattern of overall plankton abundance, of course, is the sum of the patterns of individual species. The "ideal" pattern of Figure 5.6 is realized when the phytoplankton is dominated by species which either follow the pattern themselves or complement each other so that peaks of overall abundance *do* occur during turnovers and troughs of abundance during stratification periods. Individual species of plankton, however, actually follow their own distinctive patterns of abundance, and if (for example) a "summer bloom" species is common in a particular lake, the typical summer minimum in phytoplankton abundance may not occur. Individual species require different quantities of particular nutrients for successful growth, differ in their dependence on turbulence to keep them near the surface, and sometimes differ in their light and temperature optima. For phytoplankton, all these factors plus various biological interactions—unequal competitive abilities, differential grazing pressures by zooplankton or susceptibilities to fungal infection—help to produce a seasonal succession of species (Fig. 5.7). Usually the spring burst of phytoplankton growth is dominated by species of diatoms (a class of algae characterized by relatively heavy siliceous shells); but as warming decreases the density of surface waters and as vertical turbulence simultaneously declines with the start of summer stratification, many of the larger diatoms tend to sink out of the epilimnion. Other factors may be equally important in bringing about a summer diatom decline; these include nutrient depletion (silica reserves, particularly, seem to fall below the needs of many species), and infection by fungi, which in the English Lake Windermere sometimes decimates a dominant species of diatom, *Asterionella formosa* (Canter and Lund, 1951). In an experiment directly illustrating the biological consequences of stratification, Lund (1971) artificially brought about summer turnover of a pond; the consequence was an atypical bloom of a species of diatom that usually dominates the spring maximum but is rare in summer. Lund believes that the turbulence and perhaps also the nutrient renewal associated with this anomalous turnover produced the bloom.

Certain non-diatom components of the phyto-

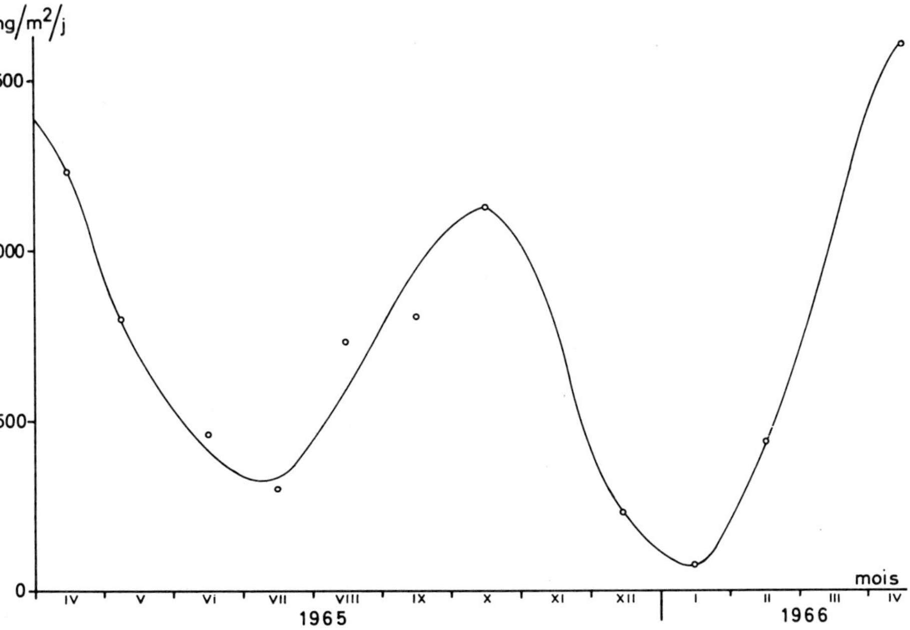

FIG. 5.6. Seasonal changes in overall phytoplankton abundance are responsible for these changes in daily rates of photosynthetic production in Lake Geneva, Switzerland. The "ideal" pattern of peaks at spring and fall overturn is realized here (see text). (From C. Serruya: *Les depots du lac Leman en relation avec l'evolution du bassin sedimentaire et les caracteres du milieu lacustre*, D.Sc. Thesis, University of Geneva, 1969.)

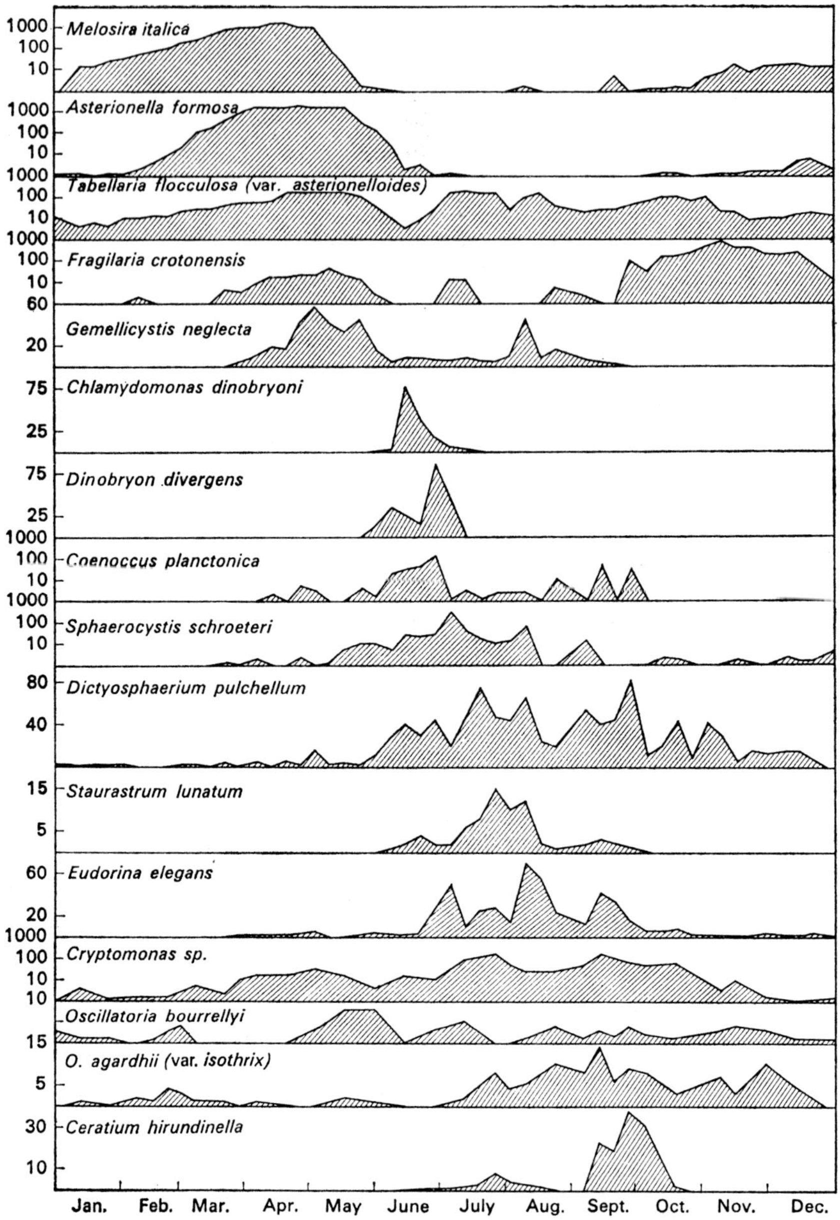

FIG 5.7. Seasonal succession of phytoplankton species in Lake Windermere, England, in 1965. Values are in number of cells or filaments per milliliter of water. Several species show generalized patterns, but most are seasonal specialists. Counts are in numbers of cells or filaments per milliliter. (From Macan, 1970, who used data of J. W. G. Lund. (T. T. Macan: *Biological Studies of the English Lakes*, New York: American Elsevier Publishing Co., Inc., 1970.)

plankton may be called summer species, because they often run counter to the rule of low summer plankton density and reach individual peaks at this time. Warming of the water aggravates the problem of sinking, since both the density and the viscosity of the medium are reduced; summer species of phytoplankton must be relatively successful at coping both with this problem and the problem of nutrient depletion. Species of the genera *Dinobryon* and *Ceratium* are frequently summer dominants; they possess flagella and can swim upward to maintain themselves in the well lit surface waters. Late in the summer, buoyant, oil-containing blue-green algae often become most prominent; the nitrogen-fixing ability of many blue-green species is probably a special advantage at this time of nutrient shortage, when nitrate and organic nitrogen supplies are meager. At the time of fall turnover diatoms usually return to prominence; and in winter, motile flagellates (different species from those of summer) often dominate the relatively sparse phytoplankton beneath the ice.

Seasonal species patterns also are evident within the zooplankton, and these often may be related to seasonal changes in the preferred food items of different zooplankton species. Sometimes, however, seasonal patterns may reflect shifting competitive advantages mediated through differences in buoyancy. For example the buoyant, jelly-enclosed crustacean *Holopedium* usually is most prominent in late summer; at this time, when sinking problems are greatest, it may be better able than most zooplankton species to remain in the eiplimnion, and hence may be relatively free from competitors for food.

For some species of zooplankton, the seasonal changes that have excited greatest attention are those of form and size rather than abundance. The summer generations of many zooplankton species contain smaller individuals than do the generations of other seasons. Reduced size may simply represent stunting caused by a sparse food supply, but it may also be a genetic adaptation which counters the reduced density and viscosity of the summer epilimnion by reducing sinking rates. Seasonal changes in size, however, are less dramatic than the seasonal changes in shape **(cyclomorphosis)** that characterize certain species of crustaceans and rotifers, as well as a few algae such as *Ceratium* (Fig. 5.8). Theories abound as to the causes and adaptive significance of these cyclic changes in morphology. One of the earliest theories saw cyclormorphosis as yet another form of adaptation to offset sinking tendencies: the helmets and spines that characterize these species in summer serve to increase the surface:volume ratio and hence the frictional resistance to sinking. A later theory holds that the summer morphology confers greater protection against predators, perhaps especially the planktivorous juvenile fishes which are abundant at this time.

Seasonality of larger organisms. Among

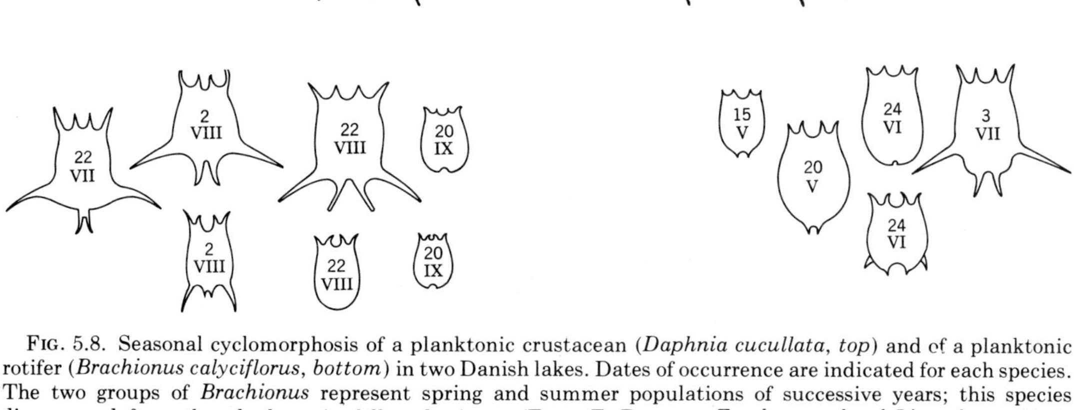

FIG. 5.8. Seasonal cyclomorphosis of a planktonic crustacean (*Daphnia cucullata, top*) and of a planktonic rotifer (*Brachionus calyciflorus, bottom*) in two Danish lakes. Dates of occurrence are indicated for each species. The two groups of *Brachionus* represent spring and summer populations of successive years; this species disappeared from the plankton in fall and winter. (From F. Ruttner: *Fundamentals of Limnology*, Ed. 3. Toronto, Ontario: University of Toronto Press, 1963; and G. E. Hutchinson: *A Treatise on Limnology, Vol. 2.* New York: John Wiley and Sons, Inc., 1967; after Wesenberg-Lund. (Hutchinson reviews this subject is great detail. Reproduced with permission.)

plankton and other short-lived organisms, seasonality often involves the actual disappearance of a species at some seasons and its predictable reappearance at others. This usually is not a matter of actual extinction and reintroduction; rather, such species pass unfavorable seasons as dormant, resistant stages (*e.g.* spores or eggs) on the bottom.

Organisms of intermediate size, such as the midge larva *Chaoborus*, may pass unfavorable seasons in a condition of arrested development. Under the photoperiodic influence of short days, *C. americanus* enters what is termed a diapause phase; during the winter it remains active but grows very little, and does not molt at all. Development is resumed in spring, triggered by long days (Bradshaw, 1972). Among still larger animals, growth and activity probably lag in winter, but the most conspicuous seasonal patterns—aside from those related to breeding—are often spatial. Many species of fish and invertebrate move to deeper water as winter approaches, and return to the shallows in warmer weather. Others, such as trout, show seasonally changing distribution patterns related to special needs. Trout require cool, well oxygenated water, and in winter and spring they find this everywhere in a typical north temperate lake, and are distributed quite generally. As summer stratification develops they retreat to the hypolimnion, but if oxygen becomes scarce there, late summer may find them concentrated within quite a narrow layer of the thermocline, this being the only part of the lake where both temperature and oxygen levels remain tolerable.

Short Term Patterns in Lakes and Ponds

Diural physicochemical changes. Superimposed on the seasonal stratification patterns of a lake are smaller diurnal changes. At the top of the epilimnion in summer, "microstratification" develops on calm days, due to absorbtion of the day's heat by water very near the surface. At night this warm layer usually cools and becomes thermally homogeneous with the rest of the epilimnion. However, a sequence of unusually hot days and nights can strengthen this secondary stratification and divide the epilimnion into two semi-independent layers. A summer temperature profile occasionally shows a double thermocline, the upper one created in the manner just described, and ordinarily much less pronounced and persistent than the lower one.

Diurnal chemical changes are associated with the metabolic processes of the biota. In layers sufficiently shallow to receive adequate light, photosynthesizing plants and phytoplankton absorb CO_2 from the water during the day and give off O_2. At night, oxygen production and CO_2 uptake cease, and respiration (which occurs both day and night) becomes the dominant process, lowering dissolved O_2 levels and raising dissolved CO_2. Thus, these dissolved gasses exhibit reciprocal diurnal patterns, O_2 rising during the day and diminishing at night and CO_2 doing the opposite. In soft waters, distinct day-night changes of pH result from CO_2 changes if the latter are pronounced. Hard waters are well buffered by the presence of bicarbonate ions, which minimize fluctuations in both CO_2 and pH by the reversible reaction $HCO_3 \rightleftharpoons CO_2 + OH^-$. In such situations bicarbonate acts as a reservoir from which additional CO_2 is withdrawn when the CO_2 in solution is being used in photosynthesis, and into which CO_2 is converted when respiration adds more of this gas to the water.

These biologically produced diurnal cycles of water chemistry are obviously more pronounced in fertile lakes, where photosynthesis and respiration are intense, than in infertile lakes, where these processes may be barely measurable. In shallow, fertile lakes and ponds, conditions sometimes arise in which poor light penetration and short-lived thermal stratification combine to bring about rapid deoxygenation of most of the water column in the course of a single day: only in a narrow surface layer is photosynthesis occurring, while beneath this layer oxygen is being exhausted by rapid respiration. If this stratification breaks down at night, the thin surface layer of oxygenated water is mixed with the nearly anoxic waters beneath, resulting in very low oxygen levels throughout the whole water column (Fig. 5.9). Fishes in such a water body can survive during the day by remaining near the surface, but may succumb to oxygen lack during the nightly turnover period.

Nonuniform distribution of plankton. We should expect the distribution of larger organisms in a lake to be patchy: different species prefer different conditions, and will be concentrated in regions favorable to them. Plankton, too, exhibit a patchiness of distribution, but a

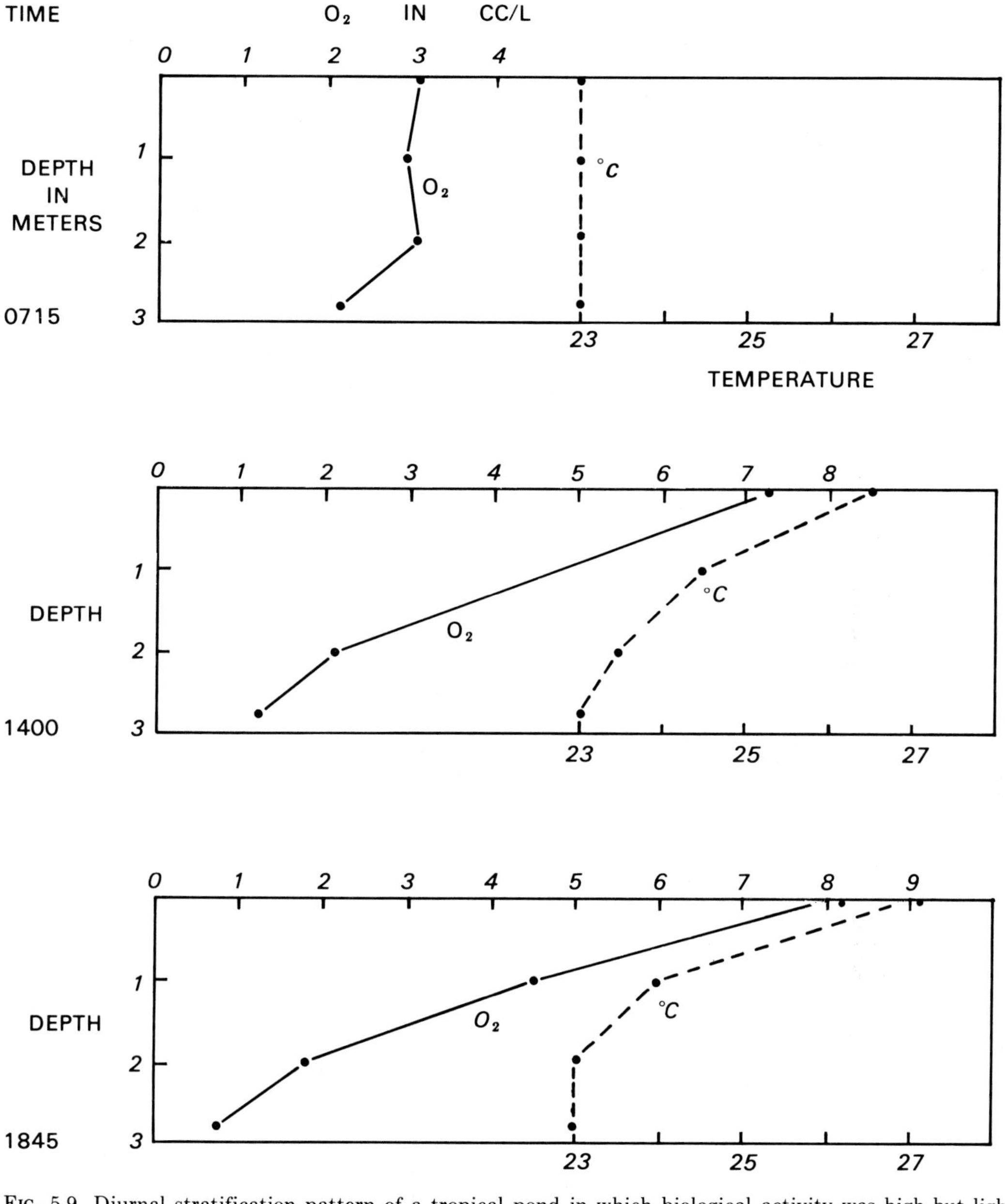

FIG. 5.9. Diurnal stratification pattern of a tropical pond in which biological activity was high but light penetration poor. The stratified condition at day's end probably will be broken down by nocturnal cooling of surface water, returning the pond to the condition shown at the beginning of the day. (Data of author.)

patchiness which often is much less permanent and predictable, being caused less by habitat preferences than by the vagaries of wind and current. One sometimes finds dense concentrations of plankton in one corner of a lake and much sparser numbers elsewhere; or the plankton may be concentrated near the surface in one part of a lake and distributed more evenly in the water column in other areas. Particularly in the case of phytoplankton, these differences may be due simply to water movements.

In the case of zooplankton, however, the phenomenon of vertical migration also produces short term vertical patchiness of a more predictable sort. Numerous species of zooplankton have sufficient powers of locomotion to permit considerable maneuvering within the water column, and many of the microcrustaceans exhibit a common tendency to rise toward the surface at night, and to return to greater depths by day. (This pattern occurs in the oceans as well.) The function of these migrations is rather mysterious. Perhaps nocturnal pilgrimages toward the surface allow these species to feed on epilimnetic phytoplankton in relative freedom from predators which hunt by sight. Descent by day would then be interpreted as retreat from lighted regions where predator risks are high. Whatever the adaptive explanation (Chapter 25 of Hutchinson (1967) contains a thorough discussion), the interesting daily migrations of zooplankton can be followed quite easily if a lake is sampled at specific depths over a 24-hour period (Fig. 5.10).

Communities and Niches of Lakes and Ponds

I would also emphasize how fantastically complicated the lacustrine microcosm is likely to be . . . It would seem likely from the several hundred species of diatoms and of insects known from certain lakes that a species list of the order of a thousand entries may not be unusual. This probably means that in the course of a season at least a thousand somewhat different ecological niches may for a time be recognizable. Most of this diversity is associated with the shallow marginal waters in which the bottom can form a solid substratum for attached aquatic plants.
G. E. Hutchinson, (1964) *The Lacustrine Microcosm Reconsidered*

The habitat divisions of a lake. Considering its three-dimensional character and its temporal variability, one is not surprised that a great many species can be accommodated in a lake. A lake of moderate depth contains at least three major types of habitat—the littoral, the pelagic and the profundal floor—and each is characterized by one or more well differentiated communities. (Compare these divisions with those of the ocean in Figure 2.19.)

The **littoral,** or shallow inshore region, extends outward to the depth at which light becomes inadequate to support the growth of plants and photosynthetic algae on the bottom. Its great biological diversity, alluded to in the quotation above, derives largely from three conditions.

1. The ability of large and small photosynthetic organisms to grow on the bottom. Not only rooted plants but a diverse assemblage of algae live on the littoral floor, resulting in a more species-rich producer level and a greater variety of food types for consumer and decomposer species than occurs in the pelagic region or on the profundal floor.

2. The many microhabitats afforded by the existence of large aquatic plants. A single water lily, for example, often supports a diverse subcommunity. Invertebrates (Fig. 5.11) are only part of this subcommunity; the leaf and stem also support a slimy film of microorganisms, the algae and protozoa that comprise the **aufwuchs.** A German term with no exact equivalent in English, aufwuchs includes not only epiphytic microorganisms, but similar growths covering rocks and other hard substrates.

3. The spatial heterogeneity of the shallow water habitat. Shores may be very exposed, very sheltered, or somewhere in between; and substrates may, in consequence, range from large rocks to sand or fine ooze. Each substrate type, in turn, provides a variety of

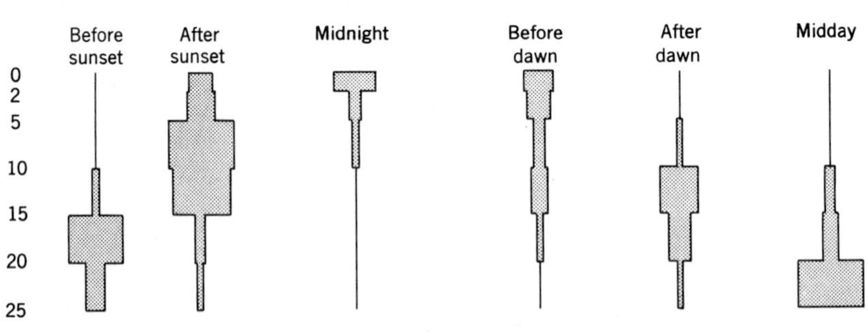

FIG. 5.10. Vertical migration of a population of young *Daphnia lonjispina* in Lough Derg, Ireland, during an 18-hour period. The width of the bar indicates the number of individuals found at each depth. (From G. E. Hutchinson: *A Treatise on Linnology*, Vol. 2. New York, John Wiley and Sons, Inc.; after Southern and Gardiner.)

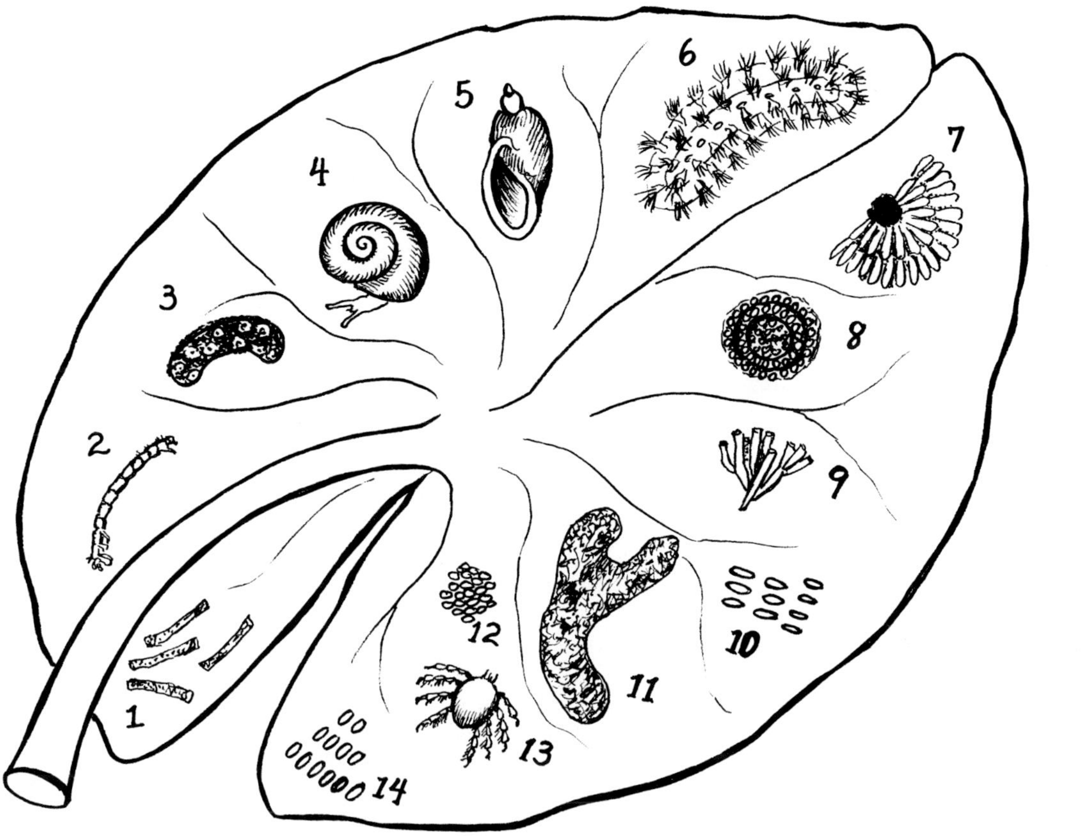

FIG. 5.11. Typical components of the microcommunity of a lily pad: (*1* and *2*) midge larva and larval tubes; (*3*, *4*, and *5*) snails (*Graulus* and *Physa*) and egg mass; (*6*) colony of bryozoans; (*7*) eggs of beetle (*Donacia*); (*8*), eggs of caddisfly (*Triaenodes*); (*9*) tube-dwelling rotifers (*Melicerta*); (*10*) eggs of damselfly (*Enallagma*); (*11*) fresh-water sponge (*Spongilla*); (*12* and *13*) water-mite and eggs; (*14*) eggs of whirligig beetle. (Redrawn from A. H. Morgan: *Field Book of Ponds and Streams*. New York: G. P. Putnam's Sons, 1930, with permission.)

microhabitats (Fig. 5.12). In addition, spatial heterogeneity is introduced by the increasing depth encountered with increasing distance from shore. This depth-related heterogeneity is clearly reflected in the depth zonation of aquatic plants that often characterizes littoral areas. Restricted to the shallows are the emergents, only the lower portions of which are submerged. Further out is a zone in which plants with floating leaves are concentrated; and deeper yet, extending to the edge of the littoral, is a zone of fully submerged plants. A similar zonation, related to the depth-determined light gradient on the littoral floor, characterizes bottom-living algae, and species separation within the aufwuchs coating a water lily stem also may occur with depth. Gradients of wave action also occur with depth, and contribute to the heterogeneity of the littoral region.

The **pelagic region** is the region of true plankton and open water fishes. It is considerably less diverse than is the littoral, yet we have seen how sharply it sometimes is differentiated into environmentally distinct water masses. Although these water masses do not themselves mix, many pelagic organisms migrate freely between them. In fact, the most fundamental division of the pelagic habitat from a biological standpoint may not be the division into epilimnion and hypolimnion, but into a **euphotic zone** where photosynthetic organisms can flourish and a **dysphotic zone** where they cannot. The boundary between euphotic and dysphotic zones lies at the depth of the lower edge of the littoral region. This depth depends on the transparency of the water, and does not coincide in any consistent way with the boundaries of the epilimnion, thermocline or hypolimnion. In Crater

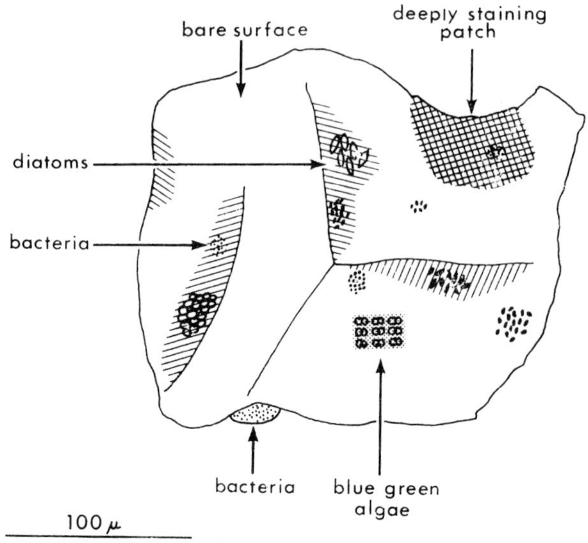

FIG. 5.12. Localized distribution of microorganisms on a single sand grain. Some species occur only in such sand-grain microhabitats. (From P. S. Meadows and J. G. Anderson: *Nature, 212:* 1059–1060, 1966. Reproduced with permission.)

Lake, Oregon, the euphotic/dysphotic boundary lies 100 m or more below the surface and is far beneath the thermocline (Larson, 1972); in Lake Lacawac, Pennsylvania, the boundary is at about 6-m depth, *i.e.* within the late summer thermocline (Fig. 5.3); while in lakes made turbid by mud or dense phytoplankton growth the euphotic zone may extend less than 1 m below the surface.

Lying beyond the lower edge of the littoral zone, the **profundal floor** is the dysphotic fraction of the lake bottom. The community of the profundal floor, like that of the oceanic abyssal floor, is completely heterotrophic except for dormant phytoplankton which may be resting on the bottom. This community includes bottom-dwelling species of fish and the micro- and macroorganisms that inhabit the sediments. Profundal sediments are more homogeneous than those of the littoral region, typically being soft and finely textured, and the species inhabiting these sediments are far fewer.

Strictly speaking, these habitat divisions—littoral, pelagic and profundal floor, together with euphotic and dysphotic zones—are not distinguishable in a pond; for a limnologist defines "pond" as a body of standing water that is completely littoral in nature.

Niche separation. The pelagic region is an environment in continuous flux. Short term as well as seasonal changes in water chemistry, temperature, turbulence and other factors ensure that the pelagic community can be an equilibrium community only in a very dynamic sense. This must be appreciated if we are to understand the seeming defiance of the principle of competitive exclusion which Hutchinson (1961) has termed "the paradox of the plankton;" for in the pelagic region we typically find a plankton of numerous species, seemingly with similar basic requirements and coexisting in a superficially homogeneous environment. Of course this environment is *not* homogeneous, particularly in the time dimension, and instead of only a few planktonic species prevailing in the competitive arena, there is a constant shifting of competitive advantage so that many types persist, though with fluctuating fortunes and seasonal optima. All planktonic diatoms need dissolved silica, for example, but Kilham (1971) has shown that diatom species differ in how much dissolved silica they need for maximum growth; thus, as the epilimnion becomes progressively depleted of silica during the summer, a progression of diatom species may be expected. Diatoms and other phytoplankton also differ in light requirements; "sun" and "shade" species exist in water, just as in a forest. Different species thus possess advantages at different levels in the water column. In vertically mixing waters, sun and shade species are carried up and down together; there is no real spatial separation between them, but first one species, then another uses the available light most effectively. When the water column becomes stratified, spatial separation of sun and shade phytoplankton may become a reality; in Figure 5.3, for example, the oxygen peaks within the thermocline are probably the work exclusively of shade-tolerant phytoplankton, perhaps belonging to but a single species which, because of its density characteristics, has been concentrated at a particular level in the thermocline's steep density gradient.

Unstable populations of numerous phytoplankton species make possible a diversity of herbivorous zooplankton and indirectly promote variety higher in the pelagic food chain. Perhaps even more than phytoplankton, zooplankton tend to partition the water column vertically; even those that migrate come to rest at different depths, and so reduce the problem of competition.

Although the environment of the profundal floor is more stable, here too there is considerable opportunity for niche specialization. For example, species of bottom-dwelling midge larvae differ in the degree and duration of anoxia that they can tolerate; this promotes a separation of species by depth, and restricts those least tolerant of deoxygenation to the littoral region. Within the littoral itself, our discussion of habitat diversity already has suggested many possible modes of niche separation.

Fishes and other long-lived aquatic organisms exhibit certain forms of niche separation not yet described, and it is worth considering briefly how fishes partition the environment. They differ, of course, in habitat preference and also in overall size, shape, mouth form and other morphological features (Fig. 5.13). These morphological differences are related to sometimes subtle differences in diet and method of hunting; and there are dietary specialists and dietary generalists among fish species, just as in other groups. Varied morphology and behavior, together with habitat differences, thus effectively restrict interspecific competition. Furthermore, like most long-lived animals, adult fishes and their fry differ in diet and also often in habitat, thereby minimizing competition at the *intraspecific* level. Juveniles of different species sometimes seem to display a more clear-cut niche separation than do the adults (Fig. 5.14). Possibly it is more important that niche overlap be minimized among the young of these species than among the adults, which exist in lower numbers and perhaps can share the same resources without exhausting them.

Lake Types, Long Term Succession, and Eutrophication

Minnesota and New England contain more natural lakes than do all the tropical and warm-temperate regions of the world. Why? Because the vast majority of the world's lakes were created by glacial processes. One of the surest signs of once glaciated landscapes is an abundance of lakes.

Of course, lakes can be created by many other agencies: the damming of valleys by landslides or lava flows (Mountain Lake, Virginia); the cratering or collapse of volcanoes (Crater Lake, Oregon and many smaller examples); major earth movements (Lake Tahoe and Reelfoot Lake in the United States, as well as Siberia's Lake Baikal and the large lakes of East Africa); meteorite showers (believed by some to be responsible for the Carolina "bay lakes"); solution of limestone bedrock; and the isolation of river meanders to create oxbow lakes.

Glacial lakes alone were formed in a variety of ways: by the scouring of shallow basins in bedrock (most Minnesota lakes); by glacial deepening of pre-existing valleys (the large lakes of Switzerland, as well as the Finger Lakes of New York and several of the Great Lakes); by melting of ice blocks embedded in moraines (the "kettle hole" lakes of Cape Cod); and in mountain regions, by morainal damming of valleys or ice-gouging within glacial cirques.

But in recent centuries, the most active agent of lake and pond formation has been not glaciation but man. In striking contrast to forests, salt marshes and certain other beleagured habitat types, lakes and ponds are becoming not less but more abundant at the hand of man. A certain type of lake—the clear, crystalline, oligotrophic type together with its pure water inhabitants—is admittedly suffering through human activities, but no one can claim that lakes and ponds are dwindling in absolute number.

If most of the world's lakes and ponds were created either by glaciers or by man, it follows

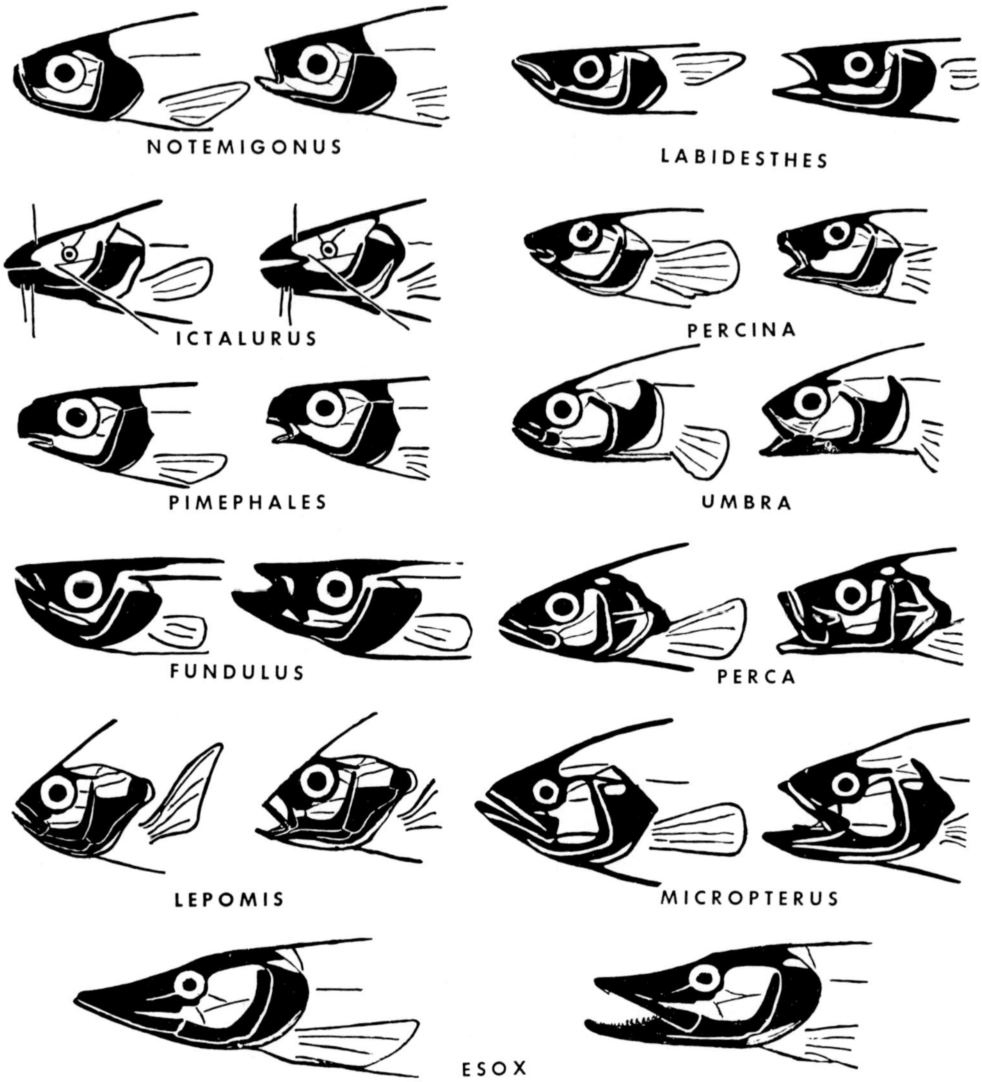

FIG. 5.13. Diversity of mouth form among fishes reflects different foods and modes of feeding. For example, *Notemigonus* is a scoop-mouthed mid-water and surface feeder on zooplankton and small insects. *Ictalurus* feeds mainly on midge larvae and small molluscs in the sediment. *Pimephales*, a smaller bottom-feeder, subsists on organic ooze and small invertebrates. *Lepomis* is an extremely generalized predator, primarily of insects. *Esox*, a large predator, captures other fish. All these genera were found to coexist in one small lake in Ontario. (From A. Keast and D. Webb: *Journal of the Fisheries Research Board of Canada, 23:* 1845–1874, 1966. Reproduced with permission.)

that they are geologically recent features of the landscape. In fact, any lake is a transitory phenomenon since it serves as a catchment basin for sediment which eventually fills it up. Some shallow glacial basins have been through the whole process already; having begun as lakes or ponds, they have progressed to shallow bogs or marshes and have finally become moist forests. Deeper glacial basins are in earlier stages of this process, a form of primary succession that is termed **hydrarch succession.** We know that the same filling-in process occurs in reservoirs, often much more rapidly than in natural lakes and ponds. The rate of filling depends both on the rate of influx to the basin of terrestrial sediment and on the fertility of the water, which affects the rate at which organic sediment is produced within the lake.

Classic limnologic theory associates the process of hydrarch succession with that of **eutrophication,** or progressive nutrient enrichment. The most common use of this term today is as a synonym for water pollution, more specifically those forms of pollution that introduce mineral or organic nutrients into a water body and fertilize it to undesired richness. But as initially developed, the term eutrophication referred to the supposed progression of a lake from oligotrophic to eutrophic conditions. This was believed to be a natural part of the process of hydrarch succession. According to this theory, a lake basin that is successionally young is one that has not yet progressed far toward a filled-in condition. Typically, the lake in such a basin, if unpolluted, is relatively deep, clear and infertile—in short, oligotrophic. Oligotrophic lakes vary in absolute depth, of course, depending on the depth of the basin they occupy. Clarity is partly a consequence of depth, but chiefly of infertility. An oligotrophic lake supports a relatively meager plankton crop (dense plankton is a cause of turbidity in more fertile lakes), and even the fishes are few in number compared to the fishes of later successional stages. Because organisms are few, an oligotrophic lake does not become markedly stratified chemically even when it stratifies thermally. Biological activity is not sufficiently intense to deplete the epilimnion of nutrients and the hypolimnion of oxygen. But why should successionally young lakes be infertile? partly, says theory, because the nutrients being supplied by inflow are meager relative to the large volume of the young lake. In addition, the depth of the water column may make mixing of nutrients into surface waters difficult even during periods of turnover.

As its basin fills with sediment, a lake diminishes in both volume and depth, and for the reverse of the reasons suggested above it may tend to increase in fertility and to support more life. This is the theoretical process of natural eutrophication. A eutrophic lake thus is thought to represent a relatively mature stage of devel-

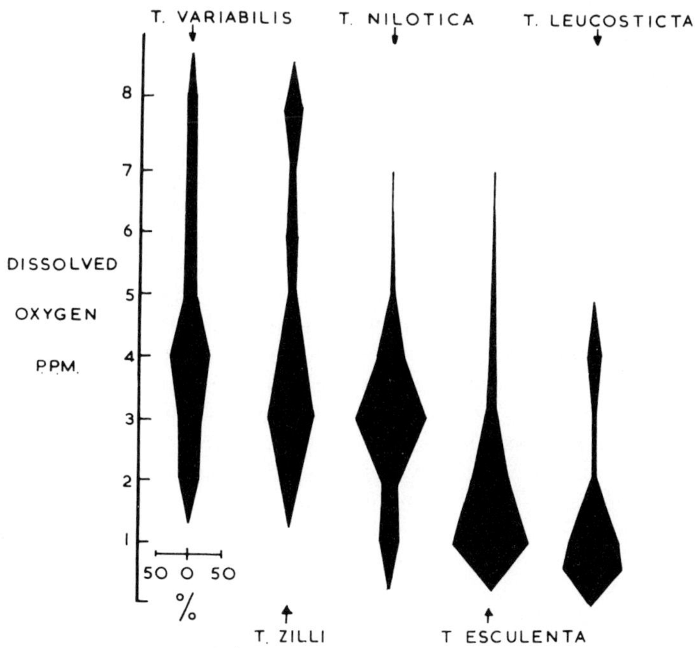

FIG. 5.14. "Species flocks" of closely related fish species pose a challenge to the Principle of Competitive Exclusion in several tropical lakes. More than 100 species belonging to one genus occur in Lake Malawi, Africa, for example. In the case of a smaller species flock, the Tilapias of Lake Victoria, careful studies of the juveniles reveal that the different species are segregated within different parts of their shoreline habitats by differing tolerances to high temperature and to low oxygen (shown here). The shallowest, most protected shoreline areas experience these factors at their most extreme. (From R. L. Welcomme: *Revue de Zoologie et de Botanique Africaines* 70:1–28, 1964. Reproduced with permission.)

opment; typically it has accumulated considerable depths of sediment, has lost some of its earlier clarity, and during the summer develops considerable nutrient deficits in the epilimnion and oxygen deficits in the hypolimnion. Judging from Figures 5.3 and 5.5, Lake Lacawac has reached the eutrophic stage, even though it remains completely unpolluted.

Not all limnologists today believe in the theory of natural, successional eutrophication. They suggest instead that the fertility of a lake is a function largely of the nutrient supply from outside, and that the shallowing of a lake during hydrarch succession is not in itself a major enriching influence. By this reasoning an oligotrophic lake is oligotrophic simply because it exists in a nutrient-poor drainage basin, and it will remain infertile throughout the successional process unless artifically enriched by man. Similarly, a naturally eutrophic lake like Lacawac may never have been oligotrophic. In essence, the proponents of the classic theory claim that the successional process in a lake is self-enriching; the sceptics doubt that this is so.

No one doubts the reality of *artificial* eutrophication, however. Introduction by man of growth-stimulating substances often increases a lake's fertility far beyond its natural levels. Lakes in an extreme condition of eutrophy sometimes are termed hypereutrophic; many examples now exist throughout the world, from the green temple ponds of India in which clothes are laundered to Sylvan Lake in Indiana and Lake Monoma in Wisconsin, both recipients of municipal sewage.* In fact, if we examine nearly any lake whose drainage basin harbors many humans, be they urban or agricultural, we see at least some degree of artificial eutrophication. The specific nutrients responsible for eutrophication, as well as the possibility of reversing it, will receive further attention in Chapter 10. We may consider here, however, some of the biological consequences of this process.

Biological changes during eutrophication involve not just numbers but also kinds of organisms. Oligotrophic lakes support a quite different flora and fauna than do eutrophic lakes. Table 5.3 shows the change in fish types that has accompanied accelerated eutrophication in Lake Erie; equally marked changes have occurred in the plankton and bottom organisms. Contrary to the claims of the popular press, Lake Erie is not "dead;" it is simply changed biologically.

The species changes that occur during eutrophication do so for much the same basic reason as they do during forest succession. The environment becomes modified, in important part by the inhabitants themselves, so that it is less suitable for them than for other species. As eutrophication proceeds, planktonic species more adept at monopolizing the larger nutrient supply probably outcompete species possessing lower intrinsic growth rates which suit them to the low nutrient levels of oligotrophy. Explanations in terms of competition are only part of the answer, however. Often the changed physicochemical environment simply becomes unliveable for oligotrophic species. My colleague Ken John found this to be the most likely explanation for the decline of cisco (a species of whitefish, related to trout) in Lake Mendota, Wisconsin. Cisco once were so abundant in Mendota that residents of Madison organized cisco-dipping parties (complete with beer and bratwurst) to catch the fish when they entered shallow water in late autumn to spawn. By the 1950s, cisco were no longer common in the lake, though fair numbers sometimes were found dead along the shore in late summer. Gill-netting studies by John and other students at the University of Wisconsin's limnological laboratory revealed the probable reason. In winter, cisco were netted throughout the 30-m depth range of the lake, but in late summer the whole population was squeezed into the middle part of the thermocline, where the water was both fairly cool and fairly well oxygenated. A fish of clear northern lakes, the cisco requires both low temperatures and abundant oxygen. Eutrophication, linked to the growth of Madison, had made the hypolimnion untenable as a summer retreat for this fish; now, with increasing fertility, summer deoxygenation was spreading upward through the thermocline, gradually eliminating the cisco's last refuge at this season.

STREAMS AND RIVERS

The environment of a small, swift-running stream is very different from that of a lake or

* Naturally hypereutrophic lakes also exist, examples being the green flamingo lakes of East Africa. The term is not simply a synonym for overpolluted lakes, though most hypereutrophic lakes have achieved their state via pollution.

TABLE 5.3

Changes in the Commercial Fish Fauna of Lake Erie, 1925–1965

Information taken from Beeton (1965).

Species	Period of Decline	Notes
A. Species formerly important in the lake but now uncommon:		
Cisco	1925 to present	Former catch averaged 9 million kg/year; catch in 1962 was 3,200 kg.
Sauger	Since before 1945	Former catch was 500,000 kg/year or more; catch in early 1960s was down to about 1,000 kg.
Whitefish	Since before 1948	Former catch was 1 million kg/year or more; catch in 1962 was 6,000 kg.
Blue pike	Since mid-1950s	Former catch averaged 6.8 million kg/year; rapid decrease to 450 kg in 1962.
Walleye	Since 1956	Increased during 1940s and 1950s to a catch of 7 million kg in 1956; decreased in early 1960s to less than 450,000 kg.
B. Species now more common in the lake: Freshwater drum, Carp, Yellow Perch, Smelt		Total catch from Lake Erie remains as high as formerly (about 23 million kg/year); Increased catches of fishes in this group have offset declining catches of fishes in group A.

pond; the environment of a slow-moving river is much less different. Unlike the swift stream, a large river like the Ohio may stratify thermally and may even contain many typical lacustrine species. Within a continuous stream-river system, therefore, one encounters marked ecological contrasts; and perhaps the best introduction to the ecology of running waters is a chronicle of the changes that occur between the source of a typical stream and its terminus in the ocean.

Journey to the Sea

Our typical stream rises in hilly terrain from cool groundwater seepages. In its headwater stretches it flows rapidly for the most part, down a steep, wooded valley; occasionally it passes through open meadows. Because of its turbulent flow it is never stratified, and it seldom freezes even in midwinter, except along the edges. Oxygen is always abundant because the shallow, mixing waters present a large and constantly renewed surface for diffusive exhange with the atmosphere. Carbon dioxide, too, remains near saturation levels at all times; dissolved minerals vary somewhat more over time, depending on the changing proportions of groundwater and surface runoff that make up the flow. Because it flows mainly in shadows and receives continual infusions of cool groundwater, this headwater stream remains cool year-round; but the meadow stretches are more variable in temperature than the wooded stretches. In the meadows the stream warms rapidly by day and cools quickly at night, owing to its exposure to sun and sky, its small volume, and its intimate interaction with atmosphere and substrate.

Gradient, and consequently current speed, tend to decrease progressively downstream, but the trend is not a smooth one. In the stream's upper reaches there are rushes and pauses, rapids and pools; and well downstream, the now lazy river may make a final rush across a fall line as it descends from piedmont to coastal plain. The stream-turned-river traverses flatter country, usually, and is generally warmer in summer than it was upstream, though remaining cooler in summer than the lands on either side. Except at the fall line, the rocky, current-scoured substrate of the headwater rapids has now become sand or soft silt. But the current remains strong enough to carry the finest sediments in suspension; light penetration consequently is limited, and aquatic plants grow only in shallow water. In its deepest and slowest stretches the river may stratify thermally in summer and the oxygen content of its dark lower waters may fall. But these stratification phenomena are seldom as pronounced as in a lake, and they are dissipated by turbulent mixing wherever the flow quickens.

A continuum of intergrading communities occupies this changing aquatic habitat. The headwaters harbor brook trout, darters, and the nymphs of stoneflies and certain mayflies, together with other organisms which share a common need for constantly cool temperatures, constantly high oxygen, or both. Many of these organisms are conspicuously adapted, morphologically or behaviorally, in ways that equip them to combat or escape from the swift current. Also found here are animals whose way of life *requires* moving water; filter feeders such as the caddis larva *Hydropsyche* and the black-fly larva *Simulium*, which strain their food from the passing flow. In headwater pools the fauna is somewhat different. The bottom is often silt-covered and harbors clams and burrowing species of mayfly and midge; above them cruise suckers and dace.

Plants are unimportant in our shaded headwater stream, but dead leaves from the overhanging trees are stacked in masses against projecting rocks and logs, and carpet the bottoms of quieter pools. In the meadow reaches algae more thickly coat the submerged rocks and water plants are more abundant. But aside from dislodged aufwuchs organisms, no true suspended plankton exists in these headwaters. Most of the headwater community, in fact, consists of bottom-dwellers: organisms that cement themselves to rocks or hide beneath them, or that live in masses of decaying leaves.

Farther downstream a true plankton starts to develop. The current now is less, and the water mass is sufficiently old for colonizing planktonic species to have multiplied. Animals that are found also in lakes begin to appear: dragonfly and damselfly nymphs, mud-dwelling oligochaetes and perhaps sunfish and bass. For the most part these animals are ill adapted to the swift headwaters but are able to live here. Still further downstream, catfish and carp cruise slowly through the oxygen-poor bottom waters in search of worms and midge larvae. The community inhabiting this broad, sluggish river is as dissimilar to the community of the headwaters as it is to the marine communities that lie ahead.

Current and Substrate: Interlinked Determinants of Stream Communities

Differences in current and in substrate are the primary agents of community zonation in a flowing water system. Other important determinants, such as changing conditions of oxygen or of turbidity, often are clearly a function of changing current speed and current-related turbulence. The nature of the bottom, too, is partly a function of current speed. Rapids and riffles are erosional habitats, their rocky bottoms scoured by swift water; pools and the lower reaches of sluggish rivers are depositional habitats, where slackening currents drop some of their suspended load. Thus if we seek the most fundamental cause of biological differences within a stream, it must be current.

One often is uncertain, however, whether the influence of current is direct or indirect. Trout, for example, probably are excluded from the downstream reaches of many streams by low oxygen levels, elevated turbidity, high temperature, or some combination of these. Trout can live perfectly well in still water if it is clear, cool, and well oxygenated. Thus the influence of current on the distribution of trout in a stream would seem to be primarily indirect. But there is probably a direct component as well; current constantly renews the supply of oxygen in the vicinity of the trout's gills, preventing the formation of a stagnant, oxygen-depleted film of water on the respiratory surfaces of the fish when it is not moving, and for this reason the actual concentration of dissolved oxygen that a trout requires in a stream is probably less than the concentration required in a lake.

The situation is even more complicated when, in the case of bottom organisms, one tries to separate the direct influence of current from the influence of substrate. Even in the swiftest headwater reaches current varies greatly, not just along the course of the stream but also laterally and vertically at a single place (Fig. 5.15). At the bottom, friction between water and substrate slows the flow considerably, and beneath rocks or in eddies the force of the current may be negligible. What, then, determines which organisms live in torrential habi-

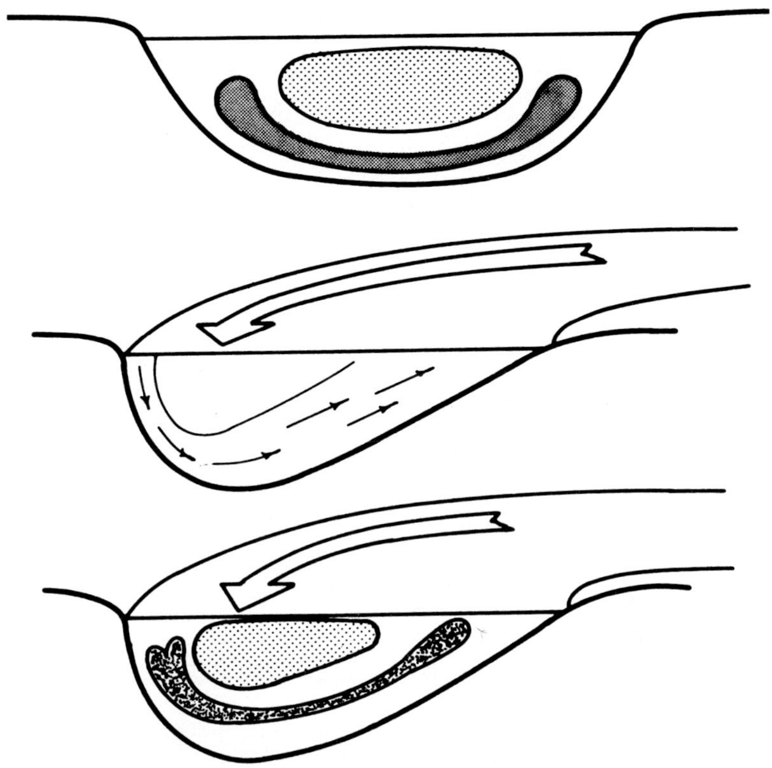

FIG. 5.15. Patterns of current variation at two points in a stream. (*Top*) In a straight stretch the swiftest current is just below the surface in the center of the stream. Below this fast "core" is a layer of turbulence. Relatively quiet water lies immediately adjacent to the bottom and along stream edges. (*Middle* and *bottom*) At a curve the swift core and turbulent layer are assymetrically displaced. Active erosion occurs at one side of the stream and cross-eddies deposit fine material in quiet shallows on the other side. From W. H. Amos: *Limnology: An Introduction to the Fresh Water Environment*. Chestertown, Md: LaMotte Chemical Products Co., 1969.

tats? Are torrential species those having unusual abilities to withstand or escape from strong currents? or are they simply those having an affinity for the clean, rocky substrates of swift water? It is not easy to answer these questions by observation, but they can be investigated by fairly simple methods (Fig. 5.16).

Regardless of which acts more directly, current and substrate together conspire to provide a remarkable range of microhabitat conditions within a single macrohabitat such as a short stretch of rapids. The bottom fauna is usually much more varied and abundant in a stretch of rapids than in an adjoining pool, because in the rapids the range of current and variety of substrates is much greater. Perhaps, too, it has been easier for many groups of bottom organisms to evolve adaptations to the erosional habitat of rapids than to the depositional habitat of pools, where finer sediments accumulate and shift unpredictably, posing to the inhabitants a constant threat of burial.

In rapids, the distribution of invertebrate types on and under a single stone is worth careful study, as are the adaptations correlated with preference for particular habitats (Fig. 5.17). Insect species that expose themselves to the current often possess ventral adhesive devices, or cement themselves in stone cases to the rocks (the cases probably serve to protect their possessors as much against predators as against the current). Other species are markedly flattened, and as they move across a rock they present little resistance to the current; but many such species are typically found in crevices or beneath rocks, suggesting that flattening is primarily an adaptation to escape the current rather than to withstand it. Behaviorally, too, many stream invertebrates seem to follow a strategy of escaping the current: a majority of

FIG. 5.16. Colonizing sampler for stream invertebrates that tests the relative importance of current speed and substrate type as habitat factors. Trays containing stony substrates of different size (*bottom*) are placed within the chute and are colonized by drifting invertebrates. After several days the trays are examined and species distributions are noted. Current can then be changed (by adjusting movable vanes at front of apparatus) and the process repeated. (Designed by J. M. Vasievich for an undergraduate project.)

species are negatively phototactic, which leads them—at least during the day—to seek dark places such as the undersides of rocks. But here again we may ask what is the primary function of this trait: has photonegativity evolved primarily as a response that keeps these species out of the current, or as one that keeps them out of sight of predators?

Systematic or quantitative sampling of a habitat as heterogeneous as the rapids of a stream is a very difficult task. Even when one uses standardized collecting equipment like the Surber square-foot sampler one can easily gain false ideas of the relative abundance of different species, or even of the general abundance of all organisms. Not only are some microhabitats more difficult to sample than others, but some species are much more easily dislodged and collected than others. At the least, a collector should not simply turn over rocks and be satisfied with what washes into the net after such disturbance; he will find much more if in addition he picks up individual rocks and, with forceps, picks off what still clings to them.

It is difficult also to obtain measurements of current speed that are real as far as the bottom organisms are concerned. One would like to measure the currents to which bottom dwellers actually are subjected in their microhabitats. But the simplest measure of current speed—determinations of the rate at which a floating object moves downstream—provides only the roughest index of the currents of bottom microhabitats. Relatively simple devices can be bought or made that do measure flow near the bottom, and while these are a great improvement over measurement of surface flow, none are as precise as one might like. The problems here are akin to those of obtaining useful micrometeorological measurements in a forest.

Environmental conditions in pools and in the quiet stretches of rivers are less heterogeneous than in rapids, and the problems of sampling and environmental measurement are fewer. But here one must pay heed to subtle gradations of substrate size; to a burrowing species, the difference between sand and silt is vast. Substrates finer than gravel can be characterized quite accurately by passing them through a series of geologist's sieves of graduated mesh size.

Zonation of Polluted Streams

The communities and conditions we have been describing are those of relatively clean

FIG. 5.17. Characteristic invertebrates of the rapids of a cool, well-oxygenated stream. (*1*) *Synagapetus* (Trichoptera), larva in a case of sand grains; (*2*) *Ancylus* (Gastropoda); (*3*) *Liponeura* (Diptera): (*a*) pupa, (*b*) larva; (*4*) *Helmis* (Coleoptera): (*a*) larva, (*b*) imago; (*5*) *Silo* (Trichoptera), larva; (*6*) *Simulium* (Diptera): (*a*) pupa, (*b*) larva; (*7*) *Baëtis* (Ephemeroptera), nymph; (*8*) *Rhithrogena* (Ephemeroptera), nymph; (*9*) *Perlodes* (Plecoptera), nymph; (*10*) *Rhyacophila* (Trichoptera): (*a*) pupal case, (*b*) larva; (*11*) *Epeorus* (Ephemeroptera), nymph; (*12*) *Planaria alpina* (Turbellaria); (*13*) *Philopotamus* (Trichoptera), larva in its catching net. (From F. Ruttner: *Fundamentals of Limnology*, Ed. 3. Toronto, Ontario: University of Toronto Press, 1963. Reproduced with permission.)

streams. Some forms of industrial pollution poison virtually all life in a stream, but pollution by organic wastes exerts a selective effect, producing a series of highly altered communities that are as interesting to study—if not as pleasant—as the communities of unpolluted streams.

Organic sewage that enters a stream represents not poison, but an enormous influx of nutrients. For some distance below the point of addition, the stream harbors greatly enlarged populations of organisms that thrive on this rich fare and can tolerate the chemical effects of their own metabolic exuberance. But eventually the extra nutrients are exhausted and clean water communities reappear. Organic pollution of streams is thus a somewhat localized phenomenon, and the water remains polluted only if further doses of sewage, entering at intervals downstream, keep it heavily enriched in nutrients. At least potentially, pollution of streams by organic sewage is a far less serious problem than pollution by toxic industrial effluents. No biological cleaning process eliminates the latter, and if life returns downstream it is only as a result of dilution of the toxins. Such water remains dangerous for human consumption as well, whereas water polluted only by biological wastes can, when sterilized, safely be used again by humans.

The zoned communities of a stream that occur downstream from a source of organic pollution are a striking illustration of the organismic concept of communities. Each community clearly recognizable as an entity, and the sequence of communities that one finds is highly predictable, as are the organisms in each community. But the individualistic concept also is well illustrated: species overlap between successive communities, and their pattern of distribution is a classic continuum (Fig. 5.18). The gradients that determine this continuum are gradients primarily of oxygen and of nutrients. Oxygen levels are lowest not immediately below the point of sewage entry but farther downstream, since oxygen is consumed less by the sewage itself than by the respiratory metabolism of organisms multiplying in the rich nutrient supply. Sewage fungi (genus *Sphaerotilus*), coating harder substrates with a white slime, are the first conspicuous organisms downstream from a major sewage source. In the same stretch, hemoglobin-rich Tubificid worms protrude in multitudes from the softer sediments, sometimes appearing as an undulating red carpet. Further downstream a green mantle of the branched alga *Cladophora* replaces *Sphaerotilus* on the rocks, and the communities become more diversified with the addition of midge larvae, isopods and perhaps carp. Gradually, in this zone of recovery, such species lose their predominance and piece by piece the diverse clean water community is reconstituted. The striking lack of diversity in the pollution communities of streams will receive comment in the next chapter; low diversity also characterizes many communities unaffected by man if the natural physicochemical environment is extreme.

Because both chemical trends and biological changes are so dramatic, the pollution-recovery sequence in a stream is a classic ecological lesson in much more than the lurid, "isn't-it-awful" sense.

Seasonal Patterns in Streams

The most marked seasonal changes in lakes and ponds are those brought about by thermal stratification. But small streams do not stratify thermally or chemically, and their temperatures change relatively little throughout the year. Perhaps in large, slow-moving rivers we can find somewhat muted seasonal changes similar to those of lakes, but should we expect to find pronounced seasonal changes in a small headwater stream?

If we say no, we are thinking too narrowly. Seasonality actually is pronounced in many small streams, but is of a type quite different from that in lakes. Seasonality in streams is primarily a consequence of the seasonality in adjacent terrestrial ecosystems; in this sense, a stream is less an independent entity than is a lake, and more part of a combined aquatic-terrestrial system. Minshall (1969) notes, for example, that particular biomes are not only terrestrially distinct, but are characterized by particular types of community in the streams flowing through them. Stream communities in coniferous forest regions are very different from stream communities of desert or grassland regions, and quite distinct even from those of deciduous forest.

Why should this be? To illustrate, let us consider a small stream flowing through a deciduous forest. Consumers and decomposers

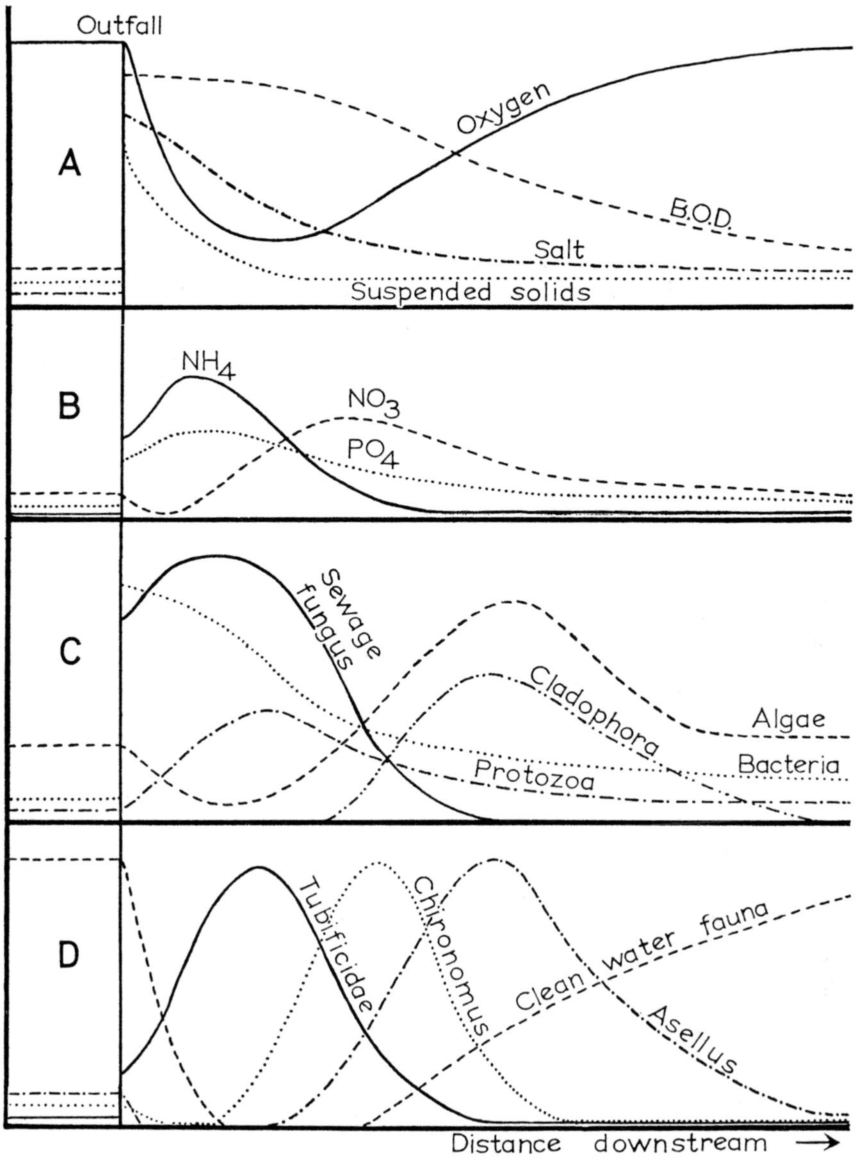

FIG. 5.18. Degradation and recovery of a stream below a sewage outfall or other organic effluent. (*A*) and (*B*) show physical and chemical changes. (BOD, the biological oxygen demand, is an index of abundance of aerobic bacteria and of the nutrients needed to sustain them.) (*C*) and (*D*) show patterns of characteristic microorganisms and invertebrates, respectively. (From H. B. N. Hynes: *The Biology of Polluted Water*, Liverpool, England: Liverpool University Press, 1960. Reproduced with permission.)

dominate the community of such a stream. Some algae are present, but such producer organisms are not really abundant; in summer, shade from overhanging trees limits the light available for photosynthesis by aquatic producers, and even in winter the stream is shaded during much of the day by the sloping banks on each side. The communities of a small forest stream are, in fact, heterotrophic communities, similar in this respect to the communities we have described in the aeolian zone, the oceanic abyss, and the profundal regions of lakes. In these stream communities, too, the food base is primarily organic matter imported from elsewhere, and in a stream flowing through deciduous forest there is a decided seasonality to the

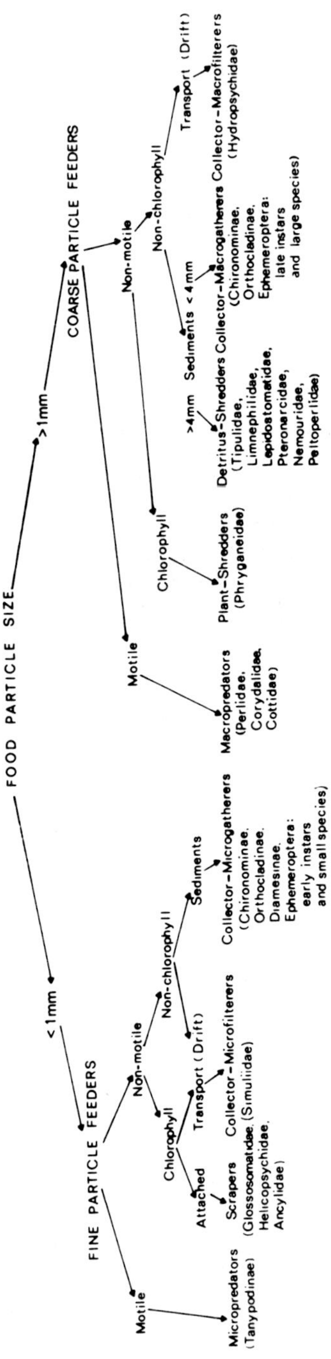

FIG. 5.19. Feeding categories of insects and fishes in a small forest stream. Fine particle foods include tiny animals, algae, bacteria, fungi, and fine organic sediment (often the feces of other organisms). Coarse particle foods include larger animals, green plants, and dead leaf matter ranging in size from whole leaves to small fragments. The variety of food types and feeding modes is remarkable. (From K. W. Cummins: *Bioscience* 24: 631-641, 1974.)

size of these imports. A great pulse of organic matter arrives in late fall, in the form of dead leaves; Vannote (1969) measured yearly leaf input into three Pennsylvania streams at between 200 and 800 g/m^2, and most of this input came in the fall. The leaves tend to lodge in masses against obstructions, become waterlogged, and remain as a food source until well decomposed in spring.

These natural leaf packs are colonized by a variety of aquatic organisms including bacteria, fungi and many types of insect larvae (Fig. 5.19). Carbon and nitrogen leachates from the leaves also promote rapid bacterial growth in the flowing water itself (Wetzel and Manny, 1972). Within the leaf packs, species succession of insects and preferences for particular leaf types are quite easily studied. Investigators at the Stroud Water Research Laboratory in Pennsylvania prepare a number of fresh leaf packs in the fall of both single and mixed leaf types, submerge them in a stream for varying lengths of time, and then retrieve them for study of their populations and degree of decomposition. (The packs are prepared either by sewing leaves together or by encasing them between pieces of wire mesh.) These studies have shown that besides microorganisms, three trophically different groups of insects invade leaf masses (Vannot, 1969). A herbivorous group, comprising cranefly larvae and certain species of larval caddis fly, mayfly and stonefly, colonize new leaf masses rapidly and build up large populations by midwinter; because they favor certain leaf types, *e.g.* willow over oak, they are more abundant in some leaf masses than others. A second group, also quick to invade new leaf masses, feeds on tiny organic particles—plant detritus, insect feces, bacteria, fungi and diatoms—that are trapped between the leaves. Included in this group are other species of immature mayfly and stonefly, together with midges. Predators comprise the third group, which ordinarily delays invasion of leaf packs until populations of the first two groups have reached considerable size. These populations often are very substantially reduced after the appearance of the predators, prominent among which are still other species of stonefly.

It is apparent that the seasons of maximum biological activity and food availability are nearly opposite in a deciduous forest and in a stream flowing through it. The stream would differ sharply, too, from a nearby lake in abundance of winter food, though both would share summer food shortages. Many stream-living insect larvae are quite long-lived, taking a year or more to mature, and some species enter dormancy or semi-dormancy during summer. The timing of dormancy varies among species: there is mutual adjustment such that potential food competitors are not all active during periods of scarcity. These mutual adjustments of life cycles are a form of temporal niche separation analogous to that seen in deciduous forest plants.

Streams in other settings may show quite different seasonality. In coniferous forests there is no autumn bounty of dead leaves, and seasonal patterns are probably quite muted. In open habitats, stream communities are more autotrophic; here algae and aquatic plants are more abundant and much more important as a community base, and summer may be the period of greatest activity. But to streams traversing open farmlands, the timing of plowing and fertilizer application, as well as the type of fertilizer used, can be controlling influences. Farmers frequently apply fertilizers in late winter, and the nutrients in these fertilizers enter the stream at different rates. Most fertilizer phosphate that reaches a stream does so quickly, in surface runoff during the first rains after application. Fertilizer nitrogen arrives at a stream more slowly, via groundwater movements; very little is carried in surface runoff (Nelson, 1969). The seasonal implications therefore differ for streams running through fields treated with phosphorus alone, nitrogen alone, or with a combination. Here, as in the case of all small streams, the intimate relationship of aquatic patterns to terrestrial influences is apparent.

In streams we thus see opportunities for niche specialization with respect to current, substrate, food and time. There are, of course, other dimensions to the niches in flowing water, but the possibilities inherent in these four dimensions alone go far toward explaining the remarkable diversity of life in unpolluted flowing waters. A small stream may harbor as many as 270 species of insects alone, and the number usually increases as the stream becomes larger. (Minshall, 1969).

Short Term Changes in Streams

Running waters are much more subject to severe, temporary environmental change than are most lakes and ponds. Cases are well known in which small streams in open country have been cooled so suddenly by a summer thunderstorm that fishes succumbed to thermal shock. But such events are much less common than the changes in current velocity, turbidity and water chemistry that affect relatively small streams during any severe rainstorm.

In deserts, many streams flow only during and after storms. Even in regions with abundant rainfall, major storms may account for remarkably high percentages of yearly flow and yearly sediment transport. In the Broad River in Georgia, during a year considered fairly normal, 90% of the particulate load carried by the stream was moved during 20% of the year, and 50% of this load was carried during 2% of the year (Kennedy, 1963). In the Conestoga Creek watershed of Pennsylvania, the week-long floods associated with Hurricane Agnes carried 52% of the total 1972 flow and 49% of the total suspended sediment (Kochel and Dorsey, 1972). In streams that carry appreciable sediment loads, organisms of clean, hard substrates probably *require* the scouring action of storm flows if they are to persist. Without periodic floods, the rocky substrates would soon be buried in silt and their habitats would vanish.

Flow volumes and suspended sediment loads are much more variable in some streams than others, depending on the vegetation cover of the landscape. This is true also of water chemistry. In forested regions there may be little change in water chemistry even with fairly large variations in flow. In watersheds having much erosion streams may become more concentrated chemically during storms; while in watersheds consisting largely of bedrock, the opposite may occur. The chemical changes produced by storms usually are not sufficient to cause major alterations in stream or river communities, although great biological disruption resulting from salinity change may occur in estuaries receiving swollen rivers. In streams and rivers, the more important short term influences are storm-induced changes in current and sediment load. But surveys by my students before and after Hurricane Agnes indicate that even these influences are less pronounced than one might expect, at least in streams where the gradient is not great.

MARINE AND FRESHWATER ENVIRONMENTS: CHEMICAL AND OSMOTIC CONTRASTS

One thinks of lakes and streams as being dilute; one thinks of the ocean as a place of dissolved salts and smarting eyes. This image glosses over a wealth of chemical variability in the case of inland waters. "Fresh" waters can be hard or soft, depending on whether they contain relatively high or relatively low amounts of dissolved Ca^{++} (or Mg^{++}) and HCO_3^-. Although the concentrations of these ions vary greatly in lakes and streams, they usually exceed the concentrations of Na^+ and Cl^-, the ions most abundant by far in sea water (Table 3.2). Some closed basins, however, contain lakes in which Na^+ and Cl^- not only are the dominant ions but exceed the quantities found in the sea.

Much of the variability of world water chemistry can be understood in terms of the relative importance of three major processes: precipitation, solution of mineral ions and evaporation (Gibbs, 1970). The most dilute lakes and streams in the world possess a chemistry resembling that of rainwater; their chemistry, in fact, largely is determined by the ions in atmospheric precipitation. Water bodies of this type usually occur in regions of low relief where the soils are old and heavily leached. A number of lakes and rivers in the moist tropics, such as that giant tributary of the Amazon, the Rio Negro, are of this type. Other "rainwater" streams and lakes occur in regions where the soil is thin and the bedrock very resistant to solution. Certain lakes in the Adirondacks are interesting examples; they have been made quite acid in recent years by rainwater enriched in the pollutant SO_2, which oxidizes to SO_3 and then dissolves to form H_2SO_4 (Likens and Bormann, 1974). These Adirondack lakes are very dilute, and do not contain enough bicarbonate to buffer the acid rainwater.

"Groundwater" lakes and streams stand in contrast to those we have just described. Here the major influence on water chemistry is the rocks and soils of the drainage basin. Solution of ions from the substrate largely determines the chemistry of the majority of lakes and rivers in temperate regions, so they would be included in the groundwater class. The hard or soft nature of these waters, for example, reflects the abundance of limestone relative to other, noncalcitic rocks in their basins.

Evaporation and crystallization, the third major control of water chemistry, are dominant in the world's oceans and in the inland waters of most arid regions. Evaporation leads to concentration of solutes in the water, but when concentrations become too high the less soluble compounds crystallize out as precipitates, usually leaving sodium chloride (which is highly soluble) as the predominant substance in solution. Most of the highly ionic waters of the world fall into this class, as do desert streams and rivers which, though quite dilute compared to the sea or to closed lakes, are ionically rich compared to streams in regions of higher rainfall. Waters of this class in the United States include the Rio Grande and Pecos Rivers and Great Salt Lake in Utah.

Precipitation, solution of substrate, and evaporation-crystallization of course act together to determine the chemical nature of any water body; we have been speaking above of the dominant influence, not of the sole influence when we assigned water bodies to specific categories. In some river systems each process may become dominant in turn. My colleague Ken John found clear cases of this when conducting fisheries surveys in Wyoming. In these surveys, the streams were sampled with an electric fish shocker, which ordinarily stuns fishes unlucky enough to swim between two electrodes placed in the water, and allows them to be easily collected. But in Wyoming's high mountain streams the shocker was quite ineffective; the meltwater from high snowfields was too low in ions to conduct a current effectively between the electrodes. Farther down these watercourses the shocker worked well: the streams had become better conductors as ion-rich groundwater was added. At still lower altitudes the streams emerged onto the dry Wyoming plateaus and evaporation became the dominant process, raising the water's ion content still further. Eventually the shocker was again ineffective; the water now conducted electricity so well that fishes between the electrodes ceased to be the path of least resistance, and the current went around instead of through them!

The remarkable chemical variety of the world's inland waters is a major reason for the remarkable biological variety that exists among them. Aquatic life is sensitive not just to nutrient differences but also to osmotic differences. For example, the body fluids of bony fishes contain more dissolved matter than does fresh water but less than salt water. Fishes living in water of a given solute content must be able somehow to counter the tendency for water to diffuse either from the medium into their body fluids (as would be the case if the medium were more dilute than their body fluids) or vice versa. Since osmotic problems differ in waters of differing solute content, it is not surprising that dilute lakes and streams contain different plants and animals than do ion-rich lakes and streams. In fact, a biological continuum of changing species parallels the osmotic continuum between dilute and ionically concentrated waters. To some extent we also can differentiate among the biotas of calcium carbonate waters, sodium bicarbonate waters, sodium chloride waters, etc., depending on which ions are predominant. Thus an aquatic ecologist often recognizes strong biological similarities between waters of similar chemistry. For example, I have found that particular species of diatoms characterize sodium bicarbonate lakes of approximately the same osmotic concentration wherever such lakes occur—in East Africa, Hungary or Mexico. On the other hand, watersheds differ so widely in the proportions and total amounts of ions that they deliver to their lakes and streams that the total result is great biological variety among the world's inland waters. When one considers also the short term and seasonal changes, both chemical and nonchemical, that characterize *individual* lakes and streams, it is not surprising that every limnological excursion, be it to a new water body or to the same one for the hundredth time, becomes something of an adventure.

SUGGESTED READING

References and texts

AMOS, W. H. The Life of the Pond. New York, McGraw-Hill Co., 1967. Full of interesting information, though written for the layman.

BARDACH, J. Downstream: A Natural History of the River. New York, Harper and Row, 1964. Another good book at a popular level.

HUTCHINSON, G. E. A Treatise on Limnology. New York, John Wiley and Son. Volume I (1957) deals with physics and chemistry, Volume II (1967) with fauna and flora, emphaaizing plankton. *The* reference book on lakes.

HYNES, H. B. N. The Biology of Polluted Waters. Liverpool University Press, 1960. A very useful short book.

HYNES, H. B. N. The Ecology of Running Waters. Toronto University Press, 1970. A fine reference, and one of the only ones, on streams.

LEOPOLD, L. B. Water: a Primer. San Francisco, W. H. Freeman & Co., 1974. A good short paperback emphasizing the geology and physics of stream processes.

MACAN, T. T. Freshwater Ecology. London, Longman Group Ltd., 1974 (2nd edition). A paperback by an Englishman at the famous Windermere freshwater laboratory.

ROLICH, G. A., ed. Eutrophication: Causes, Consequences, Correctives. Washington, National Academy of Sciences, 1969. See also references on this subject at end of Chapter 10.

RUTTNER, F. Fundamentals of Limnology. Toronto University Press, 1963 (3rd edition). A much used introductory text by a famous Austrian limnologist. Very readable.

WETZEL, R. G. Limnology. Philadelphia, W. B. Saunders Co., 1975. A comprehensive new textbook on lakes.

Methods and identification manuals

EDMONDSON, W. T., ed. Ward and Whipple's Fresh Water Biology. New York, John Wiley and Son, 1959 (2nd edition). Mostly for identification.

NEEDHAM, J. G., AND P. R. NEEDHAM. A Guide to the Study of Freshwater Biology. San Francisco, Holden-Day, Inc., 1962. A useful paperback for identification and collecting methods.

PENNAK, R. W. Freshwater Invertebrates of the United States. New York, Ronald Press, 1953. A widely used and valuable identification manual.

SCHWOERBEL, J. Methods of Hydrobiology. Oxford, Pergamon Press, 1970. Excellent.

TARAS, M. J., A. E. GREENBERG, R. D. HOAK, AND M. C. RAND, eds. Standard Methods for the Examination of Water and Wastewater. Washington, American Public Health Association, 1971 (13th edition). A voluminous work on chemical methods in fresh water.

WEBER, C. I., ed. Biological Field and Laboratory Methods for Measuring the Quality of Surface Waters and Effluents. Cincinnati, United States Environmental Protection Agency publication 670/4-73-001, 1973. Extremely useful manual on field and lab methods for microscopic and macroscopic organisms and analysis of data.

mud and estivate in molted skins that surround them like cocoons, helping to retain moisture as the soil dries. Sometimes they do not reappear for more than a year. These are the main elements, but not the only ones, in the spadefoot's adaptive survival kit (see Bentley, 1966; Whitford, 1969; Schmid, 1969).

The riddle of desert survival is especially dramatic in the case of aquatic species, but deserts are, at least to man, awesomely inhospitable environments and we can fruitfully explore the riddle for any species that thrive there. Indeed, ecologists working in physically or chemically extreme environments of any sort cannot help being fascinated by particular species and their adaptations. This chapter is an introduction to the ecology of extreme environments. We will approach the more general subject through a consideration of deserts, and like many ecologists before us, we will spend considerable time pondering how specific organisms manage to find the desert so congenial. Community characteristics and relationships deserve attention too, however, and in the later part of the chapter we will turn more directly to these.

Almost any environment can be considered extreme in some sense, or for some organisms; but deserts are not physically harsh places at all for some species. Are they only extreme in human eyes? Can we ever be purely objective when we use this word? Perhaps not, although the communities of habitats that are very cold, very hot, very dry, very polluted, or simply very variable usually share certain characteristics, such as low numbers of species. This simplicity, when we find it, seems to result from physical or chemical conditions too harsh for any but highly adapted species, specially equipped for existence at one of the ecological if not always geographical fringes of the biosphere.

DESERT AND POLE

Burning desert and ice-bound arctic—can one imagine habitats more opposite? Yet in what they require of their inhabitants, these environments are similar even in their differences. Both are low in available moisture; the absolute drought of deserts is matched in polar regions by the physiological drought of ice and snow. Extreme temperatures, too, characterize both desert and pole; though the extremes are opposite, the adaptive response is remarkably similar in some ways. Many animals hibernate through the arctic winter, and most plants become dormant. Many animals estivate through the desert summer, and many plants become dormant. Fur and feathers of many animals turn white in the arctic winter; in some animals, similar changes occur in the desert summer. A comparative approach, beginning with a discussion of heat exchange processes and thermoregulation, may be the clearest way to pinpoint the adaptive responses of living things to high desert temperatures; as a bonus, we learn something about adaptation to cold. Not all deserts, of course, are excessively hot. But most are hot at least some of the time, and since for many organisms the problems of heat and moisture are interlinked, specific thermoregulatory adaptations are very common in deserts.

Exchanges of Heat between Organism and Environment

If the environment becomes hot enough all life is killed, but what is excessive to some species remains tolerable to others. One avenue of adaptation to hot environments is simply enhancement of the physiological ability to endure high body temperatures. The same is true for cold environments. But most organisms do more than simply endure high or low temperatures in the environment. They thermoregulate, by controlling to some extent the rates at which they are warmed or cooled.

Terrestrial organisms obtain external heat by radiation and conduction.* They may also rid themselves of heat by these processes, as well as by evaporation. When water at the surface of a plant or animal evaporates, the heat of vaporization is transferred from the organism to the atmosphere. Convection, a fourth heat transfer process, does not serve as a direct means of heat exchange between organism and environment but it greatly affects the rates of conductive and

*Radiation is a wave phenomenon: a warm body radiates heat from its surface toward a cooler receptor even if that receptor is millions of miles away, so long as the medium can transmit the radiation. Air is sufficiently rarefied to serve as a good radiation transmitter, but liquids and solids are not. Conduction of heat is accomplished by warm molecules exciting adjacent molecules; hence solids, being dense concentrations of molecules, are the best conductors and gases the worst. Convection is the bulk transfer of warm gases or liquids by turbulent flow.

evaporative exchange, being responsible for the rates at which warmed or moistened air is removed from the vicinity of the organism.

How is thermoregulation accomplished? largely by modifying the rates at which these processes of heat exchange occur. In addition, particularly in warm-blooded animals (homoiotherms), rates of internal heat production may be varied. Much thermoregulation is accomplished behaviorally. A bird may flap its wings when overheated, thereby increasing its rate of heat loss by conduction and convection. If cold, the bird may close its wings and ruffle its feathers, trapping a stagnant layer of air next to its skin. Air is a very poor conductor of heat, and unless the warm layer next to the skin is constantly removed by convection, conductive heat loss to the atmosphere is very slow. Heat conduction through solids is much faster, however, and an animal can quickly gain or lose heat by conduction alone if it presses its body against rocks or soil markedly warmer or cooler than itself.

Animals, though not plants, can alter their *radiant* heat flux simply by controlling their exposure to the sun. During hot days in the Arizona desert, the collared peccary (a wild relative of the pig) minimizes radiant heat uptake by staying in the shade. If the night remains hot the peccary cools by sleeping in the open and radiating heat to the clear, cold depths of space. The rates of both conductive and radiant heat exchange depend importantly on the temperature difference between organism and environment, but in the case of radiant heat, "environment" is whatever lies in the path of the heat waves approaching or leaving the organism. A peccary in the sun always will absorb more radiant heat than it loses, because solar temperatures are always hotter than body temperatures. In daytime shade the peccary may still have a net radiation income if the leaves beneath which it is resting are warmer than its body, but of course the income is much less than if the animal were in the sun. If it remains beneath the trees at night, the rate of which the peccary radiates heat is directly proportional to the temperature difference between its body and the foliage overhead. This temperature difference is much less than that between the peccary and the night sky, which becomes the radiation target when the animal sleeps in the open. Hence, an animal which becomes overheated during the day can cool itself more rapidly during the night if it seeks open areas.

The daily heat balance of the peccary is broken into its major components in Figure 6.2. This animal is a homoiotherm, and metabolic heat production is a much larger element in the whole picture than it would be for a plant or poikilothermic animal. In winter the peccary produces more metabolic heat than in summer; it keeps warm partly by becoming more active. Thus, although the animal is considered a physiological thermoregulator, behavior influences nearly all components of its heat balance. The magnitude of the various components also is influenced by other nonphysiological properties, some of which vary seasonally in adaptive ways. These include the thickness of the insulating coat (it is sparser over much of the animal in summer), and its color. A dark object is both a better absorber and a better emitter of radiant heat than is a light object. In the case of the peccary one therefore finds adaptive significance in the summer lightening of the coat, when many hairs become tipped with silver, and in its winter darkening. The light summer coat reflects more of the incoming radiant heat than does the dark winter coat (while at the same time, because of its sparseness, permitting greater evaporative loss of body heat). Absorption of radiant heat by the peccary is thus minimized in summer, at the same time that the potential for evaporative heat loss is maximized; while in winter, when the animal can use extra heat, radiant heat absorption is maximized while evaporative and conductive heat loss are minimized.

The seasonal color changes of arctic birds and mammals are the reverse of the color changes in the peccary, but are understandable in the same terms. A snowshoe hare is brown in summertime, and can absorb considerable radiant heat when basking in the sun—a summer advantage in the Arctic, though a disadvantage in the desert. But should not the hare turn even darker in winter, thus maximizing its radiant heat uptake? No, because there is so little radiant heat available from outside during the short winter days that dark fur would gain very little additional heat for the animal. Dark fur would, in fact, be a disadvantage because it would act as an efficient radiant emitter of the heat produced internally. White fur, on the

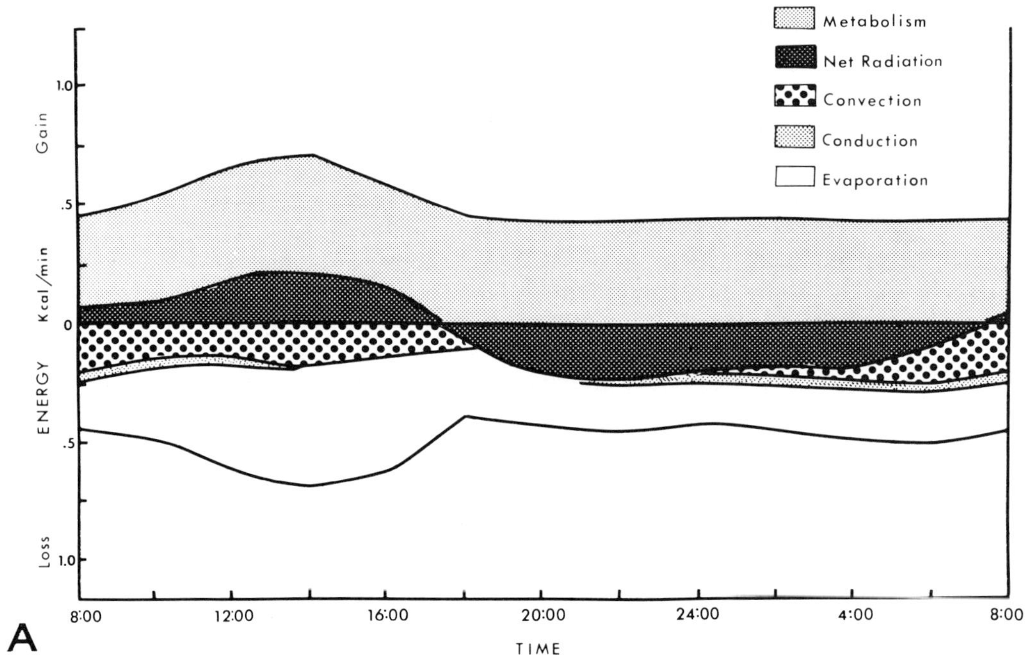

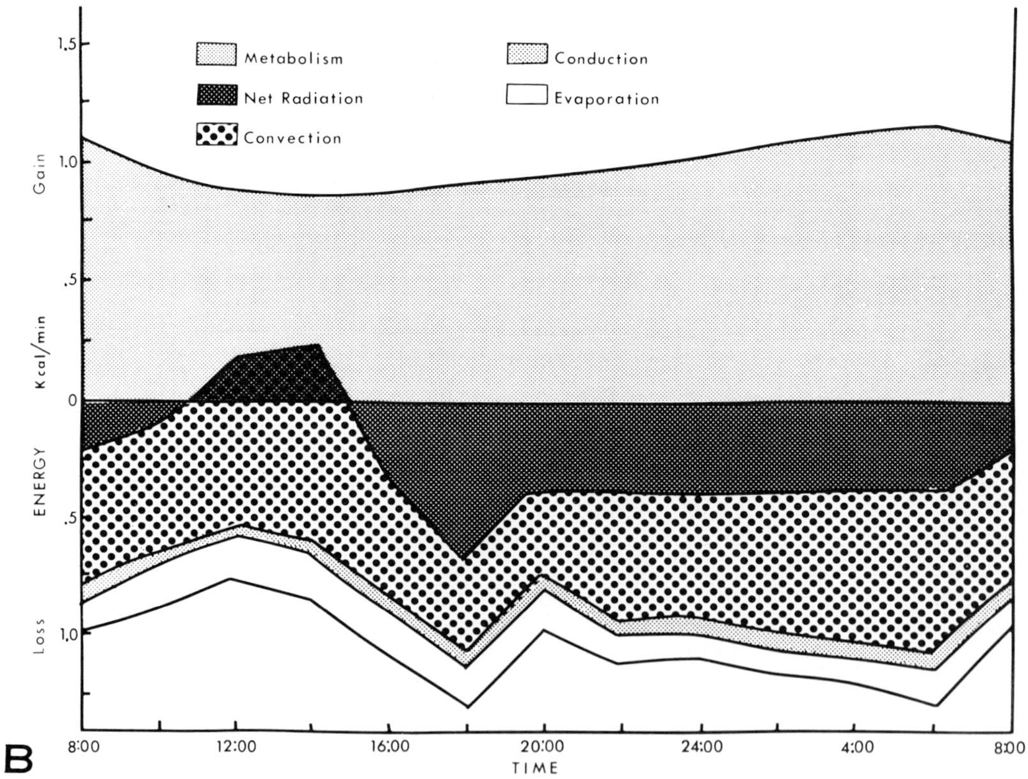

FIG. 6.2. Components of heat balance in the collared peccary on a summer day (*A*) and on a winter day (*B*). The animals may exhibit net gains or losses of heat during parts of each day, but gains and losses overall must balance. The metabolic and behavioral events described in the text may be followed by careful inspection of these graphs. (From S. M. Zervanos and N. F. Hadley: *Ecology*, *54:* 759–774, 1973.)

contrary, reflects much of the animal's metabolic heat back into the body and is a precious winter asset. The seasonal color change of arctic animals is often explained, of course, in terms of its seasonal camouflage advantages, but it is equally useful to the animals' heat balance.

Body Size, Rates of Heat Exchange, and Limits of Homoiothermy

Both the problems and the mechanisms of maintaining tolerable internal temperatures differ between large and small organisms. Two interlinked causes are responsible for these differences. First, when heat is exchanged with the environment by any of the processes we have discussed, the exchange necessarily takes place at the *surface* of the organism. Second, small bodies have proportionately more surface area than do large bodies (*i.e.* they have a high surface:volume ratio), and all internal regions are relatively close to the surface.

These facts mean that small organisms exchange heat with their environments much more rapidly than do large organisms under the same conditions. A lizard can warm itself rapidly by basking on a sunlit rock; crocodiles also bask, but their temperatures rise much more slowly and they cool more slowly at night.

Several considerations follow from the heightened heat exchange capacity of small organisms. (*a*) If they are small poikilotherms like the lizard, their temperature tends to follow closely that of their environment; while the temperature of large poikilotherms like the crocodile lags behind changes of environmental temperature, often by a matter of hours. (*b*) If they are small homoiotherms, like a mouse or sparrow, they can maintain a constant body temperature only by expending proportionately more energy than do large homoiotherms like men or elephants. The large homoiotherm is, so to speak, insulated from environmental temperature change by its low surface: volume ratio. The small homoiotherm must shiver, sweat or pant harder if it is to offset the rapid gain or loss of heat that occurs whenever a significant temperature difference exists between its body and its surroundings. In other words, the small homoiotherm must work harder to maintain a constant body temperature.

These contrasts are reflected in the relationships of Figure 6.3. Why are the metabolic rates of homoiotherms inversely proportional to body size? probably largely because homoiothermy is progressively more expensive as the animals' surface:volume ratios increase. Metabolic rates are so high among the smallest homoiotherms that one may suggest than an animal must be of a certain minimum size if it is to be homoiothermic; for animals smaller than about 5 g, the metabolic cost of maintaining a constant internal temperature becomes astronomical.

Figure 6.3 also shows how size affects both the efficacy and the practicality of one possible cooling mechanism in animals—evaporation. Small organisms can cool very rapidly by evaporation, but they must expend proportionately more of their water than large ones if they use evaporation as their cooling mechanism. In deserts, where water is a scarce commodity, evaporative cooling is too extravagant for small animals to afford, except in brief emergency situations. The animal would need to expend relatively large amounts of water to keep cool if the environment was hotter than its body, and because of its small size its water reserves would be very limited. Although only homoiotherms are considered in Figure 6.3, evaporation is equally unrealistic as a cooling mechanism for small desert poikilotherms. On a hot day in the desert, a fly would have to evaporate more than twice its own weight in water each hour to keep its temperature from rising (Edney, 1967).

Bergmann's and Allen's Rules

Conservation of body heat is a prime requirement for homoiothermic animals in polar regions. Numerous adaptive stratagems promote such conservation: the color change patterns we have mentioned; thick, insulative fur, feathers or blubber; and circulatory specializations that facilitate transfer of heat from outgoing arteries to incoming veins before the blood reaches exposed areas or extremities (Irving, 1966). In hot regions, *elimination* of body heat must often be facilitated if a homoiotherm is not to overheat. In view of these opposite needs, we might therefore expect that, in addition to those adaptations just mentioned, polar homoiotherms might exhibit tendencies to limit their ratio of surface to volume, whereas homoiotherms of hot environments might exhibit tendencies to maximize this ratio. Such tendencies were long ago recognized in the generalizations that are known as Bergmann's Rule and Allen's Rule. The first may be stated as follows:

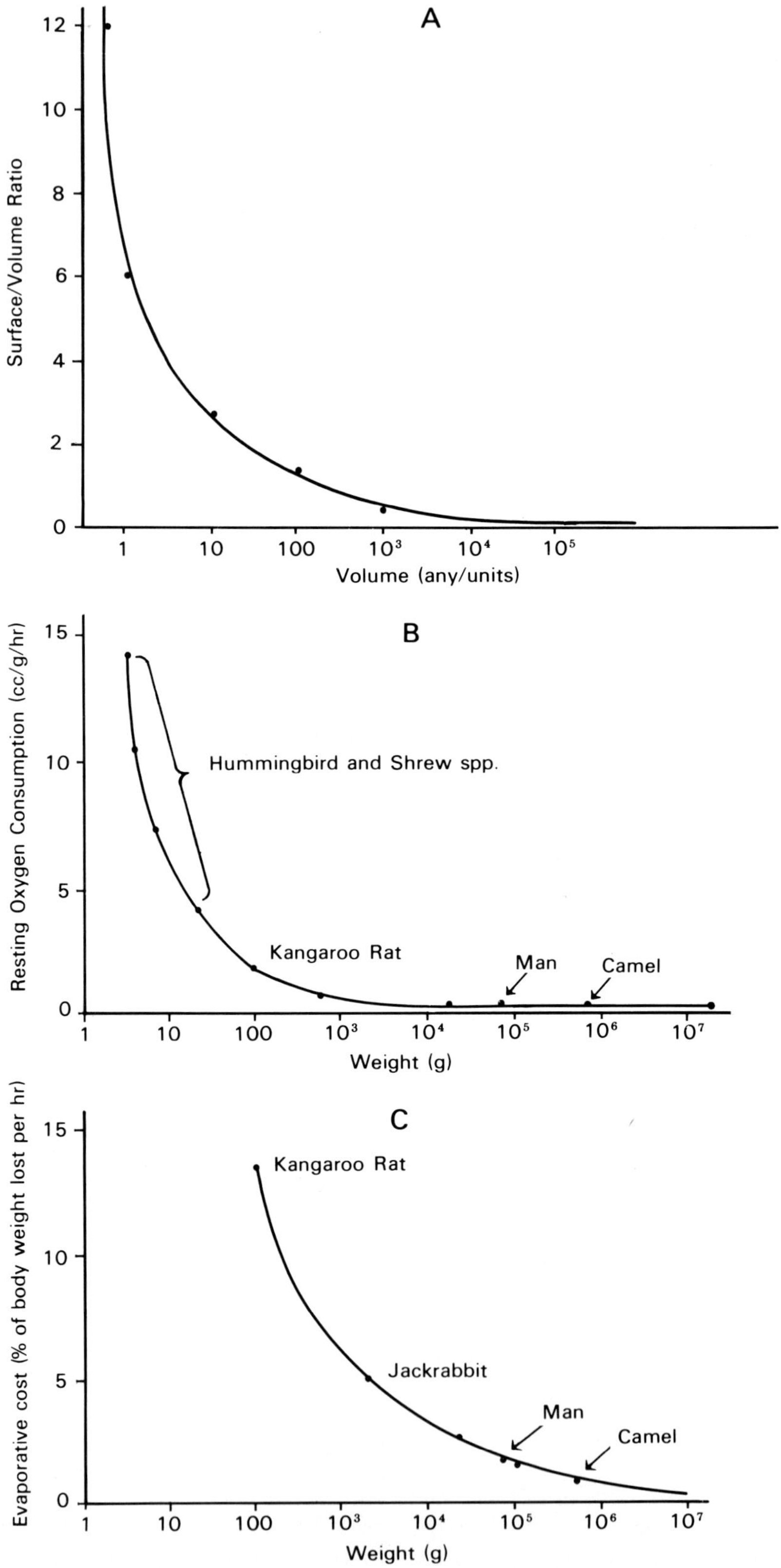
A
Surface/Volume Ratio
0
2
4
6
8
10
12
1
10
100
10³
10⁴
10⁵
Volume (any/units)
B
Resting Oxygen Consumption (cc/g/hr)
0
5
10
15
Hummingbird and Shrew spp.
Kangaroo Rat
Man
Camel
1
10
100
10³
10⁴
10⁵
10⁶
10⁷
Weight (g)
C
Evaporative cost (% of body weight lost per hr)
0
5
10
15
Kangaroo Rat
Jackrabbit
Man
Camel
1
10
100
10³
10⁴
10⁵
10⁶
10⁷
Weight (g)

TABLE 6.1

Old World Bears and Bergmann's Rule

The species of bear inhabiting warm regions are relatively small, and there is a progressive increase in size of species inhabiting higher latitudes. This is true both of the Old World and the New World. In North America the several varieties of black and grizzly bear tend to exemplify Bergmann's Rule at the intraspecific level as well—though there are exceptions such as Richardson's grizzly, an atypically small variety of the Alaskan tundra. Such exceptions demonstrate that other factors (such as inadequate food supplies, perhaps, in the case of Richardson's grizzly) may work at cross purposes to Bergmann's Rule. (The extinct European cave bear, common in central Europe during the last ice age, was larger than any living species.)

Species	Typical Weight of Adult	Geographical Range
	pounds	
Malayan sun bear (*Helarctos malayanus*)	100	Malaysia and Indonesia
Sloth bear (*Melursus ursinus*)	200	India and Ceylon
Asiatic black bear (*Selenarctos thibetanus*)	300	Temperate and subtropical Asia
European brown bear (*Ursus arctos*)	600	North and Central Europe, North and Central Asia
Polar bear (*Thelarctos maritimus*)	900	Arctic regions; recorded to Lat. 88° N

> In species of birds and mammals which range over considerable latitudes or altitudes, individuals in the colder parts of the range tend to be larger, and individuals in the warmer parts of the range tend to be smaller.

The adaptive explanation of this tendency is clear when we remember the relationship between size and rate of heat exchange. Bergmann's Rule is followed by many homoiothermic species. The white-tailed deer, for example, reaches larger size in the northern parts of its range than in the south, and in subtropical southern Florida it is represented by a dwarf race, the key deer. Table 6.1 provides evidence, using bears, that Bergmann's Rule sometimes applies not just to individuals or races within the same species, but to groups of related species.

Allen's Rule states that:

> Among birds and mammals, limbs and other extremities are shorter and more compact in cold regions than in warm ones.

Allen's Rule is simply an extension of Bergmann's Rule to cover body shape as well as overall size. Extremities such as ears, nose and legs are small structures relative to the main torso of the animal, and each such extremity therefore has the propensity to exchange heat more rapidly with the environment than does the more massive torso. In cold regions, therefore, one finds chunky animals in which protrusions from the main body regions are minimized; in very hot regions, conversely, the extremities of homoiotherms are often accentuated, allowing them to more easily dispel the heat from their "metabolic furnace." Snowshoe hares and jackrabbits make a striking contrast in this regard, as do foxes (Fig. 6.4).

Though not always followed, both rules apply also to man (Schreider 1964). The people of south China, for example, are small and slender; those of north China are large and husky. Of all the world's people, Eskimos and Andean Indians most closely approach the spherical shape that, for a given volume, has the lowest possible surface area, whereas the spindly Nilotic tribes of tropical Africa represent the most extreme departure from sphericity.

FIG. 6.3. Interlinked relationships involving size. (*A*) Surface: volume ratio vs. volume (for cubical shapes; other shapes would yield similar but not identical curves). (*B*) Metabolic rate (as measured by rate of oxygen consumption) vs. weight of homoiotherms. (*C*) Relative evaporative rate vs. weight, for homoiotherms hypothetically employing evaporation to maintain constant body temperatures on a hot desert day. (Adapted from Pearson, 1953, and Schmidt-Nielsen, 1964.)

necessarily to the level of external temperatures but enough to significantly reduce the temperature differential between themselves and the environment. The rate of heat loss is thereby reduced, and since an animal's metabolic rate also is reduced during this hypothermic period, an animal needs to feed rarely if at all during its period of dormancy—a corollary advantage, since available food is usually limited.

The estivation of certain desert homoiotherms during seasons of great heat or drought confers similar advantages. The Mohave ground squirrel, for example, retreats to its burrow in August and remains there till March, much of the time in torpor. Its body temperature during torpor drops close to that of the burrow (which remains relatively cool even in late summer), and the animal lives largely on fat reserves. Its ability to do this is advantageous, since the season of torpor is one of food and moisture shortage. The underground sojourn is ended in March, when leafy, water-rich foods become more abundant (Bartholomew and Hudson, 1961).

Only relatively large homoiotherms can employ seasonal strategies of hibernation or estivation. Small birds and mammals simply cannot store enough fat to tide them over a long period of torpor. In the desert we therefore find Mohave ground squirrels but not smaller rodents passing the season of stress in torpor. In the northeastern United States, similarly, groundhogs enter true winter hibernation; gray and red squirrels experience partial torpor, but need food periodically throughout the winter; and mice, shrews and small birds remain active through the cold season. The perpetual voracity of shrews is renowned: they must feed frequently or they starve. This is due not just to their high metabolic rates, but to their size-limited capacity to store fats and carbohydrates. Although such small animals cannot endure long periods of torpor, some enter torpor for a few hours. Hummingbirds, for example, become dormant at night, allowing their body temperatures to drop as air temperatures fall. Their metabolic rates also drop dramatically (Fig. 6.6); torpor permits these minute homoiotherms to sleep through the night without starving.

It seems paradoxical that homoiothermy, an evolutionary adaptation that allows animals to remain active in thermally rigorous environments, should be virtually abandoned by many

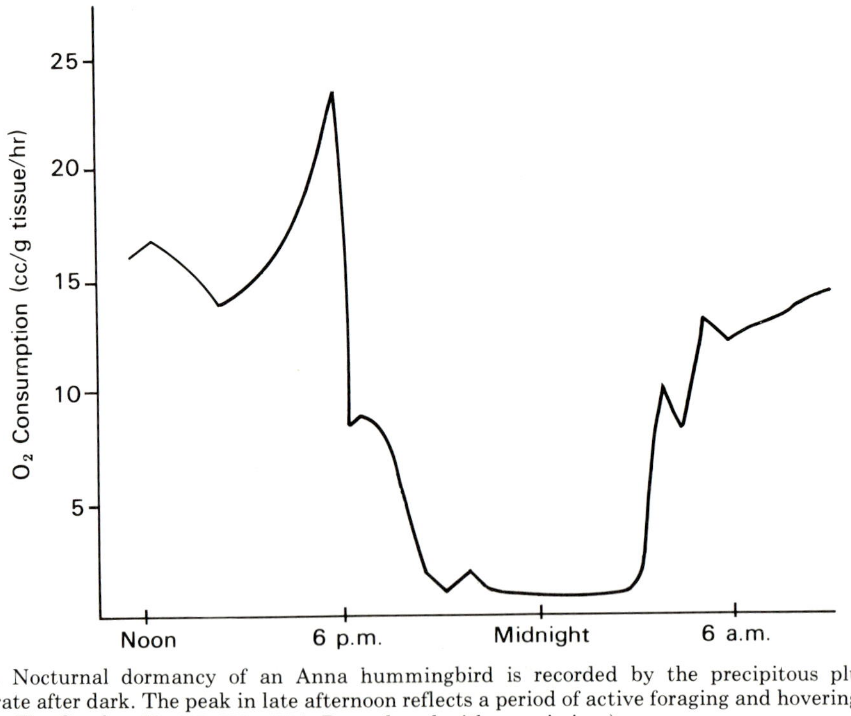

FIG. 6.6. Nocturnal dormancy of an Anna hummingbird is recorded by the precipitous plunge in its metabolic rate after dark. The peak in late afternoon reflects a period of active foraging and hovering. (From O. P. Pearson: *The Condor, 56:* 317–322, 1954. Reproduced with permission.)

homoiotherms when temperatures become too extreme. But the paradox is understood when one realizes the excessive cost of homoiothermy under such conditions, and we see that the ability to relax strict homoiothermy becomes a further form of adaptation to extreme heat or cold.

LIFE IN DESERTS

Introduction

We now have a groundwork to understand the adaptive ecology of desert organisms. We see that homoiotherms and poikilotherms must be treated in part as separate categories; the problems facing each are similar, but the possible solutions are not identical. Small and large organisms differ too, not only in the range of solutions available to them but also to some extent in the problems they face.

The world's deserts differ both in degree of aridity and in temperature characteristics. In the United States there are two main desert types. Sagebrush-shadscale deserts occur in arid, seasonally cold areas such as the Great Basin and the northern Rocky Mountain states, and cactus-creosote bush-yucca deserts are found in warm, low-lying parts of the Southwest (Fig. 6.7). Each of these types may be subdivided. The hot southwestern deserts, for example, include the Mohave region of southeastern California, the Sonoran region of southern Arizona and northwest Mexico, and the Chihuahuan region of southern New Mexico and north-central Mexico. The Mohave is driest, the Sonoran warmest, and the Chihuahuan slightly moister and cooler than the two other regions, and each boasts a flora and fauna having unique elements. Our discussion will center on these hot southwestern deserts, where the problems of heat and moisture are most closely linked and, in extreme areas, most acute.

Desert communities often are surprisingly varied, particularly with regard to angiosperms, arthropods, rodents, reptiles and birds. One should not go to the desert expecting to find a diversity of mosses, ferns, amphibians, land snails or primates, but these groups have a few representatives in arid lands. The variety of desert life, however, largely springs from diversification within those groups that find the desert particularly hospitable. Several such groups seem "preadapted" to deserts. For example, nearly all reptiles and arthropods, desert-dwelling or no, possess integuments that hinder the evaporative loss of body water; and together with birds, they excrete urinary products in solid form, minimizing the excretory loss of water. Such characteristics undoubtedly made it easier for these groups to colonize deserts, but there is no evidence that impermeable integuments and solid urine evolved in these groups as specific adaptations to desert climates.

Numerous instances of convergent evolution (Fig. 6.8) make clear, however, that desert conditions *have* called forth many specific adaptations. Deserts exist in several biogeographic realms, and the rigors of these similar habitats have provided striking examples of similar solutions by unrelated species. Whether one is considering the thorniness or succulence of cactus, aloe, euphorbia and yucca; the small size and exaggerated ears of the foxlike Saharan fennec and the American kit fox (Fig. 6.4); or the curious morphology and urine-concentrating ability of the kangaroo rat, what one observes is the way in which the desert environment has cast its denizens into adaptive molds. The sorts of adaptations one should expect are various: they may be peculiarities of morphology or physiology, of behavior or diet, or of timing within the life cycle.

Desert Climates

The biosphere's highest terrestrial temperatures occur in deserts. Air temperatures have reached 58° C (134° F) in the Sahara and 57° C in Death Valley, California. Unshaded soil surfaces at such times are much hotter, probably 70° C or more. But even hot deserts cool dramatically at night; the daily temperature range is very great (Fig. 6.9), and desert organisms must cope with these fluctuations, not simply with the heat. While always low, rainfall is highly erratic from year to year. Much of a year's rain may come in a single cloudburst; one year may see two cloudbursts, the next none, so annual totals often differ by more than 100%.

Deserts are characteristically windy places, because the strong solar heating by day causes large convective air movements. In many deserts, too, few barriers impede the air currents.

FIG. 6.7. (*A*) Cold desert on the norther Arizona plateau. The San Francisco Peaks, in the background, intercept the moisture of the prevailing westerly winds and create a rain shadow on this, their leeward side. Typical plants are saltbush (foreground), snakeweed (center), and a grass, probably *Hilaria jamesii*. The bushy, globular growth form is found in many deserts; it may help plants retain humid, transpired air in the vicinity of the leaves, thus lowering evaporation rate. Fine soil has been largely removed by wind, leaving a stony "desert pavement." (Photograph by author.) (*B*) Hot Sonoran desert of southern Arizona. In the foreground a paloverde tree is seen closely associated with young saguaro cacti. This phenomenon is so common that desert ecologists suggest a "nurse tree" relationship in which young saguaros are protected or otherwise benefited by the paloverde's proximity. (Photograph by author.)

FIG. 6.8. One of the classic examples of morphological convergence involves the gerbil of Asia Minor (*left*), the kangaroo rat of southwestern United States (*center*), and the jerboa of North Africa (*right*). These unrelated rodents of sparsely vegetated deserts each have evolved a jumping mode of rapid locomotion and the balancing tail, enlarged hind legs, and reduced front limbs that accompany this locomotory mode. The enlarged, hairy feet of kangaroo rat and jerboa permit leaping in soft sand. (Drawing by Sarah Allan.)

Wind accentuates the problem of desiccation faced by desert organisms.

Desert soils are either scanty or unstable. The absence of dense ground cover subjects the land to denudation by wind and occasional rainstorms, and in many places the substrate is reduced to bare rock or coarse pebbles; elsewhere it is unstable dunes. Where they are deep enough, desert soils are often quite fertile: there has been little leaching away of nutrients. But in some low lying areas, salts have accumulated by runoff and evaporation to the point where most plants cannot cope with the osmotic problem.

These are characteristics of the desert macroenvironment. If they were true everywhere in the desert, life there would be far more limited than it is. But even the harshest desert harbors numerous microenvironments where conditions for life are less severe. There are riverbeds and oases where permanent water exists, either at the surface or not far below; some Saharan oases even harbor crocodiles! Fierce daytime temperatures can be avoided by burrowing species, whose subterranean retreats are both cooler and more humid than the world above. If we were to list desert organisms in terms of relative ability to evade desert extremes, plants (being sedentery) would be at the bottom of the list and small animals at the top, with large animals between. Many small animals, disadvantaged by size in ways we have described, survive in deserts primarily because of their ability to exploit microenvironments that are relatively cool and moist.

Perennial Desert Plants

Table 6.2 lists a number of adaptive modifications that are found in desert plants. No one species possesses all; but each adaptation in the list is quite common among desert perennials. Desert annuals, on the other hand, possess very few special features of morphology or physiology; we will turn to them presently. Here, a few autecological sketches will suggest the variety of survival strategies among perennial species.

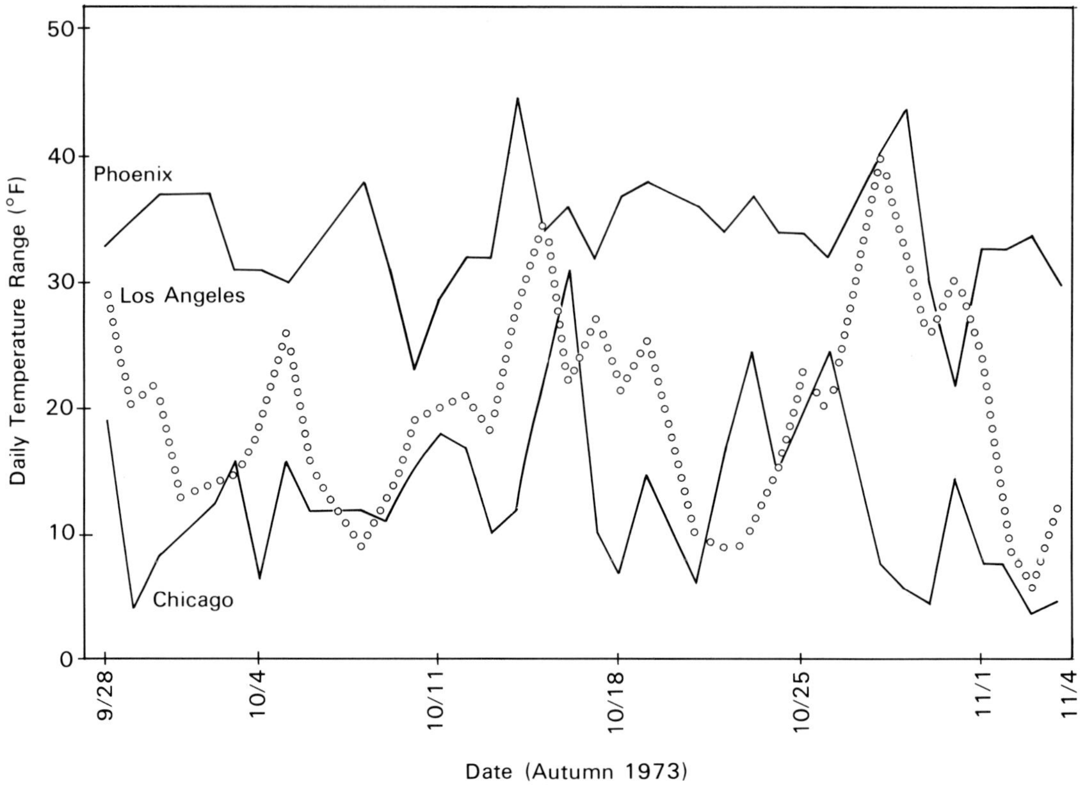

FIG. 6.9. Low atmospheric moisture allows abundant radiant energy to reach the desert surface by day, but also permits strong back-radiation at night, so deserts experience pronounced day-night changes of temperature. In this figure the daily temperature range at an inland desert locality is compared to the daily range in coastal California and in the Midwest during a 47-day span in the autumn of 1973.

Saguaro and other succulents—The saguaro cactus (*Carnegia gigantea*) shares with other succulent species the trait of water storage; but in this giant of the cactus family the ability is particularly impressive:

> These monster cacti, sometimes as high as fifty feet, sometimes weighing as much as two tons, and sometimes living as long as two hundred years, have no real taproots at all. Just below the surface of the soil, a flat, disk-like network spreads for yards around them; when a rain comes they quickly take up the water from a wide area, swelling visibly and sometimes absorbing as much as a ton of water from one rain. After that they may go a year, if necessary, without taking in water again.
>
> Joseph Wood Krutch (1951), *The Desert Year*

The saguaro thus can grow in soils that are shallow, rocky and nearly always dry; it does not depend on sources of permanent water underground, but on its ability to soak up dew and occasional rains. Not just the root system of saguaro but also its stem (and that of certain other species of succulent) is specialized for rapid uptake of water: it is pleated like an accordion, allowing for considerable expansion without rupture of the tissue. Of course, the pleating of the stem serves equally for noninjurious shrinkage as a water hoard is used up.

If it must live for indefinite periods on its own water stores, the saguaro must be parsimonious in water use. Saguaro and other cacti dispense altogether with leaves, the prime avenue of water loss in most plants, and their surface:volume ratios, already vastly reduced by this stratagem, are diminished even further by the bulky stem shape. Barrel cacti and young saguaros are nearly perfect spheres. Succulents such as agave and aloe, which retain their leaves, also have low surface:volume ratios because their swollen leaves are appreciably three-dimensional; but saguaro and barrel cacti are champions in the surface:volume sweepstakes. Unlike deciduous desert perennials, saguaro and other succulents remain photosynthetically active year-round, since the water necessary for photosynthesis comes from an internal reservoir. In many succulents, photosynthesis is carried on with minimal water loss by an ingenious reversal of the usual stomatal rhythm. The usual pattern is for the stomata (leaf pores) to be open by day, allowing CO_2 to diffuse to photosynthesizing internal tissues, and to be closed by night when CO_2 is not needed. But these succulent species are biochemically adapted so that they can stockpile CO_2 for later use; during the day they carry on photosynthesis with the stomata closed, having absorbed sufficient CO_2 the previous night when the stomata were open (Fig. 6.10). How does this minimize water loss? High temperatures during the day would promote rapid evaporation if the stomata were open; evaporation is considerably lower if the stomata are open during the cool night hours instead.

But if daytime evaporation is minimized, how do these succulents keep cool beneath the blazing sun? The saguaro, for one, does not present a broad, nearly horizontal surface to the overhead sun as does, say, a spreading elm tree. Instead, like the lombardy poplar (Chapter 1), the saguaro accents the vertical. In addition, its succulent habit and low surface:volume ratio retard internal warming. Like a cool melon in a warm melon patch, a saguaro's internal temperature lags behind that of the environment, partly because of its high water content (cf. Table 2.1), partly because the surface for heat exhange is minimized. Small succulents are more vulnerable to heating than is the saguaro; Gates et al. (1968) have found that in the morning sunlight the flat blades of prickly pear cactus reach temperatures well above those of the air, and these researchers speculate that prickly pear has evolved a special tolerance to high internal temperatures. But the blades of prickly pear typically are oriented so that minimal surface is perpendicular to the mid-day and afternoon sun; clearly, tolerance of high temperature is only part of the thermal strategy of this cactus.

Plants having a limited surface area for light reception might be expected to grow quite slowly, and this is true of most succulents. Saguaros take perhaps 10 years to reach a height of 2 cm, and 20 to 50 years to achieve the modest height of 1 m (Niering *et al.*, 1963); in this and in many other characters they are the antithesis of desert annuals, as we shall see. In one trait, however, saguaros do resemble desert annuals: these cacti produce as many as 40,000 seeds per plant, of which only a tiny fraction usually survive. It probably is more common among desert perennials for large seeds to be produced, containing ample food reserves for the seedling when it germinates in its harsh environment. This tendency is well exemplified by the pinyon pine, whose seeds, large enough to have been a staple food of Shoshone and

TABLE 6.2

Adaptive Modifications of Desert Plants

Partly from R. H. Whittaker, 1975, *Communities and Ecosystems*, p. 169, New York, The Macmillan Co.

1. Deep or wide-ranging roots for effective uptake of water.
2. Ability to store water in tissues.
3. Morphological features that reduce water loss from leaves, including waxy cuticle, surface hairiness, sunken stomata.
4. Limitation of leaf surface area by reduction of leaf number or complete absence of leaves, or by loss of leaves during periods of drought.
5. Use of stem surface for photosynthesis.
6. Reversed day-night cycle of stomatal opening; other controls on water loss mediated through control of size of stomatal opening.
7. Tolerance of tissues to desiccation; some club mosses and ferns can recover from nearly air-dry conditions.
8. Ability to take up water from relatively dry soils when tissue fluids have been concentrated by desiccation.
9. Use of biochemical pathways that permit photosynthesis with reduced water loss and at unusually high temperatures.
10. Possession of small leaves that do not heat appreciably above air temperatures.
11. Orientation of leaves or of whole plant in ways that minimize solar heating of tissues.

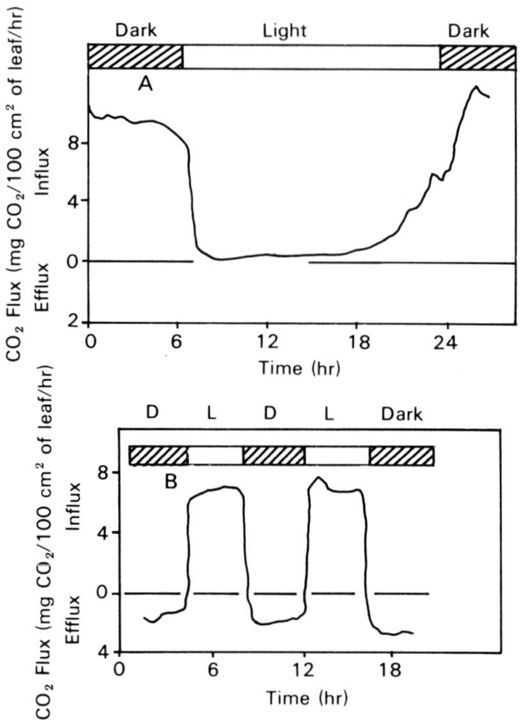

FIG. 6.10. Reversed stomatal rhythm of desert succulents is evident in *graph A*, which shows the carbon dioxide flux of an agave plant under light and dark conditions. In the dark, CO_2 is taken up rapidly from the atmosphere by a special carboxylation mechanism known as crassuleic acid metabolism (CAM). When lights are turned on, this uptake ceases abruptly; the stomata have closed. The CO_2 flux of a nonsucculent (tobacco) is shown in *graph B*. CO_2 is taken up only during light periods, and in the dark there is a net *loss* of CO_2 via respiration. Returning to *graph A*, we see that no loss of CO_2 occurs at any time in the agave cycle. Respiratory CO_2 is photosynthetically refixed when the lights are on and, when the lights are off, it is stored by CAM for later photosynthetic use. Thus succulents like agave need not replenish respiratory CO_2 losses from the atmosphere and can operate with less stomatal exchange with the atmosphere—hence less water loss. This saving is over and above the saving achieved by the day-closed, night-open stomatal pattern. (From T. F. Neales, A. A. Paterson, V. U. Hartney: *Nature, 219:* 469–472, 1968. Reproduced with permission.)

Navajo, contrast strikingly with the light, wind-dispersed seeds of pines in moister regions. Survival of saguaro seedlings, however, probably depends less on stored nutrients than on the occurrence of an opportune shower. Species perpetuation thus may be best assured by production of many light seeds, virtually all of which may succumb in some years, but goodly numbers of which may survive in others.

Paloverde and mesquite—These two species of perennial shrub (Figs. 6.7*B* and 6.11) possess small leaves that are shed during times of water stress, and are most abundant in the neighborhood of watercourses. They are desert **phreatophytes,** plants which, unlike the saguaro, rely on relatively permanent underground water sources. Some phreatophytes, such as cottonwood, require water continually and grow only along the largest and most reliable watercourses. Mesquite (*Prosopis juliflora*) and paloverde (*Cercidium microphyllum*) are not so restricted; they simply become photosynthetically dormant when their water supply fails. Dormancy is perhaps too strong a word for paloverde; chloroplasts in the green stem remain active even when the leaves are lost, but stem photosynthesis is probably inadequate for indefinite survival (Klikoff, 1967).

The massive taproot systems of these shrubs (Fig. 6.11) are as remarkable in their way as is the root system of saguaro. But because they have few shallow roots, mesquite and paloverde respond hardly at all to the light rain showers that sustain shallow-rooting species (Klikoff, 1967). In fact, paloverde and saguaro often grow together in dense clumps (Fig. 6.7*B*): in a land of little water, they compete for this common necessity hardly at all.

Small leaves are a common feature of desert perennials, except for succulent species, and deserts dominated by small-leaved species sometimes are termed microphyll deserts. The tiny leaves or leaflets of species like mesquite and paloverde are probably a mode of adaptation to desert heat; although small leaves heat rapidly, they remain very close to the air temperature (rather than rising above it, as do the blades of prickly pear), since they are as quick to pass heat on to their surroundings as they are to absorb it (Gates *et al.*, 1968). With their access to water, paloverde and mesquite may maintain leaf temperatures below those of the air by evaporation, but certain other small-leaved species cannot afford evaporative cooling. Many such species lose their leaves in summer, perhaps responding less to water shortage than to thermal stresses greater than the leaves can tolerate.

The seeds of mesquite and paloverde are

relatively large, providing nourishment for the rapid growth of an initial taproot. The seeds of paloverde and several other species have exceptionally hard coats, and germinate only when the coat has been ruptured by mechanical abrasion (Went, 1955). The most common cause of such abrasion is the occasional desert rainstorm, which tumbles the seeds roughly downslope or down the course of a wash. In the case of plants like paloverde, which grow best on uneven ground or along watercourses, reproductive success obviously is furthered by the requirement for abrasion. The seeds are produced without regard to rainfall, but germinate only when a storm has cracked them—and soaked the ground.

This brief look at perennial desert plants by no means exhausts their interest. Every species possesses its own kit of adaptations. For example, creosote bush, the shrub that dominates the most arid of American deserts (Fig. 6.12), lacks many of the specializations so far mentioned, but thrives with a different set. Nor have we considered Death Valley's *Tidestromia*, whose remarkable heat tolerance is a center of current attention among desert physiologists. *Tidestromia* actually does *best* at leaf temperatures of 47° C (117° F) and photosynthetic rates drop

FIG. 6.11. Root system of a mesquite plant has been exposed by erosion at the edge of a desert wash. The deep-plunging taproot of this species is known to penetrate as much as 53 m beneath the surface. (Photograph by author.)

FIG. 6.12. Widely spaced creosote bushes dominate America's most arid hot deserts. Like saguaro, creosote bushes possess spreading surface root systems, and the spacing of the plants may result from competition for moisture. The morphological specializations of this species seem comparatively minor, but it grows in habitats too dry even for cacti. Note the paucity of other plant species in this community. Such community simplicity is a sign of extreme conditions. Many careful studies (e.g., Chew and Chew, 1965; Barbour, 1968; Oechel *et al.*, 1972a, 1972b) have revealed subtle adaptations of creosote bush, but its ability to live in the most extreme deserts remains somewhat mysterious. (Photograph by author.)

rapidly when leaf temperatures drop below 44° C (Bjorkman *et al.*, 1972; Bjorkman and Berry, 1973). Like several desert plants, *Tidestromia* employs the C_4-dicarboxylic acid pathway of photosynthesis, a metabolic pathway that seems to be a biochemical adaptation to hot, arid climates (it facilitates CO_2 uptake at higher temperatures and less evaporative expense than does the common Calvin cycle of photosynthesis).

Annual Desert Plants

For the most part they are small, like the herbaceous plants of a garden, and they burst suddenly into a riot of blossom. It is they which are chiefly responsible for the desert's sensational spring . . . , and it must be strangely like that of the arctic, where the shortness of the warm season produces the same result as the shortness of the wet one. In both regions the plants rush from bud to seed with astonishing rapidity. Of the prudence, the foresight, and the thrift of the paloverde or the cactus these desert prodigals seem to know little and care less.

Joseph Wood Krutch (1951), *The Desert Year*

Desert annuals really should be termed ephemerals; "annual" is a misnomer for such short-lived plants. These species typically possess none of the water-obtaining, water-conserving, or desiccation-tolerant traits of desert perennials. Rather, desert annuals escape periods of water stress as dormant seeds, and restrict growth and reproduction to the desert's infrequent periods of adequate moisture. Even seeds, of course, cannot remain viable if completely dry, but they are the most drought-resistant stage in the life cycle of any plant.

Among desert annuals, the term "rare" assumes a meaning somewhat different from its meaning in other communities. Two botanists from the University of Arizona, Barbara and Arthur Phillips reported to a recent symposium on the ecology of the Grand Canyon that during the wet spring of 1973 "rare" annuals appeared in profusion in the canyon's inner gorge. Obviously there was no rarity of the seeds of these species, only a rarity, in normal years, of conditions suitable to their growth.

In this respect, rare desert annuals are like the rare weeds of a mature forest. Seeds of both are present continually but may remain dormant for long periods, often for 20 years or more. "Common" desert annuals may be more regular and profuse in their appearance only because they are less exacting in their germination and growth requirements. But whether common or rare, the primary adaptation of most desert annuals appears to be precise control of the conditions under which their seeds will germinate. A series of studies by Went (1955) and others showed that in Death Valley, the seed of annual species normally would not germinate unless at least an inch of rain had fallen within a relatively short period. Went believes that the seeds possess a chemical inhibitor which must be leached away by water before germination can occur, a neat moisture-measuring device which helps to explain the high success rate of many desert annuals once germination occurs. Success here is defined as growth through the flowering and seed production phases, and it is a relative term: the size attained by individuals, and the amount of seed they produce, is closely related to how much *extra moisture*, beyond that needed simply to trigger germination, is present at the start of growth (Went, 1973). Thus the pool of seeds lying dormant on the desert surface from year to year is replenished unevenly; while representing the accumulation of a number of seasons, it may be *largely* composed of the fruits of a single season in which growth was unusually lush and seed production lavish.

The success of desert annuals is promoted by their rapid growth rates, allowing maturation and seed production before exhaustion of the initial moisture supply. But topography can sometimes affect success considerably. In the Australian desert, Mott (1973) found that annuals germinated after rain in approximately equal numbers on slopes and on flatter terrain beneath the slopes, but there was considerable difference in survival. Presumably because drainage was more rapid, survival on the slopes was always lower than on the flats, and success rates on the slopes were sometimes less than 10%.

The scanty precipitation of some deserts is concentrated into distinct winter and summer "rainy seasons." In Western Australia, where this is the case, the annual plants are divided into distinct winter and summer floras: annual grasses predominate after the summer rains, and annual dicots after the winter rains (Mott, 1972). This separation is caused by differing characteristics of the seeds: those of species forming the summer flora have a high-temperature requirement for germination, while those of the winter flora have a low-temperature requirement. A temperature trigger is effective because the seeds of most desert annuals remain unburied on the desert surface (some, indeed, will not germinate if buried), and experience the full seasonal fluctuations of temperature.

Small seeds are the rule among desert annuals; widespread dissemination is advantageous, and wind is the most common agent. A variant of this common dispersal pattern is displayed by the many convergently evolved types of tumbleweed, which break off at the base when mature, and because of their globular, bushy growth form are blown easily across the land, scattering seeds as they go. A number of desert annuals produce larger seeds which nonetheless are dispersed widely (and inadvertently) by animals; burs or other holdfasts ensure that seeds of this type are carried abroad, while at the same time discouraging the carriers from eating them. Burs also help, eventually, to anchor the seeds to the ground. A more specialized anchoring device characterizes the seeds of certain other species; when wet, these species' seeds produce an adhesive slime that attaches them to the soil (Van der Pijl, 1972).

Widespread dispersal is not characteristic of all desert plants, be they annual or perennial. Reproductive "parochialism" has evolved in some, presumably favored by the likelihood that if a desert plant succeeds in producing seed, it occupies a favorable microhabitat. If this is true, its offspring may fare best if they do not wander very far. In some species, the seeds are retained in a heavy seed head, ensuring that they ordinarily will germinate near the parent; the seed head serves further as an absorber of moisture when it rains, benefitting the enclosed seeds in this way also. Still other desert plants schizophrenically employ both dispersal strategies: they produce two types of seed, one detachable and easily transportable, the other clinging grimly to the dead parent and germinating *in situ* during the next season of growth. This strategy is perhaps particularly common in sand deserts, where the dead parent frequently becomes buried after death and the still attached seeds are thus naturally planted. (Van der Pijl, 1972).

Poikilothermic Animals: Arthropods and Reptiles

Both arthropods and reptiles produce relatively little metabolic heat and essentially dry wastes, and both typically have highly impermeable skins. These are "preadaptations" to deserts. But the arthropods, being so small, are less able than any other group of desert animals to afford evaporative cooling. Perhaps it is not surprising, therefore, that desert arthropods, prominent among which are scorpions, spiders and insects, have evolved unusual abilities to tolerate high body temperatures. Some species are able to endure body temperatures up to 50° C, considerably outdoing terrestrial animals, including most arthropods, of other habitats (Edney, 1967; Hadley, 1972). Most desert arthropods also can survive a considerable degree of desiccation, and when dehydrated they can produce even drier excreta than at other times. When extremely hot, desert arthropods may resort briefly to evaporative cooling by opening wide the spiracular valves leading into their respiratory chambers. Because of their small size, arthropods cool very rapidly by evaporation but can ill afford it, and a commoner tactic is to take refuge from the desert heat in cool, relatively moist microhabitats. In addition, the life cycles of many desert arthopods are adjusted so that eggs or other resistant stages are present during the hottest or driest part of the year.

Case histories of the beetle *Eleodes armata* and the cockroach *Arenivaga* provide interesting contrasts because the way of life of these desert insects is nearly opposite. *Eleodes* is diurnal and a surface dweller, *Arenivaga* nocturnal and primarily a burrower. *Eleodes* has a highly impermeable integument; that of *Arenivaga* is rather permeable.

Eleodes is colored a jet black, a trait which would seem maladaptive in a species exposed to desert sunlight on hot days. But even hot days in the desert often are cool when they start: a basking *Eleodes* can warm quickly to temperatures optimum for activity, then retreat to shade during the heat of the day. Overheating also is prevented by an air-filled dorsal cavity which creates an insulating gap between the tough integument and the soft abdomen beneath (Fig. 6.13). It is thought that *Eleodes* controls the thickness of this insulative layer by raising and lowering its abdomen. On a cool morning the abdomen could be warned rapidly if it were pressed against the wings of a sunbathing beetle; as the day warmed, the abdomen might gradually be moved away from the wings, creating an insulative layer of appropriate thickness to the degree of heating. The dorsal air cavity of *Eleodes* undoubtedly serves also to restrict the loss of moisture from the spiracles, since the air in the cavity is much more humid than that outside.

Arenivaga is a tunneler in sand dunes. Seldom during the day does it encounter humidity as low as does the surface-dwelling beetle, and it is protected also from the heat when it is underground. At night *Arenivaga* comes to the surface and feeds on organic debris. Although the air is cool at this time, it is still dry, and dehydration may drive these cockroaches underground at intervals to replenish their moisture reserves. This they can do even in the absence of liquid water; females and nymphs of the species have the nearly unique ability to absorb moisture from humid air. Edney (1967) reports that this occurs when the air has a relative humidity of 82% or above. At such times the permeable integument of *Arenivaga* becomes a positive asset, by facilitating moisture uptake. But when, one might ask, is air ever that humid in the desert? To a human, the answer is "almost never," but to an animal beneath the soil surface the answer is "frequently." In the desert one even finds species of burrowing isopod, gill-breathing crustaceans which would seem particularly prone to desiccation. A few centimeters beneath the desert surface, they usually find sufficient humidity to sustain them.

Like arthropods, desert reptiles often are small enough to make use of the crevices,

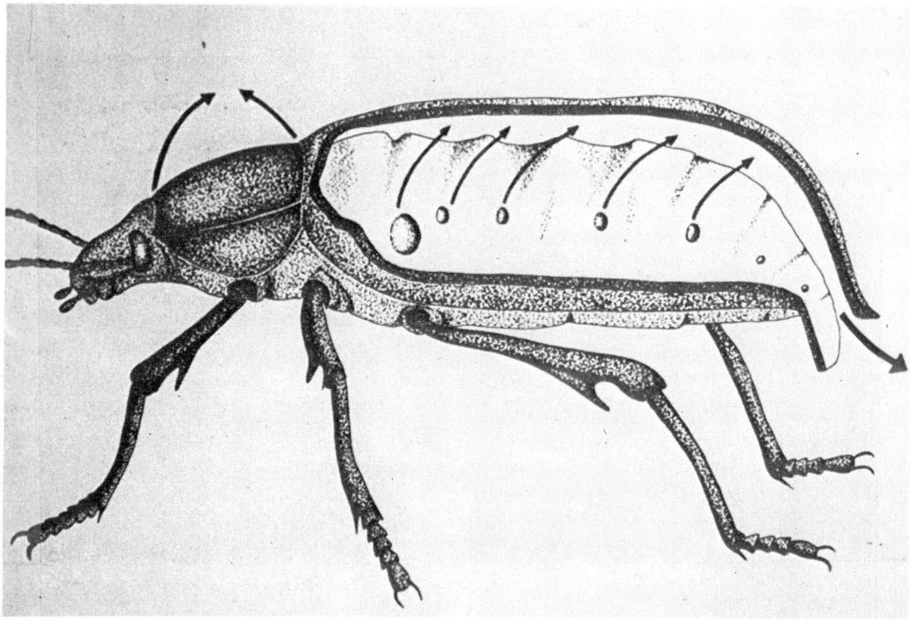

FIG. 6.13. Diagram of *Eleodes armata* showing dorsal cavity and probable route of respiratory water loss from the spiracles. (From G. A. Ahearn: *Journal of Experimental Biology, 53:* 573–595, 1970.)

burrows and recesses that provide respite from extreme heat and dryness. In addition, reptiles are experts in the art of behavioral thermoregulation (Bogert, 1959). A basking lizard can radically alter its heat balance simply by changing the orientation of its body toward the sun. In early morning it warms itself rapidly by lying broadside to the incoming rays: later it may face the sun, minimizing the amount of body surface exposed to direct insolation; and at mid-day it may retreat to the shade of a plant or burrow into sand. The conductive heat balance can similarly be controlled simply by altering the degree to which the body makes contact with the substrate. Many agile lizards seem to scamper on tiptoes across hot rocks or dunes; but if they are hot and can find a cool recess, they press their bodies against the substrate and maximize conductive heat loss.

Certain physiological abilities also are involved in the thermoregulation of reptiles, though these are not really analogous to the metabolic thermoregulation of homoiotherms. For example, the familiar lizards known as "horned toads" are able to alter their skin color. When exposed to the sun they lighten if they are too hot and darken if they are cool, thereby altering their absorption properties with regard to radiant heat. As a general trait, we might expect reptiles of hot deserts to possess lighter coloration than the reptiles of other biomes, thereby facilitating reflectance of the sun's radiant heat. This, in fact, is what one finds (Hutchinson and Larimer, 1960).

Desert reptiles drink water when it is available, but some are able to survive nearly entirely on the moisture in their food. A remarkable specialization for water uptake characterizes the Australian lizard *Meloch horridus* (Bentley and Blumer, 1962). The skin of this lizard is thrown into many tiny longitudinal folds. If the animal discovers a small pool or seepage, it crawls into it and, in effect, blots up the water. The liquid is drawn by capillarity into the skin furrows, and is then pulled toward the head by a mucus-rich, hydrophilic saliva which is extruded from the sides of the mouth. This saliva, with its load of water, is periodically swallowed and more is produced to maintain the flow toward the mouth. Despite its impermeable skin, the lizard thus is able to obtain water from pools too shallow for normal drinking.

Desert Rodents

Among small desert mammals, the rodents are such a successful and adaptively versatile group that we may turn to them for examples of nearly all the problems and survival strategies that one finds among small homoiotherms in the desert. We already have described the seasonal torpor of the Mohave ground squirrel; appreciably smaller rodents, like the desert pocket mouse, are probably incapable of long term torpor, but hummingbird-like they may conserve energy (and in the desert, moisture as well) by brief daily periods of torpor. Tucker (1966) has shown this to occur in pocket mice during periods of drought and food shortage. Many species of rodent, however, do not escape desert stresses by dormancy but manage to satisfy their needs even during the most extreme seasons of the desert year while remaining fully active. Two which we will consider in detail are Merriam's kangaroo rat (*Dipodomys merriami*) and the antelope ground squirrel (*Citellus leucurus*).

The kangaroo rat employs a stratagem unavailable to plants but extremely common among small desert animals as its sole apparent means of coping with desert heat. It restricts its activity to the night-time, and rests comfortably in a cool burrow by day. This pattern of activity greatly alleviates the animal's moisture problems as well, but moisture problems remain of great interest since Merriam's kangaroo rat thrives in America's driest deserts even when drinking water is absent for months at a time. In fact, laboratory studies indicate that it need not ever drink. Animals ordinarily obtain some of their water by drinking, some as juices in the food they eat, and some as an end product (oxidation water) of the metabolic breakdown of their food. Oxidation water is created even from the driest of foods; the breakdown of 1 g of dry carbohydrate, for example, results in the formation of 0.6 g of water as one of the end products. The water balance of any animal can be determined by measuring how much water it obtaines by these three means and how much it loses by evaporation or excretion. In a classic, beautifully conceived study of water balance in the kangaroo rat, Knut and Bodil Schmidt-Nielsen (1953) provided the model for many later inquiries into how desert animals solve their water problems, and probed the severity of conditions that the kangaroo rat could endure. Table 6.3 shows that at 0% relative humidity and a temperature of 24° C, kangaroo rats limited to a diet of dried seeds have a negative water balance: they lose more water than they gain, and cannot survive indefinitely. At the same temperature but at a relative humidity of 50% the rats get more water than they need from the same diet because of the reduction of evaporative water loss. In fact, with the other two parameters held constant, income and outgo of water just balance at a relative humidity of 10% (Schmidt-Nielsen, 1964).

But evaporation would be affected by temperature, and the laboratory temperature of 24° C seems unrealistically low for a hot desert summer. Armed with their data, the Schmidt-Nielsens returned to the field to investigate the microclimate that kangaroo rats actually expe-

TABLE 6.3

Water Balance of Dipodomys merriami in the Laboratory

A diet consisting solely of dried barley can sustain Merriam's kangaroo rat in the laboratory, when temperature and humidity conditions are favorable. The table below indicates the calculated water available and the measured water losses of animals kept on this diet over a 5-week period. If an animal is to survive indefinitely, water loss cannot exceed water intake. The experiment was run at a temperature of 23.9° C (75° F). Data from Schmidt-Nielsen (1964). R.H. = relative humidity.

Water Income			Water Outgo		
	At R.H. of 0%	At R.H. of 50%		At R.H. of 0%	At R.H. of 50%
	g			*g*	
Drinking	0	0	Urine	14	14
Water in food	0	13.5	Feces	2.5	2.5
Oxidation water	53.7	53.7	Evaporation	44	25
Total	53.7	67.2		60.5	41.5

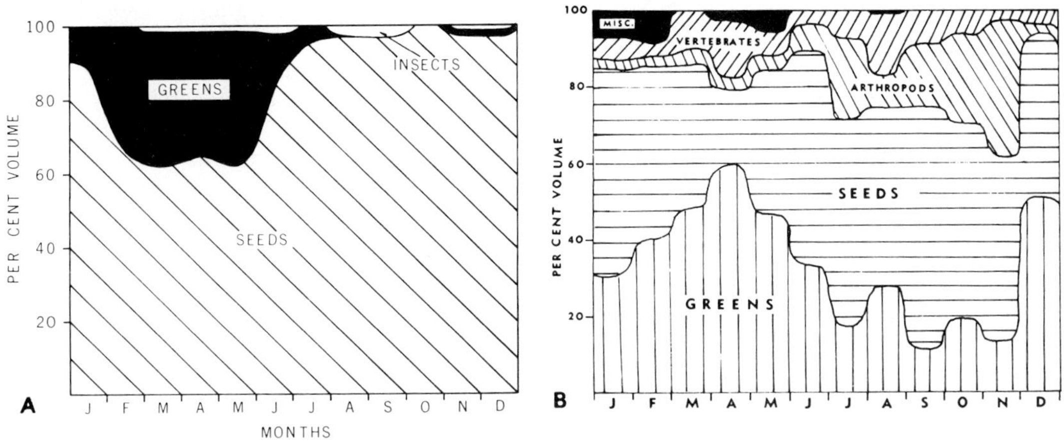

FIG. 6.14. Seasonal composition of the diet of two desert rodents: Merriam's kangaroo rat (*A*) and the antelope ground squirrel (*B*). (From W. G. Bradley: *Journal of Mammalogy, 49:* 14–21, 1968, and W. G. Bradley and R. A. Mauer: *Journal of Mammalogy, 52:* 497–507, 1971.)

rience. A captured animal, its tail loaded with micrometeorological instruments, obligingly revealed the the secrets of its burrow in early summer: a temperature range of 24–31° C, and relative humidities of 30 to 50%. The Schmidt-Nielsens calculated that if it remains in its burrow by day and exposes itself to surface conditions (16–24° C temperatures and 15–40% humidities) only at night, a kangaroo rat *is* able to live only on its meager income of oxidation water, at least during periods of normal water need. But during the reproductive season females lose considerable water via lactation, and thus need more. The Schmidt-Nielsens did not investigate water balance during lactation, and it is not known how much the water requirement of nursing females is elevated; but the breeding season occurs during the spring, when leafy annual plants are available to form a juicier supplement to the normal diet of dry seeds (Fig. 6.14 *A*). The timing of breeding, and the concurrent dietary shift, can be viewed as components of the whole adaptive complex of these animals with regard to water.

We have not yet explained the remarkable parsimony of the outgo columns in Table 6.3. Water losses of 40 to 60 g over a 5-week period are so small as to suggest that the kangaroo rat possesses unusual powers of water conservation. This species' nocturnal habits and humid burrow are important, but cannot explain the parsimony achieved in the Schmidt-Nielsens' laboratory experiments. In fact, the kangaroo rat constricts every normal avenue of water loss. Its kidney is remarkably efficient in reabsorbing water; the kangaroo rat, like all mammals, excretes a liquid urine, but one remarkably concentrated in salts and urea, wastes being flushed away with very little water expenditure. Fecal water loss is likewise minimized: small quantities of very dry pellets are produced. In processing the same amount of food, white Norway rats lose 5 times as much water in the feces as do kangaroo rats (Schmidt-Nielsen, 1964). Loss of water as sweat is low in all rodents due to a paucity of sweat glands, but the kangaroo rat also minimizes evaporative water loss during respiration. This derives from the fact that the animal's snout is relatively long and its nasal passages unusually narrow and subdivided. Whenever the outside air is significantly cooler than the rat's deep body temperature (normally about 38° C), inhalation sets up a thermal gradient within the nose, the tip being cooler than the upper nasal passage. Inhaled air is progressively warmed as it travels toward the lungs, and within the lungs itself it becomes saturated with moisture from the respiratory surfaces. On the outward trip, however, air travelling through the long, narrow passages of the snout becomes progressively cooled, and its saturation capacity decreases. The result is that much of the moisture from the lungs condenses on the walls of the nasal passages and is recovered. Air leaving the nostrils is still saturated with water, but the saturation level is low since the air is relatively cool. This mechanism of water conservation is much more effective, of course, during the cool night-time hours than it would be during the day, and the kangaroo rat's nocturnal habits thus work to its considerable advantage. It has been shown that when the outside air has a temperature of 30° C and a humidity of 25%, a kangaroo rat conserves 54% of its outgoing respiratory water by nasal condensation; when the outside air is 15° C and 25% saturated, the saving is 83%. Man, by contrast, recovers very little moisture by nasal condensation because his nasal passages are broad and relatively short and exhaled air is expelled almost at lung temperature (Schmidt-Nielsen, 1972; Jackson and Schmidt-Nielsen, 1964).

The second of our desert rodents, the antelope groundsquirrel, possesses many of the water-conserving traits we have described for the kangaroo rat; its kidneys, for example, produce a urine nearly as concentrated. But the antelope groundsquirrel differs from most desert rodents in being diurnally active. It challenges the desert sun with a recklessness that is undoubtedly costly in terms of water and that invites investigation of possible thermal adaptations. Studies by Bartholomew and Hudson (1962) and by Bradley (1968) set forth the special problems faced by this animal, and the solutions it has evolved.

Essential to the survival kit of the antelope groundsquirrel, as of the kangaroo rat, is a burrow. To the groundsquirrel it is a haven of moist air and cool soil that is sought repeatedly even during the daytime activity period. During the groundsquirrel's forays beneath the sun its body temperature rises rapidly, but periodically it seeks its burrow, flattens its body against the cool floor, and just as rapidly dispels the heat it has absorbed (Fig. 6.15). Most above ground activity occurs in the morning and later afternoon; following southwestern tradition, the groundsquirrel retires for a siesta during the hottest hours. But severe heat stress is nonetheless a problem during the hours of activity. The typical groundsquirrel travels up to 170 m away from its burrow in search of food (Bradley, 1968), and shade is often scarce within this foraging range. Physiologically, however, the antelope groundsquirrel is equipped to minimize the heat problem. It tolerates body temperatures unusually high for a mammal—up to 44° C or slightly more—and its extremely broad thermal neutral range (9.4° C) extends nearly this high. Thus, the fluctuations in body temperature that occur during each foray away from the burrow ordinarily cost the animal little

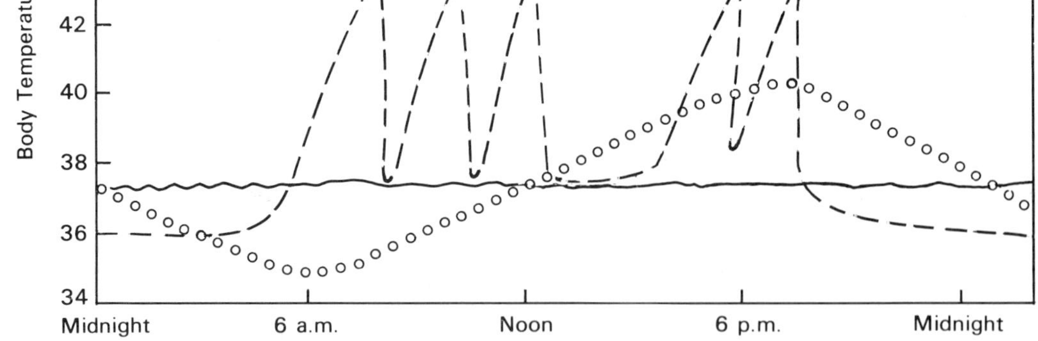

FIG. 6.15. Typical 24-hour body temperature patterns in the camel, antelope groundsquirrel, and man. Each is discussed in the text. Camel and groundsquirrel are relaxed homoiotherms, man a strict homoiotherm. (Adapted from G. A. Bartholomew: *Symposia of the Society for Experimental Biology, 15:* 7–30, (1964.)

in terms of water used for cooling. Indeed, the groundsquirrel runs such a high temperature that on cool days it can unload some of its body heat by conduction or thermal radiation even when outside its burrow, and so prolong its food-gathering journeys. On very hot days it may resort to evaporative cooling when its body temperature nears the lethal limit. Though the animal has no sweat glands, it spreads saliva over its face and paws and is cooled in this way. But the bulk of evaporative water loss undoubtedly occurs during respiration; because it breathes the hot daytime air on the desert's surface, the antelope groundsquirrel can recapture little of the outgoing moisture from its lungs, at least during its daytime activity periods. How are these water losses replenished? probably largely in the diet (Fig. 6.14*B*). Far more than the kangaroo rat, the antelope groundsquirrel is a dietary generalist, and foods containing considerable water—greens, insects, vertebrate carrion—are eaten throughout the year. Because its kidneys are able to produce a highly concentrated urine, the groundsquirrel may sometimes exploit mineralized desert pools, drinking the saline water and excreting the excess salts; but frequently such pools are unavailable and the juices and oxidation water of its food are the groundsquirrel's only sources of water. Its generalized diet serves it very well in this regard. In an environment where many food types are erratic in abundance and where water is scarce, we might expect many species to be dietary generalists.

But a specialized diet may be just as adaptive if it is reliable and provides water. The white-throated wood rat (*Neotoma albigula*) seems in many ways to be very poorly adapted to the desert, but its specialized diet of cacti (90% of its food intake during the driest months of the year) provides all the water it needs to survive (Schmidt-Nielsen, 1964). The wood rat resembles the collared peccary, another cactus-eating desert mammal, in having a "leaky kidney" unspecialized for water conservation. Both these mammals, instead, possess metabolic specializations that detoxify the poisonously high oxalic acid levels in cactus tissue and allow them to use its stored water.

Large Mammals in the Desert

The bedouin and his camel, together with a few date palms, populate anyone's mental picture of the hot deserts of the Old World. It is appropriate that we consider these mammalian partners together. Moreover, it is instructive: despite certain similarities the adaptive contrasts between camel and man can hardly be more dramatic.

Both are large, diurnally active species, and they share the consequences of their size: reduced rates of heat exchange with the environment; greater internal water reserves than possessed by small animals; a resultant potential for considerable dependence on evaporative cooling; the ability to travel considerable distances; and the inability to utilize many of the microhabitats where small animals find refuge from desert extremes. But if we probe the riddle of desert survival by these species, the explanations we uncover are vastly different.

The camel is the subject of another classic study by the Schmidt-Nielsens (*cf.* Schmidt-Nielsen, 1959 and 1964). Like the antelope groundsquirrel, this mammal is a "relaxed homoiotherm" which conserves energy and moisture by letting its temperature float over a range of some 7° C (Fig. 6.15). But because of its large size it warms and cools much more slowly than does the groundsquirrel, and hence can spend longer in the sun without overheating. Indeed, the camel ordinarily can spend the whole day in the sun without serious stress and without excessive water loss; only near the upper limit (about 41° C) of its thermal neutral range does it begin to sweat.

Size alone is not responsible for the camel's slow thermal changes. As in other mammals (except man), the fur traps air and provides an insulating boundary layer that retards both heat exchange with the environment and evaporative water loss. But a camel's fur cannot be too thick; if it were, a mammal this large would be unable to rid itself of internally generated heat.

Although not the equal of the kidney of many desert rodents, the camel's kidney nevertheless qualifies as a desert kidney: much water is reabsorbed from the urine, and the final product is a waste rich in salts and small in quantity (Table 6.4). The camel's feces, too, are very low in moisture. In other respects, including its prickly disposition, the camel is not unlike a cactus: it tolerates remarkable levels of dehydration (up to 30% of its body weight), and when water again becomes available it swells like a saguaro. The Schmidt-Nielsens have

TABLE 6.4

Maximal Urine Concentrations of Selected Mammalian Species

The listed figures are, in most cases, the highest recorded for each species, and hence give a good measure of their relative abilities to concentrate urinary output and reabsorb water from the kidney tubules. The urine electrolytes are of course salts, of which sodium chloride is usually the most common. The urine: plasma ratio is a measure of how many-fold the fluid filtered from the blood has been concentrated by the time it is excreted. Some of the values are approximations based on related measures. From Schmidt-Nielsen (1964) and sources quoted therein; data on the collared peccary (Zervanos and Hadley, 1973) have been added.

	Urine			Urine:Plasma Osmotic Ratio
	Urea	Electrolyte	Osmotic concentraction	
	mmol/liter	*meq/liter*	*osm/liter*	
Collared peccary (*Tayassu tajacu*)	600		1.28	4.0
Man	792	460	1.43	4.2
Wood rat (*Neotoma albigula*)	1980	714	2.7 (est.)	7 (est.)
White laboratory rat	2160	760	2.9	8.9
Camel (*Camelus dromedarius*)		1068	2.8	8
Antelope ground squirrel (*Citellus leucurus*)	2860	1200	3.9	9.5
Kangaroo rat (*Dipodomys merriami*)	3840	1200	5.5 (est.)	14 (est.)
Gerbil (*Gerbillus gerbillus*)	3410	1600	5.5 (est.)	14 (est.)
Jerboa (*Jaculus jaculus*)	4320	1530	6.5 (est.)	16 (est.)

watched dehydrated camels recover their 30% weight loss in 10 minutes at the water trough; in the case of large individuals, the quantity drunk during such an orgy may exceed 100 liters.

The camel's enigmatic hump is largely fat. It may be useful as extra insulation against the overhead sun, but it certainly does not serve as a useful water source. To metabolize fat and recover its oxidation water requires a greater than normal intake of oxygen, resulting in heightened respiratory losses of water from the lungs and no net water gain for the animal (Schmidt-Nielsen, 1964). Probably the most important function of the hump is realized in times of food shortage, when it can be tapped as a reserve source of energy.

Now let us turn to man. In the Arizona desert there is a dustly little settlement named Why?. The name hardly could be more apt, for man unquestionably is a morphological and physiological misfit in the desert. Unlike the camel he is a strict thermoregulator (Fig. 6.15), spending energy and water extravagantly to maintain a body temperature that is not only virtually constant but also relatively low when compared to other desert homoiotherms. Man sweats early on a desert day, and continues long after the camel has stopped sweating in the evening. He has no insulative coat of hair, and lacks the camel's tolerance of dehydration. Men seldom recover from water losses exceeding 10 to 12% of body weight. It is not simply water loss, but an associated overheating, that causes human death from dehydration: the blood, robbed of water, becomes more viscous; its flow slows, and it no longer is as effective in moving metabolically produced heat to the body surface for dissipation. The metabolic furnace overheats and the human dies. The blood of the camel, by contrast, remains quite fluid during moderate dehydration because water is withdrawn from other tissues; the camel thus avoids the thermal complications of dehydration.

How does man survive in the desert? His kidneys, respiratory tract and skin all leak badly, and he suffers thermal problems when short of water. Uniquely human characteristics give him a precarious desert foothold: the ability to carry water supplies outside the body; the use of camels, automobiles and other means of transport more drought-resistant than his legs; and the ability to fashion appropriate substitutes for fur. The loosely fitting bedouin burnoose, for example, provides protection from solar radiation and considerable insulation from the outside air, but at the same time it permits sufficient air circulation next to the skin to allow evaporative cooling. The bedouin prospers in the Sahara, but only with the aid of his camel and his inventions.

Desert Birds

We already have seen that certain desert birds possess special physiological adaptations to the desert. The desert poorwill (*Phalaenoptilus nuttali*), with its high thermal neutral range and abnormally low metabolic rate within that range, (Fig. 6.5) is perhaps the most remarkable example. The poorwill, a relaxed homoiotherm, lets its body temperature float upward to nearly 45° C before it begins to actively thermoregulate, and even then its self-cooling methods do not appreciably increase its metabolic rate: it simply opens its huge mouth and with a nearly effortless fluttering of the throat muscles pumps air in and out of the highly vascularized mouth cavity; the blood in these vessels is cooled by a combination of evaporation, convective air transfer, and radiation (Bartholomew *et al.*, 1962). More than other birds, the poorwill probably needs such specializations; it is a nocturnal flyer, but during the day it rests quietly on the unshaded desert surface, completely exposed to the desert sun. Perhaps it avoids predators in this way; they tend to seek shade during the day, and the well camouflaged poorwill probably is seldom discovered.

Perhaps the most common avian adaptation to deserts involves breeding periodicity. In the Australian desert, for example, a number of bird species breed erratically, displaying reproductive activity after rain regardless of the season, and failing to breed, even for long periods, during droughts. A biphasic breeding season (mid-April to mid-June and late July to mid-September) characterizes several species of the American Sonoran desert: the birds nest during the relatively cool spring and the characteristic rains of late summer, but not during the hot, dry days of early summer. The best known of these species is the road-runner (*Geococcyx californianus*). In the case of this bird, nest site selection and characteristics of the young are further specialized. Road-runner nests are so positioned as to be partly sunlit, but to be crossed by bands of shade during at least the hottest daylight hours. The young are born naked and deeply pigmented with black; this seems to facilitate both heat uptake during the cool early morning, when the nestlings presumably seek sunlit parts of the nest, and radiative cooling from shaded nest regions during the heat of the day (Ohmart, 1973).

Aside from juvenile and reproductive specializations, however, the road-runner does not seem particularly adapted to the desert climate (Calder and Schmidt-Nielsen, 1967). Indeed, most adult birds of the desert probably do not find the climate terribly stressful. There are three reasons for this. First, their mobility allows birds to seek shade, to travel great distances to obtain drinking water and to forage for food over relatively wide areas. Only nestlings are disadvantaged in this regard; hence the careful timing of the nesting season. Second, the excretory product of birds, like that of reptiles, is a semisolid paste of uric acid, not dissolved urea or ammonia. This means that birds lose less water via excretion than even the most frugal desert mammal. Third, the normal body temperature of birds is quite high. In the desert this means that during at least part of each day, net heat flow is from the bird to the environment rather than vice versa, giving birds the same protection from overheating that was noted for the antelope groundsquirrel.

Coping with Heat and Aridity: A Summary

Reviewing what we have learned of desert species, there appear to be four general ways of coping with desert heat. A species may evolve (*a*) a tolerance for high internal temperature, (*b*) ways, often behavioral, of escaping the extremes of desert heat; (*c*) ways of avoiding the rapid uptake of heat, despite exposure to it; or (*d*) self-cooling mechanisms and/or mechanisms that reduce the production of metabolic heat internally. Evaporative cooling is employed particularly by plants and large homoiotherms. Plants, of course, scarcely can avoid direct sunlight if growing in the open (though wilting and the resultant drooping of leaves is avoidance behavior of a sort); but certain plants, at least, have access to ample water for evaporative cooling. To animals, nonevaporative self-cooling mechanisms may be quite important: conductive self-cooling by the antelope groundsquirrel in its burrow, or radiative cooling from the mouth of the poorwill or from the huge, vascularized ears of a jackrabbit when it is sitting in the shade.

General methods of coping with moisture stress are (*a*) an evolved tolerance of dehydration; (*b*) highly efficient mechanisms of exploiting deep, small or ephemeral moisture supplies; and (*c*) highly evolved abilities to conserve internal water.

These major avenues of adaptation have numerous interesting byways and branches, but the broad avenues themselves are limited in number. In our many examples we have been witnessing numerous variations on a few basic themes. A desert autecologist thinks first of the themes, then of which themes are most feasible and of most benefit to the species he is studying. Then he seeks the species-specific adaptive complex, the intermeshing byways that together form his species' survival "strategy." Since the problems of heat and moisture are so closely linked in warm deserts, one cannot think separately of the adaptive specializations to these two stresses. The best studies of desert adaptation have been able to show how the tools in a species' survival kit function together as a combined strategy. Such studies provide marvellously clear insight into the achievements of natural selection: successful species molded to their environment.

UNRECOGNIZED DESERTS

If one lives in a moist climate, it is not necessary to travel great distances to study desert adaptations. Arid habitats can be found nearly everywhere. The interior of a house, for example, is an extremely dry habitat, at least to organisms deprived of moist food and unable to turn on the tap. The house mouse, in fact, is a species with desert origins and the water-conserving abilitis of a desert rodent (Fertig and Edmonds, 1969). Common domestic insects such as flour beetles and mealworms occupy habitats more arid than many true deserts: and the firebrat (*Thermobia domestica*), frequently found near furnaces and hot pipes, tolerates temperatures as high as 50° C while resisting evaporative water loss with unusual effectiveness (Edney, 1967).

Bare rocks, tree branches and sand dunes are similarly arid wherever they are found. These substrates are either too impervious or too porous to retain water, and only drought-resistant species can exist on them. Indeed, many plants and animals are common both to coastal dunes and true climatic deserts. Along the Carolina shore one finds species of armadillo and prickly pear cactus; and in a reverse twist, a tumbleweed now common in American deserts (the "Russian thistle," *Salsola kali*) is actually a seaside plant introduced from the Old World.

Studies such as those of Amos (1959) emphasize the many factors that are common to coastal dunes and inland deserts: aridity, shifting substrate, intense daytime heat, frequently strong winds and, in places, saline soils. There are differences as well, of course, one being the salt spray that is such a potent factor on exposed coasts. The fascinating zonation of vegetation in sandy coastal areas is the consequence of landward gradients of salt spray intensity and soil instability, together with a topographic diversity that produces sharp constrasts in soil moisture and exposure to wind (Fig. 6.16). Nowhere are terrestial microhabitats more clearly defined or more rewarding to study than on an exposed sandy coast.

THE COMMUNITIES OF EXTREME ENVIRONMENTS

Species Diversity

With few exceptions, the communities of extreme environments contain relatively few species, and the number of species drops progressively as conditions become more extreme (Fig. 6.17 and table 6.5). This statement is as true of environments made extreme by man, such as polluted streams, as of naturally harsh environments such as the aeolian zone of mountains or the upper rocks of wave-battered coastlines (Figs. 2.1 and 3.6). Not only are such communities species-poor, but in each trophic level one or two species tend to be vastly more common than any others. Most mathematical measures of community diversity are based on both these elements, the number of species present and their relative abundance (see Appendix). If two communities contain the same number of species, the one in which dominance is muted and resources more evenly shared is rated the more diverse. In extreme environments, both aspects of diversity are typically low.

Many desert communities, however, are surprising in this regard. Near Tucson, Arizona, Whittaker and Niering (1965) found the desert

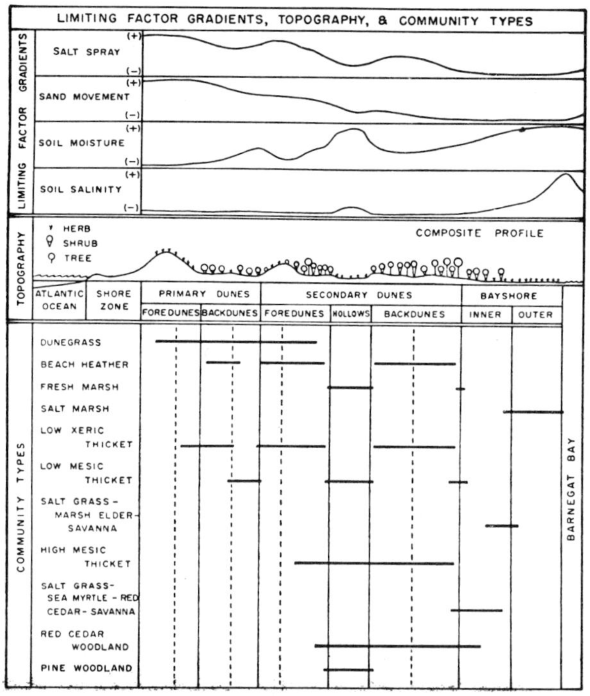

FIG. 6.16. At Island Beach, a barrier island on the New Jersey coast, W. E. Martin described 11 different plant associations in the half-mile distance separating beach and bay sides of the island. The locations of these associations, and the sharp environmental gradients responsible for such diversity, are shown here. (From W. E. Martin: *Ecological* (*Monographs, 29:* 1-46, 1959.)

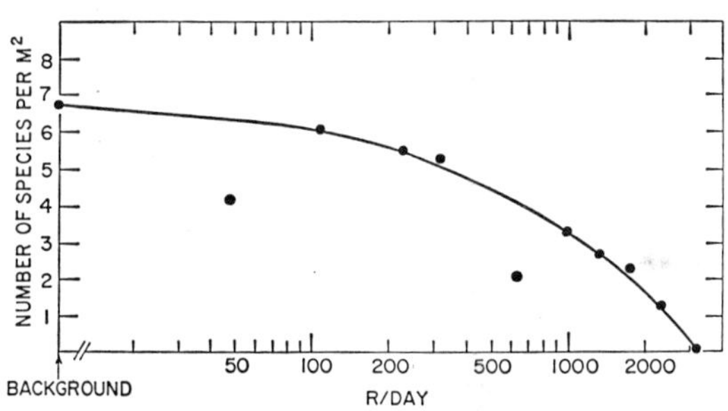

FIG. 6.17. At the Brookhaven National Laboratory, ecologists have studied the sensitivity of individual plant species and whole communities to radiation stress. This was done by placing ionizing sources in a recently abandoned field and in a forest and recording effects along the gradient leading to each emission source. Diversity of both old field and forest plant communities decreased as radiation stress increased. Shown here is the old field pattern. Radiation is expressed as roentgens per day. (From G. M. Woodwell: *Science, 156:* 461-470, 1967. Reproduced with permission from the American Association for the Advancement of Science.)

vegetation to be more diverse than that of any other biome in temperate America, including eastern forests. Is the desert, after all, not an extreme environment? Perhaps, as desert misfits, our perception is distorted. But the topographic variety near Tucson provides a considerable range of both temperature and moisture conditions, none of them as hot or dry as some other regions of the Sonoran desert. Perhaps we should simply recognize that some deserts are less extreme and/or more locally varied than others. In Death Valley, community diversity doubtless is quite low. But even here, there might be more microhabitats and more species than we would initially predict.

Trophic Relationships and Community Stability

If communities are typically simple in extreme environments, it follows that the food web also is relatively simple. Each heterotrophic species has fewer alternate foods, and often is heavily dependent on a single source of supply. As a result of this simplicity the community is poorly buffered: the collapse or increase of one species population has serious consequences for others (Fig. 3.23). Although the community may be in long term equilibrium, the populations of its constitutent species

TABLE 6.5

Distribution of Attached Algae along a Thermal Gradient

In the outflow stream of Sleeping Child Hot Springs, Montana there is a sharp, more or less constant thermal gradient. In this stream one sees evidence of partitioning of niches with respect to upper tolerable temperature. In addition, two species can be seen to *require* warm water but to differ in lower tolerable temperature. Note the reduction of species that is encountered at increasingly extreme temperatures. From Kullberg (1971).

Taxa	Temperature in degrees Centigrade																							
	52.0	51.0	50.5	50.5	50.0	49.8	49.5	49.0	48.0	42.0	40.0	39.5	39.5	37.0	36.5	36.5	36.0	36.0	36.0	36.0	35.8	35.5	35.3	34.5
Chroococcus minor	x	x	x																					
Phormidium truncatum	x	x	x	x	x	x	x	x	x	x														
Mastigocladus laminosus	x	x	x	x	x				x	x	x	x	x		x		x		x	x	x		x	x
Oscillatoria geminata	x	x	x	x	x	x	x	x	x	x			x			x	x	x	x	x	x	x	x	x
P. tenue					x	x	x	x	x	x														
Ulothrix subconstricta					x	x	x	x	x	x														
Stigeoclonium attenuatum										x	x	x	x	x	x	x	x	x	x	x	x	x	x	x
Mougeotia sp.										x	x	x	x	x	x	x	x	x	x	x	x	x	x	x
Lyngbya digueti										x													x	
Fragilaria sp.										x		x	x	x	x					x	x			
Achnanthes sp.											x	x		x	x	x	x	x	x	x	x	x	x	x
Nitzschia sp.											x		x			x		x			x	x	x	x
Pseudanabaena oblonga											x	x	x								x	x	x	x
Oedogonium sp. (12.5 μ)											x	x	x	x	x	x	x		x	x	x			x
O. sp. (25 μ)														x	x									x
Navicula sp.															x	x	x	x	x	x	x	x	x	x
Oscillatoria limosa															x									
Spirogyra sp. (38.5 μ)																x								
Rhizoclonium fontanum																	x	x	x	x	x		x	x
Nodularia harveyana																		x	x	x	x	x	x	x
Microcystis holsatica																			x	x				
O. princeps																				x				
O. tenuis																				x				
Chroococcus turgidus																				x				
Calothrix thermalis																					x			
Cosmarium obtusatum																						x	x	x
Amphora sp.																							x	
Epithemia sp.																							x	
Microcystis pulverea																							x	
Phormidium jenkeianum																							x	
Anabaena sp.																								x

fluctuate dramatically. These fluctuations often are initiated by the unpredictable character of extreme environments. In the summer of 1973 the desert around Phoenix experienced a rainless spell more than twice as long as any previously recorded; a few months later, parts of America's driest desert, the Mohave, were blanketed with 2 feet of snow. Events such as these directly and dramatically affect the abundance of certain species, and indirectly influence the abundance of other species linked trophically to the first group. Even in relatively diverse desert communities, particular species often depend heavily on particular food sources. Wood rats specialize on cacti, road-runners on lizards. In at least one part of the Mohave desert, up to 76% of the diet of Merriam's kangaroo rat consists of the seeds of one plant species, *Erodium cicutarum* (Soholt, 1973). Can one doubt that the fortunes of the rat are closely linked to those of *Erodium*? Even if alternate food sources can be exploited in emergencies, it would be surprising if the rat population in this locality did not rise and fall with every major change in the abundance of *Erodium*.

We should look further, for co-evolutionary consequences of food web simplicity. A species heavily preyed on by another is under strong selective pressure, and it would not be surprising if it evolved specific defenses against its predator. The predator is under equally strong selective pressure; only those individuals able to thwart the prey's defenses are likely to survive. The wood rat, as a cactus specialist, has successfully penetrated the morphological and chemical barriers thrown up by its prey, but perhaps it has not really won the war. Even wood rats spice their diet with other foods, perhaps of necessity diluting their high intake of oxalate; and given this respite, the cactus population survives. In cold deserts we find a similar case: the volatile oils of sagebrush inhibit bacterial activity in the digestive processes of the mule deer. The deer eat quantities of sagebrush nonetheless, but must consume approximately equal quantities of plant foods lacking these oils if they are to avoid acute indigestion (Nagy and Tengerdy, 1968).

Higher on the trophic scale, co-evolution seems similarly common in extreme environments. Desert rodents and rattlesnakes supply a remarkable example (Webster, 1965). Rattlesnakes, unlike most desert reptiles, are active at night. We may suspect that their nocturnalism is an evolutionary response to the nocturnalism of most desert rodents. Rattlesnakes possess, in fact, a remarkable means of hunting warm-blooded prey in the dark: pits below the eyes contain cells sensitive to the infrared heat radiating from desert mammals at night. But the kangaroo rat, and certain other nocturnal rodents, have evolved enlarged middle ear cavities that provide great auditory sensitivity to the low frequency noises produced by a rattlesnake as it prepares to strike. The outcome of a hunt, then, is not easily predictable. Rattlesnake cannot be wholly responsible for the ear specialization of the kangaroo rat. Owls, too, are a major nocturnal predator of desert rodents, and they too produce low frequency sounds as they swoop toward their prey. The rat is adapted to both.

Competition and Niche Specialization

How important is competition in the communities of extreme environments? Is not competititon between species likely to be minor, simply because many potential competitors cannot cope with the physicochemical harshness of the habitat? Work wth desert plants has convinced at least one authority (Went, 1955 and 1973) that this is so. Among desert rodents, however, interspecific competition has been considered a process of great importance (Rosenzweig and Winakur, 1969).

Went is perhaps right in his contention that desert plants do not live or die, ordinarily, by competition. Other agencies—herbivores or the vagaries of desert weather—may be more direct agents of plant mortality. But Went's argument implies that competition has never been important in shaping the desert plant community, and this conclusion would certainly be wrong. Everywhere in the desert we see plants specialized in ways that enable them to *avoid* competition for the moisture they all need. In some extreme environments, particularly where "extreme" refers to some resource in short supply, the community we see is one molded by competition. In such environments the consequences of direct competition have been so severe that the selective pressures favoring specialized, competition-evading habits have been very strong. What better explains the separation of annual desert plants into two groups: species that germinate only after summer rains and others that germinate only after winter rains? Desert annuals display few other competition-reducing traits, but this simple temporal strategy reduces interspecific pressures considerably. Among perennial desert plants, shallow- and deep-rooting habits sometimes confer nearly complete freedom from competition for moisture between two species; indeed, shallow-rooting saguaro often grows better in close association with deep-rooting paloverde than it does alone (Fig. 6.5*B*). Plant species having the same root characteristics, on the other hand, tend to differ conspicuously in some other niche parameter. In Arizona it is clear to the casual observer that shallow-rooting saguaro is more common in the coarse soils of slopes than the finer soils of flat outwash areas; shallow-rooting creosote bush shows opposite preferences, but is excluded in turn from the lowest depressions where salts accumulate, and where shallow-rooting saltbush finds its niche. Deep-rooting paloverde and mesquite minimize competition in similar fashion; the former can tolerate coarse soils and hence is at home on rocky slopes, while the latter is restricted to the finer soils of outwash plains and desert creek beds.

Among desert animals, we have noted already how the wood rat avoids competition from most other rodents by specializing in a cactus diet. A similar food specialization was discovered by Kircher *et al.* (1967) in studies of desert fruit flies. *Drosophila pachea* lays its eggs quite specifically in the rotting stems of senita cactus. Of nine coexisting species of *Drosophila* in the Sonoran desert, this species alone was found able to tolerate the alkaloids contained in this cactus: it has become a senita specialist, and thus avoids food competition with the other fruit flies.

Many species of desert rodent coexist in the same area, and undoubtedly they compete to some degree for the same food. But Rosenzweig and Winakur (1969) have described habitat preferences that serve to reduce the level of competition. Areas of sparse, open vegetation are characterized by any of several species of kangaroo rat (*Dipodomys*). Dense, low vegetation is the preferred habitat of certain species of pocket mouse (*Perognathus*), while other species of pocket mouse, together with wood mice of the genus *Peromyscus*, are most at home in dense, bushy vegetation. All three sorts of vegetation may be part of a very local topographic mosaic, so each species may find its preferred habitat within the same acre or two of desert. We are left, however, with the question of niche separation among the rodents of each subgroup: for example, there are 21 species of kangaroo rat in western North America, and up to four of these may coexist in the same local habitat. Where such species complexes have been studied carefully, competition-reducing differences usually have been discovered. In the Owens Valley desert of California, for example, the two most common species of kangaroo rat may occupy burrows beneath the same saltbush yet compete hardly at all with each other for food: *Dipodomys merriami* eats the seeds of the saltbush, and *D. microps* the leaves. Kenagy (1972) describes the beautiful morphological and behavioral specializations of the latter species, which carefully strips away the outer, salt-laden tissues of the leaf and feasts on the "sweet" inner tissues.

We have seen in earlier chapters how competition frequently is reduced in communities by the partitioning of dominant environmental gradients among species having different optima, or simply different tolerance limits. In the desert one can think of several gradients simply involving moisture: moisture-holding capacity of the soil, depth to permanent water, average frequency of showers, average nightly dewfall, potential evaporation during the hot season, and so forth. Desert species probably separate along each of these gradients, not simply along some single, all-encompassing, "moisture gradient." The situation is not as simple here as in the hot spring of Table 6.4, where temperature gradation by itself appears adequate to separate the niches of many algal species. In the desert, as in the hot spring and all other community types, active, interspecific competition is less because of gradient separation; but we may suspect that more direct competition in the past is what drove the species toward the different optima and specialties we now discern.

Ecological Succession

Should we expect true, biologically mediated, ecological succession in the communities of extreme environments? Let us consider the history of an abandoned field near Flagstaff, Arizona (McDougall, 1966). Flagstaff lies not in the desert but on a pine-forested plateau. It is neither especially hot nor especially dry, but the seasonal and daily temperature range is large and moisture is erratic. The field in question probably once was a meadow, dominated by the perennial grasses which typically supplant the pines in areas of deep, fine soil near Flagstaff. It was cultivated for some years before 1941, then sown to tall yellow clover and abandoned. Fifteen years later, when McDougall initiated his study, the field still contained considerable clover, mixed with annual, biennial and perennial weeds, but little native grass. Over the next 10 years there were marked year-to-year changes in the vegetation, but no identifiable progression. Clover virtually disappeared for 2 years, then returned. Two perennial native grasses advanced to importance (the approach of climax conditions?); then they regressed. By 1973, 32 years after abandonment, there was still little sign of anything that could be called true ecological succession (McDougall, personal communication to the author); the vegetation that dominated the field in any year seemed largely to be determined by such factors as duration of snow cover and the pattern and amount of spring and summer rain. McDougall feels that biological factors may play significant roles as well, particularly on a short term basis, but the enormous and unpredictable year-to-year changes in the plant composition of this field seem to represent a situation in which the community is not determining its own destiny so much as being reconstituted again and again by the vagaries of weather. If the perennial native grasses ever became permanently re-established as dominants, McDougall feels that it may be through sudden population explosions during favorable years rather than through any orderly successional sequence.

What is true of the Flagstaff plateau probably is even more true of the more arid parts of Arizona—and of the world. In the desert, active competition by plants for light is minimal, and the vegetation usually is too sparse to modify the soil in major ways. Whatever species replacement occurs over time is probably the result largely of weather changes, not biological agencies, and such changes are as likely to be reversed as to proceed. Unnaturally heavy grazing, where it occurs, seems to be the only significant biological agent of community change. In the Sonoran desert, for example, heavy grazing by cattle leads to the replacement of grasses by prickly pear and mesquite. The removal of ground cover and increase of prickly pear in turn encourage the proliferation of kangaroo rats and wood rats. Jackrabbit populations also increase, and reduce the incidence of paloverde by browsing on the saplings. The protection afforded to young saguaros by paloverde is thereby reduced, at the same time that the wood rat population is increasing; and saguaros, young and old, fall victim to the depredations of this rodent. Depending on the degree of grazing, then, desert communities assume quite different characters; and when they change due to grazing they pass through somewhat predictable stages, as just described.

Such grazing-induced sequences are perhaps the only example of true ecological succession in deserts, and they are usually, perhaps nearly always, associated with human interference. Examples of ecological succession in other extreme environments are similarly rare. Aside from the dominance of nonbiological agencies we may implicate, in truly extreme environments, the low pool of species available to constitute a successional replacement pattern. In Atlantic salt marshes, for example, two species of the genus *Spartina* overwhelmingly dominate the vegetation; apparently very few macroscopic plant species are able to survive daily baths of salt water, occasional doses of much fresher water from rainstorms, and a highly anoxic medium for the roots. If the *Spartina* is killed by some agency, the marsh will probably lie barren until *Spartina* itself returns: there are no other species available to constitute a vegetational succession. The same fate is all too evident in the simple mangrove community of coastal Vietnam, which was decimated in the Vietnamese war and is undergoing little in the way of successional recovery.

Like these saltwater communities, the montane fir forests of the Adirondacks and New England are nearly a monospecific vegetation type. Here Douglas Sprugel (1974), in a doctoral thesis for the Yale School of Forestry, has recorded a cyclic pattern of community decay and regeneration that forms an interesting contrast to the noncyclic seral progression that one associates with ecological succession. Perhaps such patterns, or their equivalent, are not uncommon in simple communities subjected to physicochemical stress. The balsam fir forests are characterized, at least on exposed, windward slopes, by elongated patches of dead trees which represent slowly moving waves of death. These waves seem to be driven by the death of older trees which become exposed (by the death of upwind neighbors) to intolerable wind desiccation, ice deposition and abrasion during winter. Most of the trees killed each winter are at the downwind edge of the dead band, where the protection afforded by other trees is least. Thus the wave of death moves progressively downwind, and is followed by a wave of regeneration as young firs, encouraged by the increased light, begin vigorous growth. On Mount Whiteface Sprugel found that the waves move about 1.8 m per year, and that a given area experiences a wave of death and regeneration at approximately 60-year intervals. He thinks that a wave begins when some local disturbance flattens a small patch of trees in an old stand; trees left standing at the downwind edge of this patch are then exposed to winter's full fury. True succession is not completely suppressed in these fir forests: when the overstory dies, light-demanding weeds such as raspberry and goldenrod spring up in the open patches, along with young paper birches and firs. But thanks partly to the selective feeding of snowshoe hares, firs soon prevail in the second growth. The overarching pattern is one of death and rapid return by the sole forest dominant.

Human Impact in Extreme Environments

The communities of extreme environments seem at first glance to be fragile things: typically they contain relatively few species and these may be living close to their biological tolerance limits. One might speculate that a slight push by man would set in motion a chain reaction, and even if the disturbance ended recovery might be slow.

If we consider the effects of overgrazing in the American Southwest, this indeed seems the case. In the 1880s cattle ranching became big business in the less extreme desert areas. Under this onslaught the relatively sparse vegetation cover shrank, and a severe drought from 1891 to 1893 compounded the pressure on the land. The result was serious erosion of topsoil and the cutting of innumerable small canyons and arroyos (Niering *et al.*, 1963). Cattle stocks were cut back as the range deteriorated, but the erosional damage was done, and its effects on the natural communities of the Southwest still are seen, decades later. Many areas, of course, have remained major rangelands, and the vegetation of many thousands of acres has become dominated by prickly pear and mesquite, species unpalatable to cattle. The face of the desert has indeed been changed by man, and as humans pour into the Southwest it will be further changed.

The same fragility may characterize tundra communities, where growth of plants is so slow that ecologists fear the scars left by builders of the Alaska pipeline will not heal for centuries. But it probably is untrue to generalize that the communities of extreme environments are *particularly* vulnerable to man. To the degree that desert communities, for example, are naturally subject to fluctuating stresses, they may be more resilient to the pressures imposed by man than are the communities of more stable and benevolent environments. Speculations of this sort, and other points brought out in this chapter, raise some fundamental questions about whether communities evolve different characteristics in different environments. We will turn to these questions in the next chapter.

SUGGESTED READING

Bartholomew, G. A. The roles of physiology and behavior in the maintenance of homeostasis in the desert environment. In G. M. Hughes, ed. *Homeostasis and Feedback Mechanisms*. New York, Academic Press, 1964.

Bartholomew, G. A., and J. W. Hudson. Desert groundsquirrels. *Scientific American* (Nov., 1961). Interesting approach in terms of niche separation.

Bjorkeman, O., and J. Berry. High-efficiency photosynthesis. *Scientific American* 229: 80–83 (Oct., 1973). Biochemical adaptation to the desert.

Cloudsley-Thompson, J. A., and M. Chadwick. Life in Deserts. Philadelphia, Dufour Editions. 1964. Deals with plants and invertebrates as well as vertebrates.

Hesse, R., W. C. Allee and R. P. Schmidt. Ecological Animal Geography. New York, John Wiley and Sons, 1951. A well known text and reference in adaptive ecology.

Jaeger, E. C. The North American Deserts. Stanford University Press, 1957. A descriptive and comparative account.

Krutch, J. W. The Desert Year. New York, Viking Explorer paperbacks, 1960. This book and the next are fine nontechnical reading by a perceptive naturalist.

Krutch, J. W. The Voice of the Desert. New York, Morrow Paperback Editions, 1971.

Maloiy, G. M. O., ed. Comparative Physiology of Desert Animals. New York, Academic Press, 1972. A symposium volume that provides a good introduction to the literature.

Schmidt-Nielsen, K.: Desert Animals: Physiological Problems of Heat and Water. London, Oxford University Press, 1964. A notable work by a leading student of adaptive physiology.

Went, F. W. The ecology of desert plants. *Scientific American* (April, 1955). A good introduction.

chapter 7

Evolution of Communities

Let us take, for example, a grass field, which can provide seeds and insect prey for birds, and assume that the field is accessible to many bird species of a variety of morphological, ecological, and behavioral attributes. We can measure a number of variables in the resident bird community: number of species, types of species, and just how they use the field's resources. The question is: Are the values measured in any sense optimal? It can be answered by finding a similar field somewhere else, which is accessible to and has been colonized by a taxonomically or genetically different bird fauna. We now compare ecologies between fields both at the level of single phenotypes and of community makeup. If we find extensive similarities or convergences, we can infer that selection has reached optimal solutions in both fields, despite differences in history, time scale, and genetic origins.

Martin L. Cody (1974), Optimization in Ecology. *Science 183:* 1156–1184. Reprinted with permission from the American Association for the Advancement of Science.

INTRODUCTION

All biological communities have much in common in the way they are organized and the way they function. Wherever we look in nature we can be sure that certain principles are operating: the principle of resource partitioning within a food web, the principle of competitive separation of niches, the principle of mutual interdependence of species. In every community we can classify the species by type: herbivores and carnivores, autotrophs and decomposers, specialists and generalists. Communities in nature seem cut from the same fabric.

But communities are different, too. They differ in the number of species they contain and in the pattern of abundances of these species. And some communities today seem more tightly integrated and organismic than others.

Natural selection is an optimizing process. It constantly favors better adapted types in the survival sweepstakes. But "better adapted" may have one meaning today and another next year; the direction of evolution shifts as environmental circumstances change. Optimization at the community level must be extraordinarily complicated. The community unit itself is not subject to natural selection: selection molds only the constitutent species, and whatever characteristics the community as a whole assumes, it simply reflects the outcome of optimization processes among these constituent species. It is impossible to imagine this shaking-out process at the community level as being anything but cumbersome and drawn out, since each constitutent species is pushed and pulled by myriad forces, its own fortunes depending on the relative success or failure of many other species and on the vagaries of a physicochemical environment which never is completely stable. Intuitively one might doubt that the optimization process extends to the community level in any effective way.

Yet we begin to detect patterns if we compare communities of different species in similar environments. Seeking evidence for optimization, we might try the approach suggested by Cody in the quotation that begins this chapter. Comparing two fields in different hemispheres he looked for ecological equivalence not just of individual species but of community organization, and found it in astonishing degree (Table 7.1). In his words, such a degree of convergence gives "reason to believe that there is a single optimal way of dividing up the resources of this type of field, and that it has been achieved, or at least approximated to the same extent, on both continents."

What we wish to discuss in this chapter is the evolutionary process as it affects communities. What do the convergent features of some communities, and their apparent divergence from other communities, really mean? Are different communities subject to quite different molding processes dictated by different sets of physical and chemical constraints? Or is there a single pathway along which all communities evolve, such that differences among today's communities simply represent different stages in a common maturation process? Does the Arctic tundra, for example, contain fewer species than the Amazon rain forest because it is a younger community? Or is there something about the Arctic environment that dictates a species-poor community as the only type that can exist?

Talk of an evolutionary maturation process for communities is easy to confuse with the idea of ecological succession. The two are not the same. Succession is a developmental process, not an evolutionary one. It is a sequence of colonization and replacement whereby communities on a site change over time. But in succession we assume that each species already has a particular niche which determines where and when it will appear. All species in a successional sequence are preadapted entities; they do not evolve new roles during the process of succession.

Community evolution covers far greater time spans far more slowly than succession. It never ends, although it may slow as niches fill and as species evolve finely tuned balances with one another—to the extent that they do. Can we document the course of community evolution in the past? Can we identify constraints and flexibilities affecting community organization in the world at present? Starting at the beginning may give us perspective.

TABLE 7.1

Convergence in Community Structure (Vegetation Structure and Bird Species Structure) of Fields in Kansas and in Chile

From Cody, 1974.

Comparison	Kansas	Chile
Vegetation structure		
Mean height (m)	0.29	0.27
Vertical density (number of leaves)	5.10	6.23
Horizontal density (number of leaves per meter)	8.68	9.26
Area under foliage profile	32.4	39.2
Bird community characteristics		
Number of species	3	3
Average between-species overlap in		
Habitat	0.63	0.60
Feeding heights	0.78	0.89
Food and feeding behavior	0.18	0.21
Overall ecology	0.53	0.56
Bird species characteristics		
Eastern meadowlark and red-breasted meadowlark (*Sturnella magna* and *Pezites militaris*)		
Body size (mm)	236	264
Bill length (mm)	32.1	33.3
Ratio of bill depth to bill length	0.36	0.40
Grasshopper sparrow and yellow grass finch (*Ammodramus savannarum* and *Sicalis luteola*)		
Body size	118	125
Bill length	6.5	7.1
Ratio of bill depth to bill length	0.60	0.73
Horned lark and Chilean pipit (*Eremophila alpestris* and *Anthus correndera*)		
Body size	157	153
Bill length	11.2	13.0
Ratio of bill depth to bill length	0.50	0.42

COMMUNITIES OF THE PAST

If we believe that life's remarkable diversity sprang from simple beginnings, and that all species trace their lineage back to a few (or even a single) primordial species, then it follows that Earth's first communities were far simpler than those today. At first there must have been major functional gaps—whole trophic levels missing, for example—and the few species present in early communities perhaps played broader roles than do the species of modern communities.

According to current popular theories of the

origin of life, Earth's first communities contained nothing but decomposers: aquatic organisms adept at utilizing the dissolved organic compounds synthesized by lightning, heat or ultraviolet radiation from simple precursor molecules. Competition for these nutrients must have been the earliest type of community interaction. We have no precise record of when this primordial community stage was reached, but Cloud (1974) believes that the existence of simple decomposers antedates the oldest sedimentary rocks still in existence, the 3.4 billion-year-old Onverwacht beds in South Africa (Fig. 7.1).

The next great step in the history of life really made possible the trophic complexity of modern communities. We refer to the appearance of autotrophs, *i.e.* organisms able to use outside energy sources to synthesize their own organic compounds. If species having such abilities had not appeared, the abundance as well as the diversity of life never would have approached today's levels.

From the beginning, sunlight seems to have been the chief energy source utilized by autotrophs. Bacteria exist today which capture energy for biosynthesis from inorganic oxidation reactions, but geological evidence indicates that such oxidations would have been severely limtied on the early Earth by a lack of free oxygen. So the first autotrophs probably were photosynthetic autotrophs: simple prokaryotic creatures like today's photosynthetic bacteria and blue-green algae. Fragmentary geologic evidence suggests that such organisms first appeared more than 3 billion years ago (Cloud, 1974 and 1968).

When did the other trophic levels appear? Herbivory probably evolved not long after photosynthesis. Decomposers already existed, and it does not seem a difficult step for some decomposers to have begun obtaining sustenance from living autotrophs rather than waiting for them to die. If this is how herbivory arose, the first herbivores probably were simple digesters of algal tissue, not organisms with mouths and stomachs. We can imagine many possibilities here for the beginnings of parasitism, not just of herbivory.

Although they left no record, the first carnivore trophic levels may have appeared before the autotrophs and herbivores. Small carnivores could have evolved as predators of the early decomposers soon aftter the latter had evolved. But if such carnivores existed, they (like their prey) must have been very small. Only much later in Earth's history, shortly before the Cambrian period, 600 million years ago, do the first macroscopic animals appear in the geologic record. These were soft, wormlike invertebrates, and they may well have been not carnivores at all, but decomposers burrowing like earthworms through organic sediment. Rather soon thereafter, however, marking the advent of the Cambrian itself, a rich array of hard-bodied invertebrates entered the fossil record. Hutchinson (1965) believes that this sudden paleontological bonanza signals the evolution of macroscopic predators. Armor, previously an encumbrance, took on great adaptive value once some animals acquired jaws (or their equivalent) and a taste for flesh. Along with selection for armor we might expect selection for rapid size increase as an aid to defense or to offense; size increases were, in fact, pronounced in the Cambrian and succeeding periods.

Thus, 600 million years ago, the elements of community food webs apparently had been established. Much has happened to influence communities since that time, as Figure 7.1 documents; but in terms of community structure there seem to have been no fundamental innovations: merely refinement, internal diversification and perhaps development of greater balance between producers, consumers and decomposers. Even the colonization of land and the appearance of the first terrestrial communities produced no radically new aspects of community structure. The fossil record subsequent to the Cambrian indicates bursts of evolutionary diversification at the organismic level, interspersed with periods of evolutionary stagnation and major phases of extinction. But the overall trend has been toward more and more species extant at the same time. At the community level this implies refinement and subdivision of niches, and a proliferation of highly specialized interactions between coexisting species—parasitic, mutualistic, allelopathic, mimetic, etc. Broad-leaved forests illustrate the subtlety of later evolutionary processes. Few major new groups of angiosperms have appeared on the scene in the last 60 million years, and there is an illusion of evolutionary stagnation among broad-leaved angiosperm plants. But mutual adaptation, particularly at the biochemical level, actually has proceeded apace among for-

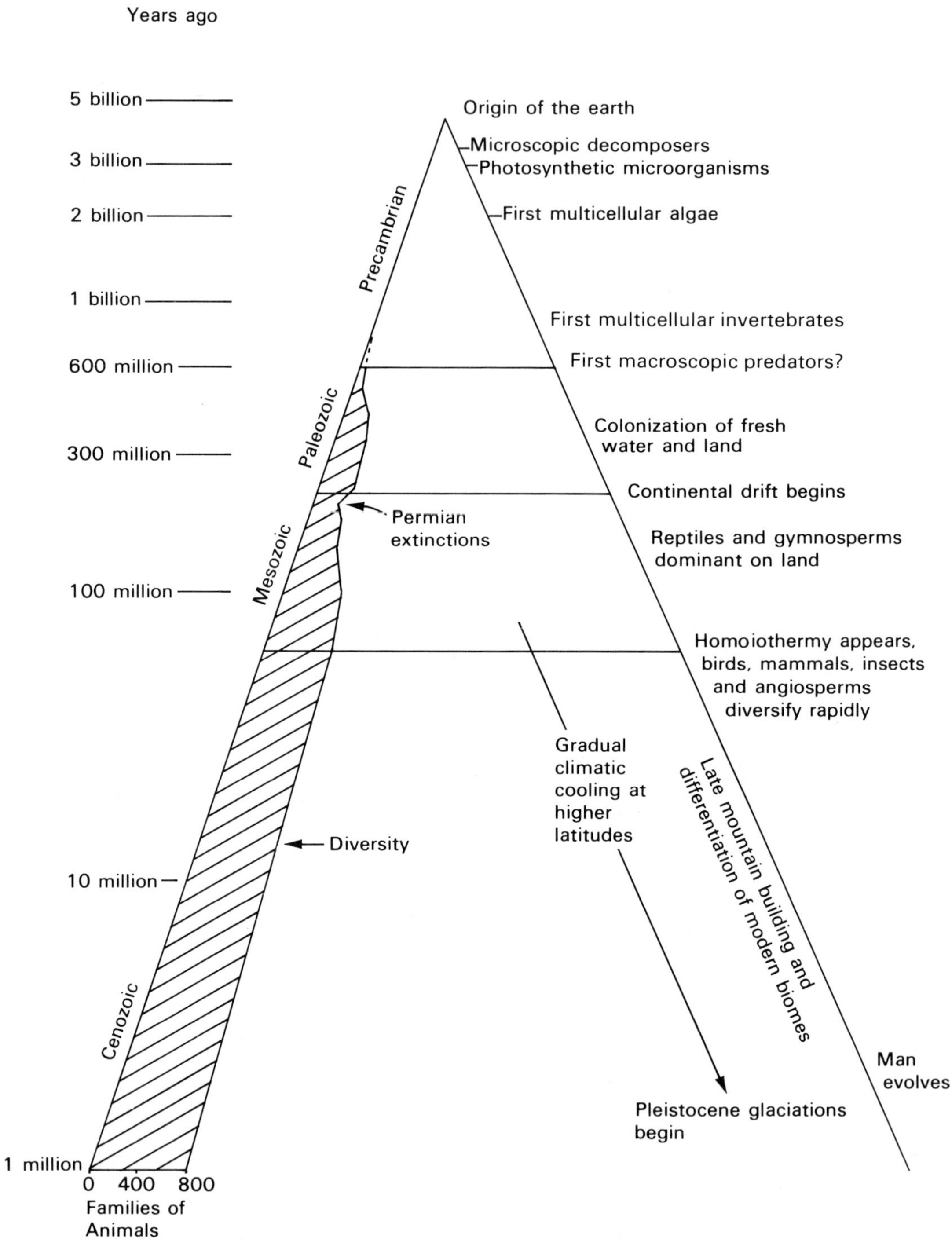

FIG. 7.1. Major events in Earth history relating to community evolution and the diversification of life. Note the logarithmic time scale: the Precambrian actually represents ⅞ of Earth's history in terms of time. As an index of diversification the number of families of animals in the fossil record (from Newell, 1963) has been plotted.

est plants, and between these plants and their insect herbivores and pollinators (Whittaker, 1970; Ehrlich and Raven, 1967; Grant, 1951).

From this description of postCambrian trends it is easy to imagine an evolutionary progression of many community types toward the organismic model proposed by Clements, and away from the individualistic model of Gleason. If there is such a tendency, however, almost certainly it is not universal. Intellectually, the community most like a superorganism *seems* more highly evolved than the loosely organized individualistic community, but very probably it is wrong to imagine *all* communities evolving toward a tightly integrated organismic model. If this were a universal trend it would spell evolutionary doom for successional types of communities, since these are loose and individualistic relative to mature communities. Indeed, the entire success of successional species derives from their individualistic, colonizing strategy, which obviously has been strongly favored in these species by natural selection. Successional communities are composed of successional species; obviously, then, we cannot suggest that the organismic model is the appropriate evolutionary model for such communities.

In the pages ahead we largely will avoid discussion of successional communities, restricting our comparisons and evolutionary speculations to equilibrium communities or to those in long term arrested succession. The organization of these communities in terms of species structure has received much attention in recent years, whereas very little information or theory has been forthcoming regarding the species structure of successional communities.

MODERN NONSUCCESSIONAL COMMUNITIES: SOME COMPOSITIONAL CONTRASTS

Rain Forest and Salt Marsh

Ecologists in the moist tropics are hard pressed simply to learn what lives in the communities they are studying. Everywhere they turn they find a new species. P. W. Richards (1973) reports that as many as 200 species of large tree may be found within 2 hectares (5 acres) of a rich lowland rain forest. The richest temperate forests, by contrast, might harbor 25 species in a comparable area. Animal contrasts are similar: rain forests of Central America contain more than 4 times as many bird species as does the North American deciduous forest, despite the relatively great diversity of bird species in the latter; and in the undergrowth of a tropical rain forest entomologists have collected 500 species of insect in 2000 local sweeps of a net. Richards concludes that "the most distinctive characteristic of the tropical rain forest is probably its great species richness."

As might be anticipated when 200 types live cheek by jowl, each tree species in a rain forest tends to be relatively rare; moreover, individuals of the same species tend to be quite widely scattered. Janzen (1970) suggests that this spacing pattern is the work of species-specific herbivores which kill the numerous seeds and seedlings near a parent tree but tend to overlook the occasional seed or seedling that is dispersed farther from the parent. Janzen sees herbivores as at least partly responsible for the richness of plant species in a rain forest, in that their pressure prevents dominance by a few species and allows many to coexist. Turning the argument around, many plant species should provide feeding niches for many types of herbivore, so a co-evolutionary proliferation of diversity can be visualized together with a biological constraint on excessive dominance by any one species.

What nonbiological factors might also facilitate the species richness of the rain forest? The habitat typically is poor in mineral nutrients, but the forest maintains very high productivity nonetheless. Perhaps this allows many species to coexist. Moreover, the climate is mild and equable. From day to day and month to month, rainfall and temperature are exceptionally stable for a terrestrial habitat. It was suggested in Chapter 3 how this seasonless climate could permit specialized "professions," such as year-round fruit eating, that are impossible in more seasonal habitats. Here too, then, may lie a partial key to the species richness of the rain forest.

Over longer periods, however, most rain forest regions have experienced marked changes in climate. The forests have shrunk, fragmented and expanded, and the fortunes of individual species have waxed and waned (Livingstone, 1975). Hence the rain forest habitat cannot be considered particularly stable over the long term. Because of their geographical location, however, certain rain forest areas like the Malay Peninsula seem likely to have experienced comparatively mild climatic fluctuations in the past, and it is these forests that have the greatest species richness. In the richest rain forests, therefore, climatic stability not just from season to season but over many centuries may have contributed to high diversity.

Let us look now at a temperate *Spartina* marsh. In terms of structure, no greater contrast to the rain forest could be found, for along the Atlantic coast of North America the vegetation of such salt marshes is essentially a one- or two-species monopoly. If there are two species co-dominant, both usually are *Spartina*, a tall species growing in relatively pure stands in the deeper water and a short species growing in relatively pure stands in shallower areas. A few other plants occur, but the dominance of *Spartina* is overwhelming (Table 7.2). The animal component of such a marsh consists of a number of insect, crab, mollusc, annelid and bird species, but here too the diversity is relatively low.

Like the rain forest, the salt marsh is a highly productive community, but fluctuating water levels and salinities make the physicochemical environment of the salt marsh rigorous. Few angiosperms have been able to adapt to marine habitats of any type, and in a salt marsh the variable salinity and the anoxic character of the substrate impose special rigors. Perhaps this is why so few plant species are found in the salt marsh; and we can understand why *Spartina*, a plant which can tolerate the rigors, manages —with rich nutrients and few competitors—to flourish so luxuriantly. The relative paucity of animal species, in turn, probably derives as much from the low diversity of plant foods as from the physicochemical rigors of the habitat.

TABLE 7.2

Dominance of Spartina in a New Jersey Salt Marsh

The data below are derived from small samples of vegetation (10 × 10 cm quadrats) harvested at 30-foot intervals along two transects in an intertidal salt marsh. Most of the variety is at the edge of the actual marsh, where *Spartina* abuts other communities. From an unpublished undergraduate study by Karen E. Daley.

	Number of Samples Containing Species	Total Dry Weight in All Samples
		g
Transect A (11 samples)		
Salicornia sp.	1	0.35
Distichlis spicata	1	2.4
Spartina alterniflora	11	152.3
Transect B (18 samples)		
Phragmites communis	1	0.9
Salicornia sp.	1	2.6
Distichlis spicata	4	18.3
Spartina alterniflora	16	227.7

Hot Spring and Abyssal Floor

Let us add the communities of two more habitats to our comparisons. A hot spring, of course, is a thermally extreme environment, and at the highest temperatures in a hot spring very few species are found (Table 6.5). Moreover, at any one temperature dominance by a single species often is pronounced. In community structure the hot spring therefore is analogous to a salt marsh. In seeking to explain this similarity we look for environmental commonalities, and immediately recognize the commonality of physical or physiological stress. A notable dissimilarity is that one habitat (the hot spring) possesses great environmental constancy, whereas the other (the salt marsh) experiences markedly fluctuating conditions, but this difference apparently is not important in determining the community structure of one or both habitats, judging from the similarity of the two communities.

In its environmental constancy the hot spring habitat is much more like the abyssal floor than it is like the salt marsh. The deep sea floor in most places experiences very little change in temperature, water movement, pressure or substrate features. But here, unlike the hot spring, there is a wealth of species. These are drawn, it is true, from a relatively small number of phyla, but it is a diverse community nonetheless. Even though the abyss receives a starvation ration of nutrients, a more species-rich community exists here than in many shallow marine habitats much better supplied with resources. Howard Sanders (1969), a student of abyssal communities, suggests a reason for this: environmental constancy over a great period of time has allowed species to adapt very closely to one another on the deep sea floor, and unusually narrow niches are the result.

But hot spring habitats also have experienced

environmental constancy over long periods; at least this habitat *type* is ancient, even though we do not know the age of individual hot springs. Some authorities even believe that the first life on Earth evolved in hot springs. Why, then, are hot spring habitats so species-poor? We must conclude from our comparative analysis that in hot springs the physical stress of high temperature overrides those environmental attributes that favor a species-rich community. Sanders probably would agree with this assessment: he believes there are two broad classes of community, and would assign the abyssal community to one and the hot spring community to the other. In the first class, which he calls **biologically accommodated communities,** stability of the physicochemical environment has allowed interspecific adaptation to proceed to great lengths, resulting in species-rich associations. He calls the second class **physically controlled communities;** these, like the salt marsh and hot spring, are much less rich in species, according to Sanders, because the physicochemical regime demands adaptive priority and prevents the evolution of closely balanced and integrated species interactions.

The two types of community Sanders recognizes actually intergrade along a continuum. Somewhere along this continuum, he suggests, all communities will fall. The examples we have used happen to lie near the extremes.

Small and Large Islands

Complicating the influences already suggested is the effect that habitable area seems to exert on the number of species in a community. The so-called **species-area relationship** is the empirical finding that the larger the habitable area, the more species-rich the community (Figs. 7.2 and 7.3). This relationship, which so far has been demonstrated for parts of communities (*e.g.* bird or reptile faunas) rather than whole communities, is clear only when one compares communities of the same type, *e.g.* rain-forested islands of different size. Other constraints on species richness cloud this area effect if one compares, say, islands of different size, some bearing rain forest, others grassland, and others desert. One must first group them by habitat, then look for a species-area relationship. The fact that such a relationship usually can be found suggests that geographic extent sets an upper limit to the number of species that a community of a particular type can harbor.

In Figure 7.3 a species-area relationship is evident even though there is a good deal of scatter about the regression line. The explanation usually offered for such imprecision is that the scatter represents different degrees of **species packing.** Communities falling on the regression line are thought to be at the long term species-area saturation point, *i.e.* their available niches are fully occupied or "packed." Communities whose points lie above the line are thought to be oversaturated and unstable, and those below the line to be undersaturated (capable of absorbing more species).

Except for isolated islands, terrestrial communities usually seem to be "packed full," judging from the small degree of scatter one

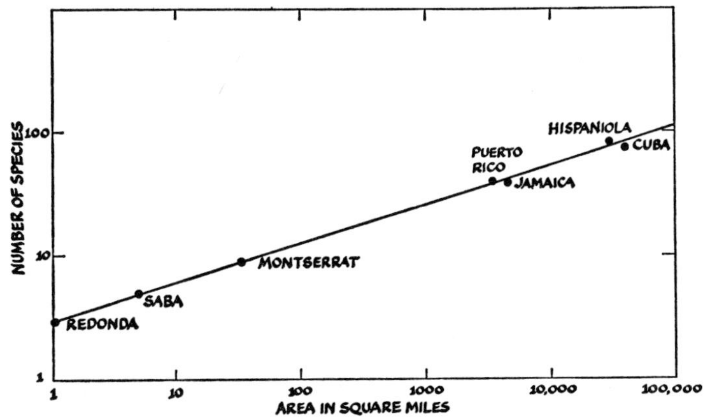

FIG. 7.2. Species-area relationship: amphibians and reptiles of the Caribbean islands. (From R. H. MacArthur and E. O. Wilson: *The Theory of Island Biogeography*, Princeton, N. J.: Princeton University Press, 1967. Reproduced with permission.)

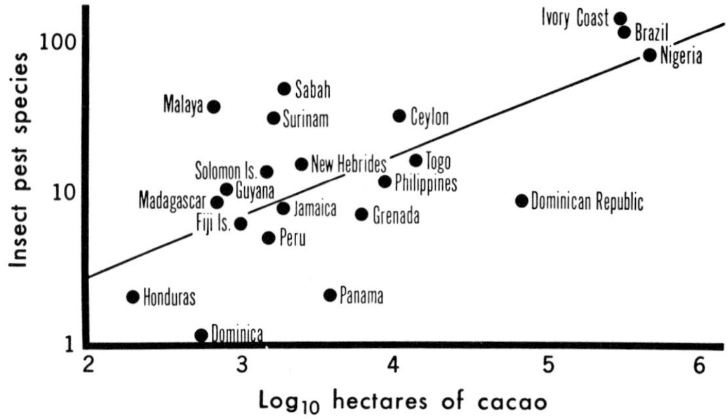

FIG. 7.3. Insect pests of cacao: a species:area relationship in which the areas plotted are the amounts of land under cacao cultivation in each country. Scatter of points may represent nonequilibrium conditions, since cacao cultivation is relatively recent in some countries (see text). (From D. Strong, Jr.: *Science, 185:* 1064–1065, 1974. Reproduced with permission from the American Association for the Advancement of Science.)

usually finds around a species-area regression line. A few community types, however, *e.g.* lakes, may normally be undersaturated (Barbour and Brown, 1974) because they are relatively youthful habitats where diversification and colonization are not yet complete.

Why should geographic extent affect a community's species richness? One can understand that geographically cramped habitats might not meet the niche requirements of certain community members, such as large mammals, but it is hard to apply this explanation to insects or plants, which also show species-area relationships. A cause that is applicable to all groups of organisms is suggested by MacArthur and Wilson (1967), who believe that a different balance between species immigration and species extinction prevails on small and large islands. (An "island" here refers to any isolated patch of habitat, be it an island in the geographic sense or a patch of forest in a sea of prairie.)

But why should the balance between immigration and extinction be a function of geographic area? As we shall see in Chapter 15, extinction rates are relatively high when species populations are small, as they are in communities of small geographic extent. Extinction rates are lower in large communities. Thus, the equilibrium between new species invading a community and species being eliminated from it is shifted progressively in the direction of extinction as one passes from larger to smaller communities of a given type. In communities of any size there is an equilibrium between these processes, but the result of this area-related shift is fewer species in smaller habitats (Fig. 7.4).

MacArthur and Wilson's idea of a community continually being altered by introduction and extinction of species is a view that does a certain violence to the organismic concept of the community. Introductions and extinctions seem, however, to be continuing phenomena in all communities, and MacArthur and Wilson's explanation of the species-area relationship has gained wide acceptance.

SPECIES RICHNESS: A SUMMARY OF CURRENT THEORY

What do the foregoing comparisons suggest about the factors controlling a community's species richness? There appears to be an area effect, a stress effect, an environmental stability effect, and a synergistic biotic effect, the last named involving a feedback among species such that their activities create niches for more species. These effects sometimes work at cross purposes, sometimes additively. They all should be considered when one seeks to explain the species richness of a particular community or community type.

Having reached these conclusions, it would be well to review briefly and more formally the major hypotheses that have been advanced regarding species diversity, of which a major component is species richness. The other component, as pointed out in Chapter 6 and the Appendix, is degree of dominance. For example,

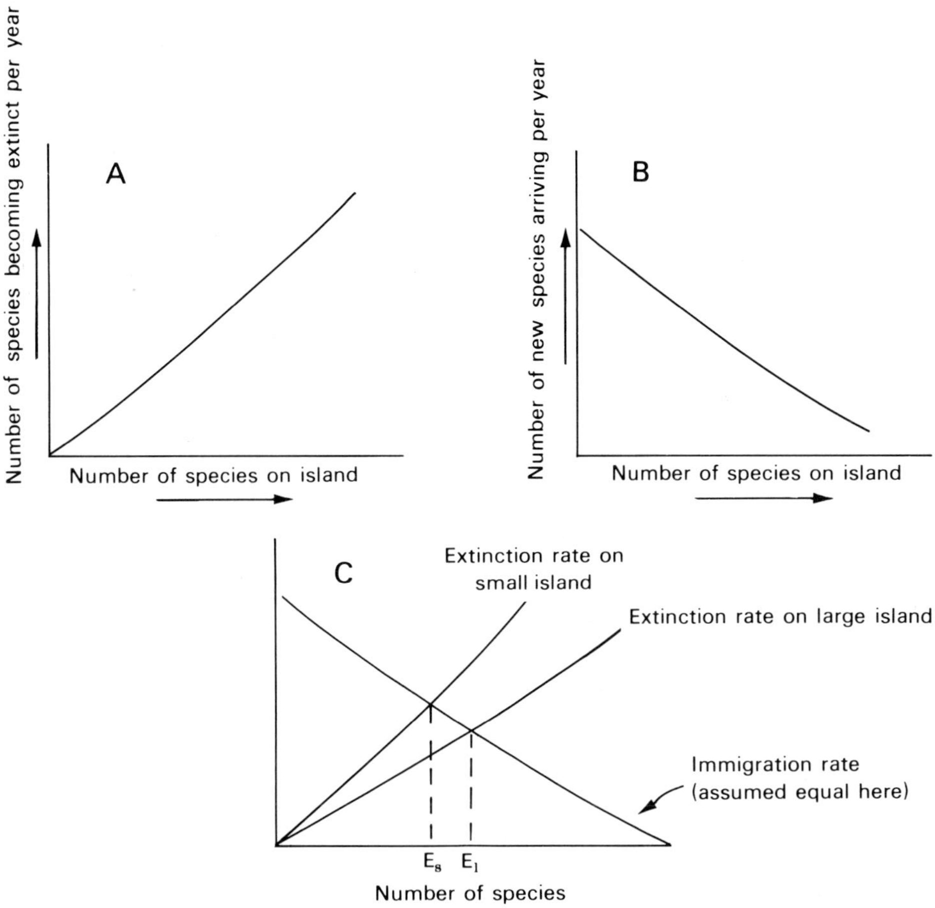

FIG. 7.4. Balance of species immigration and species extinction determines the number of species in a "saturated" community, according to MacArthur and Wilson's theory. *Graphs A* and *B* show theoretical extinction and immigration functions separately; they predict that the more species there are in any community, the more extinctions there should be and the fewer new species immigrants there should be. The extinction curve rises mostly as a matter of probability, but perhaps also because of more competition as the number of existing species rises. The immigration curve declines simply because more and more of the species available from source areas already are present. Plotted together (*graph C*), the crossing of these curves defines the equilibrium number of species (E_s, E_l) in any habitat. But if the extinction curve is steeper for geographically limited habitats (small "islands"), then the equilibrium species level will be less in such habitats (E_s) than in larger ones (E_l).

if two communities possess equal species richness, *i.e.* contain the same number of species, one nonetheless is considered less diverse than the other if it is characterized by disproportionate dominance of one or a few species. We have spoken so far in this chapter only of species richness because we wish to speak separately of dominance in a later section. The theories to be presented here can be considered for our present purposes theories of species richness because simple species richness normally correlates strongly with diversity.

Pianka (1966b) reviews a number of the hypotheses regarding the causes of differing species richness.

1. The *productivity hypothesis* suggests that only in resource-rich, highly productive habitats can communities rich in species evolve. The real world, however, furnishes both contradictions of this hypothesis (*e.g.* the resource-poor, species-rich abyssal floor) and exceptions to it (*e.g.* the productive but species-poor salt marsh). We should not discount the productivity hypothesis completely—perhaps the high

productivity of the rain forest does facilitate species richness there, and the minuscule productivity of the aeolian zone may mitigate against species richness on mountaintops. But overall, the productivity hypothesis seems very weak.

2. Hypotheses involving *physical influences or stress* include what Pianka calls the competition hypothesis, and relate to Sanders' classification of physically controlled and biologically accommodated communities. It seems quite likely that increasing degrees of stress (physical control) underlie the latitudinal gradients in species richness that have been shown to broadly characterize terrestial communities. In general, tropical communities are species-rich, temperate communities intermediate, and polar communities species-poor. In Pianka's words, stress hypotheses of species richness would suggest that "natural selection in the temperate (and polar) zones is controlled mainly by the exigencies of the physical environment, whereas biological competition becomes a more important component of evolution in the tropics." As already noted, the assumption behind this hypothesis is that species in unstressed communities are able to adapt more closely to each other; niches therefore can be defined more narrowly, and more species can coexist. Latitudinal gradients in diversity or species richness seem good evidence for the importance of stress, but we can find much more evidence elsewhere as well. In hot springs and deserts, for example, we find gradients of species richness that correlate directly to degree of physical severity in the habitat (*cf.* Table 6.5).

3. Biological influences obviously are implicit in the stress hypothesis of species richness: where biological influences are slight, and physical influences strong, species richness theoretically is low. Biological influences also are implicated in two other hypotheses regarding species richness, the first of which emphasizes the role of predators (Paine, 1966; Spight, 1967). This *predation hypothesis*, touched on earlier with regard to the rain forest, suggests that herbivores, if really effective, prevent domination of a habitat by one or a few plant species and give other plants room to flourish. The increased plant diversity encourages more species of herbiovore, which in turn provide niches for more species of carnivore, etc. This hypothesis is unlike the stress hypothesis in that it suggests that species-rich tropical communities have become that way not because of accentuated interspecific competition but rather because of low competition (due to effective predation). Reduced competition, according to the predation hypothesis, allows more species to have a place in the community and promotes evolutionary synergism between different levels of the food chain.

4. The *habitat heterogeneity hypothesis* would occur to anyone who looked closely at the microenvironmental differences and corresponding species preferences that occur in a community. Obviously if a community encompasses a wide range of microenvironments it should have more species than a community containing only a few microenvironments. We can talk here about contrasts in moisture, slope, wind, light, etc., but we must also include biotic contrasts when we talk about habitat heterogeneity. Different kinds of organisms in a community represent qualitatively different food sources and competitors. The number of kinds of plants, for example, should directly influence the number of kinds of herbivores that one finds in a community. In addition, a diverse vegetation, say, provides a wide range of physically different microhabitats and may increase both the horizontal and vertical heterogeneity of the environment in a spatial sense. For example, MacArthur and MacArthur (1961) found that the number of bird species in terrestrial communities is related closely to the vertical extent and distribution of foliage (the "foliage height diversity")—much more closely, in fact, than to the number of plant species that are present. The example of the red-eyed and yellow-throated vireos in chapter 4 illustrates the way in which foliage height diversity can affect the number of bird niches in a community. In a very similar finding regarding the important role vegetation plays in habitat heterogeneity, Pianka (1966a) found that he could accurately predict the number of lizard species present in different desert habitats if he measured the diversity of size and form in the local vegetation. To lizards as well as to birds, the variety of growth forms in the vegetation seemed a more important determinant of species richness than the actual number of plant species that were present.

5. A simple *time hypothesis*, suggesting that

the species richness of a community chiefly depends on the amount of time it has had to diversify, often is deduced from the historical patterns we described earlier in this chapter. However, it is not at all clear that all communities, even nonsuccessional ones, will continue to diversify over time: we may recall the ancient but undiverse hot spring community. Obviously time is needed simply for species to arise, and species richness may be further facilitated by time as communities become more tightly integrated. The Serengeti grazing species described in Chapter 3, for example, must have required time to attune their niches so finely to one another. But time alone will not bring such a community about; other factors also must be permissive if real species richness is to arise.

A combination of several theories obviously may be needed if we are to account for the differing species richnesses of different communities. Sanders' ideas (1968 and 1969) represent such a combination, though perhaps an insufficiently broad one. His *stability-time hypothesis* invokes time and environmental stability as major factors, acting under the constraint of stress. It includes the ideas of physically controlled and biologically accommodated communities as the two ends of a continuum, with greater species richness possible toward the biologically accommodated end. The stability-time hypothesis is perhaps deficient in that it does not recognize habitat heterogeneity as an influence on species richness. This is understandable, since the ocean floor communities with which Sanders works demonstrate how much diversity is possible in a very homogeneous habitat. But we need not doubt that habitat heterogeneity is a real influence on species richness in most communities, and we can add the area effect demonstrated by MacArthur and Wilson.

Where does this leave us, if we are asked to suggest the most likely reason why a particular community is species-rich or species-poor? I would suggest that we consider first how harsh or how mild is its abiotic environment; and second, how varied is its habitat. In most cases, though not all, other determinants of species richness will be less important than these.

MODELS OF RESOURCE PARTITIONING

We now turn to the question of dominance. When comparing nonsuccessional communities one frequently is struck not just by differing species richnesses but also by apparent differences in the way available resources are partitioned among the species present. In some communities, even some containing many species, dominance is a very evident phenomenon. In other communities space or food seems to be shared more evenly. We can rephrase our earlier question about optimization: do these contrasts in dominance reflect differing stages of community evolution along a single pathway, or can selection favor more than one mode of resource partitioning?

Robert Whittaker, a leading student of community diversity, has developed graphical methods of depicting a community's species richness and dominance pattern simultaneously, and furthermore for easily comparing different communities in these aspects (Fig. 7.5). As in the case of species richness, dominance studies usually focus on component groups in a community, like birds or plants, rather than entire communities.

Among plants, even in an area as geographically restricted as the Great Smoky Mountains, Whittaker found enormous contrasts in both species richness and dominance when he looked at communities occupying locally different habitats. If we look again at Figure 7.5 we will note that the extreme patterns are connected by many intermediates, so that in terms of dominance the idea of a range of evolutionary possibilities may be more appropriate than a concept of mutually exclusive alternative modes of resource partitioning. Nevertheless, attempts to analyze dominance have resulted in several distinct, alternative models of resource partitioning (Whittaker, 1969), three of which we will examine in the pages that follow.

1. One famous model of resource partitioning is the *broken stick model* (random niche hypothesis) of Robert MacArthur (1957). MacArthur postulated that in any community there is a finite amount of "total niche space," and that the way this is divided up among the species present is a matter of chance. He envisioned the total niche space as a stick of finite length onto which points are cast at random such that the stick is divided into segments of different length, one segment for every species present. Some segments will be large (large species niches), some will be small. The implication of this model is that species do not have a poten-

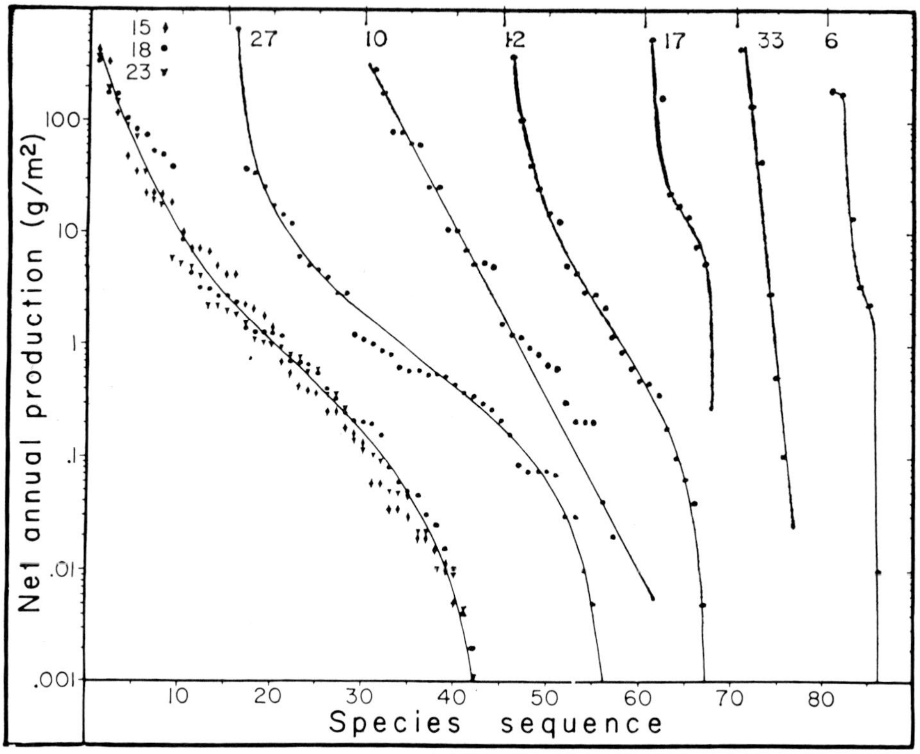

FIG. 7.5. Dominance-diversity plots for different plant communities in the Great Smoky Mountains. For visual clarity only the *left-hand curve* is shown originating at point *1* in the species sequence. The numbers on the abscissa are correct only for this curve, since the origin of the other curves has been shifted. However, each plot otherwise is done in the same way. On each curve every point is a species, and the shape of the curve indicates the degree to which the community's total photosynthetic production is monopolized by a few species or shared by many. Three different communities (*15, 18, 23*) were found to follow the left-hand pattern. Each of these contained about 40 species, dominance was relatively muted, and lesser species were scattered evenly along the curve of decreasing "importance." By contrast, community *6* contained only six species, two of which were co-dominant and only one of which was very rare. (From R. H. Whittaker: *Science, 147:* 250–260, 1965. Reproduced with permission from American Association for the Advancement of Science.)

tial for secondarily enlarging their niche spaces at the expense of others, *i.e.* of competing their way to dominance. Rather, whatever differences exist in niche size are predictable by the laws of probability, given a total niche space that must be divided among *n* species.

In testing such a model one first determines the number of species and of individuals belonging to each species in a sample from a real community. Usually, as already suggested, one samples only a subsystem that can be treated as a well integrated unit in its own right, such as the birds, plants or mammals of a community. (Root (1973) describes such subsystems as "*component communities*" of a larger "*compound community.*") From the total number of species and of individuals that are enumerated, one can calculate the expected abundances according to the broken stick model. Then the actual distribution of species abundances is compared to the predicted distribution (Fig. 7.6).

2. A second model, known as the *niche preemption hypothesis*, was developed in the 1930s in Japan, by I. Motomura and his colleagues. Here the supposition is that total niche space (a finite entity, as in the broken stick model) is divided up by unequal competition. Species A pre-empts as much space as it can and becomes dominant; species B pre-empts as much of the *remaining* space as it can, and assumes second importance; species C, in turn, pre-empts what it can of the space still remaining, leaving a still smaller fraction for subordinate species to

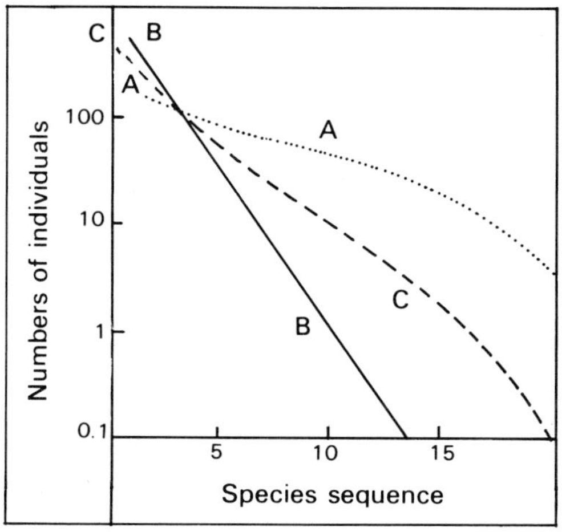

FIG. 7.6. Given a sample of 1000 individuals in which 20 species are present, the broken stick model will predict a curve of relative productivities (or abundances, if numbers of individuals are being compared) having the shape of *A* on a dominance-diversity plot. The niche pre-emption model will predict *curve B*, and the lognormal model will predict *curve C*. Predicted slopes will be different for samples of other sizes or other total species numbers, but the broken-stick model always predicts more even abundances among species than do the other two models, and the niche pre-emption model always predicts the most unequal abundance pattern. *Curve B* here predicts that, if niche pre-emption is really taking place in the manner theorized, you should not find as many as 20 species in a sample of 1000 individuals. (Modified from R. H. Whittaker: *Science 147:* 250–260, 1965, with permission from the American Association for the Advancement of Science.)

divide in the same fashion. After this pecking order process has gone to conclusion, the least dominant species gets the fraction no others can claim. According to this model of resource partitioning, a strict geometric series of species abundances (or "importances") would be predicted in real communities (Whittaker, 1969, and Fig. 7.7).

3. A third well known hypothesis is the *lognormal model* of Frank Preston (1948). This is more of a generalization than a true model: Preston discovered a pattern in real species abundances which he termed the lognormal pattern, then developed hypotheses (and a consequent model) which might explain the observed pattern. Here the assumption is that neither chance nor competition for dominance predominate in determining species abundance but rather that abundances are determined by many variables acting differently on each species. The result, in any given habitat, is that some species are much more common than others. In samples of any sort that contain variability, the variability often takes the form of a bell-shaped normal curve, with intermediate values being more abundant than extreme values: this is true, for example, when one is graphing heights or weights of human individuals or the distribution of grades on an exam. In the case of species abundances in a community, Preston found that by grouping the species of a sample into logarithmic "octaves" of abundance, a similar bell-shaped curve could be generated. Relative abundances of species cover such a wide range that a logarithmic scale is convenient, but Preston also believed it to be natural. Because sample sizes differ, observed commonness or rarity of species is a *relative*, not an absolute matter; hence it is best to talk about species being twice as common, half as common, etc. as others. Thus Preston used a $\log_2$ scale, in which the numerical range within each successive octave is twice that of the previous octave (Fig. 7.7). All species falling within a particular octave are considered to have roughly the same degree of commonness.

How does each of these models fare when its predictions are tested against reality? The broken stick model, of low competition and minimum dominance, is on the whole unsuccessful in describing real species abundances. It does not describe the relative abundance or importance of plants in any community so far studied, and successfully describes relative abundances in only a few animal groups, such as microcrustaceans in the littoral zone of lakes and territorial birds during the nesting season. In the latter case the broken stick model seems to fit more by accident than because the birds really are behaving according to its assumptions. In the case of territorial birds, *intra*specific competition is strong and keeps any species from becoming overly abundant; thus *inter*specific competition is minimized. The broken stick model's assumption of random niche size clearly is faulty here, but since each species prevents itself from becoming very abundant the distribution pattern comes close to that predicted (for other reasons) by the broken stick model.

Among most groups of organisms in most communities niche space apparently is *not* partitioned in the broken stick manner, *i.e.* in a

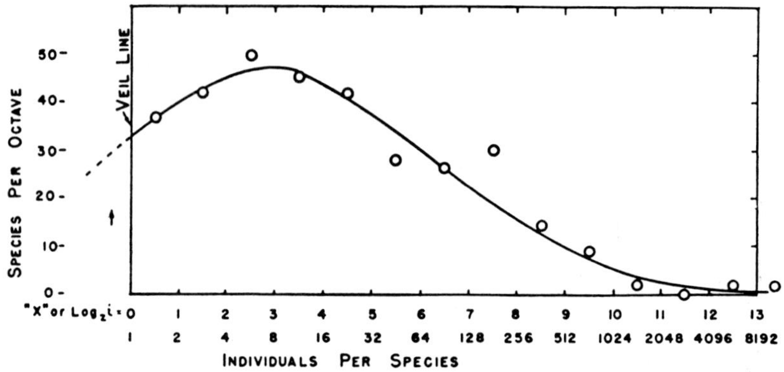

FIG. 7.7. In a huge sample of moths (56, 131 specimens, caught over 4 years at a light trap in Maine), relative species abundances follow a lognormal pattern. Note that the numbers per interval along the abscissa are not arithmetically equal but form a logarithmic series to the base 2. Log scales other than base 2 sometimes are used; the curve still is bellshaped. Always, however, it is truncated at the "rare end"—theoretically because some extremely rare species in the community were missed completely in the sampling. (From F. W. Preston: *Ecology, 29:* 254–283, 1948; after Dirks.)

random way. Rather, it is hammered out in a determinate way. The pre-emptive niche hypothesis, which suggests interspecific competition as the hammer, *does* successfully predict abundances of at least some groups in many communities, particularly species-poor communities of rigorous environments. Whittaker (1965) has shown, moreover, how the preemptive niche model can be modified rather simply to account for the co-dominance of species in plant associations like the spruce-fir timberline forest of the Great Smokies. Here the strict geometric progression of species abundances is marred by the presence of a second species nearly as abundant as the first, but otherwise the geometric pattern is quite apparent (Fig. 7.5, curve 6).

In the majority of natural communities, however, the lognormal model seems to best describe species abundances. This is true particularly of communities that are species-rich. As Whittaker (1969) notes, lognormal abundance patterns characterize component communities as different as the trees of a tropical rain forest, the diatoms of streams or lakes, and the microarthropods of soil. In the Great Smoky Mountains, the species-rich plant associations of lower elevations follow the lognormal model but the species-poor associations of high elevations mostly do not.

What a lognormal distribution means in ecological terms is rather vague. It presumably implies a multifactorial determination of niche sizes in a community, with no single process or factor standing out above all others. The prevalence of the lognormal model in nature is not unexpected, given the complexity of any environment. But if one encounters a community in which species abundances are lognormal, it does not follow that the niche determinants of any given species are too complex to unravel. McNaughton and Wolf (1970) have shown that relatively localized subsamples from lognormal communities often fail to fit the lognormal pattern of the whole community, suggesting strong local influences on abundance that may be revealed by intensive study.

EVOLUTIONARY IMPLICATIONS AND POSSIBLE TRENDS

Species Richness and Dominance

As we have seen, both species richness and dominance vary greatly among present-day communities. This variety is confusing, but if we recognize a distinction between biologically accommodated and physically controlled communities, the patterns we have recognized may suggest patterns in community evolution. Let us recapitulate.

Biologically accommodated communities are subject to relatively low levels of abiotic stress. They tend to be species-rich and low in dominance; the term "biologically accommodated" implies that intricate mutual adaptations of community members are relatively prominent.

The idea of biological synergism appeared several times in our discussion of such communities. This is the idea that a biological force, *i.e.* predation, may promote an increase in the number of species of one type (here prey), which in turn will promote diversification of other species (here particularly the predators), and so on in an ascending spiral. It seems possible that species richness may feed on itself to produce more species richness, at least in biologically accommodated communities.

This idea is strongly espoused by Robert Whittaker (1969):

> I conclude that species diversity of plant communities is an evolutionary product, subject to self-augmentation through time. There is no ceiling or saturation level for the diversity which results; the chemical differentiation of the higher plants implies virtually unlimited potentialities for the addition of different species.

Whittaker concludes that the same potential for unlimited diversification characterizes insects, a group of animals which, in fact, contains more species than any other. But Whittaker is not sure that all groups of organisms have the same potential:

> . . . plants and insects may differ from birds. The niche responses of birds are not to chemistry but to community structure and broad food categories in which the possibilities of niche division are limited. Bird communities can thus be saturated, determinate in their evolution of diversity; but plant and insect communities may show indefinite, indeterminate evolutionary increase in diversity.

One finds support in the historical record for the idea that communities are evolving toward greater species richness. Presumably the increase in number of families and species that has characterized the past (Fig. 7.1) is still continuing, and we can suspect that it is occurring chiefly in biologically accommodated communities.

But is not a concept of evolutionary increase in species richness contradicted by the species-area relationship? After all, this relationship suggests that communities become saturated with species and can hold no more. One can overcome this stumbling block if one makes the assumption that saturation levels themselves are changeable. As evolution proceeds and biological accommodations between species become more specialized, perhaps saturation levels will rise. There always will be a species-area relationship, but the line on graphs like Figure 7.2 may shift upward as community evolution proceeds. In spite of Whittaker's arguments, however, it cannot go up forever. MacArthur and Levins (1967) point out that there *is* an ultimate limit to a community's species richness, even in the case of groups such as plants and insects. Since resources are limited in any community, more species implies fewer individuals of each species, but as we shall see in Chapter 14, no species can persist in a community when its numbers become too low.

Species richness and dominance are to some degree inversely related; and if biologically-accommodated communities are tending toward an increase in richness, they probably are progressing toward a reduction of dominance, *i.e.* toward a more equal sharing of resources. The tropical rain forest, with its many species, none of which are very abundant, exemplifies a relatively advanced stage in this hypothetical process. We must be careful, however, in suggesting that species-rich, biologically accommodated communities are evolving toward a particular abundance structure. The lognormal abundance structure of such communities today is perhaps not so much a pattern toward which communities evolve, as one from which they cannot escape to evolve something more distinctive. Lognormal abundance structure may simply be a least common denominator, achieved in communities which are too complex and cumbersome to be "going anywhere" in terms of evolving an optimal abundance structure.

None of these trends or considerations need necessarily apply to physically controlled communities, where mutual adaptation of the members is (in Sanders' view) limited by the need to accommodate to abiotic stresses and fluctuating conditions. Species here cannot become so specialized relative to one another, and perhaps this limits more stringently, or even rules out, any evolutionary tendency toward greater species richness or decreased dominance. This point is speculative, but an "evolutionary destiny" for physically controlled communities is quite difficult to imagine. Possibly in some instances the species of physically controlled communities may be able to evolve superior adaptations to the stresses they now face, so that the stresses become less controlling. If so, such communities will move toward the condition of biologically accommodated communities. We must remember that today's communities exhibit a full range of intermediates between the extreme biologically accommodated type and the extreme physically controlled type, and this suggestion may well be valid for many intermediate communities.

Competition

Here again we should distinguish, despite intermediates, between biologically accommodated and physically controlled communities. Competition can occur either at the *intra*specific level or at the *inter*specific level. In the case of biologically accommodated communities it is possible to suggest that each form of competition tends to become less important as community evolution proceeds.

Such a suggestion should not be surprising: it follows logically from the preceding discussion. If biologically accommodated community types tend toward a structure of many species, none of which are very abundant, then competition among members of any species should be relatively low. This is particularly true if some outside agency, such as predation, is keeping a species' numbers down. Fedorov (1966) suggests that intraspecific competition is nearly absent among plants of the tropical rain forest. Likewise, at the interspecific level the idea of mutual adaptation within biologically accommodated communities implies an evolutionary minimization of mutually harmful interactions such as competition. As such communities evolve, then, competition should become less important and interactions beneficial to associating species should become more important.

In Chapter 6 the ideas of Frits Went (1955 and 1973) regarding competition were considered rather naive. Perhaps, however, we should consider naive only his assumption that interspecific competition *never was* important in communities like the desert and the rain forest. Perhaps, as Went states emphatically, interspecific competiton *now* is a relatively minor force in these communities. If so, we can view the present situation as the latter stages of an evolutionary trend.

In physically controlled communities, however, an evolutionary trend toward reduced competition may be lacking or weak. Conformity of many stressed communities to the niche pre-emption model of species abundance suggests that interspecific competition remains strong in such communities. One might think that species adaptively occupied with physical stresses might not have time for competition, but one could better argue from the other side: since such species are unable to adapt closely to one another, competition remains important. Sanders (1969), in fact, argues as follows regarding physically controlled communities: "To the degree that species do not make subtle distinctions about the environment, as in the physically unpredictable or high stress habitats, competition or predation can be extreme. Thus biologic effects are often drastic or catastrophic." We can suggest, then, that an evolutionary trend toward reduced competition probably is absent in physically controlled communities but is present in biologically accommodated ones.

What Is Being Optimized in Community Evolution?

As we end this chapter we must grapple with the questions that began it. Is natural selection at the level of individual species converging to some optimization at the community level? Evolutionary optimization at the species level involves self-perpetuation: for opportunistic species, self-perpetuation through high fecundity and colonizing ability, and for equilibrium species self-perpetuation through lower fecundity, lower resource demands, and staying power of individuals. One thing being optimized at the community level, therefore, certainly is self-perpetuation. But what optimizing strategy or strategies are being followed?

Physically controlled communites may appear nonsuccessional, but many of them really are communities in arrested (or, conversely, continual) succession. Their species often follow the opportunistic strategy, particularly if the stresses they must face are fluctuating stresses. The go-for-broke strategy of the opportunist allows species in such environments to recover and recolonize rapidly after setbacks. In a few physically controlled communities, like those of hot springs, stress is constant rather than fluctuating and we can expect an equilibrium strategy to characterize most species. This should be the strategy in most biologically accommodated communities too.

Consider what we have said about biologically accommodated communities: they typically are species-rich, and they are communities in which an equilibrium strategy of self-perpetuation is favored among most species by selection. These two characteristics set the stage for an idea that has permeated ecological thinking in recent years: the idea that community stability is linked to community diversity, and is inherent in it. The converse of this idea is that communities of low diversity are unstable. But these ideas are overgeneralizations, and they carry certain false connotations. After all, hot spring and salt marsh communities are very stable, even though very undiverse.

What, then, is the relationship between diversity and stability in communities? If diverse communities in reality are unusually stable, it is not basically *because* they are diverse, but because the species in them tend to be equilibrium species and hence are maintained at relatively constant levels. Diversity and stability both characterize biologically accommodated communities, but the first does not necessarily cause the second. To some extent, indeed, we must define what we mean by stability if we wish to claim that these communities are stable. In a species-rich community no one species is all-important, so even if a given species is decimated the community as a whole will not be seriously altered, whereas in a species-poor, physically controlled community there may be seismic repercussions if a major species collapses. But recovery in a physically controlled community is likely to be rapid too, because of its species' opportunistic strategy. The stability of biologically accommodated communities like the rain forest is, in part, a fair-weather stability, not a resilience in the face of real adversity. If one stressed a rain forest unduly the whole community might collapse; certainly it would be relatively slow to recover after serious disruption, because of the equilibrium strategy of its member species. Present-day ecologists emphasize the fragility of species-rich communities like the rain forest and the East African savanna as much as they do the diversity-stability dogma. Fragility and stability sound like contradictory ideas when applied to the same community, but they really are not if one separates the several possible meanings of "stability." One *cannot* discern an evolutionary trend in biologically accommodated communities toward stability if what one means is resilience toward stress. One *can* discern a trend toward stability in such communities if by stability one simply means an equilibrium strategy among the species of these communities, along with the muted population fluctuations that such a strategy implies.

In Chapter 6 it was suggested that the communities of extreme, physically controlled environments may be fragile also. To the extent that recovery of such communities after disturbance is impeded by slow growth, due to inadequate moisture in the desert, for example, or low temperature in the tundra, such communities are indeed fragile. But many physically controlled communities are not so limited, and these are relatively robust in the face of human disturbances: they are accustomed to the disturbances of nature, and their species are specialists at rebounding and recolonizing.

Other directions of community optimization have been suggested. Might there be an overall trend toward maximal use of available resources? Is there, in addition, a conservationist principle at work, *i.e.* a trend toward maximal use of resources on a *sustaining* rather than a short term basis? Or might optimization involve a power principle, as H. T. Odum has suggested (Odum, 1971; Odum and Pinkerton, 1955)? Power is useful work per unit time, and if one believes Odum, selection favors, at the species level and at the community level as well, not those systems which necessarily are most conservative or efficient in resource use, but rather those which command the greatest useful energy (efficiently or no) per unit time. This sounds much more like an opportunist strategy than an equilibrium strategy, and we might suggest that it applies to physically controlled communities but not biologically accommodated ones.

We will not try to deal here with questions of resource optimization. When we encounter ideas of optimization at the level of energy capture or resource use we have moved from questions of community structure to questions of ecosystem function. Before we try to answer such questions we should look at those functions more directly. That is the task of upcoming chapters.

SUGGESTED READING

Optimization

Cody, M. L. Optimization in ecology. *Science* 183: 1156–1164 (1974).

Wiens, J. A., ed. Ecosystem Structure and Function. Oregon State University Press, 1972. See especially Whittaker and Woodwell's paper on community evolution.

Historical

CLOUD, P. E. Evolution of Ecosystems. *American Scientist* 62: 54–66 (1974). Concentrates on the very early history of life and of the environment.

EHRLICH, P. AND P. RAVEN. Butterflies and plants. *Scientific American* 216: 104–113 (1967). Co-evolution of special relationships is stressed.

NEWELL, N. Crises in the history of life. *Scientific American* 208: 76–92 (1963).

KURTEN, B. Continental drift and evolution. *Scientific American* 220: 54–64 (1969).

Species richness and dominance

LOUCKS, O. L. Evolution of diversity, efficiency and community stability. *American Zoologist* 10: 17–25 (1970). One of several pertinent papers in this issue of the journal.

LOWE-MCCONNELL, R. H., ed. Speciation in Tropical Environments. London, Academic Press, 1969. A valuable symposium on tropical communities.

MACARTHUR, R. H. Geographical Ecology. New York, Harper and Row, 1972. A summarizing of MacArthur's work, and an entry to the literature.

PIANKA, E. R. Evolutionary Ecology. New York, Harper and Row, 1974. A textbook that emphasizes questions treated in this chapter.

RICHARDS, P. W. The tropical rain forest. *Scientific American* 229: 58–67 (Dec., 1973). A good introduction to the species-rich rain forest community.

WOODWELL, G. AND H. SMITH, eds. Diversity and Stability in Ecological Systems. Brookhaven Symposia in Biology No. 22. Upton, N. Y., Brookhaven National Laboratory, 1969. A very important symposium volume.

PART III

Ecosystem Processes

FIG. 8.1. Giant epiphytic pitcher plant in a Malayan cloud forest. This species has no access to moist soils, and the quart-size jugs capture water for later use by the plant, in addition to serving as insect traps. (Drawing by Sarah Allan.)

chapter 8

The Study of Ecosystems

To see a world in a grain of sand . . .

William Blake (1757–1827)

A NATURAL MICROCOSM

An investigator could approach the study of pitcher plants at several levels. As an autecologist he might travel to many countries, investigating the striking habitat contrasts among members of this insectivorous family (Fig. 8.1). In the United States, mention of pitcher plants brings to mind a wet sphagnum bog, but the dry habitats of many tropical species might make our investigator wonder: did the strange, juglike leaves evolve initially as insect traps, or merely as water-storing organs?

As a community ecologist, our specialist might become fascinated by the rich yet relatively simple aquatic community within each "pitcher" (Table 8.1). These little communities already have demonstrated great potential for investigations of how the survival strategies of coexisting species are integrated and interrelated (Addicott, 1974; Maguire, 1971).

Our ecologist would realize also that each pitcher represents not just a leaf, or a flask full of organisms, but, like Blake's grain of sand, a little world or microecosystem. The pitcher plant microecosystem is sustained, like our giant global ecosystem, by inputs and internal transfers of energy. It is sustained also by nutrients, imports of which (in the form of leaf exudates and trapped insects) must at least equal the amounts exported (absorbed by the leaf, perhaps, or lost in the bodies of emerging mosquitoes); if imports to the ecosystem are less than exports, it will run down. As an ecosystem ecologist our student, hopefully not yet old and gray, would direct his questions toward these processes, for an ecosystem is not just a community but a resource-processing unit. How much energy enters this aquatic microcosm over the course of a day, week or season in the form of sunlight and insect prey? How much energy leaves? What gains and losses of specific elements occur, and to what extent do they balance over short periods and over longer ones? Within the microcosm itself, what are the transfer routes of energy and nutrients, and what degrees of hoarding and recycling characterize each species population? In the little world of the pitcher plant it might be relatively easy to answer these questions quantitatively. What an ingenious investigator could learn by studying the pitcher plant microcosm might be as close as one could ever get to fully understanding the processes of a natural ecosystem. No one, it seems, has yet applied the ecosystem approach to pitcher plants, though it frequently has been applied to artificial microcosms in the laboratory.

DRAWING THE LINE

Few would differ regarding the boundaries of a pitcher plant ecosystem, but boundary disputes might be common when defining other ecosystems for study. One virtue of the pitcher plant system is its distinct and relatively sharp borders, namely the surface of the reservoir and the enclosing walls of the jug. The latter boundary, admittedly, could be blurred slightly by the burrowing habits of certain members of the community. But what passes completely through the walls, or through the water surface, can be classed without argument as imports or exports.

What of the boundaries of a pond ecosystem? They seem equally distinct, but why not include

TABLE 8.1

Members of the pitcher plant community studied by Addicott (1974)

Other natural aquatic microcosms (*e.g.* water-filled knotholes in trees) contain quite similar communities. The community includes photosynthetic organisms, organisms subsisting on decayed matter, and active predators.

I. Unicellular Organisms
- 16 species of flagellates, including both photosynthetic types (*Euglena, Chlamydomonas, Chlorogonium, Carteria,* etc.) and nonphotosynthetic types (*Cryptomonas, Astasia, Urophagus,* etc.)
- 23+ species of ciliates: *Urotricha, Colpoda, Paramecium, Vorticella,* etc.

II. Multicellular Organisms
- Several species of rotifer
- One species of mosquito: *Wyeomia smithii*

Suggested Food Chain Relationships

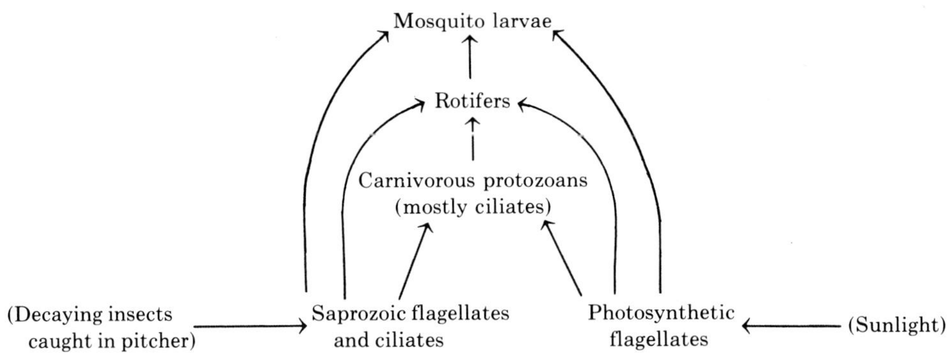

the whole watershed within the system? After all, are not nutrients that nourish the life of a pond derived nearly continuously from the surrounding watershed? And do not ducks and herons regularly remove materials from the pond and deposit them on land or in their nearby nests? Why not, then, enlarge the boundary of this ecosystem to include those adjacent land areas with which the pond is intimately associated?

Indeed, there is nothing sacred about *where* one fixes the boundary. In their studies of weathering, erosion and nutrient turnover in a New Hampshire forest, Bormann and Likens (1970) simply ignored the fact that the whole forest could (and by most people would) be considered a single ecosystem. They narrowed their focus to a single watershed within the forest and defined *this* as their ecosystem. For the purposes of these investigators it was a very satisfactory choice. It mattered little to their study if the boundary they had defined separated trees which rubbed shoulders at the crest of the ridge, or if squirrels scampered freely back and forth between these trees. The lateral limits most useful to a study of soil nutrients and their fate were those defined by drainage patterns. For vertical limits Bormann and Likens chose, with equal freedom, the top of the forest canopy and the lower limit of root penetration. They thus carved their ecosystem's third dimension from a continuous column of air and a continuous depth of substrate which could have been carved differently. Someone studying energy flow through the animals of this forest would be more troubled about those squirrels than were Bormann and Likens, and might choose different horizontal borders. The animal ecologist's ecosystem might comprise parts of two watersheds, perhaps being delimited by a pair of streams and by other natural features serving as partial barriers to animal movement.

For the purposes of study, then, an ecosystem can be delimited in almost any way convenient to the interest of the investigator. The ecosystem concept is most useful if it is molded to the situation being investigated.

ECOSYSTEM CONTRASTS

Size

The ecosystem concept applies equally well to the leaf reservoir of a pitcher plant, to a single lake, watershed or meadow, or to the whole world, the global ecosystem. Microbiologists have profitably used the ecosystem concept in studying the rumen of a sheep, which of course is merely one part of a digestive tract. The rumen is clearly delimited, has its own community of microorganisms, and constantly imports and exports nutrients. Another biologist might study a whole sheep as an ecosystem; still another, the pasture together with its flock. Earth's global ecosystem encompasses innumerable smaller ecosystems, like boxes within boxes. The same processes can be followed in each, although the contrasts in size and complexity are enormous.

Degree of Autotrophy

An autotrophic organism is self-nourished with respect to carbon: it manufactures its own compounds from carbon dioxide or bicarbonate in the air or water. A heterotrophic organism is other-nourished: it can use only the carbon that exists as more complex compounds in the tissues or products of other organisms. There are degrees of autotrophy: a pitcher plant is partly heterotrophic because it feeds on other organisms, but it is partly autotrophic because it performs photosynthesis.

We speak in similar terms of autotrophic and heterotrophic ecosystems, and of their degree of autotrophy. An autotrophic ecosystem contains both auto- and heterotrophic organisms, but is self-sustaining from the standpoint of carbon fixation. Its autotrophs produce enough food and capture enough energy in their tissues to support both themselves and the heterotrophic organisms in the system. But there are ecosystems for which this is not true. For example, cave ecosystems support only heterotrophic organisms and depend completely on imports of carbon that was biologically fixed elsewhere. Similarly, a forest stream is a largely heterotrophic ecosystem: though it contains photosynthetic plants and algae, its animal populations are sustained only by sizeable additional imports of terrestrial leaves. In fact, ecosystems vary from the condition of complete heterotrophy to the condition of overbalanced autotrophy, in which more organic matter is produced by photosynthetic organisms than can be utilized by the resident consumers and decomposers (Fig. 8.2). In such ecosystems organic matter may accumulate (for example, as humus, peat or organic lake sediment) or it may be more or less lavishly exported (as in the case of agricultural croplands from which harvests periodically are removed).

Openness

All ecosystems are open systems, using imported solar energy as the power source for their dynamic processes. Autotrophic ecosystems use solar energy directly in their biological manufacturing processes; heterotrophic ecosystems import most of it indirectly, already transformed into the chemical energy of organic matter.

It is useful to distinguish between energy and matter when studying ecosystem processes, even though in many cases transfers of matter are vehicles for the transfer of chemically trapped energy. As a rule one speaks of "energy flow" through ecosystems, because once it actually is expended a given quantum of energy is not reusable. The energy captured by plants flows through an ecosystem in a series of essentially one-way steps, some energy becoming degraded to heat and lost at each step and a variable but usually small amount being exported as chemical energy in the bodies of emigrating animals, or as eroding humus. Matter, however, is not used up but may be exchanged again and again between the components of an ecosystem. In opposition to the flow of energy we therefore witness a "cycle of matter" within an ecosystem (Fig 8.3), the cycle of most interest being from nonliving to living and back to nonliving compartments of the system. But some matter is lost from the system via processes such as erosion; and some stops cycling even though it remains, because it has been trapped in some natural sink such as the anaerobic sediments of a peat bog or deep water body.

All ecosystems then, are open systems, but they exhibit enormous contrasts in *degree* of openness. "Spaceship Earth," our global ecosystem, imports virtually all its energy from the sun but is essentially closed with regard to matter. Aside from meteorites no new matter enters this system, and the only losses are light

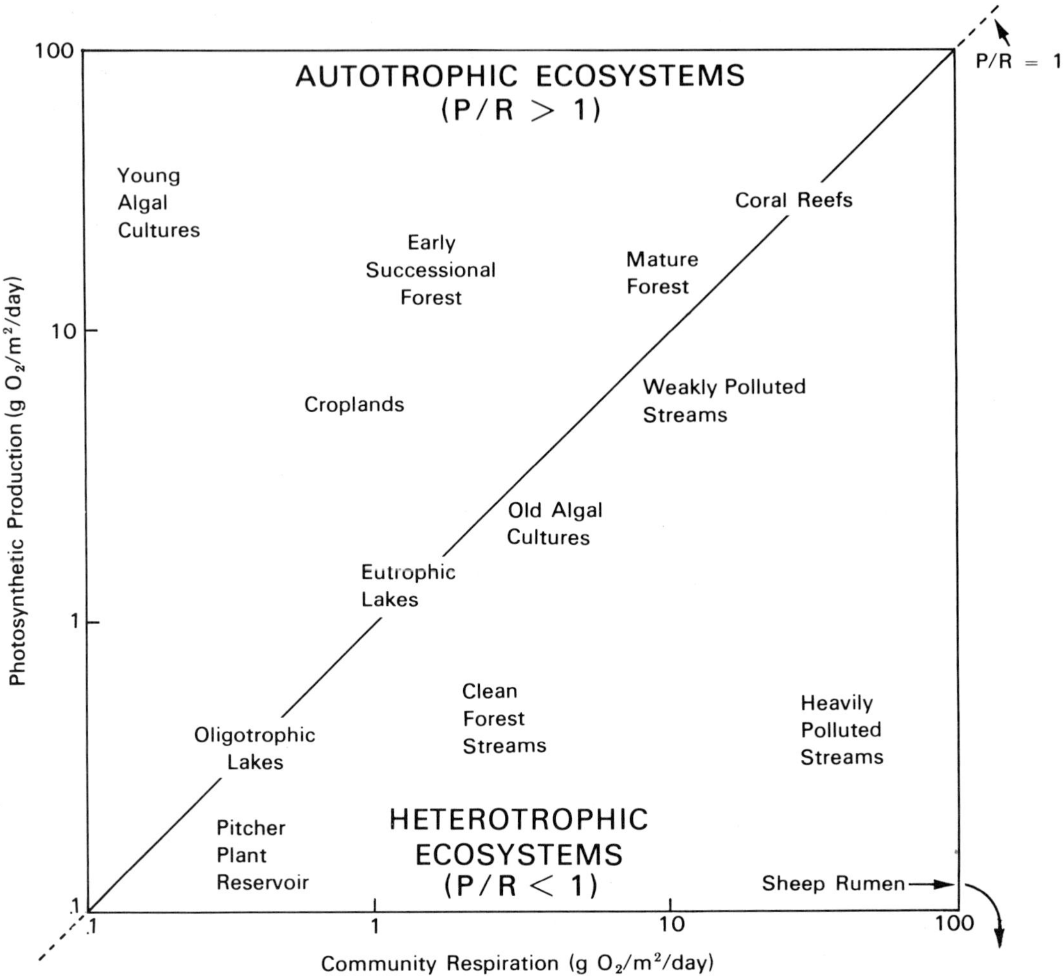

FIG. 8.2. Ecosystems differ in the degree to which they are autotrophic or heterotrophic, as this graph makes clear. The *diagonal line* indicates balanced self-sufficiency, in which the ratio of organic production to respiratory consumption is unity. (From H. T. Odum: *Limnology and Oceanography, 1:* 102–117, 1956. Adapted here with permission.)

elements which still may be escaping from Earth's atmosphere. On the other hand, the rumen ecosystem of a sheep is incredibly open with respect to *both* energy and matter (Brock, 1967). The rumen fluid, the "culture medium" of this ecosystem, is constantly being lost and renewed: the system's volume is only about 5 liters, and the flow rate averages 7.5 liters per day. With the departing medium go millions of microorganisms together with the wastes they have excreted. If the system is to continue functioning it must be replenished by new raw materials and energy (which come neatly packaged in the sheep's food), and by the unceasing growth of the bacterial population. Certain elements that support the rumen's bacterial growth are deficient in the sheep's diet of cellulose, but are supplied by the sheep itself. Nitrogen and phosphorus, for example, are secreted in the sheep's saliva, used by the rumen bacteria, then recovered in the intestine where the microorganisms that have been flushed out of the rumen are digested. Nitrogen and phosphorus thus cycle fairly efficiently in the sheep as a whole, and the rumen is kept supplied, but there is relatively little cycling of these or other elements within the rumen itself.

Let us consider the openness, or leakiness if you wish, of other ecosystem types. Stream and river ecosystems obviously experience much less

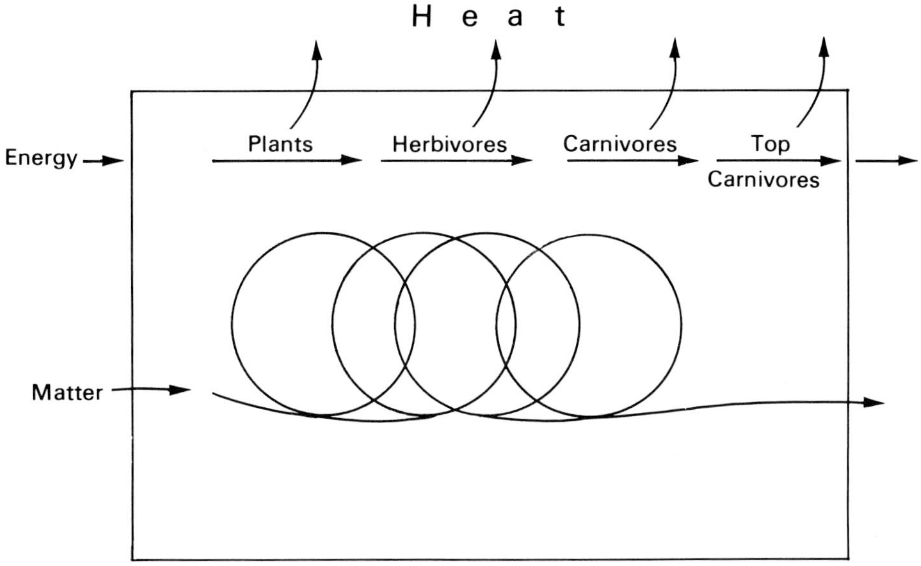

FIG. 8.3. Energy flows through an ecosystem and is dissipated. It is not recycled. Matter entering an ecosystem may be captured and reused many times, though eventually it too may be lost from the system. For simplicity, energy transfer to decomposers is not shown.

internal recycling than do standing water ecosystems, where the same atoms may be used again and again; but along the course of a stream there is efficient reuse of nutrients (and extraction of energy) due to the turbulent circulation. Matter is retained within a lake for much longer periods, but often it tends to cycle less rapidly. When organisms die, sink and decay in a lake, their atoms may not immediately reenter the living portion of the cycle; indeed, they may be more or less permanently sequestered in accumulating sediments. This, too, is leakage of a sort.

The leakiness of terrestrial ecosystems is partly a function of the type and degree of vegetation within the system (Fig. 8.4). Leakiness is expressed obviously through the mechanism of particulate erosion, and less obviously through the leaching away of materials. Besides the vegetation cover (which is related to a region's precipitation regime), such factors as soil type and hilliness help to determine the degree of openness of a terrestrial ecosystem. For example, in Chapter 4 we noted the rapid leaching action that occurs in the porous sands of the New Jersey Pine Barrens. This ecosystem is leakier than one would expect if reasoning only from its vegetation cover (which is ample).

Human activities generally act also to make terrestrial ecosystems more leaky. For example, erosion rates in croplands and overgrazed pastures may be as much as 1000 times the rates that characterized the native forests or grasslands they replaced. These losses must be added to the nutrient losses inflicted on agricultural landscapes simply by the harvesting and removal of crops. Does one wonder that most farmlands need repeated doses of fertilizer if they are to continue to be high yielding?

THE STEADY STATE ASSUMPTION

When an ecologist measures the rate of phosphorus turnover in a marsh or photosynthesis in a forest he usually assumes that his finger lies on the pulse of a steady state ecosystem. If he could not repeat his measurements at some other time with much the same result, his study would lose much of its value.

Within a steady state system there may be continual change and turnover, but imports and losses are balanced. Physiologists often make the steady state assumption when studying the metabolic processes of cells or organisms, and thereby are able to calculate input-output budgets:

> The body remains more or less constant in the quantity and quality of its elements . . . even though

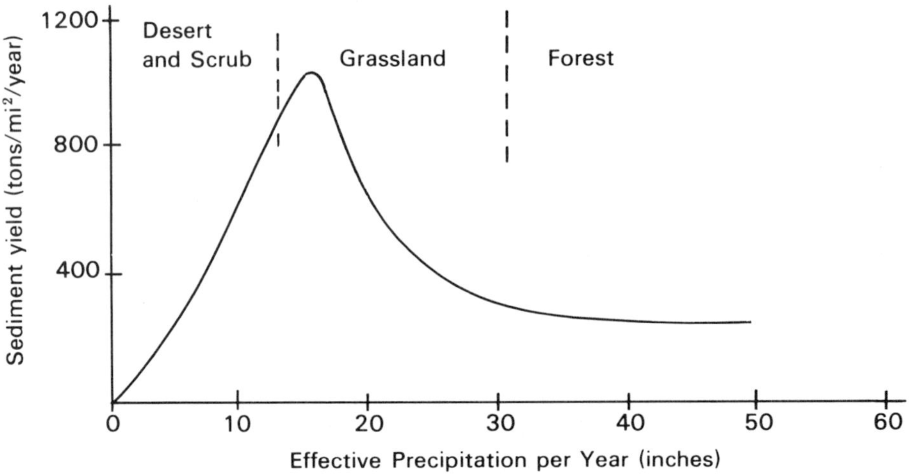

FIG. 8.4. Erosion plotted against precipitation does not yield a continually rising curve, as one might expect. Rather, erosional losses are greatest in regions of relatively low rainfall. Ample precipitation encourages a more complete vegetation cover that retards erosion. Since data from warm and cool localities were used in this study, yearly rainfall figures were corrected for differing amounts of evaporation, yielding a scale of "effective precipitation." (From W. B. Langbein and S. A. Schumm: *Transactions of the American Geophysical Union* 39: 1076–1084, 1958.)

many cells and tissues are turning over continuously. Food enters, wastes leave, yet everything remains the same.

Thomas Brock (1967), *The Ecosystem and the Steady State*

But the steady state assumption must be applied with care to the metabolic processes of ecosystems. For example, photosynthetic energy capture fluctuates both seasonally and diurnally in a temperate forest. It makes some sense to think of 12-month or 24-hour steady states, but none at all to use 6-month or 12-hour periods as our time interval. If we choose an appropriate time span, we still must recognize that energy fixation (or almost any other ecosystem process) will vary daily and yearly in unpredictable ways because of fluctuating weather conditions or other random events. To assume a steady state is to impose a simplifying symmetry on the system.

Random or nonrandom deviations from balance, such as those produced by cloudy days, day-night changes, or the cycling seasons, presumably can be accounted for in steady state ecosystem analysis. If there is an equilibrium state, departures from it will cancel each other out when a suitable time span is selected. But what of a system not in equilibrium? What of a growing organism, or an ecosystem undergoing succession? The steady state assumption is not applicable here, unless the directional changes are so slow that over a suitably short time period they are negligible. Fortunately, often they are this slow. Though ecological succession may be occurring, it simply may be ignored in the study of a forest's yearly energy budget. Similarly, seasonal succession may be overlooked in a nutrient or energy budget calculated for a single day. Brock (1967) states it baldly: ". . . successional events, although of interest to the ecosystem historian, may frequently be ignored by the ecosystem physiologist."

This is true, but the metabolic *changes* of an ecosystem during succession also are of great interest. In the next two chapters, we will first consider energy flow and biogeochemical cycling as if they were steady state processes; then we will drop the steady state assumption and look briefly at these processes during ecological succession.

SUGGESTED READING

Concept of the ecosystem

BROCK, T. D. The ecosystem and the steady state. *Bioscience* 17: 166–169 (1967). Discusses common assumptions in ecosystem studies.

EVANS, F. C. Ecosystem as the basic unit in ecology. *Science* 123: 1127–1128 (1956). A widely quoted statement of the ecosystem idea. This and the next paper have been reprinted in G. W. Cox, ed., Readings in Conservation

Ecology. New York, Appleton Century Crofts, 1974 (2nd edition).

ODUM, E. P. Relationships between structure and function in ecosystems. *Japanese Journal of Ecology* 12: 108–118 (1962). A lucid exposition of the thrust of ecosystem studies. Also reprinted in Cox (above).

Microcosms and their study

ADDICOTT, J. F. Predation and prey community structure: an experimental study of the effect of mosquito larvae on the protozoan communities of pitcher plants. *Ecology* 55: 475–492 (1974).

BEYERS, R. J. The metabolism of twelve aquatic laboratory microecosystems. *Ecological Monographs* 33: 281–306 (1963).

MAGUIRE, B., JR. Phytotelmata: Biota and community structure determination in plant-held waters. *Annual Review of Ecology and Systematics* 2: 439–464 (1971).

MCCONNELL, W. J. Productivity relations in carboy microcosms. *Limnology and Oceanography* 7: 335–343 (1962).

chapter 9

Biogeochemistry: The Nutrient Game

Full fathom five thy father lies;
Of his bones are coral made;
Those are pearls that were his eyes:
Nothing of him that doth fade
But doth suffer a sea-change
Into something rich and strange.
—William Shakespeare, *The Tempest*

THE SEA AND THE JUNGLE

Rich and strange indeed are the transformations that characterize the nutrient game. In some ways it is not unlike a game of Monopoly. Imagine that the Monopoly board is Earth, and that the board's four sides are Earth's four biogeochemical realms: hydrosphere, lithosphere, atmosphere, and biological sphere (the realm of live and decaying organisms). Like tokens in the game, molecules and ions cycle round and round this board, following a loosely prescribed course fraught with detours, even with temporary incarceration. The cycling game goes on forever, as most Monopoly games seem to do. Some tokens scamper along lightly but, like the hare, they intermittently tire. Other tokens plod like tortoises; they may even burrow into the mud for a long or short season's slumber. The workings of the game pose many riddles. Let us consider two for openers.

The Sea and Its Salt

The world ocean truly can be envisioned as a giant evaporating dish. Rivers endlessly pour into it a solution of minerals eroded from the land. The aqueous solvent evaporates, cycling back to land through the atmosphere and making its way to the sea again with another load of minerals (Fig. 9.1). Water has been at this job for hundreds of millions of years, and the quantities of salts it has brought to the ocean are enormous. Yet inland lakes which lack outlets, like the Dead Sea or Great Salt Lake, often are far more nearly saturated with minerals. These lakes have been in existence for a much shorter time than has the ocean. Where are all the ocean's salts? And what controls their levels in the sea?

The Self-sustaining Jungle

The tropical rain forest, Earth's most luxuriant biome, subsists on soils that for the most part are old, leached and infertile, soils that seem grossly inadequate to meet a persistent demand for nutrients by the vegetation. By what magic have rain forests been able to flourish through the ages on such soils, their existence finally threatened not by nutrient exhaustion but by saws and bulldozers?

Questions like these will occupy us in the present chapter. Nutrients sustain life, and their availability is determined by the rules of biogeochemistry, the science which traces movements of matter between the biological sphere and the nonliving spheres that together comprise this planet and all of its ecosystems.

RESERVOIRS, MERRY-GO-ROUNDS AND SINKS

Global Cycles and Ecosystem Cycles

First we must explore more fully what ecologists mean when they talk about nutrient

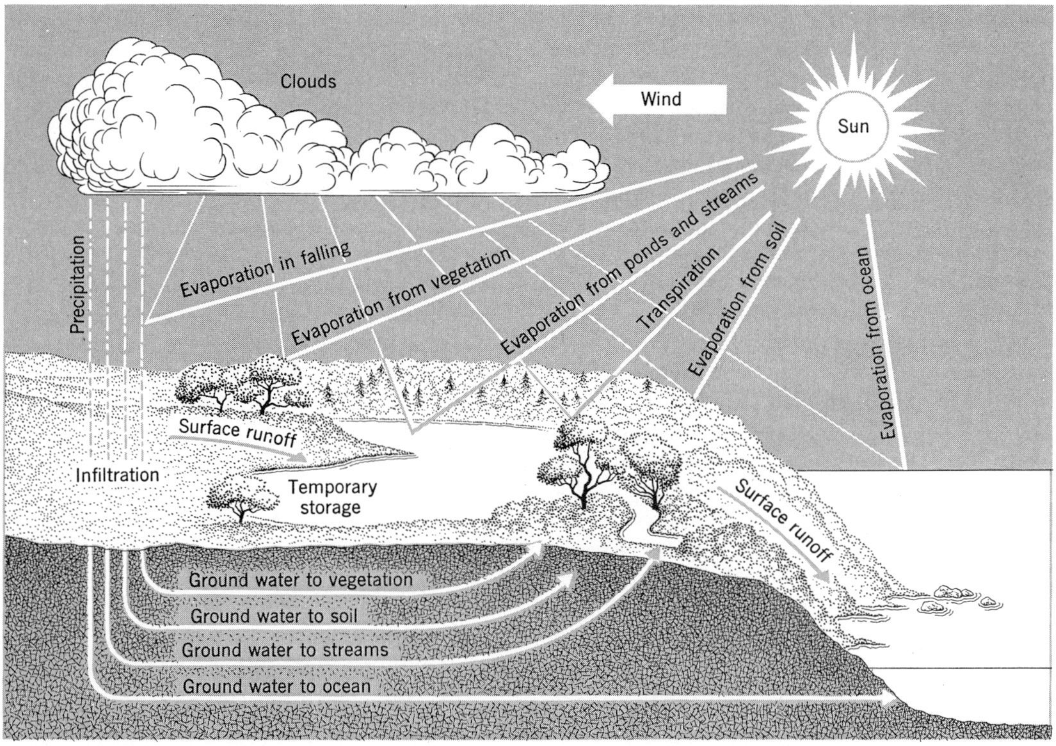

FIG. 9.1. Global hydrologic cycle, powered by the sun and gravity, is an apparent one-way conduit of minerals to the sea. (From R. F. Dasmann: *Environmental Conservation*, ed. 4. New York: John Wiley and Sons, Inc., 1976).

cycling. This term is used in both a global sense and a narrow, intra-ecosystem sense. Much confusion may be generated if one fails to realize which of these two meanings is being used in a particular instance, for a substance may be part of a closed global cycle yet may cycle hardly at all within the confines of a particular ecosystem. If we speak for example of the hydrologic cycle, we must realize that we are speaking on a global scale of water's free movement between the biologic sphere, the atmosphere and the hydrosphere. But when speaking of the water relations of a pond or forest, we must largely abandon the concept of a cycle and speak instead of imports and exports. Water molecules are highly mobile; leaving the ecosystem easily as water vapor or runoff, they are replaced by other water molecules. Water exemplifies many substances which follow closed global cycles but cannot be said to cycle appreciably within ecosystems; the molecules simply pass through. Lighter elements and volatile compounds resemble water: any ecosystem is relatively leaky with regard to these. It is chiefly in regard to heavier mineral elements that one can make the distinction between tight ecosystems, in which there is considerable *internal* cycling, and leaky ecosystems in which there is much less. When we speak of efficient cycling *within* a rain forest, for example, we obviously refer to soil nutrients, not to water or atmospheric gasses.

Figure 9.2 depicts the dynamics of any nonvolatile mineral element with respect to a single terrestrial ecosystem. Molecules of a mineral element may enter the ecosystem from above or from the sides via rainwater, inflowing streams or immigrating organisms; or they may enter from beneath through the slow weathering of parent rock. The same molecules, or their constituent atoms, eventually leave the system via outflow or emigrating organisms. It is the rock and soil of the lithosphere that constitute the main repository, the **reservoir,** of most mineral nutrients. This lithospheric reservoir may be very rich, but nonetheless it oftens feeds nutrients upward into the biological sphere quite slowly: for instance, upper soil may need to be removed by erosion if minerals newly weathered from underlying rock are to be near enough the surface for uptake by plant roots. On the other hand, downward leaching or chemical precipitation may return molecules of a mineral relatively quickly from the biological sphere to inaccessible portions (sinks) of this lithospheric reservoir. Inaccessibility obviously can involve either physical remoteness from the biological sphere or chemical unavailability, the latter occurring when a nutrient is not in appropriate form for biological uptake.

In addition to two-way transit between lithosphere and biological sphere, mineral subtances may cycle round and round within the biological sphere, being reincorporated as organic matter almost as fast as they are released by decay. This process underlies the ability of rain forests to perpetuate themselves on poor soils. Here, mineral elements circulate nearly totally within the biological sphere (where relatively large amounts are concentrated). In the humid, hothouse, rain forest environment decay is rapid. Nitrogenous compounds and nearly all mineral nutrients are taken up quickly and reused when biologically liberated. Beneath the thin humus layer, the infertile soil acts more as a simple rooting medium than as a source of new minerals or a sink for minerals leaching downward. Export of minerals by erosion is likewise minimal, judging from the dilute streams draining rain forest basins such as the Amazon.

Figure 9.3 illustrates, for contrast, the dynamics of a *gaseous* nutrient such as O_2 or CO_2 with respect to a terrestrial ecosystem. In the case of such substances, exchanges between a highly mobile atmospheric reservoir and the biological sphere may be much more rapid than the biological transfer steps, the "merry-go-round," within the ecosystem. However, free feed-in and removal from the biological sphere is not true of *all* atmospheric elements; nitrogen resembles many mineral nutrients in that its rate of movement from reservoir to biological sphere may be too low to satisfy the biological demand. Somewhere in every nutrient's cycle there are slower, rate-limiting steps; these are of crucial interest if the nutrient is in short supply to all or part of the biological sphere.

The Calcium Dynamics of a Forest

Let us now examine real figures for the movements of an element in a particular ecosystem (Table 9.1). In the Hubbard Brook Forest, moderately large amounts of calcium enter and leave the biological sphere each year but nonetheless remain within the ecosystem; the amounts being imported or exported are relatively small. Evidently this northern hardwood forest is a relatively tight ecosystem in the case of calcium, and internal cycling is a major process. The ratios of internally retained cal-

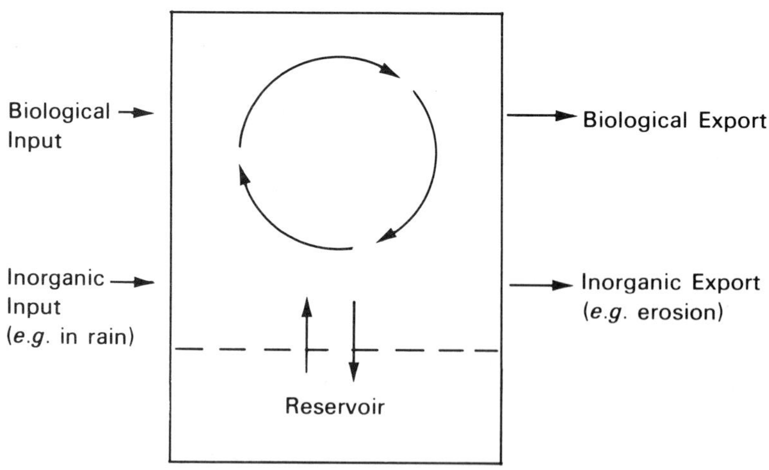

FIG. 9.2. Dynamics of any mineral nutrient with respect to a single ecosystem. The deeper portions of the lithospheric reservoir may be considered either inside or outside the ecosystem. The position of the boundary is arbitrary.

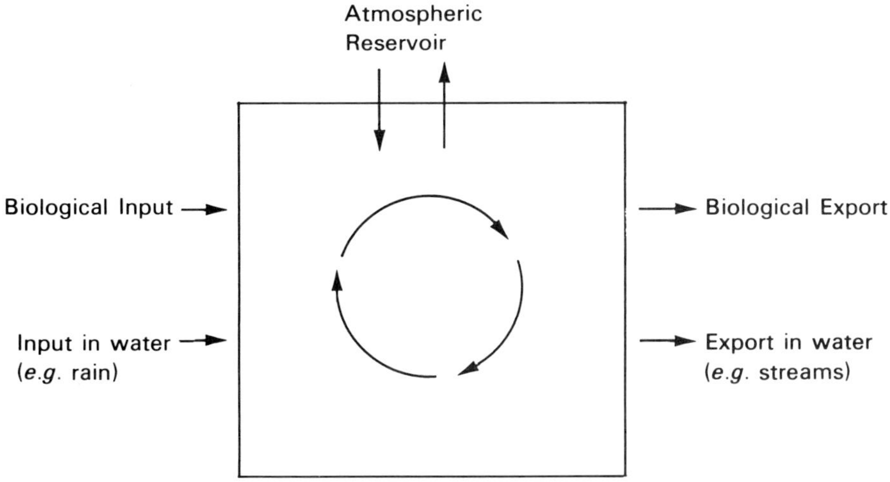

FIG. 9.3. Dynamics of any gaseous element with regard to a terrestrial ecosystem. In the case of certain nutrients, such as nitrogen, a secondary reservoir may build up in the soil.

cium to input and output would be very different in a cornfield or a stream. In leaky ecosystems such as these, the community depends on relatively massive imports of mineral nutrients. But leaky or tight, an ecosystem obviously must maintain no less than a balance between income and outgo, or its life-sustaining abilities will decline.

The Global Cycle of Sulfur

Let us look now at a balance sheet for sulfur, not in one forest ecosystem but in the great global ecosystem. Here, there are no input and output items, only internal transfers: as far as most elements are concerned, Earth is a closed system.

Sulfur is an important constitutent of protoplasm and sea water, and it moves freely through the atmosphere; its principal reservoir, however, is the lithosphere. Man has greatly increased sulfur's rate of movement from this reservoir into the atmosphere and hydrosphere, where it enters a series of relatively rapid cycling steps that include living things. In fact, man's active quest for sulfur is rapidly depleting Earth's accessible lithospheric reserves of sulfate ore, and this element may be the first of those commonly used in commercial fertilizers to be exhausted (Slack, 1970). When we speak of sulfur being exhausted we do so only in an anthropomorphic sense. Pardoxically, man is *increasing* the amount in active circulation, and in so doing is fertilizing the hydrosphere and polluting the atmosphere. But by doing this he is running out of the concentrated reserves most useful for his agricultural and industrial enterprises.

Figure 9.4 illustrates the major exchanges of sulfur between lithosphere, atmosphere and hydrosphere. In the hydrosphere and on land, organisms intercept inorganic sulfur compounds and cycle them through food chains, but for simplicity these processes are omitted from the diagram. It has been discovered that sulfur introduced into the atmosphere remains there relatively briefly: no more than a few days if it remains in the troposphere, and 1 to 2 years if it manages to reach the stratosphere. Because of this short residence time, the yearly atmospheric budget of sulfur can realistically be considered balanced: approximately equal amounts of sulfur enter the air and leave it over the course of any year. Exchanges of sulfur between the stratospheric and tropospheric compartments of the atmosphere also may be assumed to balance on a yearly basis.

The atmosphere receives sulfur from various terrestrial and hydrospheric sources and this element leaves the atmosphere in several ways. These fluxes are expressed quantitatively in Figure 9.4, the quantities for the most part being estimated from real measurements. In the case of hydrogen sulfide (H_2S) no measurements were adequate for a direct estimate of the quantity entering the atmosphere, but the assumption of a balanced overall input/output budget for the atmosphere allowed H_2S input to be calculated by subtracting all other inputs from the total atmospheric output. Such calculations are a common stratagem of the geochemist when he is confident a steady state situation exists. Here the calculation confirms the impression conveyed to our noses by tidal mud flats and crowded feedlots: impressive quantities of sulfur enter the atmosphere as hydrogen sulfide.

Nonatmospheric portions of the sulfur cycle are not in complete balance at present. There is input to the lithospheric reservoir (for example, via burial of organic sediment and precipitation of gypsum rock), but quite probably this input is less today than lithospheric output. Figure 9.4 shows how much sulfur man is withdrawing from the lithosphere simply in his fossil fuels. If we could develop economical ways to recover and purify the SO_2 that goes up our chimneys, we would not need to fear the exhaustion of sulfate ore reserves.

In addition to inputs of lithospheric sulfur to the atmosphere, some passes directly to the hydrosphere via erosional processes. Eriksson (1963) has estimated that for the period around 1960, annual riverine removal of sulfur from land to ocean was 240 million tons. Of this, 45 million tons were derived from rock weathering, 30 million from fertilizer runoff, and the remainder from the atmosphere. Terrestrial and atmospheric inputs to the hydrosphere undoubtedly have increased since 1960 beause of man's accelerating use of fertilizer and fossil fuels.

Sulfur enters the biological sphere through the uptake of soluble sulfate ions by algae and terrestrial plants. Now in organic form, this sulfur moves readily through food chains and, if liberated as sulfate by decay, is often immediately reused by plants. But biologically derived sulfate also may be reduced to hydrogen sulfide by anaerobic aquatic bacteria. These organisms play a major role in the overall cycling process of sulfur on our globe, because diffusion of biologically created H_2S into the atmosphere is one of the major ways by which sulfur is returned from water to land. Without the contribution of faraway sulfur-reducing bacteria in the oceans, terrestrial communities would be starved for this element.

Local and Global Movements of Phosphorus

In Chapter 2 a case was made for suspecting phosphorus to be Earth's master limiting nutrient. Accessible stores of this element tend to be

TABLE 9.1

Yearly calcium budget of Hubbard Brook watershed

The data below were obtained in an undisturbed forest watershed ecosystem in New Hampshire. There is no stream inflow to this ecosystem (though a stream drains it), and no groundwater influx. Animal immigrations and emigrations were not monitored, so calcium movements via this mechanism are not included; however, animal inputs and animal exports of calcium are probably in rough balance. The data are grouped into categories that resemble those of Figure 9.2. From Bormann and Likens (1970).

	kg/hectare/year
INPUT	
1. In precipitation	2.6
2. By rock weathering	9.1
OUTPUT	
1. In outflowing stream	
a. Dissolved inorganic Ca	11.7
b. Particulate organic Ca	0.3
2. Return to lithospheric reservoir	negligible
BIOLOGICALLY BOUND OR CYCLING INTERNALLY	
1. Taken up from soil by plants	49.3
2. Returned to topsoil by decomposition	49.0
3. Bound in trees and undecayed litter	2310. kg/hectare
LITHOSPHERIC RESERVES WITHIN ECOSYSTEM	
1. Available Ca in soil water or bound to exchange surfaces	690. kg/hectare
2. Ca contained in soil and rock minerals	28,550. kg/hectare
LITHOSPHERIC RESERVOIR BENEATH ECOSYSTEM (here defined as Ca in soil and rocks deeper than 61 cm)	immense

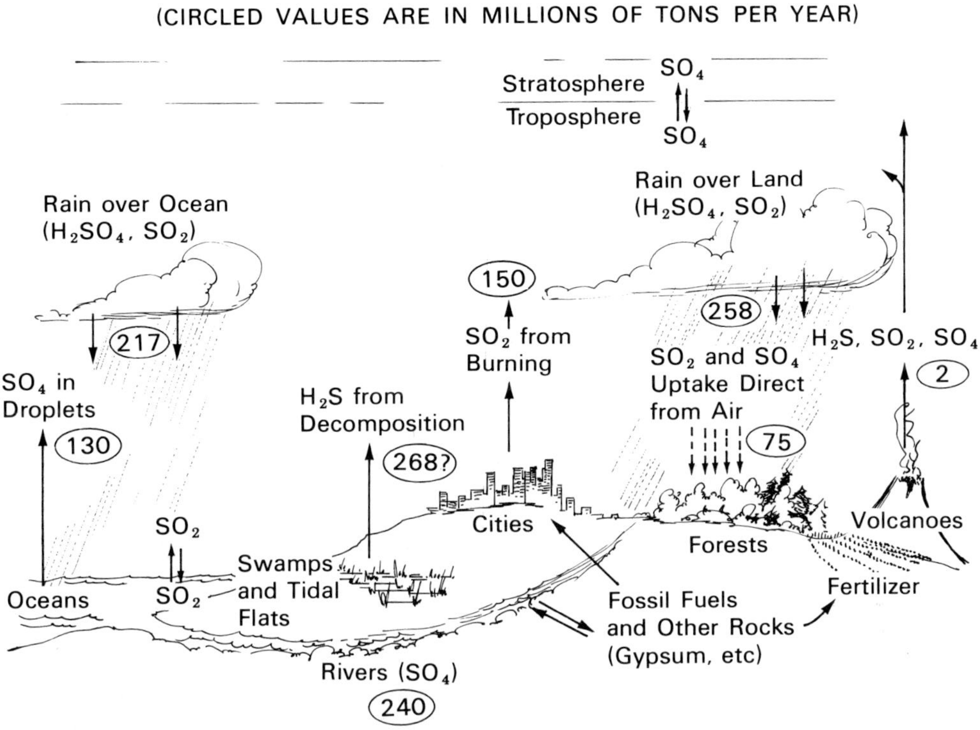

FIG. 9.4. Quantitative representation of the global sulfur cycle. Sulfur enters the atmosphere as SO_4 from breaking sea foam, as H_2S from biologic decomposition processes, and as SO_2 from combustion, especially of fossil fuels. Small contributions to the atmosphere also come from volcanoes. Sulfur leaves the atmosphere dissolved in rainwater, by diffusion into soils and vegetation, and by fallout of sulfate particles. Much of the sulfur leaving the atmosphere travels eventually to the sea, as does much land-derived sulfur. Both in water and on land, sulfur also circulates through organisms in local ecosystem cycles which are not shown in this drawing. (Slightly modified from W. W. Kellogg *et al.*: *Science*, *175:* 587–596, 1972, with permission from the American Association for the Advancement of Science.)

rare relative to its biological need, and the problem of supply is compounded by properties which impede the continued cycling of phosphorus.

There is no volatile form of phosphorus which can enter the atmosphere; thus, instead of the closed cycle between air, water and land that characterizes sulfur, the movement of phosphorus is essentially one-way, from land to sea. Thus, it is misleading to speak of a global phosphorus cycle constantly in operation. Some phosphorus *is* returned from sea to land: small islands off Peru, together with others in the western Pacific, are covered with enormously rich deposits of oceanic phosphate derived from marine organisms. Over the millenia, "recycling efforts" by marine birds have resulted in astounding concentrations of phosphate-rich excrement on these "guano islands." Modern man abets this landward movement of phosphate by mining guano for use on inland fields, as well as by directly harvesting marine organisms for food and fertilizer. But despite the efforts of man and bird, the net movement of phosphorus is from land to sea in a ratio that probably approaches 100 to 1.

Inorganic phosphorus enters the biological sphere as the soluble phosphate ion. This ion is readily assimilated by plants and by many kinds of microorganisms, and phosphorus thus is started through the food chain. But soluble phosphate ions also have an affinity for clay, and in fine-textured soils considerable phosphate may be tightly bound to mineral particles. Not only does adsorption to clay render much phosphate unavailable to organisms; this process also hastens its seaward migration. Clay is of such small grain size that it is transported easily by running water; hence clay-bound phosphate is very subject to erosive removal from terrestrial ecosystems. On reaching a lake or ocean, the clay and its passenger tend to settle out. Deep sea sediments, particularly, constitute a major sink for phosphorus; here the clay-bound fraction is withdrawn from circulation and sequestered, and dead organisms sinking from above augment this accumulation with their contribution of organic phosphorus compounds. In the fullness of geologic time these oceanic sediments may be uplifted and the phosphate thereby returned to land, thus completing the global cycle. But this is a long time for the biological sphere to wait.

The biological sphere is not simply a passive spectator watching its phosphorus go down the drain. Many aquatic species have evolved hoarding habits: when soluble phosphate is temporarily abundant they take up more than they immediately need (or may ever need), presumably storing it against a time when the supply may be insufficient. In an early experiment that revealed such biological stockpiling, Hayes *et al.* (1952) added radioactive phosphate to a lake in Nova Scotia and discovered that more than three-fourths of the amounts added were biologically removed from the water within 4 days, even though there was no immediate increase in the biomass of aquatic organisms. Many species simply increased their content of phosphorus, and in the process displayed remarkable powers of active uptake. One zooplanktonic species, on analysis, contained tissue levels of radioactive phosphate that were 40,000 times the level originally in the water. In another experiment Wilson (1972) found that when grown in phosphate-rich water a species of aquatic plant (*Myriophyllum exalbescens*) accummulated 20 times more phosphate than was sufficient to sustain maximal growth, as determined by comparative growth studies in water containing less phosphate. Such hoarding behavior by aquatic organisms minimizes the fertilizing effect that one would expect from the artificial addition of phosphate to a water body. Instead of stimulating growth of new organisms, much of the nutrient is simply taken up and stored by organisms already present. Despite this tendency, *continued* addition of phosphate to lakes or streams usually does have a major fertilizing effect, and contributes in large measure to the phenomenon known as artificial eutrophication. We will consider eutrophication further in Chapter 10.

The Global Nitrogen Cycle

Like sulfur, nitrogen cycles easily between Earth's three nonliving spheres. The atmosphere, where it is the predominant element, can be considered its principal reservoir (Fig. 9.5). But atmospheric nitrogen exists mainly as pure N_2, which is unusable by most living things. The atmosphere contains, in addition, relatively tiny quantities of ammonia (NH_3) and nitrogen oxides, the former derived from the decay of organic matter and the latter from its combustion (as well as from atmospheric fixation by lightning and photochemical reactions). Fixation, whether by lightning, chemical reactions or microorganisms, simply means the combining of pure nitrogen with oxygen or hydrogen to form mixed nitrogen compounds.

In the early part of this century man developed the capacity to fix atmospheric nitrogen by industrial processes, and since then he has been an increasingly important agent in the transfer of nitrogen from its atmospheric reservoir to the bodies of living things. But the first and still the foremost agent in moving atmospheric nitrogen into the biological sphere is a small and remarkable group of bacteria and blue-green algae. Prime examples of facilitator species, these microorganisms alone have the ability to biologically reduce atmospheric or dissolved N_2 to NH_3 (ammonia). The latter readily reacts with water to form hydroxyl and ammonium ions (OH^- and NH_4^+). Ammonium is assimilable by many plants, but often is progressively oxidized to nitrate (NO_3), another assimilable form, before it enters the food chain. The oxidation reactions usually are performed by *nitrifying bacteria* which can capture and use the energy released in these oxidations. Once assimilated biologically, the nitrogen exists as a constituent of complex protein and nucleic acid molecules; but when an organism dies and decays these are broken down again to nitrate or ammonia and are ready to re-enter the biologic cycle through plants. Some nitrate usually is tapped off at this stage by the activity of yet another group of bacteria, the *denitrifiers*. When conditions are anaerobic, denitrifying bacteria use nitrate or nitrite (NO_2) in place of pure oxygen to oxidize organic compounds, and N_2 is one of the reaction products. Denitrifying

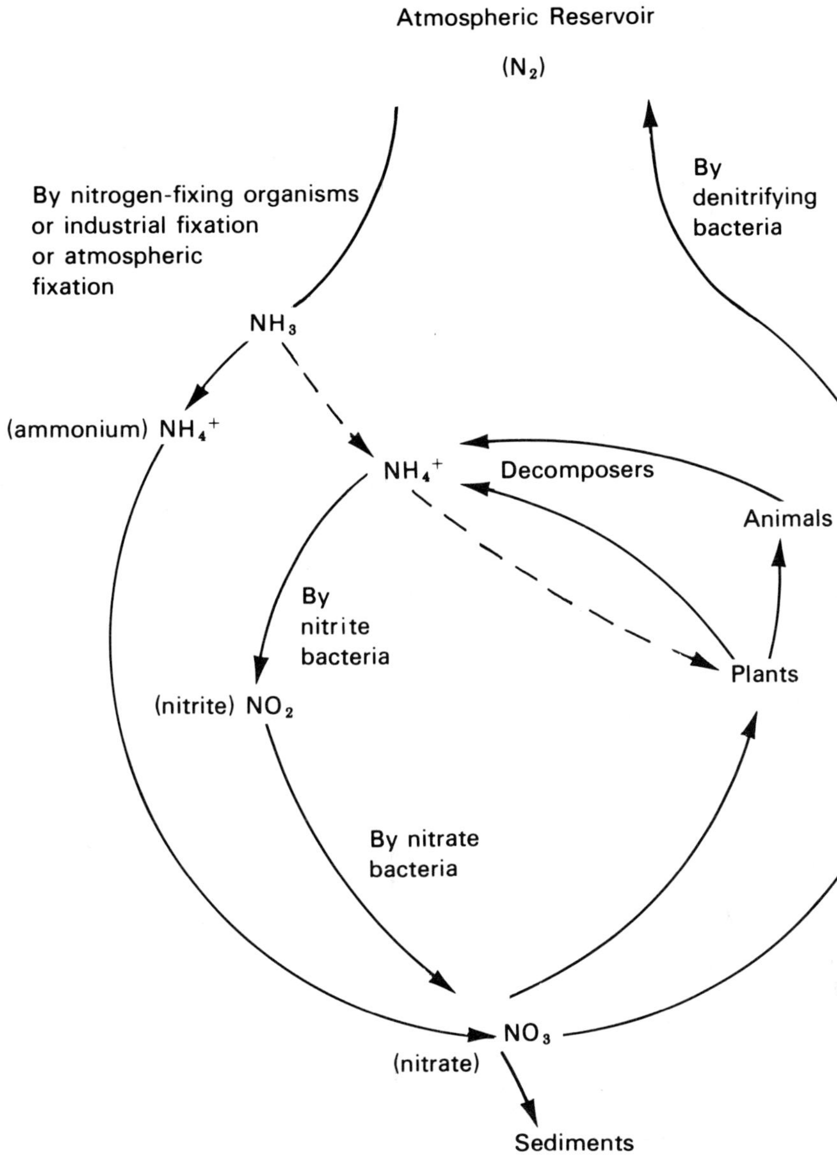

FIG. 9.5. Nitrogen cycle. In addition to the larger cycle involving the atmosphere, there is an "inner cycle" involving organisms, soil and water through which a given atom of nitrogen may move many times before it returns to the atmosphere or is removed from the cycle by entrapment in sediments. For simplicity, this diagram does not show the many forms of organic nitrogen which, as excretory or decomposition products, may join nitrate in the sedimentary sink.

bacteria actually are the chief agent by which nitrogen is returned to the atmosphere, and denitrification is a vital process in the maintenance of the atmospheric reservoir. Without dentrifying bacteria there would be a one-way flow of nitrogen from the atmosphere to the other spheres, similar to the one-way flow of phosphorus from land to sea. Sedimentation of nitrogen compounds in sinks like the deep ocean results in some leakage from the nitrogen cycle, but this loss (unlike sedimentary loss of phosphate) is partly offset by volcanic emission of nitrogen gasses trapped within the lithosphere.

These are the main steps in the nitrogen cycle (Fig. 9.5). Nitrogen moves from its atmospheric reservoir into soil, water and organisms largely through the process of biological nitrogen fixation, and returns to the atmosphere via the biological process of denitrification. The rate of fixation often seems to be a controlling process for rates of biological production: if ammonium could be fed into many ecosystems more rapidly, greater growth would result. When a nutrient plays a limiting role it often acts as a force in natural selection, and this seems to be true in the case of nitrogen. Early successional environments on land are in many cases particularly nitrogen-poor, and it has been shown that the plant species which flourish in such environments—abandoned fields, or rocks weathering to a primary soil, or slopes denuded of their organic topsoil by landslides—typically have unusually low nitrogen requirements. This is one direction of adaptation; another is the evolution by certain higher plants of elaborate symbiotic relationships with nitrogen-fixing bacteria. Such relationships, found among many leguminous species such as clover and soybeans, assure that these plants will have enough even in soils generally deficient in fixed nitrogen. Of course, the whole community indirectly benefits from such symbioses, because the newly fixed nitrogen eventually passes through food chain and decomposition transfers and becomes generally available within the ecosystem. Recognizing this, man plants legumes widely not just as a food crop but as a means to restore fertility to fallow croplands and pastures. Delwiche (1970) believes that by growing legumes, the human species has added significantly to Earth's supply of fixed nitrogen. In fact, he suggests that agriculturally and industrially, man has achieved a doubling of the rate at which nitrogen is being fixed world-wide. Since most industrial fixation is directed toward the production of fertilizer, man's efforts are seen to represent a major avenue of nitrogen flow into the biological sphere. Man himself, then, is a major facilitator species to the rest of the living world. One wonders if, through man's efforts, more nitrogen presently is being withdrawn from the atmosphere by fixation than is being returned by denitrification. No one has yet detected a decrease of atmospheric nitrogen, but the atmospheric reservoir is so great that an unbalanced drain on the atmosphere might not yet be noticeable, even if the flow is quite great.

Other Cycles

Carbon and oxygen are of supreme biological importance, and we should note at least the key features of their cycles. These substances move freely in and out of the biological sphere; unlike nitrogen, which can be fixed by only a few facilitator species, atmospheric or dissolved oxygen can be used directly by nearly all organisms, and atmospheric or dissolved CO_2 by nearly all plants and algae. Moreover, it is misleading to think in terms of a single reservoir for these elements. If asked to identify the principal reservoir of the carbon cycle, one's first answer might be the atmosphere, since that is the carbon source for terrestrial plants (Fig. 9.6). But living and decomposing organisms actually contain far more carbon than does the whole atmosphere. The hydrosphere contains even more: CO_2 is highly soluble, as is bicarbonate (HCO_3^-) and to a lesser degree carbonate (CO_3^-). And the lithosphere is the repository for abundant carbonate rocks (*e.g.* limestone) and deposits of hydrocarbons (*e.g.* coals and petroleum). For the most part these lithospheric supplies represent carbon that was once in circulation but was then sequestered in sinks. Particularly today, however, these sinks are not inaccessible. Many carbonate rocks are highly soluble, and their carbon always has been partially reclaimable via surface weathering as well as groundwater movement into surface streams. Now, in addition, human activities are returning to the atmosphere great quantities of the carbon removed from the biological sphere millions of years ago as fossil fuel. The resultant increase in atmospheric CO_2 (Fig. 9.6) would be far greater if the atmosphere retained all that it acquired. Instead, much of the added CO_2 dissolves secondarily in the oceans, equilibrating between atmosphere and hydrosphere according to the solubility constant of this gas. Of course, fossil fuel CO_2 also is withdrawn from the atmosphere by photosynthesizers, but the atmosphere should recover an equivalent amount from the biological sphere if photosynthesis is roughly in balance with respiration. Some scientists speculate however, that the total mass of living matter on Earth is increasing at present, rather than remaining stable; *i.e.* biological uptake is outstripping respiratory release of CO_2. If so, this biological imbal-

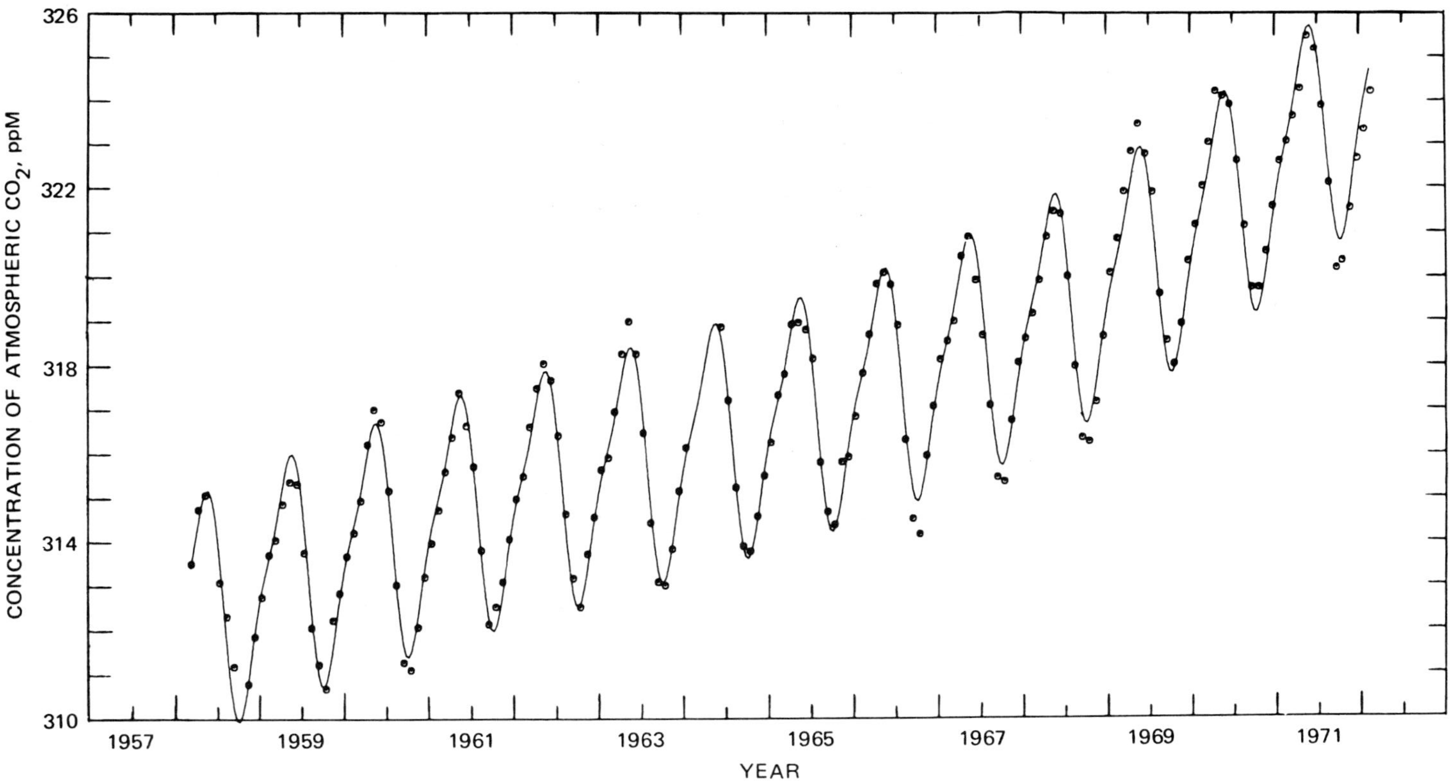

FIG. 9.6. Seasonal and long term changes of atmospheric CO_2. These changes were recorded at the high altitude Mauna Lao observatory in Hawaii, where local influences on CO_2 are slight and do not mask the larger picture. The long term trend is explained in the text. The short term oscillations reflect seasonal changes in the intensity of photosynthetic carbon fixation in the northern hemisphere, which dominates the picture because it contains the great preponderance of the world's land mass. In his concluding remarks to a symposium on carbon in the biosphere, Daniel Livingstone said of this graph that "the majestically rising seasonal cycles . . . made me feel as if I had just put my finger on the beating, living heart of the world." (From L. Machta: *Brookhaven Symposia in Biology*, *24:* 22–31, 1973. Reproduced with permission.)

ance further retards the increase in atmospheric CO_2 expected from fossil fuel use. Despite hydrospheric and biospheric damping mechanisms, however, average CO_2 levels in the atmosphere *are* rising, and the increase is sufficiently pronounced to worry many climatologists. Since CO_2 is one of the main heat-absorbing constituents of the atmosphere, the fear exists that the world will become warmer as fossil fuel carbon continues to be pumped back into circulation. Even a 2–3° C rise in global temperatures would have frightening repercussions in terms of melting ice caps at high latitudes and human heat stress in the tropics.

What of oxygen? Molecular oxygen comprises 21% of the atmosphere at present, and together with the large amounts dissolved in the hydrosphere this supply is thought to be completely derived from photosynthesis. There is no more profound example than this of how life on Earth has molded its own environment; the primitive atmosphere has been altered drastically by the activities of living things. Because rumors have been voiced that human activities are seriously depleting these biologically produced oxygen reserves, we should at least deal briefly with this possibility. Will man run out of oxygen to breathe if he continues to clear the forests, pave the prairies, dump oxygen-gobbling sewage into streams and rivers, and consume atmospheric oxygen in fossil fuel combustion? All these activities create problems, but human suffocation apparently is not one of them: calculations by Wallace Broecker (1970) show that air and ocean, the twin reservoirs of molecular oxygen, contain enormous reserves, and only in polluted lakes and streams does serious oxygen depletion result from human activity.

Living things need many elements and compounds not mentioned here. These essential substances were discussed in Chapter 2, and perusal of Table 2.2 will suggest the locations of many biogeochemical reservoirs. But rather than explore elemental cycling patterns one by one, let us see how several together relate to the riddle of the ocean with which this chapter began.

THE OCEAN: A STEADY STATE CHEMISTRY?

The ever changing yet unchanging sea: "While clay and ooze fall, like gentle snow, and rafted rocks, wrecks and other heavy debris plummet to the sea floor, manganese, iron, cobalt, nickel and copper invisibly migrate ion by ion to the deepest parts of the ocean" (Heezen and Hollister, 1971). We can imagine this process being in balance with the rate of delivery of these substances to the sea: the vast gardens of ferromanganese nodules on the floor of the oceanic sink (Fig. 9.7) well may be growing at precisely the rate at which their constituents wash in from the land, and the concentration of metallic elements in solution may be nearly constant through time.

There is truth to this notion of the ocean as a steady state intermediary between dry land and deep sediments, but the picture is more complicated than we have painted it. For example, stupendous quantities of sodium (14×10^{15} tons) are dissolved in the ocean, and submarine sediments are estimated to contain nearly as much (Livingstone, 1963). But Livingstone's calculations show that these amounts fall far short of the quantity of sodium that has been delivered to the ocean over geologic time, if long term input has been at all close to its present rate. Where is all the extra sodium that should be in or beneath the sea? The puzzle is partly resolved if we realize that an appreciable part of the sodium carried to the sea by rivers is not freshly leached lithospheric sodium but is *oceanic* sodium recycled through the atmosphere (Table 9.2). But a huge discrepancy remains, which Livingstone was able to account for only by concluding that the oceans' sodium has passed through at least four complete sedimentary cycles in addition to the inadequate atmospheric cycle—a sedimentary cycle being the age-long process of sedimentation, uplift to form dry land, and subsequent erosion and redeposition. The deep ocean and its sediments are indeed a sink, but over the long term even the sink enters the cycling process.

This conclusion does not upset the idea of an ocean in steady state chemistry, at least over the long term, although we can envision temporary departures from steady state if sedimentary uplift has been episodic rather than constant. But living things and their evolutionary history cannot be forgotten when we talk about oceanic chemistry. Sometimes biological processes have worked against the chemical steady state, sometimes for it. As an example of the

FIG. 9.7. Manganese and iron are precipitated from solution as nodules on the deep ocean floor. The technical difficulties of "mining" these enormous reserves have so far prevented their exploitation (From B. Heezen and C. Hollister: *The Face of the Deep*. New York: Oxford University Press, 1971. Reproduced with permission.)

former, consider the case of silica. Harriss (1966) has calculated that the present oceanic phytoplankton are removing dissolved silica from the medium twice as fast as it is arriving from the land. When silica-using phytoplankton (primarily diatoms and coccolithophorids) die, their shells sink; partial solution of the shells returns some silica to circulation, but large quantities of undissolved siliceous ooze accumulate on the bottom and become part of the sediment. If Harriss is correct, the dissolved silica of the oceans presently is decreasing slowly in concentration and is not in steady state. Diatoms, and probably also coccolithophorids, are evolutionary newcomers to the ocean, but are now its dominant photosynthetic organisms. They may have been reducing the sea's silica content for much of the 80 million years or so that they have been on the scene, but if silica continues to decline, these organisms should also decline in abundance. Harriss predicts that eventually a near-steady state concentration of silica may be established at a level of about 0.2 mg/liter. This is the limiting concentration of silica that the average diatom needs for growth; once dissolved levels reach this low value the populations of diatoms and coccolithophorids will no longer be living off the silica reserves of previous ages but will have their growth controlled by *present* rates of silica input. Now they will be agents of the steady state, their numbers governed by the amount of silica available while they in turn keep dissolved levels close to the limiting concentration.

Certain other dissolved substances in the ocean seem already to have attained the biologically mediated equilibrium that Harriss foresees for silica. In 1958 Alfred Redfield published *The Biological Control of Chemical Factors in the Environment*, a provocative paper which suggests that the amounts of carbon, phosphorus and nitrogen in oceanic solution fluctuate almost entirely as the result of the synthesis or decomposition of organic matter, and that these fluctuations are balanced because the processes of biological synthesis and decomposition are themselves balanced. Photosynthesis provides the oxygen that is necessary to bring about oxidative decomposition. Dissolved oxygen can, of course, be replenished from the atmosphere if it is in short supply, but Redfield believes this process is unimportant to the oceanic balance he describes. Dissolved oceanic oxygen is depleted primarily by oxidation of organic matter, and the photosynthetic creation of organic matter involves, at its first carbon-fixing step, the production of enough oxygen to oxidize it when it decays. Thus, there is an internal balance of oxygen in the ocean that is linked to the balance of the three other elements. When we note that some parts of the ocean are saturated with dissolved oxygen while others are anaerobic, and that some are far richer in dissolved phosphate than others, we are not (says Redfield) denying the steady state chemistry of the ocean as a whole; unless the balancing processes he proposes are operating at the same rate in every liter of ocean water, one would not expect (nor require, by the steady state assumption) every liter to be chemically uniform. The present steady state has probably been a slowly evolving one; eons ago, for example, there was less molecular oxygen available either in the ocean or the atmosphere than there is today, and this presumably affected the concentrations of dissolved carbon, phosphate and nitrate. But total ocean chemistry has been remarkably conservative over time; in concluding a recent review of the subject, Holland (1972) states: "It is likely that the concentration of the major constituents of sea water has rarely, if ever, been more than twice or less than half their present concentration during the past 700 million years." Perhaps the ocean is not a perfect steady state, but it is a reasonable approximation to one.

THE BIOGEOCHEMISTRY OF SUCCESSION

What happens to nutrient cycling processes when an ecosystem is disrupted and its communities undergo the process of ecological succession? In recent years a fragmentary picture has emerged of the self-repair mechanisms that tend to restore the biogeochemical equilibrium typical of a mature ecosystem. We have learned a little, too, of how the nutrient-related strategies of different species are integrated into the successional process.

Earlier in this chapter we referred to the Hubbard Brook watershed studies in New Hampshire. This program has included investigations of cut-over and regenerating forests as well as undisturbed ones. At Hubbard Brook the initial result of clear-cut lumbering is a

TABLE 9.2

Cycling of oceanic salts through the atmosphere

One avenue for removal of salts from the sea is through the air, as wind-driven spray and aerosols. As indicated by the data below, these airborne salts represent important nutrient input to terrestrial ecosystems adjoining the ocean, but are much less important farther inland. The Sunken Forest is a unique forest formation on Fire Island, New York, lying behind primary and secondary foredunes on this sandbar island. Mineral input from the soil is negligible; instead, there is a net loss through the porous sand and back to the sea. Data from Art *et al.* (1974).

	INPUT OF CATIONS				Source
	K	Na	Ca	Mg	
	g/m₂/year				
Sunken Forest (0.1 to 2 km from ocean)	0.73	14.15	0.98	1.91	Rainfall and spray
	0	0	0	0	Weathering
Hubbard Brook Forest (116 km from ocean)	0.14	0.15	0.26	0.07	Rainfall
	0.4	0.61	0.80	0.8	Weathering

	ATMOSPHERIC INPUT AS % OF TOTAL			
	K	Na	Ca	Mg
Sunken Forest	100	100	100	100
Hubbard Brook Forest	3.4	20	25	8

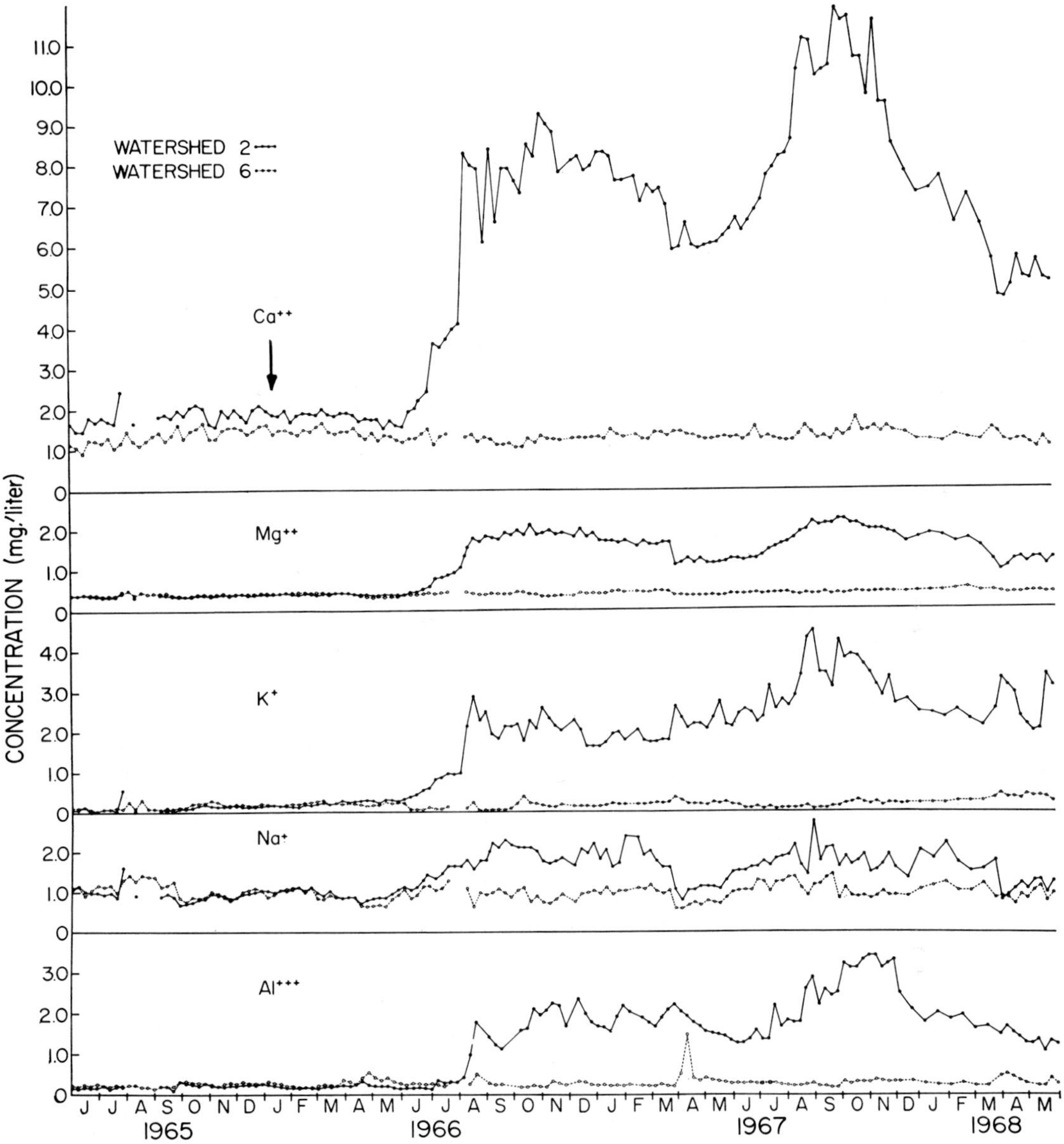

FIG. 9.8. Effect of deforestation on short term nutrient removal rates from a forested watershed, with rates from an undisturbed watershed shown also for comparison. The *arrow* shows the time of deforestation. (From G. E. Likens, F. H. Bormann, N. M. Johnson, D. W. Fisher, and R. S. Pierce: *Ecological Monographs, 40:* 23–47, 1970. Reproduced with permission.)

remarkable increase in the leakiness of the watershed: concentrations of many elements skyrocket in the outflow streams (Fig. 9.8) and the consequences of clear-cutting for soil fertility seem severe. But equally remarkable is the rapidity with which these losses are reduced as vegetational succession begins. One species of fast-growing tree, pin cherry, is a particularly important agent of nutrient conservation (Marks, 1974; Marks and Bormann, 1972). In this part of New Hampshire pin cherry rapidly proliferates in cut-over deciduous areas and establishes a dense, complete cover within 4 years of cutting. Pin cherry thickets quickly reduce the high surface runoff that follows deforestation; water instead is channeled upward through the leaves as evapotranspiration, and nutrients accumulate rapidly within the plants instead of washing out of the ecosystem (Fig. 9.9). Pin cherry thus facilitates nutrient retention in two ways: by decreasing runoff and by capturing soil nutrients in its own tissues. The leafy thickets of pin cherry also reduce summer soil temperatures, moderating what had been unusually rapid rates of decomposition and release of organically held nutrients. In this way, too, nutrient retention within the ecosystem is facilitated.

The role played by pin cherry at Hubbard Brook doubtless is played by other early successional plants in other ecosystems. There is some evidence, too, that the species of later seral stages possess characteristics that further decrease nutrient losses from an ecosystem as succession proceeds. In one intriguing example, Rice and Pancholy (1972 and 1973) report a progressive decrease in activity of nitrifying bacteria in the stages of a forest sere in Oklahoma, the decrease apparently being due to a progressive inhibition of these bacteria by the plant types of successive seral stages. Nitrifying bacteria oxidize ammonium to nitrate, so that as succession proceeds in this sere, nitrate levels in the soil fall while ammonium levels rise. For the ecosystem this is a nitrogen conserving process: either ammonium or nitrate can serve as a nitrogen source for the plants, but the positively charged ammonium ions (NH_4^+) tend to be loosely adsorbed to negatively charged colloidal particles near the soil surface, and to be retained there until taken up by the roots. Negatively charged nitrate (NO_3^-) tends to be repelled by the negative charges of the soil particles, hence is easily dissolved and removed either by downward leaching or surface runoff. One of the main inhibitors of the nitrifying bacteria seems to be tannin, a substance produced in progressively greater amounts by the vegetation of successive seral stages.

Not just communities, then, but ecosystems too seem to act sometimes in almost an organismic way during ecological succession. In the examples just cited we see ecosystems performing excellent plumbing jobs on themselves, repairing leaks in efficient and seemingly coordinated fashion. But where coexisting species are concerned, images of cooperative altruism should not be overstressed. Individuals of different species work for themselves, and though benefits may accrue to others through their activities, in many cases working for oneself means working *against* others. A second example involving nitrogen (Rice, 1971) provides a balancing counterpoint to the case cited above. In Oklahoma old fields, many early successional weeds contain substances that, when leached from the leaves, significantly reduce nitrogen-fixing capabilities of the legumes that ordinarily appear in these fields. Thus the rate of nitrogen restoration to these exhausted old field soils is slowed, and the early weeds, which generally can flourish in nitrogen-poor soils, persist longer because soil fertility increases more slowly than it would otherwise. (In earlier studies, Rice had shown that the order in which plant species invaded these abandoned fields was closely correlated to their relative needs for nitrogen.) In this example, of course, we are not dealing directly with repair of a leaky ecosystem but rather with biological retardation of the rate at which a needed nutrient is imported. But if the early weeds are relatively poor at retarding

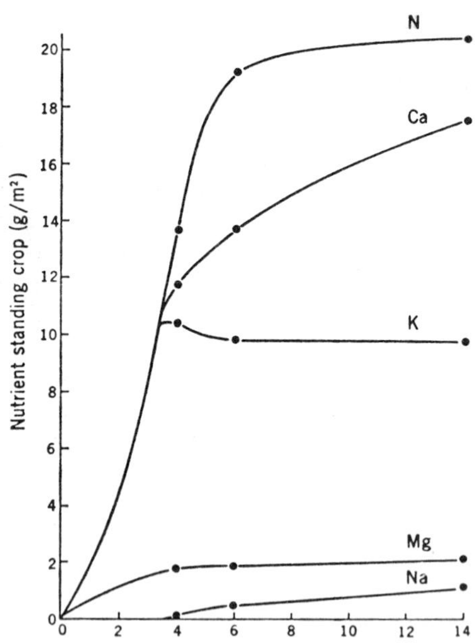

FIG. 9.9. Rapid nutrient accumulation by pin cherry: biological repair of an ecosystem made leaky by deforestation. (From P. L. Marks and F. H. Bormann: *Science, 176:* 914–915, 1972. Reproduced with permission from the American Association for the Advancement of Science.)

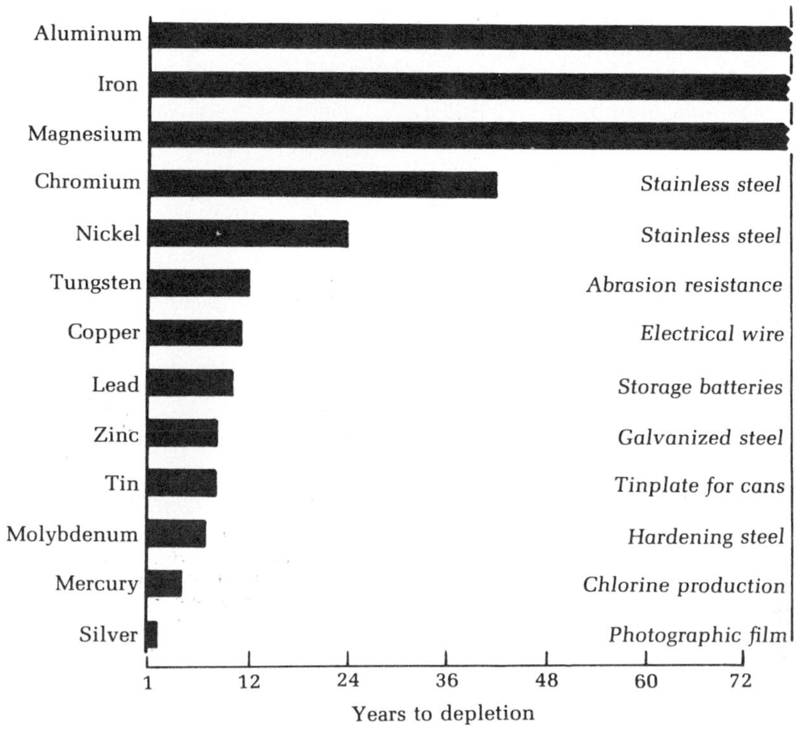

FIG. 9.10. Expected time to exhaustion of reserves of the world's major industrial metals if per capita world consumption was equal to the United States per capita consumption. (From G. Moriber: *Environmental Science*, Boston, Allyn and Bacon, Inc., 1974.)

erosion, then their persistence also implies a slowing of the ecosystem's leak repair processes.

RECYCLING AND THE ENVIRONMENTAL MOVEMENT

Clearly, the human species now plays a major role in many biogeochemical processes. In many cases—fixed nitrogen being an example—man increases the amounts available to living things. But often the long term effect of his activities is to threaten exhaustion of resources in forms that he can use himself. To modern man, essential nutrients are not just those making up his protoplasm, but also those sustaining his culture—metals and fuels as well as nitrogen for his protein and phosphorus for his bone. We have begun to recognize ourselves as prodigious wasters, and exhortations to conserve and recycle sweep the land.

Recycling consciousness is good, but if it is to help us very much we need to know what we are running out of (Fig. 9.10). Except in local instances we are *not* running out of water, because most water cycles continuously from sea to air to land. What we use and dispose of eventually returns for reuse. We *are* sometimes overtaxing or even exhausting local water supplies; in Arizona man is mining water sequestered beneath the land surface since the last ice age, and as this water rejoins the active cycle it disappears, for the most part, from Arizona. But in other places where supplies have seemed dangerously short, they could be stretched simply by tapping into the natural cycle more efficiently. New York's water shortages of the mid-60s were not real *water* shortages; the Hudson River continued to flow by unused, little effort having been made to exploit its bountiful supply. In drier parts of America irrigation makes heavy use of stream and well water, and because it transfers much of the water to the atmosphere it can seriously reduce surface and underground flows; but most municipal water is simply discharged back into streams and sent on its way when it could be purified and used again.

Is much really gained by recycling glass and aluminum containers? Definitely, if what worries us is an overabundance of trash, or the high energy cost of smelting aluminum ore. But we are in no danger of running out of the ingredients of glass, or of aluminum beer cans; if we are concerned primarily with mineral exhaustion we should emphasize the recycling of *tin* containers, as Figure 9.10 makes clear.

Man's so-called renewable resources are those substances whose natural biogeochemical cycles are complete and relatively rapid: water, oxygen, timber, cattle. Our nonrenewable resources consist of substances which are not part of closed, short term cycles, and whose supplies are therefore finite in human terms: oil, phosphate, most metals. Sometimes we must conserve even renewable resources, if consumption exceeds the rate of supply; but it is scarce *nonrenewable* resources that we must learn to do without, very soon in some cases, even if we do mend our wasting ways.

SUGGESTED READING

The Biosphere. Special Issue of Scientific American. (Sept., 1970). Many authoritative articles on the cycles of different nutrients.

BORMANN, F. H., AND G. E. LIKENS. The nutrient cycles of an ecosystem. *Scientific American* 223 (4): 92–101 (1970). A condensation of results from the Hubbard Brook study.

HOLLAND, H. D. The geologic history of sea water—an attempt to solve the problem. *Geochimica et Cosmochimica Acta* 36: 637–651 (1972). A rather technical exploration of the chemical history of the ocean.

LIKENS, G. E., AND F. H. BORMANN. Nutrient cycling in ecosystems. pp. 25–67. In J. A. Wiens, ed. *Ecosystem Structure and Function*. Corvallis, Ore., Oregon State University Press, 1972.

MARKS, P. L., AND F. H. BORMANN. Revegetation following forest cutting: mechanisms for return to steady-state nutrient cycling. *Science* 176: 914–915 (1972). The role of pin cherry at Hubbard Brook. An expanded version by Marks may be found in *Ecological Monographs* 44: 73–88 (1974).

POMEROY, L. R. The strategy of mineral cycling. *Annual Review of Ecology and Systematics* 1: 170–190 (1970). An authoritative review containing many references.

REDFIELD, A. C. The biological control of chemical factors in the environment. *American Scientist* 46: 204–221 (1958). Interesting ideas regarding the central role of marine organisms in determining the chemical composition of sea water.

VITOUSEK, P. M., AND W. A. REINERS. Ecosystem succession and nutrient retention: a hypothesis. *Bioscience* 25 (6): 376–381 (1975). Presents the suggestion that nutrient leakage from ecosystems is less during the middle stages of succession than at maturity.

WOODWELL, G. M., AND E. V. PECAN, eds. Carbon and the Biosphere. Springfield, Va., Technical Information Center, United States Atomic Energy Commission, 1973. Proceedings of a symposium on the distribution and cycling dynamics of this most central nutrient to life.

WOODWELL, G. M., AND E. V. PECAN, eds. Carbon and the Biosphere. Springfield, Va., Technical Information Center, United States Atomic Energy Commission, 1973. Proceedings of a symposium on the distribution and cycling dynamics of this most central nutrient to life.

chapter 10

Ecosystems and Energy

And as I was green and carefree, famous among the barns
About the happy yard and singing as the farm was home,
In the sun that is young once only,
Time let me play and be
Golden in the mercy of his means . . . .

Dylan Thomas, *Fern Hill*, from *Collected Poems* (1953 New York, New Directions Books)

A ONE-WAY TICKET

Some day, when the sun is old, Earth will be dark and lifeless. Our living planet requires constant transfusions of energy and it is "the sun that is young once only" that sustains us.

An ecosystem ecologist should know some physics. The Law of Conservation of Matter underlies the phenomenon of nutrient cycling discussed in Chapter 9: "Matter is neither created nor destroyed." It follows that the same atom may be used again and again, passing from one substance to another but remaining immortal.

The Law of Conservation of Energy sounds remarkably similar: "Energy is neither created nor destroyed." Moreover, just as nutrients exist in numerous chemical states that are interconvertible, so also does energy. Visible light energy, chemical bond energy, electrical energy, mechanical energy, all can be transformed one to another.

But the energy entering an ecosystem does not cycle, and the amount available for use diminishes with every transformation, and with every transfer step within a food chain. Why are the fates of matter and of energy so different? We turn to a third physical law, the Law of Entropy.* Every transfer or transformation of energy is accompanied by the production of heat, and heat quickly becomes so dispersed within the universe that it is lost as a usable entity. Dispersed heat is a degraded form of energy which cannot be converted back to other forms and cannot be used by organisms. It is an energy sink toward which all other forms of energy gravitate.

In the photosynthetic fixation of light energy, heat is a surprisingly large (and largely useless) by-product. Only part of the incoming light energy is captured as chemical energy; that which becomes heat is lost. Plants produce additional heat in the chemical reactions by which they grow and maintain themselves; this energy, too, slips away, as does the heat produced by herbivores engaged in eating and processing the plants (Fig. 10.1). Great energy losses occur with every transfer up a food chain, and typically there is very little energy available to the last links, the upper level carnivores. At Hawk Mountain in Pennsylvania, southward migrating birds of prey are concentrated in

* The Law of Conservation of Energy and the Law of Entropy are also known as the First and Second Laws of Thermodynamics.

unusual numbers by topography and wind patterns along the last great ridge of the folded Appalachians. Like many others, I have delighted in the procession of hawks and eagles soaring overhead, but as an ecologist I am impressed not with the great numbers, but with how scarce these "kings of the food chain" really are. On a good day there may be a few hundred funnelled into Hawk Mountain from a large fraction of northeastern North America, a vast area.

The Law of Entropy is not wholly to blame for the rarity of top carnivores. There are other ways by which the energy fixed at earlier steps in the food chain becomes unavailable to hawks and eagles, one being the shunting of foodstuffs from the so-called **grazing food chain** (to which these carnivores belong) to a **decomposer food chain** coexisting in the same ecosystem (Table 10.1). Different ecosystem types vary greatly in the share of available energy that passes to these two food chains. In most aquatic ecosystems a large fraction of the energy fixed by photosynthesizers continues up the grazing food chain. In terrestrial ecosystems, particularly forests, a far larger proportion of the fixed energy is shunted to decomposers (Fig. 10.2, and page 108).

BIOMASS AND PRODUCTIVITY

In energy flow studies an investigator commonly simplifies an ecosystem's complex food web to a series of trophic levels. All plants of all species in an ecosystem consitute its primary trophic level; they and they alone are capable of directly converting solar energy into the chemical bond energy of organic matter. Plants are the sole source of fixed energy for herbivores (the second trophic level) and indirectly for carnivores as well. A grazing food chain usually can be divided into four of five trophic levels; progressive energy losses with each trophic transfer preclude the existence of more steps than this.

A trophic level consists, at any given time, of a certain amount of stored chemical energy. This quantity is referred to as the **biomass** or **standing crop** of plants, herbivores, etc., and can be represented in various units: calories or kilocalories of energy, grams or bushels of total tissue, or other measures. Units of weight or volume are not themselves energy units, but can be converted to energy equivalents if a given amount of organic material is burned completely and its heat output (caloric content)

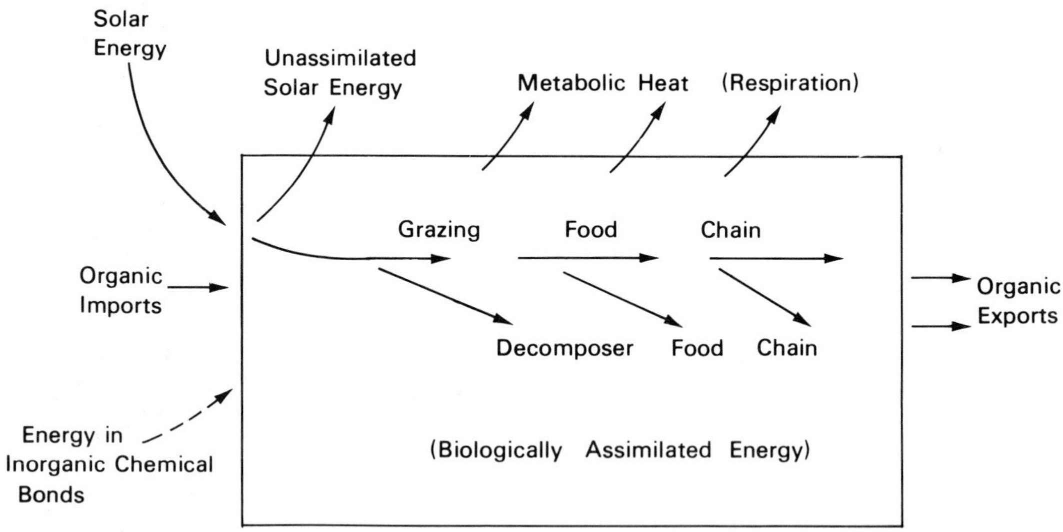

FIG. 10.1. In autotrophic ecosystems, solar energy usually constitutes by far the largest energy input. Organic imports vary in relative size, but are the largest energy input to many heterotrophic ecosystems. Some microorganisms can obtain energy from oxidation of inorganic compounds, but this generally is a very small input. Immediately lost to the biologic component of the ecosystem is unassimilated solar energy; that which is assimilated is dissipated in metabolic processes, variable amounts also being exported. The grazing food chain also loses energy at every step to the decomposer food chain.

TABLE 10.1

Food Chains in a Forest Ecosystem

The two main types of food chain in an ecosystem are the grazing food chain, based on living or recently dead foods, and the decomposer food chain based on materials that are being progressively reduced to simple molecular compounds. These compounds, in turn, serve as nutrients for plants and so feed back into the grazing food chain. Every level of the latter contributes dead materials to the decomposers. Thus these two food chains are closely linked, yet represent distinctly different pathways for energy in an ecosystem. These pathways may merge, however, as in the case of the "hybrid" food chain that is also shown below, in which decomposer organisms serve as food for carnivores. (Normally, of course, the organisms in successive levels of a decomposer food chain do not feed on each other, but represent sequential processing stations for decaying matter.) The food chains that are shown are simplified examples of direct sequences of energy flow.

Grazing Food Chain	Decomposer Food Chain	Hybrid Food Chain
Hawks	Bacteria & fungi	Weasels
↑	↑	↑
Vireos	Earthworms	Moles
↑	↑	↑
Small caterpillars	Cicada larvae	Earthworms
↑	↑	↑
Fresh leaves	Dead plant materials	Cicada larvae
		↑
		Dead plant materials

accurately determined. The biomass of a trophic level is a measure of its energy content at a given moment, but knowing the biomass tells us nothing about the *rate* at which the trophic level is *fixing or capturing new energy*. This rate is what is meant by energy flow, and is referred to as the **productivity** of the trophic level. Productivity, then is the rate of energy fixation by a trophic level, whereas biomass is a measure of its instantaneous energy content.

Trophic levels are designated primary, secondary, tertiary, etc., on the basis of their food chain distance from the initial source of energy, the sun. Thus one speaks of primary, secondary or tertiary biomasses, and for the same trophic levels speaks of **primary productivity** (the rate of photosynthetic fixation of energy by plants), **secondary productivity** (the rate of energy capture by herbivores from plants), tertiary productivity, etc. Moreover, two kinds of productivity are distinguishable at each trophic level. The first is **gross productivity,** the total amount of energy assimilated per unit time. This energy is used to synthesize new organic matter. But the organisms of each trophic level similtaneously are expending energy (and destroying organic matter) in all the energy-requiring processes of existence. Energy expended in all these ways is lumped as **respiration.** In addition to gross productivity, then, a rate termed **net productivity** can be defined. This is the algebraic sum of the gross productivity and the respiratory loss of energy over a given period of time. Net productivity can be either positive or negative (Fig. 10.3); if positive, organic matter is being created faster than it is being destroyed by respiration. In such circumstances, either the biomass of the trophic level increases or the excess production is passed on to other trophic levels. The net productivity of a trophic level, then, is the rate at which energy potentially available to the next trophic level is captured.

If the biomass of a given trophic level is large and is not diminishing, clearly its gross productivity is relatively large too; it must be sufficient to balance the considerable energy loss of this large respiring biomass. But a large biomass does not necessarily imply a large net productivity; the trophic level may be using up nearly all the energy it captures in its own respiratory processes, in which case very little is available to pass on to the next trophic level. For some trophic levels in some ecosystems it is relatively easy to measure gross productivity, net productivity and respiration (*cf.* Fig. 10.3). In other cases, net productivity or some approximation of it is both the easiest and the most practically important measure to obtain. The amount of hay harvested from a field at the end of a growing season represents the net primary production of the field over that time period (less the amounts harvested by insects and other natural consumers). To the farmer, this net figure is much more important than the values for gross production and respiration by the plants, values that would have to be determined by physiological experiments. If a published figure for biological production does not specify whether it is gross or net, it usually is safe to assume that it is the latter.

TROPHIC PYRAMIDS AND EFFICIENCY OF ENERGY TRANSFER

Three Ecosystems Contrasted

If the productivities, biomasses, or simply the individuals of an ecosystem's different trophic levels are enumerated, they easily can be compared by diagrams in which the trophic values, represented by bars of given length, are stacked one above the other (Fig. 10.4). The quantities typically decrease with each step to a higher trophic level, so that diagrams of this sort generally assume the shape of a pyramid.

Figure 10.4 depicts hypothetical trophic pyramids for the grazing food chains of forest, grassland, and oceanic ecosystems. Note that pyramids of numbers or of biomass may be partially inverted on occasion, but the Law of Entropy dictates that pyramids of productivity must always be regular. Pyramids of productivity are always steep—rates of energy capture decline rapidly with each trophic step—but the steepness of the decline varies among ecosystems. The steeper the pyramid of productivity, the less is the efficiency of trophic transfer and the fewer are the number of trophic steps possible before the available energy becomes vanishingly small.

The pyramids of these three ecosystem types represent two extremes (forest and ocean) and an intermediate form (grassland). The primary level in aquatic ecosystems consists of small, rapidly growing, easily digestible phytoplankton which pass on to herbivores a relatively large fraction of their assimilated energy. The primary biomass often is relatively low in water, because the phytoplankton may be heavily

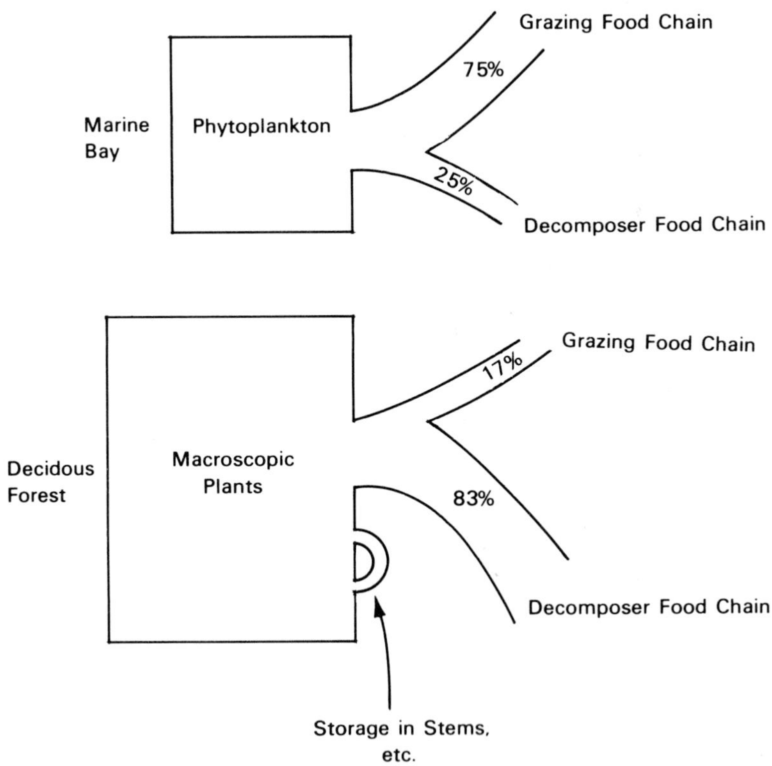

FIG. 10.2. In studies of Long Island Sound and the English Channel reported by Odum (1962), three-fourths of the energy passed on by phytoplankton went to herbivores, *i.e.* into the grazing food chain. Odum also reports studies of deciduous forests in which only 17% of energy passed on by plants went to herbivores, the rest going to decomposers. In this case the forests captured twice as much energy as the phytoplankton to begin with, but this may not always be true. The forest plants also used about one-fourth of their assimilated energy to make wood, which represents long-term storage that would eventually go largely to decomposers. (Adapted from Odum, 1962.)

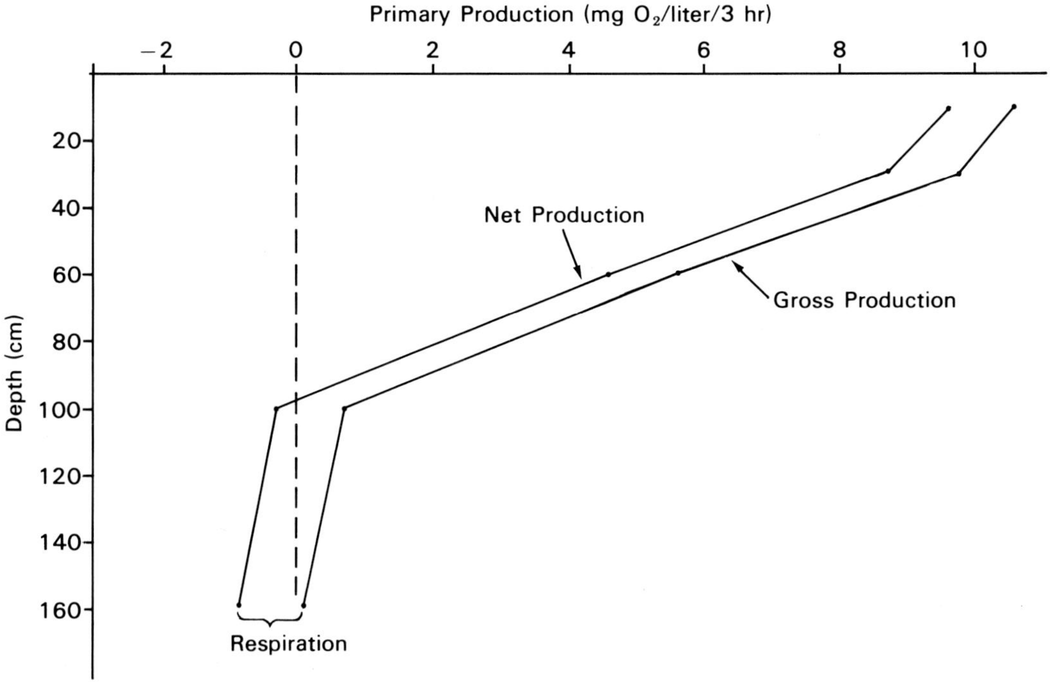

FIG. 10.3. Depth profiles of gross and net primary production in a tropical pond. The data were obtained by the "light and dark bottle method," in which paired bottles of pond water (containing phytoplankton) are suspended at different depths. One bottle of each pair is covered to exclude light. Hence respiration but no photosynthesis occurs in this bottle, the respiration rate being measured by the decrease in dissolved oxygen that occurs during the incubation period. In the uncovered bottle of each pair respiration and photosynthesis occur simultaneously, and the change in dissolved oxygen represents the net production. Gross production is easily calculated, as the net production value plus the respiration value. In this graph respiration is simply the difference between gross and net production. Respiration is approximately the same at all depths but gross and net production decline with depth due to decreasing light intensity. When gross production becomes too small to offset respiration (*i.e.*, the compensation depth is exceeded), net production becomes negative. (Data of author.)

grazed by zooplankton; but because they grow rapidly, these tiny photosynthesizers sometimes sustain a biomass of slower-growing herbivores that is larger than their own.

In a forest the producer organisms are large, long-lived, and relatively difficult to digest. To achieve an equivalent amount of production the biomass of slow-growing forest plants must be enormously larger than the biomass of aquatic pytoplankton, but the number of individual organisms is far fewer. Indeed, in a forest, there are usually far fewer trees than there are herbivores: thousands of insects can subsist on the foliage of a single tree. But forest herbivores are quite inefficient at converting plant production to herbivore production, and much plant production therefore is shunted to the decomposer food chain.

Grazing food chains usually are longer in the ocean than in a forest, and we can begin to see why when we consider efficiency of transfer. In water, no transfer step is as inefficient as the producer-herbivore step in a forest, and the energy passing up the grazing pyramid therefore is not dissipated so rapidly. Another reason for the length of grazing food chains in water is the small size of the phytoplankton and zooplankton that chiefly comprise the first two trophic levels. (This was pointed out in Chapter 5.) If there are many small herbivores at level two, this means that the carnivores at level three also can be relatively small and numerous, and additional carnivore levels can be accommodated before the last level, represented by a relatively small number of large predators, is attained. The interplay of trophic efficiency and size in determining the length of a food chain is well illustrated by grassland ecosystems. Here a much greater fraction of plant production is captured by the herbivore level than in a forest, but since many grassland herbivores are unusually large, the grazing food chain is no longer than that of a forest, often containing only three steps.

In any ecosystem, an organism's trophic position dictates to some degree whether it can afford to be a food specialist. Species low in a trophic pyramid can be epicures, as exemplified by the herbivores of the Serengeti described in Chapter 3. At lower trophic levels many types of food may be sufficiently abundant each to support its own food specialist. But to carnivores high in a trophic pyramid all food types are relatively scarce, and since upper level carnivores are usually large, energy-demanding animals, no single prey type may provide sufficient sustenance. It is striking how wide is the geographic range of many top carnivore species (Table 10.2). Forced by the Law of Entropy to be dietary generalists, top carnivore species often have found it easy to be habitat generalists as well. The dietary versatility of top carnivores often has led to conflicts with man, with unfortunate consequences for the carnivore (Fig. 10.5).

Energy Transfer Across a Single Trophic Step

What use, or nonuse, is made by a trophic level of the energy theoretically available to it, *i.e.* the net production of the preceding trophic level? Figure 10.6 diagrammatically catalogues its several fates and also illustrates, by means of a partial pyramid, how the *gross* production of any trophic level is related to the gross production of the preceding level. If these gross productions are represented by λ_{n-1} and λ_n, respectively, then

$$\lambda_n = \lambda_{n-1} - (R_{n-1} + NH_n + NA_n)$$

the other terms in this equation being defined in Figure 10.6. In the case of the step from incoming solar energy to gross primary production, the equation reduces to:

$$\lambda_1 = \lambda_0 - (NH_1 + NA_1)$$

where λ_0 is incoming solar energy, NH_1 is the portion of this energy that is not harvested by

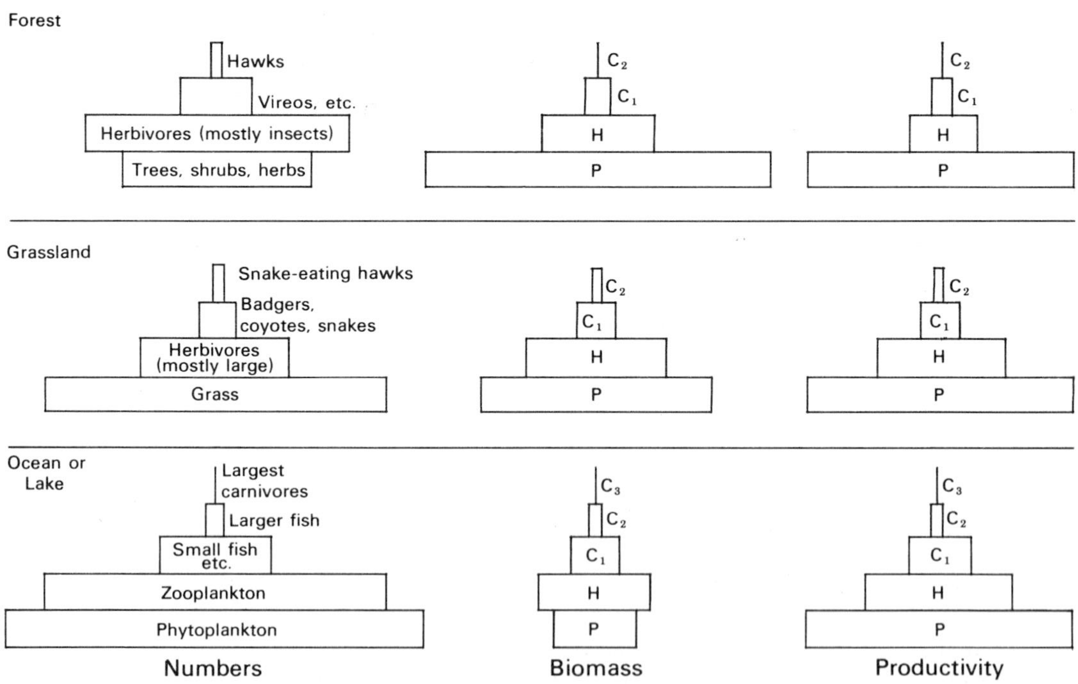

FIG. 10.4. Pyramids of numbers, biomass, and productivity for three ecosystem types. Primary productivity of each has been made equal so that there is a basis for comparison. The variations in shape of the pyramids relate to major contrasts among the ecosystem types, as discussed in the text. A logarithmic horizontal scale is used here for convenience. "Real" arithmetic pyramids become attenuated so rapidly that the upper levels are hardly visible on a diagram such as this one.

TABLE 10.2

Geographic Ranges of Top Carnivores

Many higher carnivores have very broad geographic ranges, which may be directly correlated to their lack of dietary specialization. They feed on many different kinds of prey, so need not be restricted to the habitat of any particular prey species. Homoiothermic carnivores tend to have especially wide geographic ranges because they are adaptable to a wider range of environmental temperatures than are poikilothermic carnivores. There are some glaring exceptions to this pattern (*e.g.* the Everglades kite, a hawk which is said to feed nearly exclusively on one species of snail) but the pattern is very clear nonetheless. These are some of the more spectacular ranges.

Species of Carnivore	Range
Osprey (*Pandion haliaetus*)	Every continent except Antarctica, wherever there is water containing fish. In the New World, ranges from Hudson's Bay to Paraguay.
Golden eagle (*Aquila chrysaetos*)	Throughout the northern hemisphere in mountainous areas, tundra, grassland, desert. Ranges south to Mexico, North Africa and the Himalayas.
Kestrel (sparrow hawk) (*Falco sparverius*)	Northern Canada to Mexico and Maine to California. A related Old World species ranges from Scandinavia to Japan and south to equatorial Africa.
Great horned owl (*Bubo virginianus*)	Nearly all of North America, from the Arctic south to Costa Rica.
Coyote (*Canis latrans*)	Alaska to Central America; largely absent east and south of the Ohio River, but now spreading into this region.
Leopard (*Panthera pardus*)	Africa, from the Cape of Good Hope to the Mediterranean; and across Asia to Manchuria and Java, though not in coldest or driest areas.
Mountain lion (*Felis concolor*)	Alaska to Patagonia in mountains, deserts and forests of many types.
Tiger (*Panthera tigris*)	Restricted to Asia, but found from the arid southwest to the boreal forests of Siberia, and through the tropical forests of Southern Asia to Java and Bali. Missing from the far north and the Tibetan plateau.

plants (because it is reflected, transmitted, or does not even strike foliage), and NA_1 is the fraction that is absorbed by plants as heat, ultraviolet or light radiation or is used in evaporation by foliage, but is not assimilated as the chemical energy of new organic matter.

Trophic Transfer Efficiencies in Natural Ecosystems

In a pioneering study of trophic relationships in Florida's famous Silver Spring, Howard Odum (1957) derived a number of indices of efficiency for trophic transfer processes. The term "efficiency," when expressed as a percentage, always means (output/input) $\times$ 100, but a bewildering variety of efficiencies can be calculated and the terminology of these has become quite confused. Kozlovsky (1968) reviews the subject, and provides an essential guide to the different names given by Odum and others to the same index (Table 10.3). From the community standpoint the most interesting of these efficiencies are: (*a*) the ratio of plant net production to incoming solar energy (primary AE); (*b*) the ratio of plant gross production to incoming solar energy (primary EGE); and (*c*) the ratios of gross or net productions in succeeding trophic levels (BE and NE). The studies on which generalizations about efficiency may be based are few, particularly for transfers involving higher trophic levels. Because most natural ecosystems contain many species (even if, as here, one considers only the grazing food chain), it is difficult and time-consuming to obtain accurate and complete energy flow figures for whole trophic levels.

It is clear from the evidence available, however, that plants never capture very much of the solar energy entering an ecosystem. Gross photosynthetic efficiency (AE_1) is perhaps as high as 5% in a few ecosystems, averages 1 to 1.5% in most and is as low as 0.1% in others; for the earth as a whole (including nonvegetated regions like the polar icecaps), this efficiency is estimated to be 0.24%. The efficiency with which Earth's solar input is fixed as *net* global photosynthesis (EGE_1) is only about 0.13% (Whittaker and Likens, 1973); this small percentage must sustain nearly all of Earth's heterotrophic species (the exception being the various groups of chemosynthetic microorganisms that capture energy from the chemical oxidation of inorganic compounds).

The ratios of both gross (BE) and net (NE) productivities in succeding steps of a grazing food chain are higher than those of the first step, the efficiencies ranging between 5% and 30% in different ecosystems, but typically being near 10% (Golley, 1972). Using typical figures, it can be calculated that the food energy available to top carnivores in a five-step food chain is about 0.1% of the food energy available to herbivores. It is no wonder that hawks do not darken the sky at Hawk Mountain!

Although these trophic efficiencies are always low, their variability among different ecosystems is highly significant. We already have remarked on the relative inefficiency of the plant-herbivore transfer step in forests, and seen something of its effect. Obviously it makes enormous difference to the abundance of organisms at higher trophic levels if herbivores consume 40%, 7% or 1% of the net plant productivity (Table 10.4).

What causes these important contrasts in trophic efficiency? Why should a given transfer step be more efficient in one ecosystem than in another, and within the same ecosystem, why should efficiencies differ in successive transfer steps up a food chain? Several factors contribute to these contrasts, and each is described briefly below.

Differences in digestibility. On the whole,

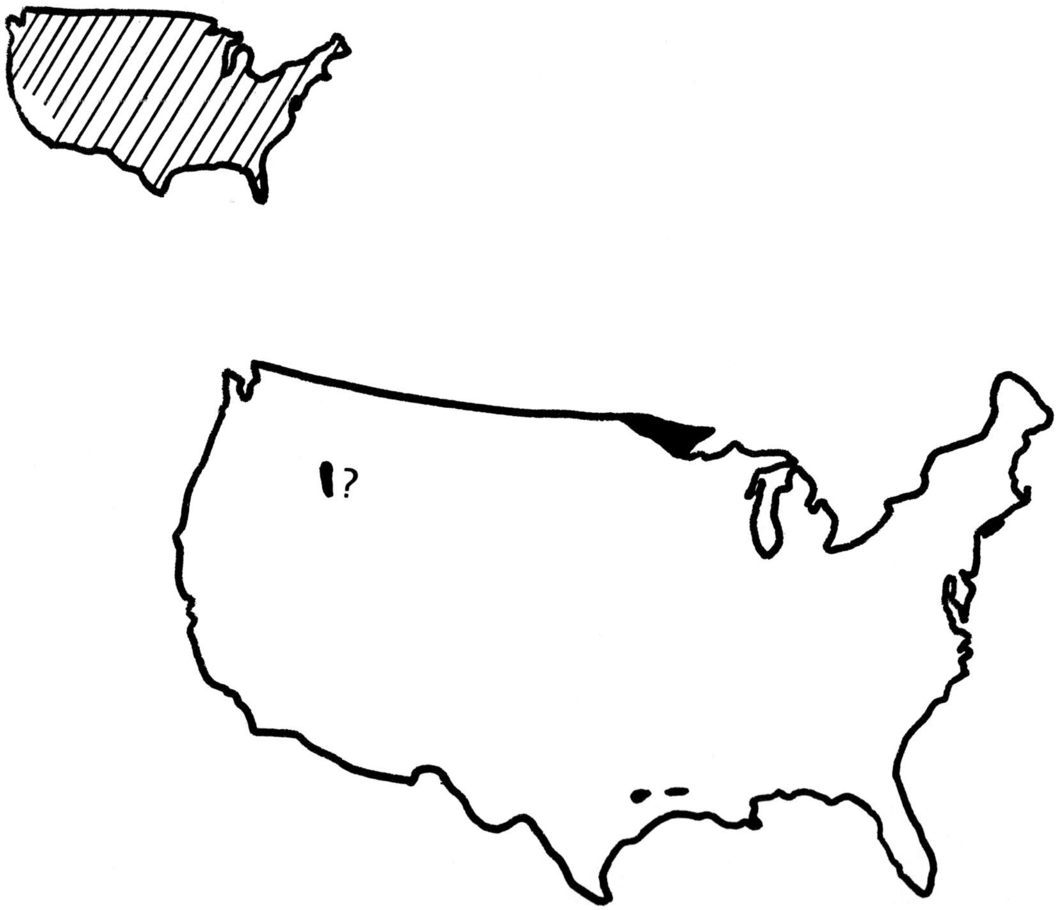

FIG. 10.5. Past and present range of the wolf in the United States, exclusive of Alaska. Wolves never existed in Hawaii, and they may have been absent from parts of presettlement California, but otherwise they were unbiquitous. In Alaska the state government was paying a $50 bounty for gray wolves as late as 1973, but elsewhere in the United States only two self-perpetuating populations remain: in the Boundary Waters Canoe Area of Minnesota and in nearby Isle Royale National Park in Lake Superior. A few "red wolves" (possibly a distinct southern species) may remain in Louisiana and Texas, and a few gray wolves may still exist in the northern Rockies.

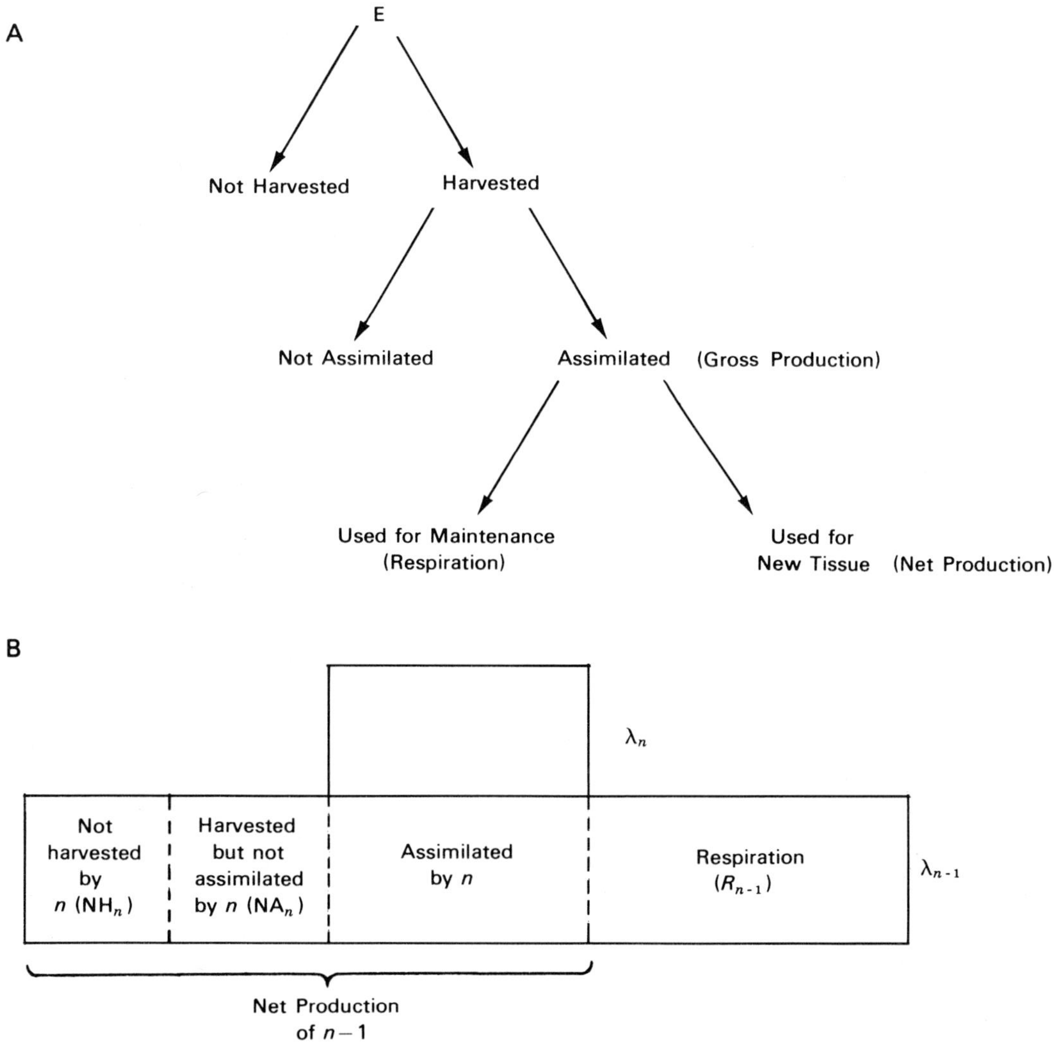

FIG. 10.6. (*A*) Partitioning of the energy available to a trophic level in terms of its actual fate. (*B*) The relationship of gross production (λ_n) of any trophic level to the gross production (λ_{n-1}) of the preceding level.

the digestibility of animal tissues is much greater than the digestibility of plants. This is the main reason why production efficiencies of animal-animal trophic steps normally exceed those of the plant-animal step. In the case of some carnivores, 75% of the food they ingest may be assimilated, *i.e.* converted to new tissue (Phillipson, 1966). By contrast, as little as 10% of the energy intake of some herbivores is assimilated; the rest passes out as feces and is shunted to the decomposer food chain (Phillipson, 1966). The huge vegetation intake of horses and elephants is out of proportion to their size and metabolic needs; their diet is so indigestible that they must ingest great quantities to extract a relatively small return.

All plant foods, of course, are not equally indigestible, nor all animal foods equally easy to digest. Among the latter, one need only contrast worms and hard-bodied insects, or tender rabbits and tough armadillos, to realize that what a carnivore eats affects its assimilation efficiency. Among plants, woody species tend to be least digestible, herbaceous land plants intermediate, and algae most digestible. But further nuances exist. For example, algae differ among themselves in digestibility: lakes differing in their phytoplankton composition therefore vary in the efficiency of the plant-herbivore transfer step; and a fish farmer wishing an optimum yield should try to manage his pond to favor digestible diatoms rather than mostly indigestible blue-green algae.

Abundance or concentration of food relative to feeders. Consider the production of beef from good and poor ranges. If the poor range is larger and therefore contains just as much food as the good range, and if both ranges are equally stocked with cattle, the beef harvest from the poor range nonetheless will be somewhat lower. Why? Because these cattle have to seek food more actively and over longer distances in order to be satisfied. More of their food intake goes into respiration and less into growth than is true of the lazy cattle on the good range.

Obviouly, too, the efficiency of the plant-

TABLE 10.3

Trophic Efficiencies

The terms and definitions below generally follow those of Kozlovsky (1968), who in turn tried to conform to the usages most common among other investigators. All fractions would be multiplied by 100 to express efficiencies as percentages. *n* denotes any trophic level.

Term	Meaning	Notes
I. AE (Assimilation efficiency)	$\frac{\text{Assimilation (or Gross Productivity) at } n}{\text{Ingestion at } n}$	Also used to describe the special ratio $\frac{\text{Gross Primary Production.}}{\text{Incoming Light Energy.}}$
2. EGE (Ecological growth efficiency)	$\frac{\text{Net Productivity at } n}{\text{Ingestion at } n}$	Ingestion here and elsewhere means food taken into the digestive tract, but in the special case of the first trophic level, ingestion means *all* incoming light energy.
3. TGE (Tissue growth efficiency)	$\frac{\text{Net Productivity at } n}{\text{Gross Productivity at } n}$	
4. BE (Biological efficiency or progressive efficiency)	$\frac{\text{Gross Productivity at } n}{\text{Gross Productivity at } (n-1)}$	
5. EE (Ecological efficiency)	$\frac{\text{Ingestion at } n}{\text{Ingestion at } (n-1)}$	A measure of food conversion efficiency that is used in some laboratory studies and agricultural computations.
6. NE (Net efficiency)	$\frac{\text{Net Productivity at } n}{\text{Net Productivity at } (n-1)}$	An efficiency not defined by Kozlovsky, but one that is probably in common use.
7. FCE (Food chain efficiency)	$\frac{\text{Ingestion at } n}{\text{Calories Available to } (n-1)}$	A measure used in Slobodkin's laboratory studies (see p. 270 and Fig. 10.7).
8. IE (Ingestion efficiency)	$\frac{\text{Ingestion at } n}{\text{Net Productivity at } (n-1)}$	A measure of how much of the net production of one trophic level actually is consumed by the next.

TABLE 10.4
Efficiency of the Plant-Herbivore Step

This is a table of ingestion efficiencies, representing the percentage of plant net production that is actually consumed by herbivores. The data on which these percentages are based are those compiled by Golley (1972), from sources cited in his review.

Ecosystem Type Number Studied	Range of Ingestion Efficiencies (IE)	Summary Comments
	%	
Tropical forests (2)	7–8.5	Except in cases of infrequent defoliation, IE for forest ecosystems is usually less than 10% for the plant-herbivore step.
Temperate forests (4)	1.5–40	
Tropical grasslands (3)	28–60	Plant-herbivore IE in grasslands is high but usually less than 50%.
Temperate meadow (1)	25–30	
Early successional fields in a forest sere (2)	1–12	
Salt marsh (1)	8	
Freshwater springs (2)	24–38	Values probably vary widely in aquatic ecosystems; most are perhaps in the range 10 to 30%.
Ocean (Georges Bank) (1)	5	

herbivore transfer step on these ranges is affected by the density of stocking. If cattle density is too low, much of the herbage is left unharvested and the transfer efficiency (any of numbers 3 to 8 in Table 10.3) also is low. Higher stocking rates will increase this particular efficiency, but if stocking is too high the plant cover will be reduced, and the efficiency of the sun-plant step will decline.

It is clear that we cannot seek to maximize particular trophic transfers without considering repercussions elsewhere in the pyramid. If cattle are being harvested at a great rate, their numbers may be too low to utilize the pasture maximally; if cattle are not being harvested fast enough (*i.e.* if the range is overstocked), their heavy grazing may be preventing maximal photosynthesis. High efficiency at one level can result in low efficiency further down the pyramid. This is perhaps a common situation in temporarily unstable natural ecosystems, for example those whose species are undergoing seasonal fluctuations of abundance. The more stable the ecosystem, the more chance there is for transfer efficiencies to come into some sort of balance throughout the trophic pyramid.

Trophic level contrasts in respiration. As we have seen, cattle scrounging a living from a poor range can pass on less of their assimilated energy than cattle on a good range, simply because they respire away more of this energy. Considering the different trophic levels, it follows that if respiratory losses are partly a function of activity, then plants (being sedentary organisms) should have relatively low respiratory losses in proportion to their gross productivity; *i.e.* they should have relatively high tissue growth efficiencies (Table 10.3). Herbivores, too, often are relatively sedentary simply because their food is sedentary. Carnivores, on the other hand, usually must capture and subdue a motile prey, and, especially if they are near the top of the food chain, they may have to spend increasing amounts of their assimilated energy merely to find their food. Moreover, homoiothermic species possessing high metabolic rates increasingly dominate most food chains as they are ascended. For all these reasons the ratio of net to gross production tends to be lower at each succeeding trophic step: relatively more energy is used for maintenance, and relatively less for growth that is potentially harvestable by the next trophic level.

This generalization should not obscure the fact that the *same* trophic level in *different* ecosystems may display quite different tissue growth efficiencies. At the producer level, for example, respiration constitutes anywhere from 20% to 80% of gross production, forests being at the upper end of the scale because their massive stem and root structures respire but do not photosynthesize (Golley, 1972).

Special considerations relating to the photoassimilation step. Since the sun-plant transfer step affects energy availability at all higher trophic levels, we must go further in explaining contrasts among ecosystems in the efficiency of this step. Several considerations are involved. First, if the amount of light-intercepting vegetation is relatively sparse, its assimilation efficiency obviously will be low. Much incoming sunlight will be absorbed not by plant tissue but by the nonliving substrate. Desert vegetation is far less efficient in this regard than is forest vegetation. In aquatic ecosystems similar contrasts exist: the water and nonliving particulate matter of an infertile lake often absorb most of the solar input and the sparse phytoplankton relatively little. Second, even when light *is* absorbed by vegetation, its utilization may not be equally efficient in different ecosystems: if incoming sunlight exceeds the saturation intensity of plants' photosynthetic apparatus, the excess will be completely wasted (Fig. 4.9). Since intensities of incoming radiation vary across the globe, so too do photosynthetic assimilation efficiencies. In some terrestrial ecosystems 90% of incoming light energy may exceed the saturation capacity of leaves directly exposed to the sun.

Considerations such as these explain why efficiencies of the sun-plant transfer step vary in different ecosystems, and also help us to understand why these efficiencies always are low relative to transfers higher in the grazing food chain. They also are low, of course, because light constitutes only half the solar energy input in the first place, unassimilable heat being most of the rest. Moreover, the photosynthetic reaction itself operates at only 20% efficiency: for every calorie of chemical energy produced, about 5 cal of light energy must be invested.

Efficiencies in a laboratory ecosystem. The studies by Slobodkin (1959) of a three-species laboratory food chain clarify certain of the variables affecting trophic efficiencies. In these studies a producer species (the alga *Chlamydomonas*) was fed to an herbivore species (the microcrustacean *Daphnia*), and a "predator" (Slobodkin himself) harvested the herbivore, varying the rate of predation in the different beakers in which the *Daphnia* were housed. The algae were not grown in the beakers, but were supplied to the *Daphnia* at different rates, much as grain is supplied to cattle in a feedlot.

Slobodkin studied the variability of two trophic efficiencies in this system. The food chain efficiency (Table 10.3, No. 7), represented here as

$$\frac{\text{Calories of } Daphnia \text{ ``consumed'' by man}}{\text{Calories of algae supplied to } Daphnia} \times 100$$

was obviously sensitive to experimental manipulations of both numerator and denominator, as was the ecological efficiency (Table 10.4, no. 5), here represented by

$$\frac{\text{Calories ``consumed'' by man}}{\text{Calories consumed by } Daphnia} \times 100$$

At low levels of predation (20 to 40% of the rate of birth, with only adult *Daphnia* being taken), Slobodkin found that he was underfishing: food chain efficiency was only 3 to 6% (Fig. 10.7). But if he increased his predation too markedly the population of *Daphnia* was reduced to levels that could not consume all the algae, particularly in those aquaria receiving large doses of algae. Now Slobodkin was overfishing, and food chain efficiency remained rather low. The *maximum food chain efficiency* that Slobodkin obtained was near 9%, and occurred in beakers to which algae were added at a minimum rate

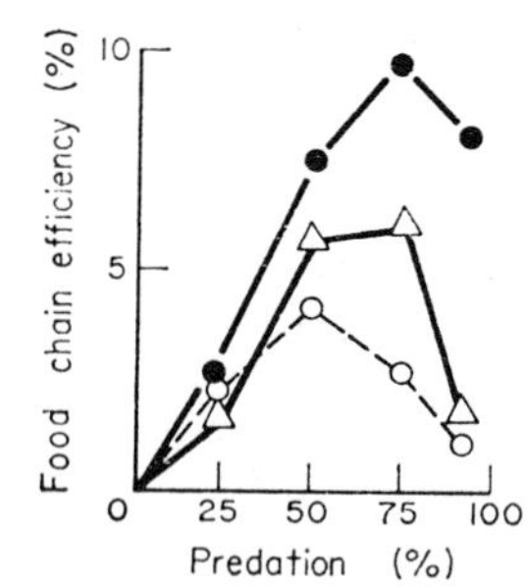

FIG. 10.7. Efficiencies of energy transfer (expressed as food chain efficiency) in Slobodkin's laboratory ecosystem. The efficiencies varied with differing rates of food supply and harvest. ●——●, lowest feeding level (1 ration per 4 days); Δ——Δ, intermediate feeding level (3 rations per 4 days); O---O, highest feeding level (5 rations per 4 days). (From J. Phillipson: *Ecological Energetics*. New York: St. Martin's Press, 1966; using data of Slobodkin, 1959. Reprinted courtesy Edward Arnold Ltd., London.)

and from which adult *Daphnia* were harvested at 65% of the rate of birth. Slobodkin was able to calculate, but not to achieve in actuality, a *maximum ecological efficiency* of 13%. This is the efficiency he theoretically could have achieved by precisely adjusting the rates of algal supply and *Daphnia* harvest so that all of the algae were being eaten and *Daphnia* were being harvested at the same rate that they produced new biomass.

The results of Slobodkin's experiments roughly agree with the "10% rule-of-thumb" that can be thought of as the average efficiency of trophic level transfers in most natural ecosystems, once the highly inefficient sun-plant step is passed.

Advantages of descending the food chain. Simplified trophic pyramids neglect the fact that some species operate at more than one trophic level. Upper level carnivores gain an obvious advantage by doing this; by eating some herbivores as well as primary carnivores, a species classed as a second level carnivore taps into the energy supply before it has dropped to a trickle. Earlier we suggested that most top carnivores need to be food generalists. Feeding at more than one trophic level is simply another aspect of this tendency.

Omnivores like man also are food generalists, and man illustrates well the benefits to be gained by descending the food chain. According to the calculations of Deevey (1956), the same amount of plant production will support a vegetarian or near-vegetarian human population 10 times that of a hunting culture subsisting on herbivores, and 1000 times that of a fishing culture in which man is operating as a tertiary carnivore. This example makes clear why the peoples of densely populated, poor countries eat so much grain and so little meat. Similarly, it is clear why beef and fish *cost* more than rice. The puzzle is why French fries outprice hamburgers at many fast-food restaurants. Have burger makers really learned how to elude the Law of Entropy?

MEASURING ENERGY FLOW

It is a considerable task to trace the fate of energy in an ecosystem. One must gather data on the growth rates and respiratory losses of every major species population in the community, and on the system's imports and exports of solar energy and organic compounds. Not surprisingly, the few attempts at complete energy flow studies have dealt with ecosystems containing relatively few species. Springs, salt marshes, cornfields and recently abandoned farmlands have been favorite subjects.

It is easiest to measure the first steps of energy flow, and many studies restrict themselves to primary productivity. Incoming solar energy can be monitored with instruments such as thermopiles (which measure total radiant energy) or pyrheliometers (which measure visible light energy). In aquatic ecosystems photosynthetic production can be measured by following the changing oxygen or carbon dioxide levels in the water. In the case of phytoplankton, for example, photosynthetic rates can be determined by the light and dark bottle method of Figure 10.3. Total daily photosynthesis by *all* organisms in a lake or pond can be calculated from measurements of oxygen or carbon dioxide changes in the water body during a 24-hour period, although this calculation is complicated by the necessity to estimate diffusional exchanges of these gases with the atmosphere. In streams and springs primary productivity can be calculated from the increase of O_2 or decrease of CO_2 between two points along the course of flow (Odum, 1956; McDiffet *et al.*, 1972).

On land, photosynthetic rates of individual plants often are measured by an infrared gas analyzer, which monitors CO_2 changes in a closed chamber containing the plant. The atmosphere is too mobile to allow one to measure photosynthesis by, say, the daily decrease of free CO_2 in a forest; the air in which changes are being studied must always be encapsulated. This means that photosynthesis measurements in terrestrial ecosystems usually are restricted to small plants or plant parts; but in one ambitious study a complete segment of Puerto Rican rain forest was covered with a huge plastic bag, allowing diurnal changes of CO_2 to be used as a measure of gross and net photosynthesis just as in a lake.

Net primary production also can be estimated by the harvest method, in which new growth over a known period of time is directly harvested and weighed. Burning of samples of the dry organic matter in a bomb calorimeter provides a measure of the energy content (calories) per unit weight of tissue.

Measurement of energy flow through higher trophic levels involves the determination of every quantity in Figure 10.6 for every species population in each trophic level, and addition of the species figures to give total figures for each trophic level. Because of the enormity of this job, investigators often limit their objective to studying the energy budget of a single species of animal in a community. The weight, food intake, fecal output, and respiration of a few individuals are measured, their assimilation and net production rates calculated from these figures, and values for the whole species calculated from a census of its population size.

The final product of a complete study is an energy flow diagram which quantitatively depicts the energy transfers within the system. More elaborate than a trophic pyramid diagram, an energy flow diagram may illustrate transfers between particular species or may simply summarize transfers between trophic levels. Figure 10.8 is a diagram of the latter type, illustrating the results of Odum's study of Silver Spring. This is perhaps the most ambitious energy flow study ever attempted.

ENERGY INTERACTING WITH NUTRIENTS: THE LIMITING FACTOR CONCEPT

Liebig's Law of the Minimum

What limits an ecosystem's overall productivity? One possibility is the amount of light energy supplied. If there is not enough, as we might suspect in the case of high latitude ecosystems, then this could limit primary productivity and indirectly the productivities of higher trophic levels.

If light input is ample, the limit to an ecosystem's productivity is set by one or several factors that limit the efficiency of the various energy transfer steps, particularly the initial conversion step from light energy to plant tissue. We have examined already several of the factors that affect the efficiency of this step, but a general category of factors that needs further attention is nutrient levels. Adequate nutrients obviously are necessary if an ecosystem is to use efficiently the solar energy available to it.

Nutrient limitation of plant growth was what concerned the German agricultural chemist Justus von Liebig, who in about 1840 formulated what has come to be known as the **Law of the Minimum** (or **Law of Limiting Factors).** The law states simply that "growth is limited by that resource which occurs in the least amount relative to its need." These are not Liebig's words—he spoke more specifically of crops and soil nutrients—but they express his idea succinctly. If for "growth" we substitute "productivity" and if we use the term "re-

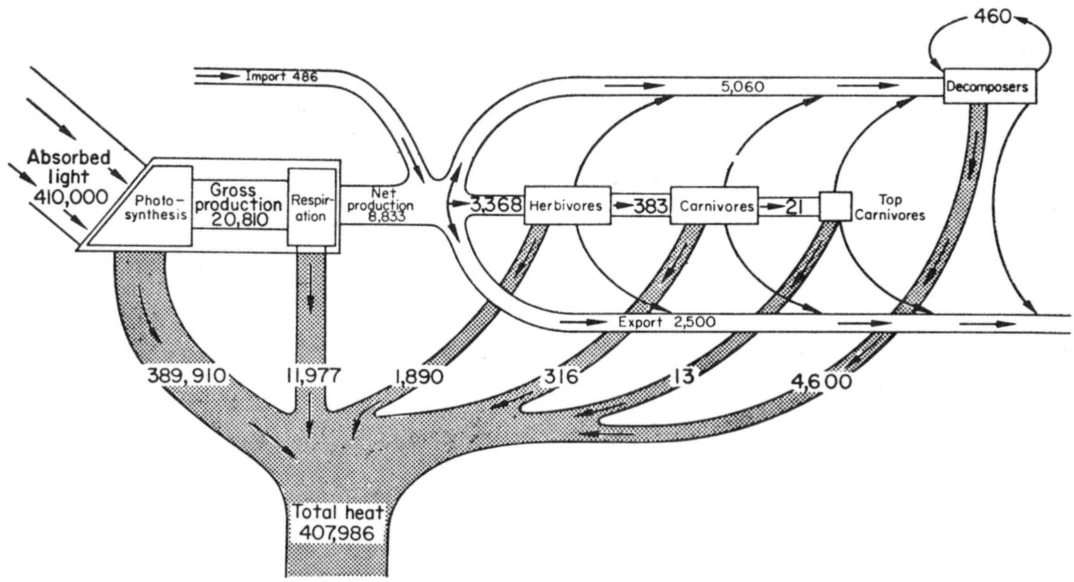

FIG. 10.8. Energy flow diagram for Silver Springs, Florida. All figures are expressed as kilocalories of energy per square meter per year. For example, the herbivores beneath 1 m^2 of water surface assimilated 3368 kcal from the plants, respired away 1890 kcal, passed on 383 kcal to carnivores, and (assuming a steady state system) contributed the remainder to decomposers and export. (From J. Phillipson: *Ecological Energetics*. New York: St. Martin's Press, 1966; after Odum, 1957. Reprinted courtesy Edward Arnold Ltd., London.)

source" broadly, we have a law of limiting factors that might be applied to any species population anywhere in the food chain.

But is Liebig's Law true? Perhaps not completely. In the first place, it implies that the productivity of a species always is determined by some resource in short supply. We will see in Chapter 15, however, that different factors seem to control different species in a community, and not all of these factors can be considered "resources." In fact, an influential and controversial article several years ago suggested that one whole trophic category, herbivores, is limited worldwide not primarily by scarce resources but by predators.

For species which seemingly *are* limited by nutrients, Liebig's idea that in any given situation one and only one resource is limiting can be further questioned. A farmer often can improve his crop yield by adding any of several nutrients. If Liebig's Law were strictly true, only one nutrient should cause more growth initially. After this had been supplied, some other nutrient might become limiting, but no more than one at a time. This is not what the farmer finds; instead, if he divides his field into several plots and adds a different nutrient to each, he gets accelerated growth in each. But if to one plot he adds all the nutrients, a synergistic effect is observed: the growth increment here exceeds the sum of smaller growth increments obtained in the plots receiving single nutrients. Synergism of this sort seems to be another example of the factor interaction described in Chapter 3 with reference to niche size.

Though often we cannot identify a *single* nutrient as limiting, yet out of several which augment growth there usually is one with a predominant effect. Thus, the concept of the minimum need not be disarded completely; often we can talk quite truthfully of a *principal* limiting factor, even if we cannot talk about a single limiting factor.

Controls of Primary Productivity: World and Local Patterns

Primary productivity has been measured in a wide variety of ecosystem types, and more comparative information is available here than in the case of secondary or tertiary productivities. Since, in addition, an ecosystem's primary productivity carries all higher productivities on its shoulders, we will restrict this discussion of limiting factors to the question of what controls an ecosystem's primary productivity.

It can be seen in Table 10.5 that there are considerable differences in the primary productivity of different ecosystem types. In attempting to detect patterns of controlling factors, let us consider terrestrial, freshwater, and marine ecosystems separately.

Terrestrial productivity. On land it is possible to draw worldwide correlations between net productivity and each of two climatic variables: temperature and precipitation (Whittaker and Likens, 1973). When these two variables are integrated as a single index, actual evapotranspiration, the correlation with terrestrial productivity is particularly convincing (Fig. 10.9). Temperature doubtless has some direct influence on productivity but we must remember that global differences of mean temperature closely reflect global differences of solar energy input. Therefore it is difficult to say whether the more important variable is air temperature or the supply of light energy. But it seems safe to say that solar energy and water are the chief determinants of terrestrial differences in global productivity.

These are somewhat unexpected findings, given the stress by many experts on the limiting nature of soil nutrients other than water. René Dumont (1969), a renowned French agronomist, has written that "lack of nitrogen constitutes the principal factor limiting the development of tropical crops." In conifer forests of Wisconsin it is possible to predict average growth rates quite accurately from regression equations based on four soil variables: (*a*) the percentage of fine particles; (*b*) the percentage of organic matter; (*c*) the available phosphorus level, and (*d*) the available potassium level (Wilde, 1970). And in many countries, Australia and New Zealand being notable examples, there exist large areas where addition of small quantities of a specific micronutrient such as molybdenum or cobalt results in enormous increases of soil fertility (Anderson and Underwood, 1959). In these "trace element deserts," Liebig's Law seems to be literally true; here there *is* a single limiting nutrient.

How can Figure 10.9 be reconciled with this abundant evidence that factors other than water and energy are enormously important to terrestrial productivity? Reconciliation is possible if we acknowledge that water and energy

the length of the growing season. Supplied by the tides with frequently renewed, nutrient-rich estuarine waters, warm-latitude salt marshes rival any of the world's ecosystem types in primary productivity. But nutrient supply affects salt marsh productivity also, the nutrient in greatest demand seemingly being fixed nitrogen (Ryther and Dunstan, 1971; Teal and Valiela, 1973).

ENERGY RELATIONS DURING SUCCESSION

Laboratory Models

Copeland (1965) witnessed a simple and interesting example of succession in a marine aquarium. Artificially illuminated at an intensity of 1500 foot-candles, the initial ecosystem was dominated by turtle grass and was balanced in the sense that community respiration

TABLE 10.6

Gross Primary Productivity of Phytoplankton in Lakes and Ponds of the Malay Peninsula

The Malay peninsula originally supported moist tropical rain forest over most of its extent. It is a region of naturally infertile soils, for the most part, and of dilute water bodies. Productivity of these water bodies tends to be quite low except when man intentionally adds fertilizers; then they may become as productive as any fresh waters in the world. These data were obtained by the author, using the light and dark bottle method (see Fig. 10.3).

Type of Water Body	Range of Daily Gross Productivity	Remarks
	g O_2/m²/day	
Natural lakes	up to 1.1	Two lakes were studied. Production in one was too low to be measured.
Man-made mining pools, reservoirs and ponds (unfertilized)	0.5 to 4.3	The least productive of these water bodies were in areas of undisturbed rain forest.
Fertilized mining pools and fish ponds	2.5 to 27	In the most productive of these waters, photosynthesis by phytoplankton was restricted to a narrow layer near the surface due to algal self-shading.

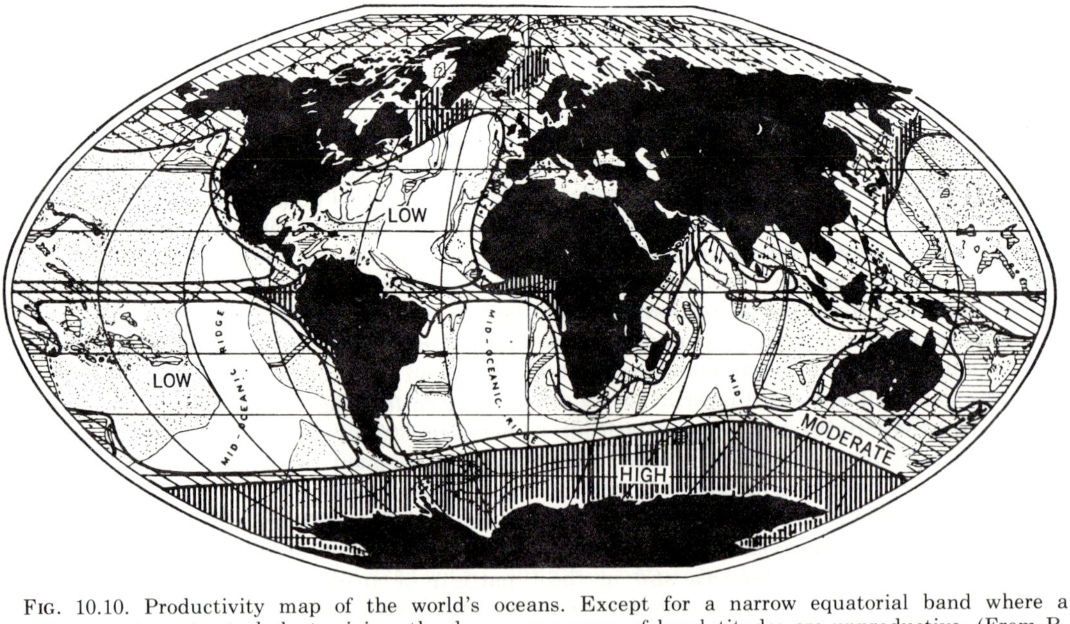

FIG. 10.10. Productivity map of the world's oceans. Except for a narrow equatorial band where a countercurrent creates turbulent mixing, the deep ocean areas of low latitudes are unproductive. (From B. Heezen and C. Hollister: *The Face of the Deep*. New York: Oxford University Press, 1971. Reproduced with permission.)

(Rt) equalled community photosynthesis (Pg) (Fig. 10.11). Such a balance, or near-balance, characterizes successionally mature, stable communities in nature.

Copeland upset this balance by suddenly reducing the illumination to 230 foot-candles, initiating the changes shown in the figure. The first overall effect was a drastic reduction in both photosynthetic and respiratory activity. The balance between the two was initially maintained. Lake and forest ecosystems, too, when a cloudy day follows a sunny one, seem to step down their whole metabolism (particularly at the producer level) and thereby buffer short term changes in energy supply without serious repercussions in the community. But in this experiment the energy reduction was permanent, and after several days the community became metabolically unbalanced. The turtle grass began to die and there was an increase in community respiration, reflecting the activity of decomposers. At the same time, however, blue-green algae began to increase, replacing the dying turtle grass as chief primary producer. Therefore photosynthetic activity also started to increase, though more slowly than did respiratory activity. Maximal metabolic imbalance, probably indicating the peak of decomposer activity, occurred about 60 days after the change in illumination. The last measurements of community metabolism, at 85 days, indicate the beginning of return to a steady state, with respiration and photosynthesis approaching the same value. Interestingly, gross photosynthesis now was nearly as high as initially it had been under the higher light regime. Just as in a maturing forest, succession witnessed the apparance of autotrophs able to function relatively efficiently at low light intensities.

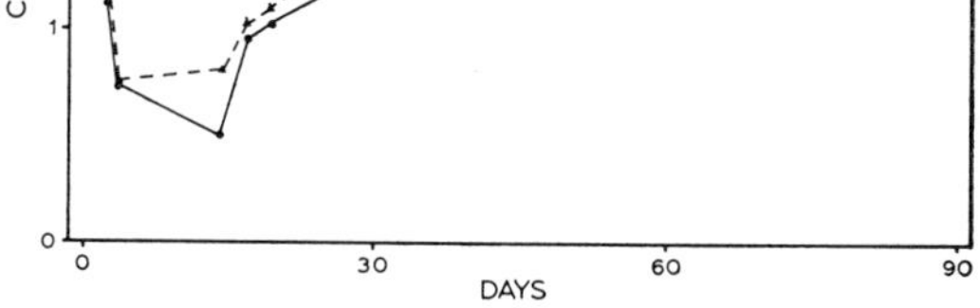

FIG. 10.11. Metabolic changes in an aquarium ecosystem following a change in illumination. For explanation see text. (From B. J. Copeland: *Ecology* 46: 563–564, 1965.)

If illumination had been reduced even lower, Copeland's ecosystem presumably would have run down rather than recovering. This is what happens during the special type of succession known as **heterotrophic succession:** energy flow through the system decreases with time because there is insufficient input and the community gradually is dissipating previously stored energy. A decomposing log is a miniature heterotrophic ecosystem illustrating this progression. Heterotrophic succession always is temporary since it culminates in extinction of the community when the stored energy supply is exhausted.*

The more usual type of succession, **autotrophic succession,** occurs in the presence of a continuing supply of solar energy to the ecosystem, and usually is characterized by a rapid recovery of the capacity to fix this energy photosynthetically.

Successional processes easily can be followed in laboratory ecosystems even smaller than Copeland's. For example, one can study the changes, both metabolic and compositional, that occur in a flask of pond water after it receives a dose of nutrients. This sequence in time is closely analogous to the sequence in space that characterizes a stream recovering from a dose of organic sewage (see Chapter 5).

Productivity and Biomass Changes during Terrestrial Succession

Figure 10.12 illustrates a typical energy flow sequence during secondary forest succession. (The reader can suggest appropriate modifications for other terrestrial biomes.) An abandoned field or forest clearing is soon revegetated by rapidly growing plant species, and these are responsible for a spurt of gross production (P_G) and a more gradual increase of biomass (B). Except initially, the increase in biomass chiefly represents nonphotosynthetic tissues (*e.g.* stems, roots, and branches) and as the bio-

* Heterotrophic succession should not be confused with a heterotrophic community like that of a cave. Many heterotrophic communities are perpetuated through time, their use of energy being balanced by continuing imports of organic matter from the outside.

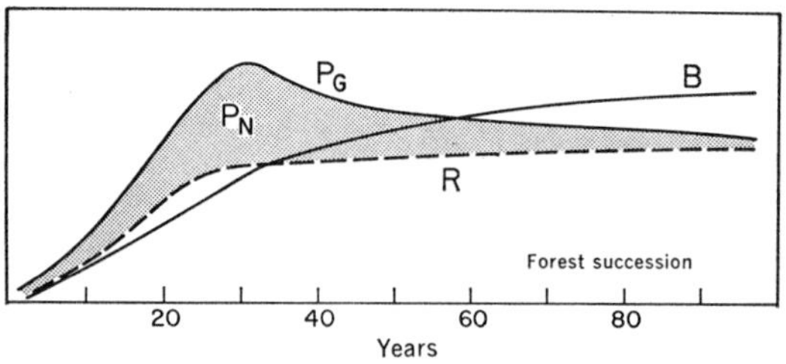

FIG. 10.12. Expected pattern of biomass and productivity change during secondary forest succession. See text for explanation. (From E. P. Odum: *Science, 164:* 262–270, 1969. Reproduced with permission from the American Association for the Advancement of Science.)

mass increases so does the maintenance cost. Hence, as succession proceeds, more of the fixed energy of gross production is consumed in respiration (R), and *net* production (P_N), the rate of increase of biomass diminishes. At climax an equilibrium theoretically is reached: gross photosynthetic production equals total community respiration, exceeding *plant* respiration only by the amount needed to sustain the rest of the community. In the equilibrium state, no net increase in organic matter is occurring.

This general scheme is illustrated by the Hubbard Brook studies of Marks (1974), biogeochemical aspects of which were described in Chapter 9. In Marks' successional stands of pin cherry the amount of foliage 4 years after clear-cutting already was the equivalent of that in a mature forest, but there was far less woody tissue to maintain. Hence net production was high and biomass accumulated rapidly. Foliage amounts and net plant production actually peaked after only 6 years, presumably indicating complete closure of the young forest canopy. Foliage amounts and net primary production declined slowly thereafter. Biomass, however, continued to increase during the 14 years for which data are available; and since a 14-year-old forest is still young, biomass would certainly continue to increase for many subsequent years. In terms of energy capture this forest recovered within 4 to 6 years of clear-cutting, long before the species succession process had run its course. Marks concludes that ecosystem maintenance processes, both biogeochemical and energetic, are repaired very rapidly after forests of the type he studied are disturbed. We may suspect that in forests lacking pin cherry, some other fast-growing tree performs its functional role.

Aquatic Eutrophication

Chapter 5 presented the classical ideas regarding eutrophication as a natural aspect of long term succession in lakes, together with a description of the biological changes associated with eutrophication. It is not certain whether progressive nutrient enrichment really is a natural feature of hydrarch succession, but *artificial* eutrophication is a process which affects many aquatic ecosystems. Having already described the habitat and species changes (*cf.* Table 5.3) that accompany eutrophication, we here wish to explore the response of ecosystem productivity to aquatic eutrophication, and the specific nutrients that seem most often responsible for this process. Several references at the end of this chapter describe the eutrophication process in detail and extend the concept to apply to the enrichment of streams and estuaries. In this extended usage, eutrophication always may be considered artificial; the idea of natural aquatic eutrophication applies only to lakes and ponds.

During eutrophication of a lake, increases of productivity do not necessarily occur as mirror images of the increases in nutrients. Lags, spurts and saturation levels intervene to mar the synchrony. As Joseph Shapiro of the University of Minnesota has stressed, the growth response of algal cultures to a nutrient increase typically follows the pattern of Figure 10.13, and there is no reason to expect the pattern in nature to be different. (Note how similar this relationship is to the photosynthetic relationship of plants to light intensity (Fig. 4.9).) When nutrient concentrations are low, the algal growth rate is very sensitive to small nutrient changes. At higher concentrations there is a saturation effect such that large changes in nutrient levels may produce very little change in productivity. Ultimately, of course, productivity in a highly eutrophic lake or pond will be limited by algal self-shading, and more nutrients will produce no increase in productivity whatsoever. Few waters unpolluted by man have reached this stage, however, and addition of nutrients usually does increase productivity.

One deduces from Figure 10.13 that the addition of nutrients should increase productivity by different degrees in different waters, depending on how rich they are before the addition. Waters that are naturally oligotrophic should respond dramatically to a small increase

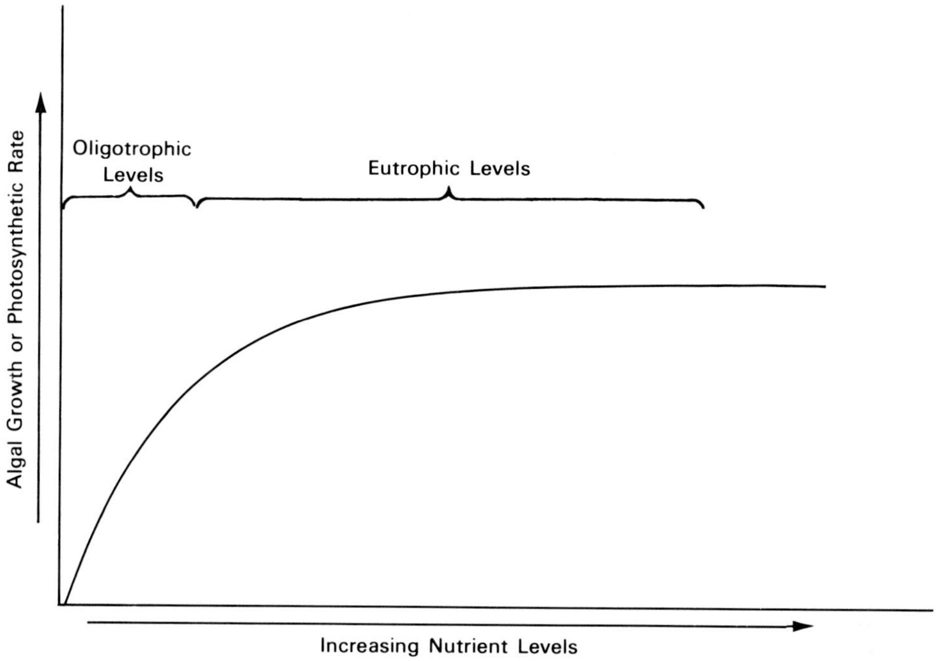

FIG. 10.13. Usual relationship of growth rate to nutrient addition in algal cultures. The effect can be produced by adding a mixture of nutrients, or sometimes by single "critical nutrients" like phosphate. Oligotrophic and eutrophic concentration ranges are indicated very roughly.

in nutrient levels if the nutrients added are limiting to start with. This prediction is the basis of fears about the impact of even small inputs of sewage into lakes such as Tahoe and Superior. The pronounced biological changes in Lake Erie in the last 70 years have resulted from quite modest changes in most dissolved substances (Table 10.7), but even in its pristine state Erie was less oligotrophic, hence presumably less sensitive to chemical inputs, than are Tahoe and Superior today.

Controversy rages regarding which specific nutrients are most responsible for artificial eutrophication, the arguments being heated and often biased by interests and economic pressures outside of science. Nitrogen, phosphorus and carbon all are pointed to as the chief culprit. Will removal of phosphate from detergents really reduce eutrophication levels in America's streams and lakes? only if phosphate is the principal limiting factor in these water bodies. Is the use of detergents that are low in phosphate but high in available nitrogen (*e.g.* nitrilotriacetic acid (NTA) detergents) likely to ameliorate the situation? Possibly in lakes and streams, but in estuaries the consensus seems to be that available nitrogen is the chief limiting nutrient, so addition of NTA to estuaries may cause an increase in eutrophication problems. Industries, municipalities and agricultural interests all have a large stake in whom the Environmental Protection Agency chooses to believe in these arguments, and even the experts are not yet unanimous in their views. But the opinion of serious investigators now leans preponderantly toward the view that phosphate enrichment is the chief single culprit in most cases of accelerated *freshwater* eutrophication. A particularly convincing example of the primacy of phosphate recently has been reported by Schindler (1974), who artificially divided an hourglass-shaped lake into two halves and, using one as the control for the other, tested the effects of a number of added nutrients. Phosphate was by far the greatest stimulant of increased productivity.

Can the effects of artificial enrichment be reversed if excessive nutrient inputs to a lake are halted? The case of Lake Washington in Seattle suggests that the answer is yes. This lake, formerly enriched by sewage from the city, has retreated dramatically from eutrophy since sewage inputs stopped (Edmondson, 1969 and 1972). The lakes studied by Schindler, too,

TABLE 10.7

Chemical Changes in the Great Lakes Since the Late 1800s

The nutrients shown below are not those nutrients generally regarded as most responsible for eutrophication, but they are nutrients for which reliable long term data are available. Total dissolved solids is a nonspecific measure of all nutrients. Present levels of dissolved phosphate and dissolved nitrate approach 27 and 640 ppb in western Lake Erie and 2 and 250 ppb in Lake Superior during seasons of peak abundance (Dobson *et al.*, 1974). In other words, they are much lower than the concentrations of other substances shown in this table, but they may have shown proportionally similar patterns of change over time. The table is based on Beeton (1965) and Ragotzkie (1974).

	Change of Individual Ionic Substances								Change of Total Dissolved Solids		
Lake	Na + K		Ca		Cl		SO_4				
Year	c1895	c1965	c1895	c1965	c1895	c1965	c1895	c1965	c1895	c1965	Per cent change
	ppm								ppm		%
Erie	7	12	32	40	7	23	12	23	140	200	+43
Ontario	7	12	32	40	7	23	13	26	140	200	+43
Michigan	4	5	32	32	4	8	6	18	130	150	+15
Huron	4	4	24	25	3	6	5	13	105	110	+5
Superior	5	4	13	13	2	2	3	3	60	60	0

reverted to their former condition when he stopped fertilizing them; here the reversion was especially dramatic because the lakes were naturally very oligotrophic. Apparently lakes, and certainly streams, are such leaky ecosystems that they cannot sustain high productivities without a continuous influx of nutrients to replace those washed or sedimented out of the system. But we must realize also that the real reason Schindler's lakes were able to return to oligotrophy is that before and after his fertilizer experiments they were on a starvation diet. These lakes lay in forested Manitoba, far from normal sources of human enrichment. The practical lesson of Figure 10.13, as Shapiro has pointed out, is that it will be very difficult, perhaps impossible, to restore many polluted waters to their pristine state because it may be nearly impossible to reduce nutrient inputs from home, industry and agriculture to levels that significantly reduce algal growth rates. Complete restoration is more than we can hope for in most cases; it will be difficult enough simply to make presently polluted waters recreationally and aesthetically attractive once again.

World Eutrophication: a Biosphere in Succession?

We saw in Chapter 9 that man is increasing the availability to the biosphere of numerous biologically important elements. These include carbon, nitrogen, sulfur and phosphorus. In this chapter we have seen that certain of these nutrients seem to be limiting or potentially limiting to productivity in many ecosystems. From such considerations Deevey (1970 and 1973) has made the startling suggestion that the global ecosystem itself is not in steady state but is in the process of "eutrophication:" additional nutrients are making possible greater growth than formerly, and Earth's living biomass may in fact be increasing. To readers accustomed to the idea that man is replacing trees by macadam and in many other ways devastating the biosphere (and he undoubtedly is, in many qualitative senses), it is perhaps difficult to believe that *quantitatively* man is stimulating more abundant life on Earth. Understandably, Deevey's idea has not gone unquestioned. But if he is right, man presently is augmenting rather than destroying the metabolism of the global ecosystem. Nonetheless, we must watch our step:

> . . . there is a unique and nearly ubiquitous compound, with the empirical formula $H_{2960}O_{1480}N_{16}P_{1.8}S$, called living matter. Its synthesis, on an oxidized and uncarboxylated earth, is the most intricate feat of chemical engineering ever performed—and the most delicate operation that people have ever tampered with.
>
> E. S. Deevey (1970)

SUGGESTED READING

ANDERSON, A. J., AND E. J. UNDERWOOD. Trace-element deserts. *Scientific American* (Jan., 1959).

The Biosphere. Special issue of *Scientific American*. (Sept., 1970). Many highly pertinent articles.

Energy and Power. Special issue of *Scientific American.* (Sept. 1971).

HUTCHINSON, G. E. Eutrophication. *American Scientist* 61(3): 269–279 (1973).

LIKENS, G. E., ed. Nutrients and Eutrophication. Special Symposium 1 of the American Society of Limnology and Oceanography. Lawrence, Kans., The Allen Press, (1972). The controversy over causes of eutrophication is thoroughly aired in this collection.

LINDEMAN, R. E. The trophic-dynamic aspect of ecology. *Ecology* 23: 399–418 (1942). A pioneering paper by the real founder of energy flow study.

ODUM, E. P. The strategy of ecosystem development. *Science* 164: 262–270 (1969). Ideas regarding ecological succession by one of the leading proponents of the ecosystem approach.

TEAL, J. M. Community metabolism in a temperate cold spring. *Ecological Monographs* 38: 169–197 (1957). A well known example of a complete ecosystem study from the aspect of energy flow.

WIENS, J. A., ed. Ecosystem Structure and Function. Corvallis, Oregon State University Press, (1972). An authoritative colloquium that includes a thorough presentation of ecosystem energetics.

WOODWELL, G. M., AND E. V. PECAN. Carbon and the Biosphere (Brookhaven Biology Symposium No. 24). Technical Information Center, Office of Information Services, United States Atomic Energy Commission, (1973). Another authoritative symposium, with articles on productivity, decomposition, and other aspects of energy flow and carbon movement through ecosystems. No. 30 in the AEC Symposium Series.

chapter 11

The City as Ecosystem and Community

The city is not an ecological monstrosity, it is rather the place where both the problems and the opportunities of modern technological civilisation are most potent and visible.

Peter Self (1973), The City and Its Environment

INTRODUCTION

The city seems a strange intrusion into a book that is focused so clearly on the natural world and its life patterns. Yet cities are the habitat of a great portion of the human species and of many other species as well. We wish some of the latter—rats, pathogenic bacteria, house flies—would find the city *less* attractive. We think of a city as an ecosystem in which one creature is uniquely dominant, but man really is no more dominant in a city than is the grass *Spartina* in a salt marsh.

Social scientists, geographers, planners and others have proposed (and sometimes adopted) ecological approaches to the study of cities and the solution of urban problems. Here we will limit our focus to those aspects of urban ecology which relate quite directly to concepts presented in other chapters. Ecologists can study a city, its climate, its metabolic processes, the adaptation and niche separation of its inhabitants, in ways sometimes identical, sometimes analogous to the ways they study the rest of nature. In a sense this chapter integrates the concerns of earlier chapters and previews topics we will consider in later pages.

THE URBAN BIOME AND ITS DIFFERENTIATION

Spacing of Towns and Cities

In a heterogeneous landscape the tendency toward symmetry in human settlement patterns seldom is very evident; rivers, highways, transportation routes and the like exert too strong an influence. But in a relatively uniform landscape the symmetry of man's settlements often is quite striking (Fig. 11.1). Once, while flying over the flat, agricultural heartland of America at night, I realized that the patches of twinkling lights below were assuming a quite predictable pattern: small patches ringed the larger ones like satellites, and even smaller dots of light surrounded the satellites. Moreover, the larger patches were repeated at fairly regular intervals. It is not difficult to explain such symmetry. In the past, rural Americans satisfied nearly all their day-to-day needs on the farm or at the crossroads general store. Within easy travelling distance a small town met regular but less frequent needs through its hardware store, weekly market and church. Larger regional settlements, more widely spaced, served much bigger rural populations which "went to the city" occasionally to purchase and sell bulk items or goods of a more specialized nature. Such was the pattern of settlement in much of the Old World and the New: a pattern dictated by how frequently something was needed, by economic considerations, and by slow modes of transportation. The pattern often remains discernible today; though rapid transport and specialized industries make it less inevitable, we live with the legacy of our forefathers.

Differentiation of the Biome

As just outlined, the needs human settlements serve are related to their size. Therefore

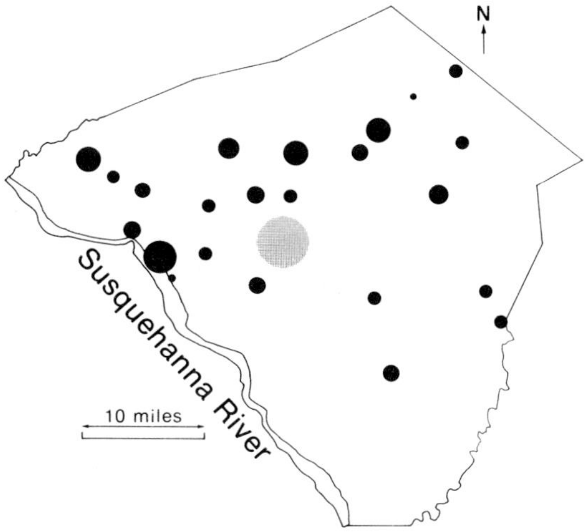

FIG. 11.1. Spacing of communities of 500 people or more in Lancaster County, Pennsylvania. The central region of this county is agricultural land of low relief and the main communities were founded and flourished according to simple dictates of distance from one another. Lancaster city, the "hub" of the county, is ringed by an inner group of small towns, then a group of larger ones. (Courtesy of Lancaster County Planning Commission.)

villages, towns and cities differ quite fundamentally in function and character. Other differences exist that are *not* size-related. In the case of cities, these other differences lend each a certain individuality and impart subtle contrasts in attitude to the citizenry. Milgram (1970) describes attempts to capture the differing flavors of three great cities through interviews with their inhabitants and visitors. Though elusive, such differences certainly exist and contribute to the diversity of the urban biome. For example, Milgram's study indicates that people's impressions of New York tend to be dominated by themes of diversity and frenetic activity, of ghettoization and rude impersonality but also of cultural richness and entertainment. London's physical features and pace are less overwhelming, and its people—tolerant, courteous and neighborly—seem more dominant in its atmosphere. Paris is quite different again: its flavor seems an amalgam of its great monuments and its cafés, its pleasant parks and its self-centered populace. Any city has its distinctive aura, compounded of its people, its industries, its amenities, its location and its age.

Age actually has much to do with urban contrasts. Keyfitz (1966) points out that the earliest cities were centers of despotism, extracting goods and services from people of the surrounding hinterlands. Often they were walled for protection and teeming within the walls. This legacy survives in the tortuous streets and close-packed buildings of many Old World cities. Only during times of strong imperial rule did cities, temporarily free from the need for protection, take on grander dimensions. Rome is a combination of the geometric, amply spaced town plan of the imperial period and the cramped, unplanned city of medieval times. The great converging avenues of Paris, on the other hand, represent later Napoleonic visions cutting through the medieval clutter of an older city.

The early cities were parasitic on rural areas: they took much and gave back little. During the Middle Ages and Renaissance, many European cities lost their despotic, parasitic character, becoming centers of craft and trade where free artisans and merchants assembled, often more or less in opposition to the feudal nobles of the countryside. No longer primarily extractive, these cities began to provide goods and services more equally in exchange for the goods of the countryside, and the spacing pattern we have described for rural America began to take shape in the Old World.

Changing modes of transportation, particularly in the past century, have left nearly indelible signatures on American cities. In the older sections of American towns and cities the streets are narrow and the buildings compressed, reflecting a time when transport was slow and vehicles small. Crowding in this case was more often for convenience than for protection. The increasing spaciousness one encounters outward from the center of an American city is a reflection less of increasing affluence—the concentration of wealth in the suburbs is a relatively new phenomenon—than of the youthfulness of these peripheral areas. Settled when trolleys, commuter trains or autos had already come into use, the newer sections of our cities almost inevitably became more spread out. Despite increasing land costs America's newest cities are its most sprawling and decentralized: greater Los Angeles, for example, is clearly a child of its freeways.

THE URBAN ENVIRONMENT: PHYSICAL AND BIOTIC ASPECTS

Cities exist on the equator and on the arctic circle. They interrupt deserts, steppes and every type of forest. Clearly the urban biome is not characterized by a particular temperature range and rainfall regime, as are "natural" biomes. But regardless of site, a city's local climate *is* measurably different from that of surrounding, less populous, areas. Cities usually are foggier but less humid than nearby suburbs and rural districts; they also are warmer, and they tend to exhibit strong internal contrasts in wind velocity. Cities inject quantities of smoke and exhaust particles into their local atmospheres, together with specific chemical compounds such as sulfur dioxide, nitrogen oxides, carbon monoxide and dioxide, and volatile hydrocarbons. The climate of cities is thus chemically as well as physically distinctive. In fact, the chemical peculiarities of urban climate are responsible for certain physical peculiarities, and vice versa.

The **heat island effect** of cities is one of their best known climatic attributes. Considerable heat is generated within factories, homes, offices and vehicles, and much of this heat is liberated into the atmosphere. At the same time, incoming solar radiation is efficiently absorbed by concrete buildings and asphalt pavements; even if initially reflected, solar radiation striking the ground or a building often is intercepted by the vertical walls of other buildings rather than escaping back to space. Furthermore, the paucity of vegetation and exposed soil limits the amount of evaporative cooling that can occur. Rain runs off quickly and the impervious surfaces of streets and buildings are soon dry. At night these surfaces, warmed during the day, radiate heat and keep air temperatures elevated. Of course, urban haze and fog sometimes reduce ground level solar warming, but the overall balance favors higher temperatures within cities than around them, and the more densely the population is packed the greater the thermal effect (Fig. 11.2). Cities typically have longer frost-free growing seasons and higher mean annual temperatures (sometimes by as much as 4° C) than their suburbs and nearby rural areas. A 3–4° difference in mean annual temperature may sound small, but in the temperate zone it is equivalent to a latitudinal displacement of roughly 300 to 400 miles (480 to 640 km).

Vertical air movements in the city exceed those of the countryside, but horizontal air movements are reduced in the city by about 25% (Landsberg, 1970). The *vertical* movements largely are consequent on heating of the lower air. It rises, cools, spreads out, and then may sink over areas adjacent to the city, to return near the ground as a "country breeze" drawn in by the low pressure associated with rising city air. Obviously the country breeze and the city air are the same air at different points in a cycle (Fig. 11.3). The reduced speed of *horizontal* air movements is a consequence of

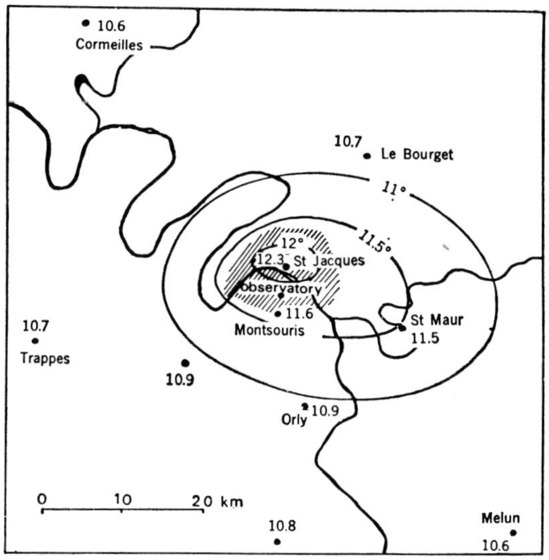

FIG. 11.2. Heat island effect in Paris, as reflected in mean annual temperatures (°C) for the period 1951–1960. Temperatures were corrected for altitude variations within the city before plotting. (From J. Dettwiller: *Journal of Applied Meteorology 9*:178–180, 1970).

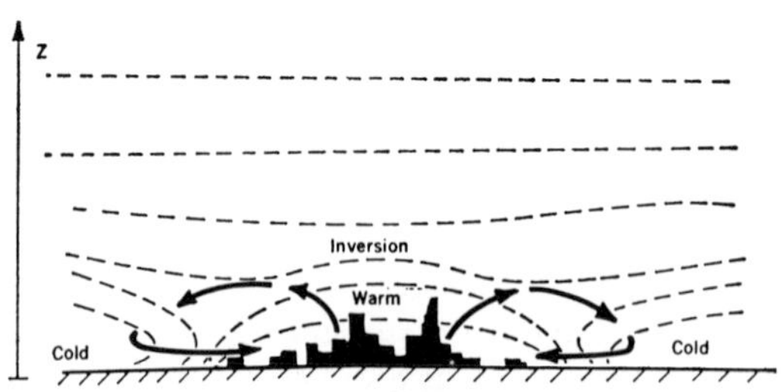

FIG. 11.3. Self-driven pattern of urban air circulation that occurs in calm weather. (From H. Landsberg: *Science, 170:* 1265–1274, 1970. Reproduced with permission from the American Association for the Advancement of Science.)

the barriers and frictional drag offered to surface winds by large buildings. Chicago's reputation as the windy city would pale if its winds were compared to those traversing the nearby lake or prairie. But wind in a city may be channeled down avenues or deflected downward as it passes over a building, so that violent gusts and sudden contrasts are commonplace on a breezy day. Cities obviously are fine places for the comparative study of microclimate, be it wind, temperature, radiation balance or a combination of parameters that is being followed.

The rising air of the city, together with its load of particulate pollutants, is responsible for much urban fog and upper level cloudiness. The minute particles in dirty air serve as nuclei for condensation of moisture, and the prevalence of fog in some cities, especially in winter, may be twice as great as in surrounding rural areas. (This is true despite the generally lower humidity of cities consequent on their greater warmth and reduced evaporation.) Once fog forms, it tends to interrupt the vertical movement of urban air and thus to perpetuate itself, because the warming rays of the sun are reflected by the top of the fog layer and do not heat the air at ground level.

The infamous thermal inversions of certain cities represent instances when the heat island effect is temporarily inactivated and the convective rise of city air is halted. In such instances a cool layer of surface air slips into the city and is trapped beneath a higher, stagnant layer of warmer air. In Denver, for example, dense air often descends the slopes of the Front Range at night, entering the city as a cool wedge beneath the warmer residual air of the previous day. The warm upper air creates a density barrier, opposing the tendency of the cooler surface air to rise as it warms. If there is a relatively slow rate of surface heating, for example because the day is cloudy, and if winds do not dispel the density pattern, the surface air will not be renewed and will accumulate high levels of urban air pollutants. Inversion patterns usually are not *caused* by cities; in fact, the convective tendencies of urban air oppose inversion, and localities such as Denver and Los Angeles would experience inversions even if no city was there. But when they occur over cities, inversions have such noxious consequences that they naturally are considered urban phenomena in the public mind.

Pollutants assault the eyes and lungs of city dwellers; noise assaults their ears. The high decibel nature of the urban biome is one aspect of its physical distinctiveness. "High decibel" also is a suitable metaphor for the biotic features that set cities apart: the rapid tempo of activity, the nerve-testing crush of traffic, above all the high density of people and the multiplicity of encounters with a multiplicity of individuals. These, even more than its physical features, make the city a special sort of environment for its human inhabitants, and for other urban animals as well.

URBAN MAN AND HIS ADAPTATIONS

Coping with Social Overload

Any distinctive environment evokes distinctive adaptations in its inhabitants. The urban environment is no exception, and in the human species we are dealing with the most adaptable of all species in the short term, nongenetic sense. Many of man's adjustments to urban life might better be called acclimation: they develop during the lifetime of the individual. These adjustments are at least partly physiological: protective responses of the lungs and eyes to air pollutants, increased immunity to the infectious diseases which assail high density populations, submersion of urban noise in the subconscious. But the most obvious adaptations of man to the city tend to be psychosocial and occupational. There probably is much truth, for example, to the oft-caricatured contrast between the rural or small town character and the urban personality: the country person warm, open, helpful, naively friendly to strangers and the urban person cold, distrustful, privately preoccupied, hardly recognizing even the presence of strangers. Of course the caricature is overdrawn: for example, country people generally are quite distrustful of strangers, but because strangers are something out of the ordinary the country person at least notices them, and his curiosity often leads to helpfulness.

The explanation usually advanced for urban-rural personality differences is that the urban personality is an adaptive response to what otherwise would be an overload of social stimuli and individual interactions. For example, the seeming callousness of city dwellers to the problems and suffering of others is explained as follows by Milgram (1970):

There are practical limitations to the Samaritan impulse in a major city. If a citizen attended to every needy person, if he were sensitive and acted on every altruisitic impulse that was evoked in the city, he could scarcely keep his own affairs in order.

City dwellers are *not* cold or callous to their circle of friends and relations, but a resident of mid-Manhattan cannot possibly relate individually to the 200,000 or so people he encounters within a short distance of his home or office. Many social characteristics of urban inhabitants—the superficiality of interactions with others (except close friends), the rudeness or at least lack of graciousness in store and subway, the sense of detachment from the surrounding human sea, the distrustfulness toward others—seem traceable, at least in part, to the impossibility of dealing on a personal basis with as many people as one encounters. In American cities the prevalence of crime obviously magnifies the element of distrustfulness: in a *New York Times* survey in 1973, 63% of those polled in New York considered crime the problem of greatest personal concern. But high crime rates are not characteristic of many European cities, and the danger of crime cannot be considered the basic determinant of the urban personality. George and McKinley (1974) state well what I have suggested as the *primary* cause and effect:

People have a limited amount of time. Dialogues, as they increase in number, become shorter in length and less profound in content. On the other hand, increased diversity of possible dialogues produces more and more interactions of high productivity; the formation of fertile subgroups is expedited. Thus, the city justifies itself, even today.

But for most urbanites, a price must be paid. Interactions become highly formalized. Hospitality is reserved for the few persons you really know. Civility itself may become a rarity in public. Decisions may represent fiat more than organic processes of growth.*

Occupational Niches and Individual Identity

Within an ecological community role specialization reduces competition among species and helps to insure the survival of each. If a species is common, further role differentiation at the intraspecific level also may be important to survival. The different foraging ranges of male and female red-eyed vireos illustrate this tendency in the deciduous forest community (page 112); further examples will be encountered in Chapter 13. Man is the most versatile of role-players, and in the city humans have greatly elaborated this process of intraspecific specialization, carving out occupational niches which are highly varied and which may be extremely narrow. Cities are not simply groups of people; they also are collections of factories, office buildings, stores, sewers, bus lines and other complex works, all requiring specialization of the populace if they are to function. That same specialization obviously eases the degree of competition faced by the city dweller:

Now the electronics engineer in Chicago, say, has to concern himself, at most, with the competition of other electronics engineers. But he does not even have to cope with them, at least in the short run. There are a hundred specialties within the field of electronics, and within each of these recognized specialties an individual practitioner, through his own tastes and capacities, can make himself unique. People in a particular plant come to depend on him. If the city is, on the one side, a jungle of potentially infinite and destroying competition, on the other, it shows a nearly infinite capacity of its members to differentiate themselves, to become useful to one another, to become needed.

Nathan Keyfitz (1966)

Thus city dwellers, particularly in industrialized countries, tend to be occupationally specialized and highly interdependent. An individual relies on many other specialists to satisfy his overall needs, and it is easy to imagine this complex web of dependence exacerbating the individual's social overload problem. The solution that seems to have evolved is for the city dweller to interact only in highly restricted ways with most of the people he deals with. The TV repairman, supermarket clerk or bus driver is simply a functional entity, not a whole person, to his client. This often is true even of people the city dweller encounters every day. One exchanges pleasantries with the elevator operator or the man at the corner news stand but seldom really gets to know him or becomes more broadly involved in his life.

Occupational niches seem essential if a city is to function and if its people are to avoid fierce competition in the job market. But obvious dangers of social fragmentation attend this

* From *Urban Ecology*, p. 35. These authors illustrate their point with an equation $D = N(N - 1)/2$ which specifies the possible number of interactive pairs, or dialogues between individuals, in populations of different size. D (the number of dialogues) is 45 for a population (N) of 10 individuals, 4950 when $N = 100$ people, 499,500 when $N = 1000$.

process. The peril is well expressed by Robert Earle:

> Aren't our cities crumbling in large part because their various departments (business, education, medicine, religion, the arts) have become so specialized, have spent so much time solely with themselves, that they cannot even understand each other's cries for help? In the city, everyone takes, no one gives.
>
> *The Chronicle of Higher Education*, December 13, 1971

The problem of overspecialization at the intraspecific level may be peculiar to man; it certainly is most acute for urban man. In the city, man's occupational niches are so diversified that they resemble the niches of different species in nonhuman communities. Yet as a *single* species, urban man retains the need for social coherence and integration at the level of the population, needs that will be explored in the case of other species in Chapter 13. As Earle suggests, it is entirely possible that cities can be too large to be viable, not just as physically functioning complexes but as socially coordinated populations. In many parts of a city, however, the residential neighborhood assumes the nature of a small community, and serves at least partly the need of humans to be known and treated as whole persons rather than as cogs. The psychological importance of the neighborhood has been recognized by many social scientists. Indeed, some large, nonWestern cities where industry is less complex and human role specialization less elaborate seem to function more as a collection of villages than as a coherent urban unit (Keyfitz, 1966).

The Urban Sink and the Culture of Poverty

In these pages we have not attempted to review the huge subject of urban socioecology, but simply to illustrate the closely analogous ways in which urban communities and other ecological communities can be viewed. To conclude this section on urban man's adaptations let us look at a specific example: Oscar Lewis' important adaptive concept of the culture of poverty. Lewis is an anthropologist who has worked for years among the urban poor of Mexico City, New York and San Juan, attempting to understand the life styles and views of slum-dwellers through close studies of individual families. He finds their values and attitudes so different from those of the middle and upper classes, yet so similar *to each other* from city to city, that he considers the urban poor to constitute a distinct, self-sustaining subculture.

What traits distinguish this urban slum culture? Of many that Lewis and his associates have recognized (*cf.* Lewis, 1966), we may summarize as follows:

1. The urban poor are largely divorced from, and indeed often hostile to, the institutions and organizations of "the establishment." They do not join the organized labor movement or political parties. They seldom make use of banks, hospitals, department stores or supposedly public amenities such as museums and libraries. The police are hated, and the church is often suspect. The slum populace is inward turning and local in outlook. Seldom having sufficient cash to buy necessities in bulk, people purchase small amounts of food at high prices, borrow at high rates, pawn their meager possessions, and make frequent use of informal exchanges of credit among neighbors. Except for such arrangements there is little neighborhood organization among adults, despite considerable gregariousness. Among youth, however, neighborhood gangs sometimes represent a temporary form of organization.

2. Marriage is often consensual rather than formalized by a legal or church weeding. Since most men have neither wealth nor steady employment, their role as provider is undermined and the uncertainty of prospect makes a loose rather than binding union attractive from the standpoint of both parents. The family tends to be centered around an authoritarian mother and tied to her relatives, not to the father's. Except as regards the family, however, male superiority is widely acknowledged.

3. Since individuals have so little control over the future, life styles tend to be spontaneous and centered on the present. There is little planning for tomorrow, and much fatalism. Childhood as a way of life is brief and poorly protected; sexual initiation tends to be early; and there is a strong preoccupation among youths and men with their masculinity.

4. These traits and values, together with strong feelings of inferiority and helplessness, tend to be transmitted by example from one generation to the next.

Many of the characteristics in this list appear negative and counterproductive, yet this culture of poverty apparently has developed independently in widely separated cities in the Western world, suggesting an adaptive process at the cultural and psychological levels that resembles convergent evolution. Lewis claims that the traits of this culture are indeed adaptive: the urban poor have very little real hope or chance of achieving success in terms of the values and achievements of the middle and upper classes, and adoption of upper class values as true goals would make life even less tolerable than it is. Instead, a new system of values and arrangements has arisen outside that of the dominant social classes. Once the culture of poverty has become established it becomes self-perpetuating and raises even higher barriers against escape from the slum existence. But if escape is unlikely anyway the system can be seen to be largely adaptive, particularly at the psychological level.

Lewis thinks the culture of poverty is not an attribute of all poor people. Rather, it characterizes the uneducated occupationally unspecialized poor of Western cities (and perhaps rural regions as well) where there is a capitalistic cash economy and a society that is stratified but upwardly mobile. The mainstream of such a society prizes the work ethic and the accumulation of wealth, and believes that material improvement is attainable by anyone who works hard, practices thrift, and keeps an eye set on long term goals. A mental corollary of this value system is that low economic status is nobody's fault but one's own; a poor man is considered poor because he is lazy, short-sighted or foolish. According to Lewis, in societies where the dominant classes hold these values the poor tend to be ignored or weakly provided for by the rest of society. It is easy to see how a culture of poverty is stimulated among those whose struggles get them nowhere, especially in cities where the unskilled poor are minimally equipped for specialized urban niches. Eventually the struggle is largely abandoned, together with the goals valued by the rest of society.

The concept of the culture of poverty has been resisted on many grounds. Social scientists attempt to explain its traits as something other than adaptation. Middle and upper classes resent Lewis' implication that their attitudes and neglect are largely responsible for the fatalistic hopelessness of the very poor; and representatives of the poor themselves fear that acceptance of this concept by government leaders will cause these leaders to question the utility of anti-poverty programs and reduce support for such programs. Some of these criticisms, of course, are not real attacks on Lewis' thesis but simply reactions to its implications. To an ecologist less familiar with urban studies than with communities of animals and plants it is hard to contest Lewis' notion that the collection of traits he has described represents adaptive response to an extreme environment. We will leave others to argue about how drastically the environment will have to be ameliorated if the urban poor are to lose the sense of futility and hopelessness that sustains the culture of poverty.

URBAN PROCESSES: OPEN SYSTEM OR STEADY STATE?

We turn now to the resources that sustain a city, i.e. to the province of ecosytem analysis. Like all ecosystems, a city imports energy and materials, and after transforming what it actually uses, it exports these commodities (Fig. 11.4). The city is a highly open system. Solar input is largely wasted: the only organic matter created in a city is the small amount photosynthetically fixed by lawns, shade trees and the weeds of vacant lots. This fixed energy goes unutilized by man. In human terms the city is a heterotrophic ecosystem: it imports its organic foodstuffs, as well as most of the energy it consumes as heat, electricity and fuel for transport. Turning to theoretically reusable resources, Table 11.1 makes clear that cycling is a minor phenomenon in cities relative to input and output. A city actually may be considered the analogue of a stream: it is a heterotrophic, extremely leaky ecosystem. Future shortages may force cities to become less leaky, *i.e.* to decrease their rates of import and increase the degree of cycling of many commodities. Water, for example, presently is used in prodigal amounts by cities because it generally is used but once, then exported downstream as waste. Adequate treatment of sewage and wastewater can make the supply usable again and again.

Similarly, cities already are facing the need to restrict the quantity of refuse they export and to find ways of reusing it, simply because surrounding areas have limited space that is suita-

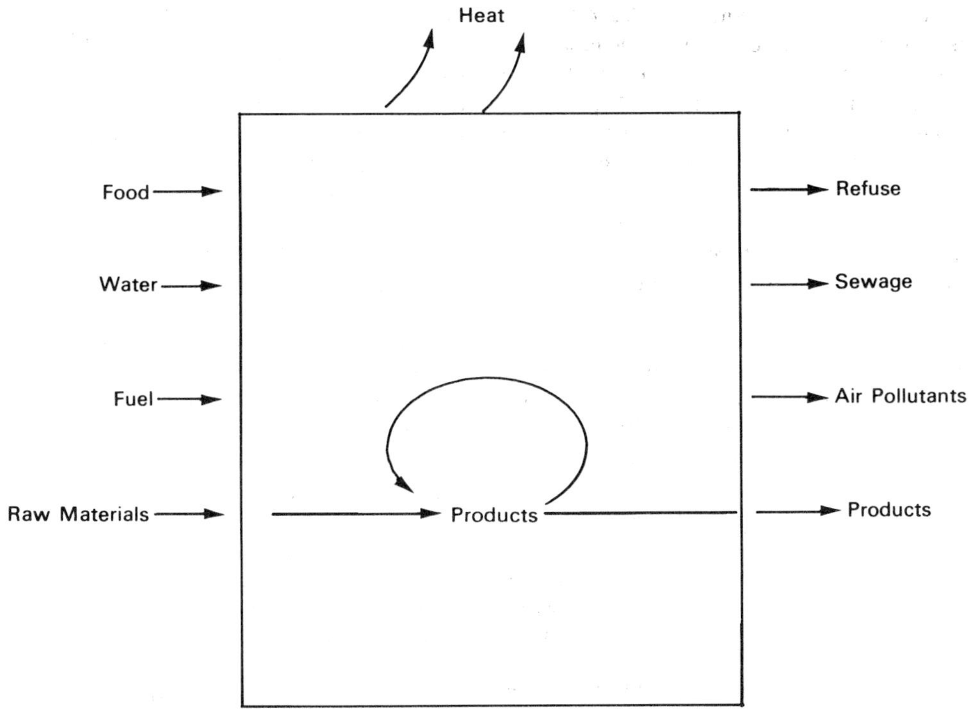

FIG. 11.4. Dynamics of an urban ecosystem. Solar input is not shown, since it plays little role in sustaining the system. Products are manufactured both for use within the city and for export from it. Food energy is used primarily by one species rather than passing through a food chain, so export is relatively great.

TABLE 11.1

Urban Inputs and Outputs

According to Wolman (1965), a typical American city of 1 million people would consume the following quantities of nutrients and produce the following amounts of effluent in the course of a single day. All quantities are expressed as tons. The major atmospheric effluent, CO_2, is not included in Wolman's budget.

Income		Outgo		
Water	625,000	Sewage	500,000	
Food	2,000	Refuse	2,000	
Coal	3,000	Soot	150	Air pollutants
Oil	2,800	Sulfur dioxide	150	Air pollutants
Natural gas	2,700	Nitrogen oxides	100	Air pollutants
Motor fuel	1,000	Hydrocarbons	100	Air pollutants
		Carbon monoxide	450	Air pollutants

ble for landfills. And the rural atmosphere, formerly easily able to absorb urban air pollutants, now is becoming dirty itself, making it more difficult for cities to rid themselves of atmospheric pollutants via natural air exchanges.

Though cities are heterotrophic systems, the modern Western city maintains a kind of balance with the countryside in terms of raw materials and products. The city is sustained by its imports, but it converts many of these into products desired by the suppliers of food and raw materials. This interdependence may be contrasted with the parasitic relationship of earlier, nonindustrial cities to outlying areas. Keyfitz (1966) points out that parasitic cities still exist in many less developed countries. After World War II many such cities grew

city). Likewise the tide of foreign migrants to most cities has slowed since Burgess' day; but it has been replaced by a movement of poor rural Americans to the inner city. Ambitious urban renewal projects today are transforming and revitalizing Burgess' zone of deterioration in many cities. Is this the start of a new successional sequence, analogous to Burgess' but updated by modern transportation methods and the satellite-like decentralizaton of urban retailing and manufacturing concerns? Perhaps a multiple-nuclei theory (Fig. 11.6) now describes most American cities better than does Burgess' older idea of an expanding bull's eye pattern. But certainly no scheme in the future will have quite the simplicity and power of his.

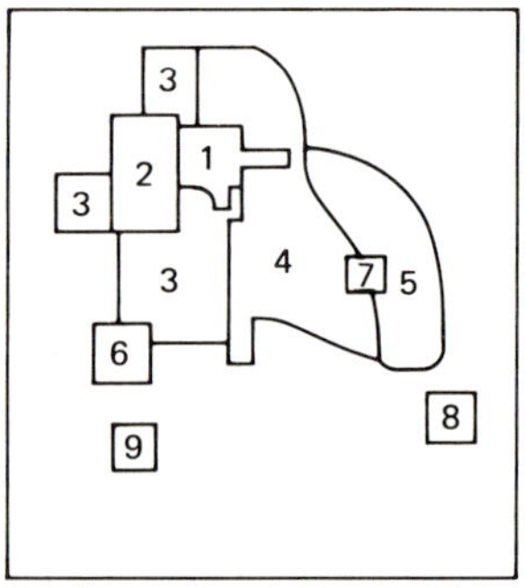

FIG. 11.6. Multiple nuclei theory of urban organization states that cities develop around more than one "center"—if not initially then at least secondarily—resulting in a much less ordered pattern than the one proposed by Burgess. In this hypothetical example the districts are as follows: (*1*) Central Business District, (*2*) Wholesale Light Manufacturing, (*3*) Low-Class Residential, (*4*) Middle-Class Residential, (*5*) Upper-Class Residential, (*6*) Heavy Manufacturing, (*7*) Outlying Business District, (*8*) Residential Suburb, (*9*) Industrial Suburb. (From C. D. Harris and E. L. Ullman: *Annals of the American Academy of Political and Social Science 242*:7–17, 1945.)

Development of Urban Heterogeneity

As succession proceeds in most ecological communities, species diversity tends to increase (Table 10.2) and the species themselves change in character from opportunistic to equlibrium types. Is this true of a developing city? There are parallels, albeit imperfect ones. Certainly occupational diversity does increase as a city grows; and in Burgess' Chicago there also were developmental trends toward greater ethnic heterogeneity and toward increasing percentages of younger residents. But in the more "mature" Chicago of today, and in many other cities across America, the two latter trends have ceased or been muted. At least in their inner parts, American cities no longer are centers of the young and middle-aged but in many cases have become concentrations of the elderly. Many older urbanites are unwilling residents trapped by low income, but others find in downtown areas the necessities and amenities of advanced years: public transit, clinics and hospitals, social and cultural opportunities. Perhaps as a city becomes over-mature its heterogeneity decreases somewhat, as does that of a forest; certainly the *increasing* heterogeneity of an actively growing city parallels the trend of earlier successional stages. Possibly, too, there is a tendency for the populace to shift toward a less mobile, more enduring "equilibrium" character as urban growth slows, but here the parallel is being stretched until it nearly breaks. Modern Americans, particularly urban Americans, are much more like opportunistic species than equilibrium species in their restless mobility, and today's cities are too much in flux for one to suggest that urban succession progresses toward stability. If any parallel to the equilibrium species appears late in urban development it must simply be that elderly segment of the human community which is content, or compelled, to live life out in this final place or this last occupation.

CONCLUSION: OTHER SPECIES IN THE CITY

A city is a place where they cut down the trees and then name the streets after them.

Anonymous

How easily one can agree with this definition! We wince as we laugh. Are ecological concepts useful in a city only to sociologists, psychologists and urban planners? What is there here for the biologically oriented ecologist eager to study something besides man?

In fact, between exotic species encouraged by man and other species that flourish unaided, much can be investigated in a city besides humanity. Traditional ecological problems lie close to every urban doorstep. Backyards, parks, playgrounds and cemeteries all are potential sites for study. Nearly any railroad right-of-way supports a well defined fringing community of volunteer plants: "weed trees" like ailanthus, mulberry and princess tree together with smaller opportunists such as pokeweed, garlic mustard and poison ivy (Fig. 11.7). Vacant lots and land disturbed by construction activity are showpieces of early succession where plantain, pigweed, smartweed, ragweed and others find their place in the sun. Little is known about most natural, city-adapted successional sequences, still less about their animal associates.

For the most part our meager understanding of such urban associations relies heavily on casual observation and anecdote. In his recollections of the London blitz, for example, Julian Huxley recalls the bitterness of devastation being eased by the music of a profusion of unlikely songbirds, which appeared in the city as tangles of vegetation overgrew the bomb craters and half-demolished buildings. It is chiefly since the start of the 1970s (and even today primarily in semi-popular publications like *Natural History*) that true studies of urban wildlife have begun to appear. In Cincinnati the remarkable success of raccoons has been documented by Cauley and Shinner (1973), whose census in a residential suburb revealed a higher density of these animals than has ever been recorded in a wild habitat. Adaptive modification of certain traits helps to explain the Cincinnati raccoons' success. For example, they have smaller home ranges and more sharply focused foraging patterns than their rural brothers. Like an urban senior citizen who habitually goes to one store for groceries, another for medicine and a third for housewares, these raccoons seem to follow set routes through sewers and storm drains to favorite provisioning points, such as garbage cans they know will be full. Interestingly, the chief agent of mortality in this population is not city traffic but canine distemper, an epidemic disease that is easily spread because of the high density of raccoons.

The raccoon study is reported in a special supplement to *Natural History* (Hartmann, 1973), which contains other examples of urban investigations: a study of squirrels in a Zanesville, Ohio cemetery; another of bird popula-

FIG. 11.7. (*A*) "Railway line community." Plants in this picture include ailanthus (*lower left*), Japanese honeysuckle (*lower right*), and ashleaf maple (*just above it*). (Photograph by author.) (*B*) Early successional weeds flourish on a mound of excavated earth. In the foreground is a tassel of foxtail grass (*Setaria* sp.). The plants just behind it (numbered 2–4) are pigweed (*Chenopodium album*), ragweed (*Ambrosia artemisiifolia* and smartweed (*Polygonum* sp.). (Photograph by author.)

tions in Washington, D. C., censused along a continuous transect through affluent, residential Georgetown, adjacent Rock Creek Park, and a concrete-dominated, low cost residential development; and several articles on city trees. Another recent study, in the new city of Columbia, Maryland, chronicles a changing bird community as agricultural land is transformed to residential development (Table 11.3). And in west Texas the Mississippi kite, a spectacular bird of prey, is reported now to prefer the city of Amarillo to the surrounding rangelands it formerly inhabited (Phelps, 1974). The kite now is quite rare on the rangelands, perhaps because mesquite-control activities have altered the landscape unfavorably for its prey or its hunting methods. But now swooping kites snatch moths from the crannies of tall buildings in downtown Amarillo.

Urban species that man intentionally fosters provide another fertile field for investigation. For example, trickling filter types of sewage treatment plant support unique food chains of microorganisms and invertebrates (Cooke, 1959). And although humans *do* clear away trees and then name streets after them, humans also plant other trees of their own choosing. Urban forestry actually is a thriving professional specialty: in Chicago the city manages a discontinuous "forest" of 800,000 trees and is constantly seeking species that will thrive under the stress of polluted air, cramped root space, rapid rainwater runoff, poor soil, and the copious stem-corroding urine of urban dogs. Presently favored tree species in Chicago include green ash, hackberry, pin oak and red oak, honey locust, Norway maple and sycamore (Andresen, 1973). Of course, urban floras must also be matched to the climate, but they usually contain more tree species than do nearby native forests, and are very different in their dominant species.

To support this last point I can cite my own rather informal censuses of the urban trees of Flagstaff, Arizona and Lancaster, Pennsylvania. As it developed, Flagstaff shouldered aside a nearly pure native forest of ponderosa pine, (which still surrounds it) and now boasts, in its older parts alone, a flora of at least 26 tree species. Ponderosa pine is among them, but by no means pre-eminent; Siberian elm, Colorado blue spruce, cottonwood, ashleaf maple, white poplar, and quaking aspen all surpass it, and apple, lombardy poplar, honey locust and black locust are hard on its heels. Why are these species popular? And how well are the aliens doing in northern Arizona's quite difficult climate? Autecological studies might tell us.

TABLE 11.3

Annual Counts of Breeding Birds in a Section of Columbia, Maryland Undergoing Development from Agricultural to Residential Land

From A. D. Geis: *Natural History Magazine*, p. 58, June–July 1974. Reproduced with permission from the American Museum of Natural History, New York.

Species	1966	1967	1968	1969	1970	1971
Decreasing						
Bobwhite	37	37	35	28	23	9
Mourning dove	17	12	11	7	5	4
Wood thrush	16	15	6	5	4	2
Eastern meadow-lark	16	13	12	10	3	6
Red-winged black-bird	9	12	8	7	2	1
Indigo bunting	14	13	9	5	3	2
Grasshopper sparrow	15	17	10	9	3	2
No change						
Cardinal	4	7	6	7	6	5
Increasing						
House wren	0	0	1	2	2	3
Mockingbird	1	4	5	11	11	10
Starling	1	8	15	13	37	32
House sparrow	0	0	3	15	26	20
Chipping sparrow	0	1	1	9	7	12
Song sparrow	4	6	9	12	14	12

Lancaster's urban forest is an intriguing mosaic of neighborhood types relating to age and available space. Where there are no front lawns the sidewalk tree flora is noticeably different from the flora in areas with lawns. Norway maple is the runaway dominant in the city as a whole, but it is especially dominant in old, no-lawn neighborhoods, with horsechestnut and sycamore as common associates. Sugar maple, elm, dogwood and holly become more prominent in neighborhoods with lawns. The small trees planted in the downtown urban renewal area give a hint of changing tastes: honey locust, ginkgo, red maple and scarlet oak are among the favored trees. Not all of them are doing well in the cramped sidewalk cubicles to which they have been confined.

Clearly, the urban biome is not just a human biome. We may take as an invitation to discovery the words of H. L. Li (1969) of Philadelphia's Morris Arboretum: "We have only scant or imperfect knowledge of how plants grow and behave in an urban environment." The same is true, to an even greater degree, of urban animals.

SUGGESTED READING

General references

DETWYLER, T.R., ed. Man's Impact on Environment. New York. McGraw-Hill Co., 1971. Excellent collection of papers, some dealing with urban environments.

GEORGE, C. J. AND D. MCKINLEY. Urban Ecology. New York, McGraw-Hill Co., 1974. Useful paperback covering many topics.

JACOBS, J. The Death and Life of Great American Cities. New York, Random House, 1961.

MCHARG, I. Design with Nature. Garden City, N. Y., Natural History Press, 1969. Ecological approaches to city and regional planning.

MUMFORD, L. The City in History. New York, Harcourt, Brace & Co., 1961.

THEODORSON, G. A., ed. Studies in Human Ecology. Evanston, Ill. and Elmford, Ill., Row, Peterson & Co., 1961. Contains reprints of several classic analyses of urban pattern, including the paper by Burgess.

Urban climate

LANDSBERG, H. E. Man-made climatic changes. *Science* 170: 1265–1274 (1970).

LOWRY, W. P. The climate of cities. *Scientific American* 217(2): 15–23 (1967).

Human adaptation to the City

KEYFITZ, N. Population density and the style of social life. *Bioscience* 16 (12); 868–873 (1966).

LEWIS, O. The culture of poverty. *Scientific American* 215 (4): 19–25 (1966).

MILGRAM, S. The experience of living in cities. *Science* 167: 1461–1468 (1970).

Urban processes

WOLMAN, A. The metabolism of cities. *Scientific American* 213 (3): 179–190 (1965). Part of a special issue on cities.

Other species in the city

COOKE, W. B. Trickling filter ecology. Ecology 40: 273–291 (1959). Describes the specialized community of organisms associated with the biological breakdown of organic wastes in sewage treatment facilities. A whole food chain is involved.

FALK, J. H. Energetics of a suburban lawn ecosystem. *Ecology* 57 (1): 141–150 (1976).

GILL, D. AND P. BONNETT. *Nature in the Urban Landscape* (1973). Baltimore. York Press. A book about wildlife in the city, with case studies of London. Los Angeles coyotes, etc.

HARTMANN, F., ed. The metro forest. *Natural History* 82 (9): 45–84 (1973). A collection of several articles.

Trees: The Yearbook of Agriculture for 1949. Washington, United States Department of Agriculture, 1949. Contains several articles relating to "urban forestry."

PART IV

Ecology of Populations

FIG. 12.1. New Jersey sand dune with beach grass (*top*), sea rocket (*center*), and beach heather (*lower right*) is fertile study ground for population ecologist and autecologist alike. Beach grass occurs in extensive dense stands on seaward dunes, aggressively spreading by runners. Sea rocket usually occurs in small clumps, reproducing only by seed. Beach heather grows in dense, well-defined pockets back of the foredune area. At the top of the picture is an enlargement of the dimorphic double seed capsule produced by sea rocket. The top half breaks off for long distance dispersal (often by floating on the sea), while the bottom half remains attached to the plant, germinating *in situ* when the parent dies and is buried by sand. Barbour (1970a, 1970b, 1972) has done extensive studies on California populations of sea rocket. (Drawing by Sarah Allan.)

chapter 12

The Population as a Unit of Study

. . . Populations of living organisms are very rarely distributed at random over the space available to them . . . when plants reproduce either vegetatively or by means of seeds there is a tendency for the offspring to be concentrated in the neighborhood of the parent plant. The same is true of animals which produce their young in litters and especially of the many forms which deposit masses of eggs thus temporarily leading to a heavy concentration of individuals within a small area. Most animals furthermore show some tendency toward active congregation.

The very large variances occurring in [non-random] *distributions greatly reduce the sensitivity of many statistical tests even to the point where it becomes difficult to place confidence in apparent changes in population size.*

L. C. Cole (*1946*), *A theory for Analyzing Contagiously Distributed Populations*

INTRODUCTION

The quotation that introduces this chapter conveys much of the taste of population ecology. The subject is a distinctive brew with a strong distributional and numerical flavor, and it is spiced for many animal ecologists with a social tang. It is a field with its own fascinations and, as Cole suggests, its own analytical difficulties.

To an ecologist the word "population" means omething quite exact. It is the local subunit of a single species: a group of individuals that exchange genes through interbreeding and usually interact in many other ways. Coexisting populations of different species constitute a community, a level of organization that, as we have seen, is also considered a unit with its own interconnecting strands and processes. Within a population, moreover, each individual itself is a functional unit at the level of the single organism. But the population ecologist is not concerned with individuals *per se*, and he is interested in the community only insofar as it molds the character and fortunes of the group he is studying. A population possesses characteristics and processes all its own, and these are what concern him.

Often these special concerns have to do with numbers, distributional patterns in space or changing abundances over time of the individuals which are the population's subunits. On most sandy beaches of North America we find one or another species of the plant known as sea rocket, a plant with a remarkable set of adaptations facilitating existence on exposed oceanic strands (Fig. 12.1). A single plant is sufficient to delight an autecologist. But a population ecologist notes other things: the sparseness of sea rocket plants relative to the beach grass covering nearby dunes; the clumping of individuals, and the separation of clumps by bare stretches of sand. Why is the average density so low? Why is it not lower? Is density usually this low, or is this population in danger of extinction? Why is the distribution of individuals patchy? Is each clump simply a cluster of offspring around a parent plant? Or are the clumps secondary phenomena unrelated to parental location but perhaps the result of enhanced survivorship among seedlings in certain microhabitats? These questions lead to counts of individuals over time, to assessments of average longevity and mortality within the population, perhaps to experiments in which seedlings are grown together and apart and survival rates compared.

If one is observing animal populations, social differentiation and interaction become another whole area of interest. Not far from a clump of sea rockets, groups of willets are playing tag with the surf, busily pecking for invertebrates as each wave recedes. How is each group of willets related to other groups? Do the groups roost in larger aggregations but break up by day to forage? Are members faithful to a single set of comrades, or is the composition of each group a fluid thing? Are all foraging groups of a certain size? If so, what is the adaptive significance of this arrangement, and why does the sandpiper that forages nearby typically do so alone?

Questions of numbers and questions of adaptation—these are what mainly concern population ecologists. Numbers are central because what happens to a population is the aggregate of what happens to its individuals, and because many of the characteristics of a population are described in terms of how many individuals play this or that role—how many are juveniles, successfully nesting adults, unmated adults, or whatever. Adaptation is central partly because it is the population, not the individual or the community, that is subject to evolution by natural selection. The population represents a pool of genetically different yet interbreeding individuals, and it is the differential reproductive success of these individuals that results in adaptive genetic change. The genetic composition of the population changes with changing environmental pressures because individuals that are more fit in a particular set of environmental circumstances ordinarily contribute more genes to the next generation, *i.e.* leave more offspring. Individuals do not change genetically over their lifetimes, but the population changes genetically with successive generations; it adaptively evolves, as we will see in Chapter 15.

In recent years many students of animal populations have become concerned with adaptation for another reason. They have realized that many of the characteristics of animal populations—the size of their functional subgroups, their particular form of social structure, the degree of cooperation among individuals, the type of mating arrangement—are characteristics that are as subject to natural selection as are the physiological and morphological traits of the species. Comparative studies of social organization have blossomed in recent years, and most of these studies have been directed toward an ecological understanding of why such characteristics as territoriality, monogamy, or food sharing evolve in some species or species populations but not in others.

Chapter 13 will deal specifically with the ecology of social organization, and Chapter 14 with those aspects of populations that have traditionally been central to ecologists—fluctuations in number, causes of fluctuation and modes of numerical regulation. In this chapter we will explore what is meant by "population characteristics," concentrating on those—such as age structure and dispersion—that can serve as useful preludes to our consideration of social ecology and population dynamics.

POPULATION CHARACTERISTICS

Just as we can recognize certain species-specific attributes at the level of the organism, so certain population characteristics also may be species-specific. As individuals, all wolves (*Canis lupus*) are warm-blooded, possess the same dental formula, and exhibit a singular shape to the skull which separates them from other canids. But wolves also mate for life, exhibit a hierarchical form of social organization, and engage in cooperative hunting. These are characteristics which all wolf populations exhibit; hence they also are species-specific. Traits such as these presumably evolved just as did warm-bloodedness, dental features and skull shape, and invite inquiry into their adaptive significance if one is studying wolf populations.

Besides social attributes, a population's characteristics include its potential and actual reproductive rates, its density, its dispersion mode, and its age structure and survivorship pattern. **Intrinsic reproductive rate** is a population characteristic that tends to be relatively species-specific. The intrinsic reproductive rate is the rate of increase achievable by a population when it is completely free of environmental constraints, i.e. the theoretical maximum growth rate. For some species, such as white oak, wolf, or man, the intrinsic reproductive rate is relatively *low* in all populations: each individual (or pair) produces only a few offspring at a time, and at relatively great intervals. Other species—sycamores, meadow mice, salmon—have a *high* intrinsic reproductive rate in all populations.

Most other population characteristics tend to be more labile within a species but still may show a degree of species constancy. Density is an example. Obviously, density may vary among different populations of a species, but this variation is not unlimited: in Chapter 3 we noted that certain species characteristically are rare in the communities they are part of, *i.e.* their population densities nearly always are low. Other species characteristically are common wherever they maintain populations, though density variations of these populations may be rather wide. Most of the population characteristics in Table 12.1 are of this semi-labile type; even such attributes as age structure vary only within relatively narrow limits among populations of a particular species. Species specificity with regard to population dispersion, social structure and mating system is quite high in many cases; but we now know numerous examples of species whose populations differ in one or more of these characteristics depending on environmental circumstances. As we shall see in the next chapter, intraspecific variations in social structure and mating system often can be related to specific ecological contexts, and comparative studies are beginning to uncover adaptive reasons for the distinct social contrasts that are apparent among populations.

DISPERSION

Turning to particular attributes, let us first ask how the individuals of a population are distributed within their habitat. Three general modes are possible: the individuals may be (*a*) clumped into one or several groups, (*b*) dispersed randomly within the habitat or (*c*) distributed relatively evenly with respect to each other. Of these three patterns, one might expect to find random dispersion if the individuals show no interest (either attraction or repulsion) toward one another or else no ability to control their distance from one another; and if, in addition, the habitat is uniform, no part being more attractive than another or more likely for "historical" reasons to be a population center. Clumped dispersion patterns indicate either a nonhomogeneous habitat, causing aggregation in certain places, or a positively social population. Even dispersion would be expected where anti-social tendencies, or nonsocial intraspecific competition, characterize the population.

We already have seen (Fig. 11.1) that dispersion patterns yield particularly interesting insights when the complication of habitat heterogeneity is minimized. If the habitat is homogeneous, random dispersion of a population indicates that the species is essentially nonsocial. Random dispersion is what one approaches if a handful of change is tossed willy-nilly on a sidewalk: some coins end up near each other, some quite isolated. This is the pattern to expect if the positioning of individuals is merely

Table 12.1

Population Characteristics

Indications are given whether each of these characteristics can be expected to vary greatly in different populations of the same species, or whether they are relatively invariable (species-specific).

Characteristic	Species-specific?	Varies Among Populations of the Same Species?
Intrinsic reproductive rate (r)	Yes	Not appreciably
Actual reproductive rate	Not very	Yes
Survivorship pattern	Largely	Somewhat
Age structure of population	Somewhat	Yes
Density	Somewhat, *e.g.* "common" *versus* "rare" species	Yes
Dispersion pattern	Largely	To small degree
Social structure	Somewhat	Somewhat
Mating system	Largely	Somewhat

a matter of chance. In a similarly homogeneous habitat, either a clumped or an even dispersion pattern indicates some organizing force affecting the population. Generally, this force is social in the first case and anti-social (*e.g.* territorial) in the second, if the population consists of motile animals.

The dispersion pattern of a population often is obvious: sea rocket plants form clumps on a relatively homogeneous beach, and gannets space their nests evenly within a breeding ground (Fig. 12.2). But one can be deceived: patterns that really are nonrandom may *appear* random if their clumping or evenness is not too extreme; and it is easy for randomly dispersed individuals to appear clumped. Opinion has differed, for example, as to whether the apparently regular spacing of creosote bushes in some desert habitats (Fig. 6.12) really is "more regular than random." Such questions can be settled by **Poisson analysis:** if sampling data are available regarding the number of individuals in equal-sized quadrats, a hypothetical random distribution can be generated mathematically and compared to the observed distribution patterns (Fig. 12.3). In the case of creosote bush there seems to be intraspecific variation in dispersion pattern: certain populations apparently *are* randomly spaced, whereas others, in very dry deserts like the Mojave, are more even than random (can the reader suggest why?). As a rule, few populations, even in apparently homogeneous habitats, prove to be randomly spaced. Cole (1946) studying forest floor invertebrates in Illinois, found only spiders to be randomly dispersed, and obviously even they would not have been so dispersed during the breeding season. Clapham (1936) found only 4 of 44 plant species in a grassland community to be randomly spaced.

FIG. 12.2. Many sea birds breed in dense social clumps, but within each clump nests are evenly spaced due to territorial behavior. These are gannets at the Cape Kidnappers rookery in New Zealand. (Photograph by author.)

In the case of plants, or of animals with limited mobility, a clumped dispersion pattern may be simply the result of where the parents happened to shed seeds or eggs; or in the case of plankton in a lake, of currents that concentrate the organisms. Although zooplankton and creosote bush offer examples of differing dispersion patterns among populations of the same species, most species actually are quite specific with regard to this characteristic: their population dispersion is either clumped, even or random, the first being the pattern characteristic of most species. Sometimes, however, one sees a hybrid of two patterns: a population which is clumped into groups, but with spacing being quite even either within the groups (Fig. 12.2) or between them (Fig. 11.1).

AGE STRUCTURE

Clues to the relative "health" of a population frequently can be obtained through analysis of its age structure. Is the population perpetuating

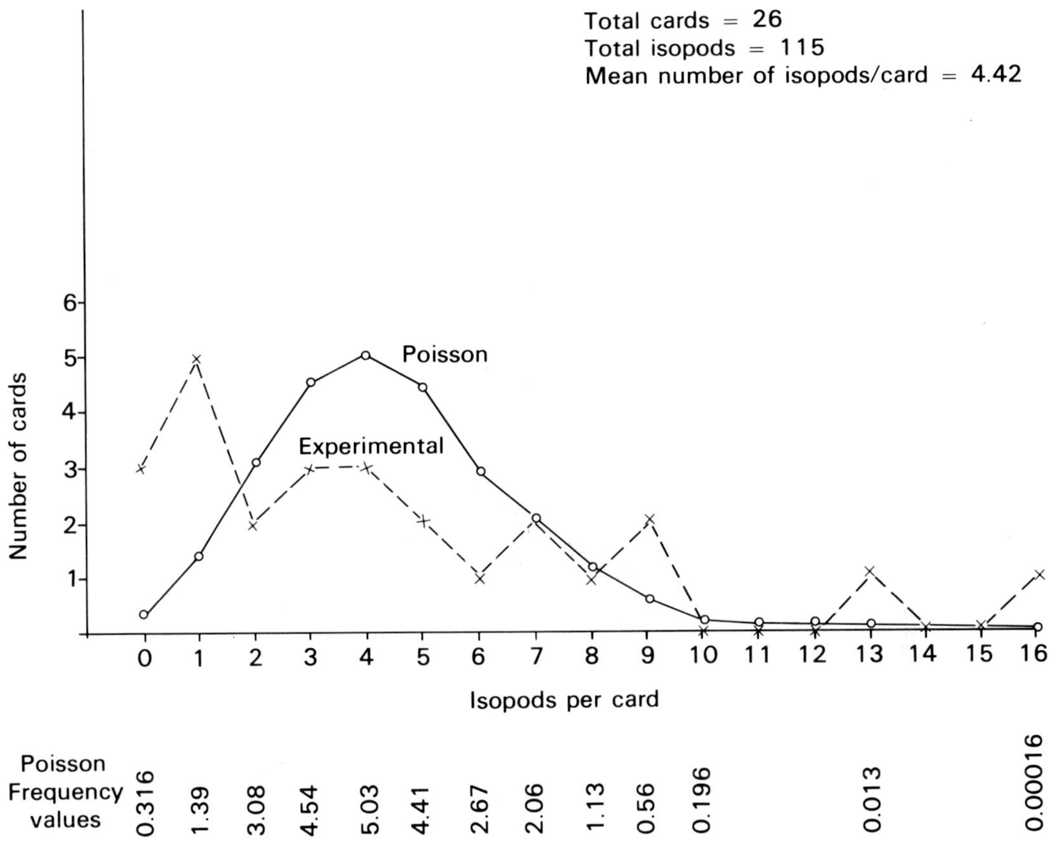

FIG. 12.3. Experiment in a college greenhouse on the aggregative tendencies of terrestrial isopods. Isopods (popularly known as sow bugs or roly-polies) seek shelter in dark places during the day, but move about at night. Here 26 identical cards were placed in a relatively homogeneous area of the greenhouse and the isopods taking shelter beneath them were counted the next day. If they simply chose any card at random, their distribution should approximate the Poisson distribution illustrated. Actually, more cards than predicted had no isopods or just one, and several cards had unexpectedly large numbers, suggesting a tendency among isopods toward mutual attraction. Given 26 hiding places and 115 isopods, the frequency with which 16 isopods should choose the same card is only 1 in 6000 if chance alone is the cause. The Poisson frequency values state how many cards one would expect to harbor a given number of isopods if they chose cards simply at random. Andrewartha (1961), Cox (1976), and others describe the method of Poisson analysis. (This example uses data from one of the author's class exercises.)

itself? If so, it must contain sufficiently large numbers of young individuals (prereproductives) to replace the present adult progenitors. Since the population suffers progressive attrition through mortality and emigration, ordinarily there must be more juveniles than adults in a healthy population. Reproduction of some species can continue throughout adult life to the end of the physiological life span. In the case of other species, reproductive activity ceases before the end of life, and it is important to distinguish a third functional group within the population: postreproductives, as well as reproductives and prereproductives. In practice, one often goes beyond this and enumerates a population in terms of individual year classes or age-related size classes (Fig. 12.4).

Age structure clearly is capable of great variation in different populations of the same species, or in the same population at different times, but there is also a degree of species specificity as regards age structure, since each species tends to vary only within limits in its reproductive potential and survivorship pattern. A species having a high intrinsic reproductive rate will also normally have a high attrition

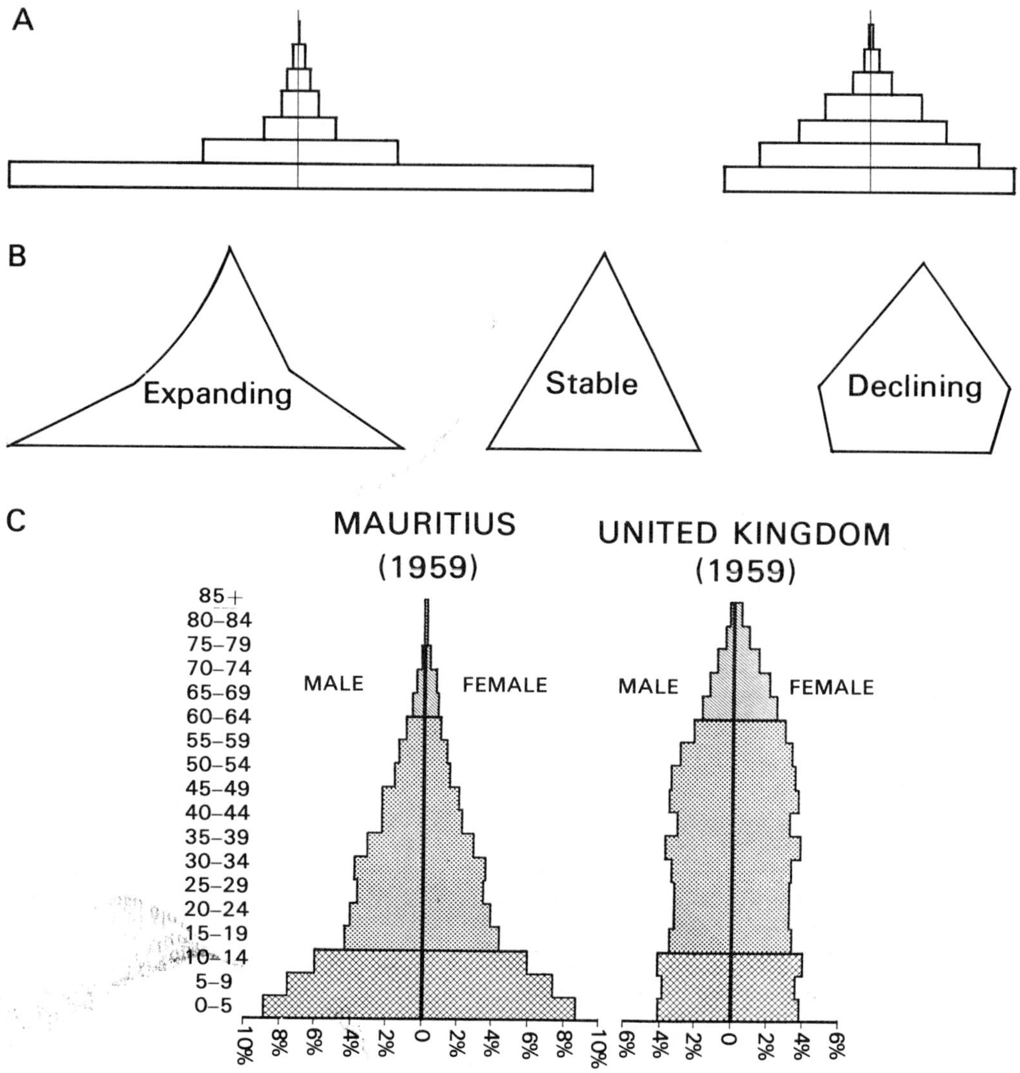

FIG. 12.4. (*A*) Broad-based and narrow-based age pyramids of two species. The species on the left has higher reproductive and juvenile mortality rates. (*B*) Contrasting pyramids of expanding, stable and declining populations of a hypothetical species. (*C*) Age structure pyramids of two human populations. Mauritius is an island in the Indian Ocean where the mortality rate dropped suddenly after World War II through public health measures. (From R. C. Cook: *Population Bulletin* 18:93–115, 1962.)

rate, and age pyramids of its populations will normally be broad-based with very acute basal angles. If a species has a low intrinsic rate of reproduction, its rate of mortality normally is low also, and age pyramids of its populations will be narrow-based and relatively steep (Fig. 12.4*A*). In other words, a species' reproductive potential normally is geared to its members' survival potential; age structure pyramids reflect this, by varying only within limits for the populations of a given species (Fig. 12.4, *B* and *C*). According to Odum (1971):

> There is evidence that populations have a "normal" or stable age distribution toward which actual age distributions are tending. Once a stable age distribution is achieved, unusual increases in natality or mortality result in temporary changes, with spontaneous return to the stable situation.

Age structure analysis obviously is useful in predicting whether a population will increase, decrease or remain relatively constant in the near future. If one knows the stable age distribution of a species, observed deviations from stability in the age structure of one of its populations will have particularly high predictive value; but intelligent predictive statements are possible even if the stable age distribution is not well known. Figure 12.5 illustrates how age structure analysis can be of economic importance in the case of commercially exploited populations; in this case it helps to diagnose the ills of the North Sea herring fishery.

Unusual age structures characterize populations in which reproductive success and/or early survival varies greatly from one breeding season to the next. For example, the North Sea herring catch was dominated for 13 successive years (1907–19) by individuals born in 1904, a year in which breeding success was much higher than normal and huge recruitment to the population occurred. (Today, as Figure 12.5 suggests, heavy fishing would not allow even a very large year-class to dominate the population for so long.) We may also note with regard to differential breeding success that populations of any species that occupy the fringes of its range, or other habitats that are only marginally suitable, are particularly likely to possess discontinous age structures dominated by only a few age classes (Fig. 12.6). Such populations may be self-perpetuating, even though conditions only occasionally permit successful recruitment of young age classes to the population. In the Arizona desert Barbour and Diaz (1973) discovered that along a gradient of decreasing annual rainfall, populations of creosote bush displayed a matching gradient in the regularity of their age structure. In the driest habitats recruitment to the population was virtually restricted to years in which slightly moister conditions prevailed, while in less dry habitats recruitment occurred every year and populations had more regular age structures. Probably it is safe to generalize that a sharply discontinuous age structure in any population indicates that the habitat is marginal for the species to which it belongs, whereas a smoothly continuous age structure indicates a more favorable habitat.

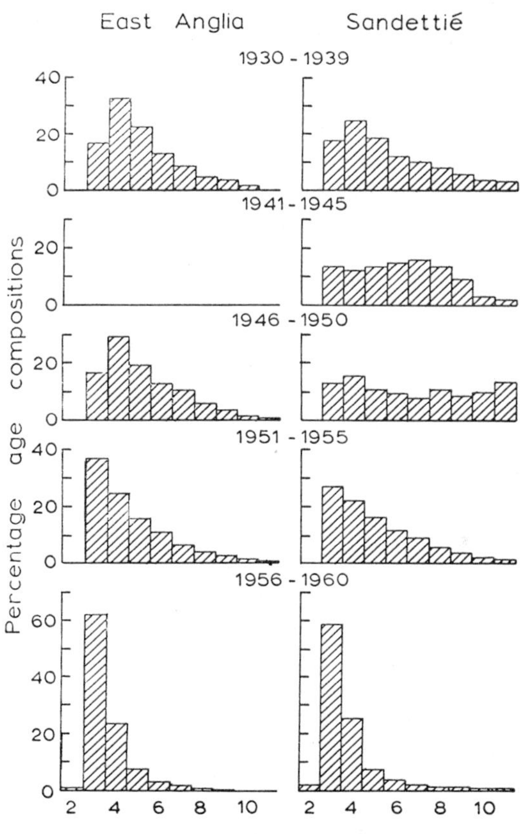

FIG. 12.5. Decline of the North Sea herring fishery is documented in these age structure diagrams of the catch over several decades in two locations. Young fish have come to dominate the catch. Most 3-year-olds are just entering their reproductive phase, so reproduction of the stock is hindred by catching fish so young. Fishing intensity now allows few fish to live to more advanced ages. (From Gulland, 1972; after a report of the International Council for the Exploration of the Sea.)

SURVIVORSHIP PATTERN

How do life insurance companies know what rates to charge their customers? When a company sells a policy to a customer of a given age, it obviously is gambling. If the customer dies shortly thereafter the company loses; if the customer keeps paying in premiums for years, the company wins. Being business enterprises, insurance companies make their bets as secure as possible by knowing all they can about their customers as individuals (health status, occupational dangers, etc.) and also about the relationships of age to life expectancy in the populaton as a whole. Much of the information regarding a population's age-related mortality and life expectancy can be summarized in a survivorship curve (Fig. 12.7).

What can be learned from such a curve? By simple inspection of Figure 12.7 one can see

FIG. 12.6. Discontinuous age structure of a ponderosa pine population near Flagstaff, Arizona. Near the edge of its range in this region, ponderosa pine reproduces successfully here only at considerable intervals. The younger pines all date from 1919, the second of two successive wet years. The first year favored high seed production, the second high seedling survival. The even-aged overstory trees probably date from a similar favorable episode two centuries or more ago. The photograph was taken by the author in 1973, 54 years after the last year of high reproductive success.

when individuals in the population are most subject to mortality; as in most human populations, mortality here is concentrated in the early prereproductive years and in the postreproductive years, and is lower during the middle age period. One also can calculate the mean life expectancy at any age Z, simply by following the curve down to the age at which the number (or percentage) of survivors has been reduced to one-half the number at age Z. It is evident, too, that one can approximate a population's age structure pyramid by using data from a survivorship curve. But this calculated age structure will closely resemble the actual age structure only if recruitment to the population is constant; if birth rate, or infant mortality, varies greatly from year to year, the age structure pyramid will show bulges and perhaps even discontinuities which do not appear on a survivorship curve. A survivorship curve describes the fate of a single age group—the "class of 1974," say—over time, whereas the age structure pyramid shows the relative sizes, at any one time, of all the age groups within the population.

Survivorship figures are far better for human populations than for populations of almost any other species; but as first pointed out by Deevey (1947), much can be learned from both intra- and interspecific comparisons of relatively crude population survivorship data. Like man, other animal species exhibit survivorship differences among their populations, or sometimes between the sexes within a single population (Fig. 12.8), and these differences almost invariably have ecological causes.

Survivorship differences *between* species can best be explored if age is expressed in relative units, say as percentage of the mean life span or of the maximum physiological life span. If this is done, survivorship curves can be constructed on the same set of axes for rabbits and humans, or for gulls and rotifers, even though their actual life spans differ greatly (Fig. 12.9). When the

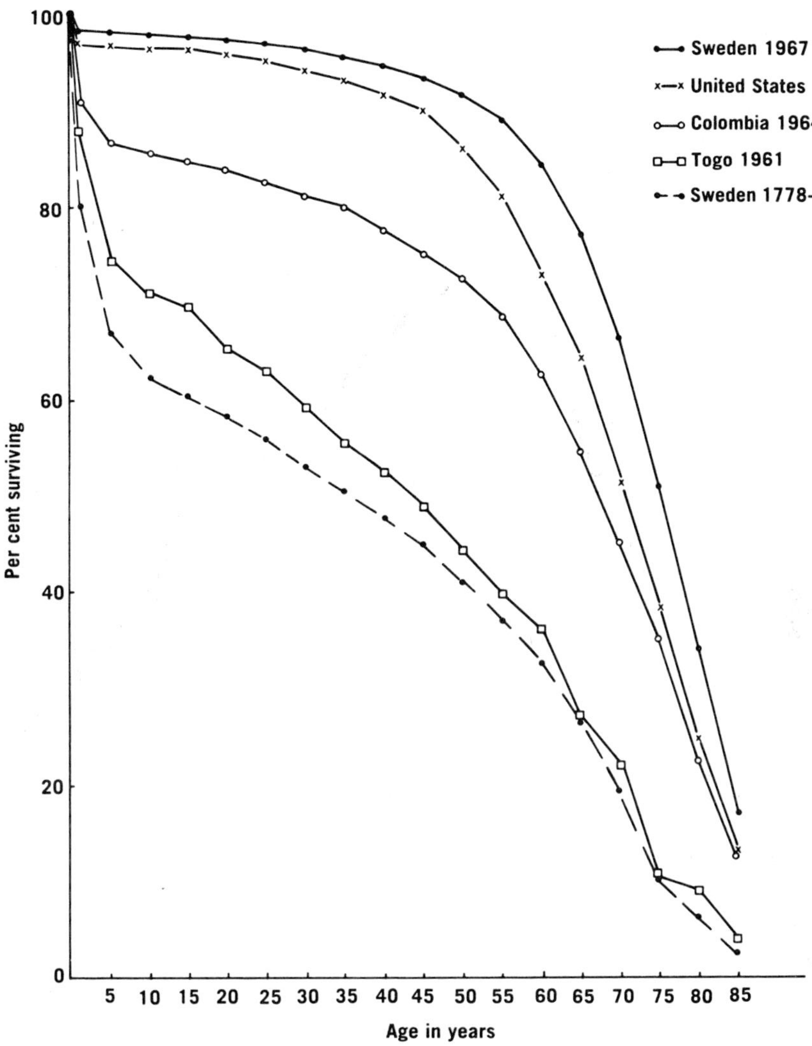

FIG. 12.7. Survivorship curves of the human populations in various countries, plotted with an arithmetic scale on the ordinate to show contrasts. If plotted logarithmically, as in the next three figures, these curves all would be highly convex and hardly separable from one another. (From A. Turk et al.: *Environmental Science*. Philadelphia: W. B. Saunders Co., 1974. Reprinted with permission. Raw data were taken from N. Keyfitz and W. Flieger: World Population: An Analysis of Vital Data. Chicago: University of Chicago Press, 1968.)

survivorship scale is plotted logarithmically, as in this figure, a further benefit accrues: a straight line relationship, if it occurs, indicates that the population is characterized by a constant mortality rate at all ages. A curve that is convexly bowed depicts a population that defers most mortality until postreproductive or later reproductive ages, whereas a concavely bowed curve characterizes a population with heavy juvenile mortality and greatly decreased mortality rates later in life. Despite such intraspecific variations as we have noted, considerable species specificity is revealed when survivorship patterns of different species are compared. For example, the bird species so far studied tend to follow the straight-line pattern; mortality is quite high and few birds live to a ripe old age, but the mortality rate does not differ markedly between young birds and old ones. Most mammalian populations have somewhat sinuous survivorship curves: concave early in life but quite convex later on. Wholly concave survivorship curves probably characterize species which produce incredible numbers of offspring, only few of which survive. Female cod lay up to 10 million eggs apiece, the American

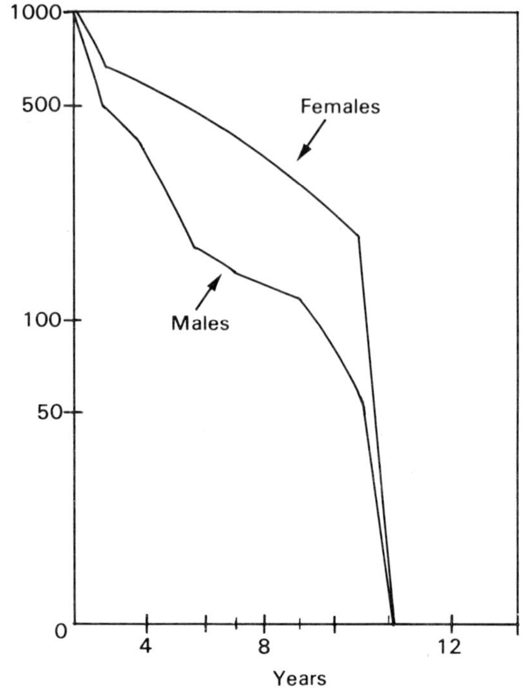

FIG. 12.8. Survival patterns of black-tailed deer in a chaparral region of California. The physiological life span of both sexes appears to be about 10 years. Males are hunted heavily from the third year onward, some being taken even as yearlings. This probably explains most of the difference between the sexes' survivorship curves. (From R. D. Taber and R. Dasmann: *Ecology 38*:233–246, 1957.)

oyster nearly as many; and though large numbers of these may be fertilized, clearly few survive to adulthood or the oceans would soon be overrun with cod and oysters.

Concave survivorship patterns seem to be common in the case of many plant species; we can expect most weed species, especially, to have deeply concave curves owing to their high reproductive rate but poor competitive ability. Concave patterns have been found, too, in a population of sea rocket (Barbour, 1972) and in 9 of 11 populations of young sugar maples studied in Wisconsin (Hett and Loucks, 1971). In the case of trees, size class analyses are much more common than survivorship studies; but when plotted as survivorship curves, size class data for trees indicate interesting differences between successional species and species of forests near maturity (Fig. 12.10).

Interesting as are comparisons between species, to the population ecologist comparisons of different populations within the *same* species carry greatest interest. Only intraspecific comparisons can suggest whether a *particular* population is numerically stable, increasing, or declining; and as we shall see in Chapter 14, the prediction and analysis of fluctuations in a population's size are central concerns among students of populations. Before turning to this

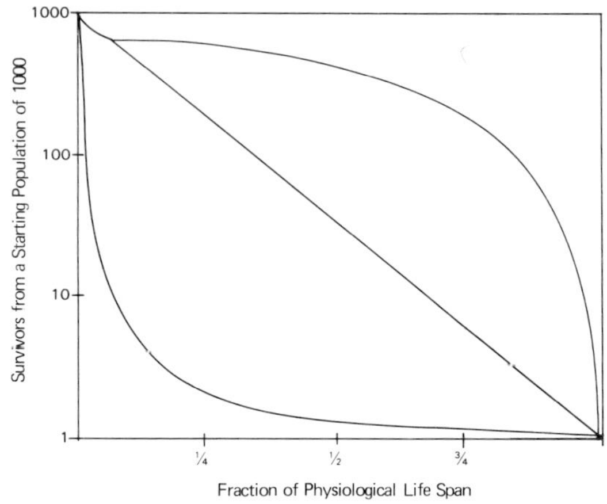

FIG. 12.9. Basic types of survivorship pattern. (Adapted from E. S. Deevey: *Quarterly Review of Biology 22*:283–314, 1947.)

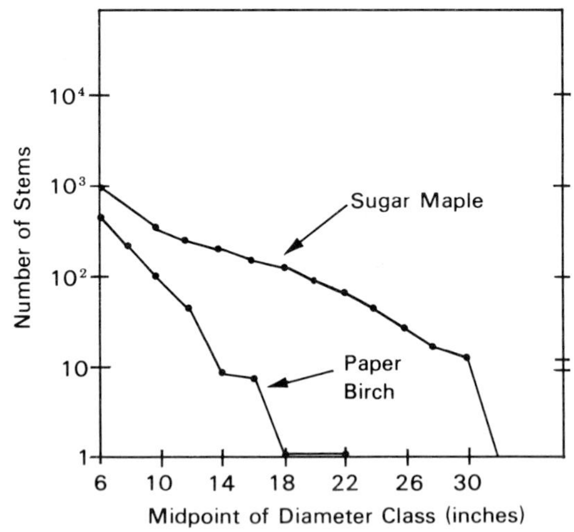

FIG. 12.10. Diameter-class data for sugar maple (an equilibrium species) and paper birch (a successional species) plotted as survivorship curves. The study area was Menominee County, Wisconsin. Maple has the more convex survivorship pattern, as might be expected of an equilibrium species. (From P. H. Zedler and F. G. Goff: Ecological *Monographs 43*:79–94, 1973.)

topic, however, we will consider some aspects of socioecology, a burgeoning new interface between animal population ecology and animal behavior.

SUGGESTED READINGS

(These are mostly references dealing with methods of population sampling and data analysis.)

ANDREWARTHA, H. G. Introduction to the Study of Animal Populations. Chicago, University of Chicago Press (paperback), 1961. A somewhat difficult book to start off with, but containing quite thorough accounts of many techniques for sampling and analysis.

BENTON, A. H., AND W. E. WERNER, JR. Manual of Field Biology and Ecology. Minneapolis, Burgess Publishing Co., 1966 (4th edition). A general manual, with a useful section on population sampling and census methods.

COX, G. W. Laboratory Manual of General Ecology. Dubuque, Iowa, Wm. C. Brown Co., 1976 (3rd edition). Excellent for methods.

DEEVEY, E. S., JR. Life tables for natural populations of animals. *Quarterly Review of Biology* 22: 283–314 (1947). A very influential paper dealing with survivorship data.

GREIG-SMITH, P. Quantitative Plant Ecology. London, Butterworth's Scientific Publishers, 1957. A standard text and reference.

LEWIS, T., AND L. TAYLOR. Introduction to Experimental Ecology. New York, Academic Press, 1967. A book that stresses methods and data analyis.

SMITH, R. L. Ecology and Field Biology. New York, Harper and Row, 1974 (2nd edition). A widely used textbook with a long appendix on sampling methods.

chapter 13

The Ecology of Social Organization

We must expect social systems to have evolved, just as structure and physiology have evolved. As the products of evolution, social systems should thus be adaptive; that is, they should be peculiarly adjusted to each environment in which they occur so as to confer maximum reproductive success . . . .

David P. Barash (1974), *The Evolution of Marmot Societies: a General Theory*

INTRODUCTION

One hundred forty years ago, Darwin's observations of finches in the Galapagos Islands propelled his mind toward natural selection (Fig. 13.1). In the mid-1960s a second ornithological visit to the Galapagos produced a brief paper which nicely outlines the concerns of this chapter. Like Darwin, J. B. Nelson (1967) dealt with evolutionary adaptation. Again like Darwin, his springboard was the careful observations of animals in the field, and their comparison with other species. But Nelson was particularly concerned with the adaptive aspects of a social system. Nelson studied frigate birds, those slim-winged, fork-tailed pirates so familiar to sailors in the tropics. Known also as man-o-war birds, these relatives of the pelican seem a natural part of the old Spanish Main as they cruise the skies in search of plunder. They roam far out to sea and subsist largely on fish, yet they never touch the water even to rest. Instead, they harass other seabirds into releasing their prey in the air, and catch it as it drops. They themselves are marvellously adapted to life in the skies; the wing span of a frigate bird far exceeds that of any other bird of comparable weight, and baggage unnecessary to flying (such as legs and feet) is reduced in size. The throat pouch familiar in the pelican is also reduced in frigate birds, but males possess an analogue in the inflatable, boldly colored sac that is used in courtship (Fig. 13.1).

Studying a breeding colony of the great frigate bird on Tower Island, Nelson became intrigued by a social behavior which he recognized as highly atypical of sea birds, and atypical even of other families of the pelican order. Eventually he realized that these social peculiarities were part of an adaptive complex associated with the frigate birds' piratical way of life (Table 13.1). Nelson believes food and feeding method are the ultimate cause of the social complex he catalogued. For frigate birds, air piracy is not really getting something for nothing: it is a skill that requires considerable time to perfect and considerable effort to accomplish. Food is hard to come by, and for the Galapagos frigate bird it tends also to be erratic in abundance, depending on unpredictable movements of fishes and other seabirds in this largely nonseasonal tropical region. Therefore (*a*) young frigate birds grow slowly, apparently being metabolically adapted to a relatively slow rate of food supply and to periods of fasting; and (*b*) only when relatively mature do the young possess the coordination and agressiveness to capture food themselves by piracy. Hence they

FIG. 13.1. Bird fauna of the Galapagos Islands makes ecologists of evolutionists and vice versa. Contrasting bill shapes and feeding modes of Darwin's finches (*A*) have an ecological basis in competitive interaction and the opportunities of the environment. The breeding system of the great frigate bird (*B*) likewise seems to be an evolutionary correlate of a particular environment and feeding mode. (Finch drawing is from an early edition of *The Voyage of the Beagle*. Drawing of frigate birds is by Sarah Allan.)

remain in the nest and are fed by their parents for many months. The social consequences of a nesting period lasting more than a year are profound, and Nelson's analysis lays them out clearly (Table 13.1). Nelson's linkage of breeding behavior to length of nesting season, and of the latter to the realities of a piratical life style in a nonseasonal environment, delineates the kind of total adaptive system that socioecologists now look for when they study animal behavior in the field.

In this chapter we will be exploring other examples of the interplay among ecology, evolution and animal sociology, limiting ourselves primarily to rodents, primates and birds. This fascinating interface of biology, initiated decades ago through the work of men like W. C. Allee, recently has exploded with activity. The prime fuse probably has been the work of John Crook, an Englishman who first excited behaviorists with his monograph on the socioecology of African weaver finches (1964) and who attracted a much wider audience through his bold comparative analysis of primate societies, pub-

TABLE 13.1

Ecological Basis of Breeding Behavior in the Great Frigate Bird

Data are from Nelson (1967).

Behavior Typical of Most Sea Birds	Behavior Typical of Great Frigate Bird	Suggested Explanation of Frigate Bird Peculiarities
1. Marked fidelity to same nest site from year to year	1. Fidelity to same breeding area, but not to specific nest sites	1. Nesting and rearing of young takes 400 days or more. Hence frigate birds breed in staggered years. and usually find their previous nest site still occupied by another pair when they return.
2. Males return to site first, aggressively establish small nesting territories, begin nest building, and display individually here.	2. Males display in groups, showing no territorial aggression or nest-building activity until after females return and pair formation occurs.	2. Group display more clearly distinguishes receptive males from males already raising young, and males' fidelity to a specific territory is weak because the previous year's nest site is not available.
3. Marked fidelity to same mate from year to year.	3. Impermanent pair bonds; usually a different mate each time breeding occurs.	3. Returning females cannot easily find former partners, since they are not at previously established nest sites.
4. Strong cooperative defense of nest site after pairing; site ownership displays; ritualized greetings when mate returns to relieve incubating partner.	4. Nest sites and even partners are frequently usurped after establishment; pair bonds and ritualized pair interactions are not well developed even on a short term basis.	4. Long periods of one-bird incubation due to difficulty of obtaining food; thus less chance of pair bond reinforcement than in most sea birds. Moreover, changes of mates and nest site in successive breeding seasons preclude long term reinforcement of these attachments. Also, waves of new breeders disrupt those still raising young from previous year.

lished with J. S. Gartlan in *Nature* (1966). The present productive ferment has been stirred also by the controversial ideas of Wynne-Edwards (1962 and 1965) on the basic functions of social behavior. We will reserve Wynne-Edwards' ideas for the next chapter since they relate particularly to population regulation, and since, in any case, most socioecologists seem to have learned more through the stimulus of disagreeing with Wynne-Edwards than through acceptance of his explanations.

WHY SOCIETIES?

Before considering how environments may mold societies, we should ask what we mean by societies and why they should be so common in the animal kingdom. Most animal species possess nonrandom dispersion patterns, even in homogeneous habitats which present no obvious incentive to clumping or to even spacing. In Chapter 12 we referred to even spacing as evidence of anti-social behavior, which often it seems to be; but in a larger perspective this behavior represents a type of social interaction too, and one that integrates the population. If only random dispersion can arise by chance, then the nonrandom dispersion patterns set up by social interaction must be the products of natural selection. Therefore we may ask first about the fundamental adaptive advantages of social organization *per se*.

Aggregations versus Societies

A society can be defined as a group of organisms that recognize and respond to one another as individuals, not as anonymous members of the same species. In a society individuals have different roles, different territories or different capabilities, and these are recognized by other members and influence their behavior toward the individual.

Social groupings probably have their origin in the simple aggregation of individuals for individual benefit. An aggregation is a grouping of organisms who *do not* recognize and respond to one another as individuals, but who may benefit from a mutual proximity. Isopods (Fig. 12.3) are representative of organisms which aggregate but do not really form societies. There is no evidence that isopods recognize each other as individuals; rather, their degree of mutual attraction seems dependent on factors of the environment such as light and humidity. Isopod aggregation tendencies increase as the relative humidity decreases, and the benefits seem easy to discern: these gill-breathing land organisms are susceptible to desiccation, and by clumping together they lower the amount of body surface exposed to dry air and decrease their individual rates of water loss. Protection against cold seems sufficient explanation for the similar aggregation tendency of many snakes, which gather into tightly intertwined masses for their winter dormancy underground. Here again, the surface area exposed by the aggregate mass is far less than would be the individual surface areas of these snakes hibernating alone; thus they can better conserve their meager output of metabolic heat. This sort of "plural equivalent of Bergmann's rule" is met, too, in many truly social organisms: honey bees (whose hives run several degrees warmer than the winter air outside), birds and bats in communal roosts, sheep huddling close in a storm. Many such activities, even in social species, may be more aggregative than social in that the "huddlers" react very little to each other as individuals.

Except for family groups such as a mother catfish and her offspring, schools of fish seem also to be aggregations rather than true social groups. A typical school consists of but a single size class, possesses no particular leader and exhibits no distinguishable role differentiation. There must be benefits to the mutual attraction of "schoolmates" or schooling would not have evolved. While there is no hard evidence as to what the advantages may be, the benefit most often suggested is that schooling fishes derive from this habit a measure of protection from predators. Simple "safety in numbers" is not a sufficient explanation, because the larger the school is, the more predators should be attracted to it; but scattering by the members in all directions (the usual response when a predator approaches) is thought to have a confusing effect that minimizes predator success. Certainly significant to any explanation of schooling is the fact that it occurs primarily among open water species rather than bottom-dwellers. Presumably the latter can find hiding places when danger threatens and they even may be disadvantaged by schooling habits if the number in a school exceeds the capacity of nearby hiding places. But open water species have nowhere at all to hide, and if schooling adds to their survival chances it will be favored by selection.

Disadvantages of Clumping

Against the benefits we have suggested must be set the possible disadvantages of clumping. A flock of birds may be better able to find food than a single individual, but will there be enough to go round when it is discovered? Considerations such as this are invoked to explain why flocking bird species are so much more common in some habitats than others. The suggestion currently put forward, and backed by mathematical modeling, is that flocking is particularly favored in habitats where ample supplies of food occur in occasional, unpredictably located patches (Horn, 1968a). Obviously, this thesis also would explain the advantage accruing to predatory fishes, such as tuna and bluefish, which typically form schools of their own to hunt schooling prey: the prey represent large, unpredictable patches of food. But in habitats where food is widely and nonpatchily distributed, it may well be advantageous for animals not to form foraging groups but to act individually.

Social termites provide a good illustration of both the advantages and the disadvantages implicit in many forms of clumping. Termites have extremely permeable exoskeletons, and hence lose moisture rapidly to dry air. Furthermore, these insects require relatively high temperatures and considerable thermal constancy; hence they are restricted to lower latitudes. But even in the tropics, except for colony-founding expeditions during the wettest time of year,

> termites survive only because their elaborate social organization enables them to build nests in which they establish the microclimate suited to their needs. They live their entire existence in the closed environment of the nest and the tunnels that connect it to sources of food and water.
>
> Martin Lüscher (1961)

Significantly, the tropical species which are most widely distributed in terms of habitat type are those which form the largest colonies and build the largest mounds. Only the large mounds are sufficiently insulated to endure the direct solar heating and dryness of tropical savannah and grassland habitats, so species building nests of more modest proportions are restricted to the forest.

Lüscher found that large mounds maintain a remarkably constant internal microclimate, with temperatures hovering within a degree or two of the optimum for termite activity (30° C), and with relative humidities never lower than 96%. But the very elements which make this constancy possible (walls 0.5 m thick) also pose the threat of asphyxiation to the 2 million or more oxygen-demanding, CO_2-producing individuals inside. Only a remarkably engineered gas exchange system overcomes this problem (Fig. 13.2). Among termites it is very clear that potential disadvantages to extreme gregariousness had to be overcome at the same time that the obvious advantages of large colony size were being realized. Recalling our discussion of human cities, the same statement obviously can be made about them. But whereas termites have proceeded no farther in colony size than is commensurate with the need to breathe, cities like Los Angeles and Tokyo seem to be straining that particular limit.

Advantages of Social Differentiation

Termites are social species with social differentiation, but in terms of the advantages and disadvantages so far discussed (which are simply examples, not an exhaustive catalogue), we have treated them and other species mentioned as undifferentiated aggregations. What advantages beyond the sort available to aggregations can be advanced to explain the evolution of societies?

Broadly speaking, the special benefits of social differentiation and individual recognition seem twofold: (*a*) social differentiation facilitates specialization of individuals for different tasks, and (*b*) together with individual recognition it reduces levels of active, intraspecific competition within the group. If individuals or subgroups are functionally specialized, as are the castes in a bee or termite society, group cohesiveness is furthered by a heightened interdependence; cooperation tends to supplant competition; and tasks are performed by specialists with an efficiency probably exceeding that of a jack-of-all-trades individual performing the same tasks. We speak of an extreme when referring to bees or termites, for in no vertebrate society is caste differentiation and division of labor so sharp and in very few societies is active cooperation a major trait. On

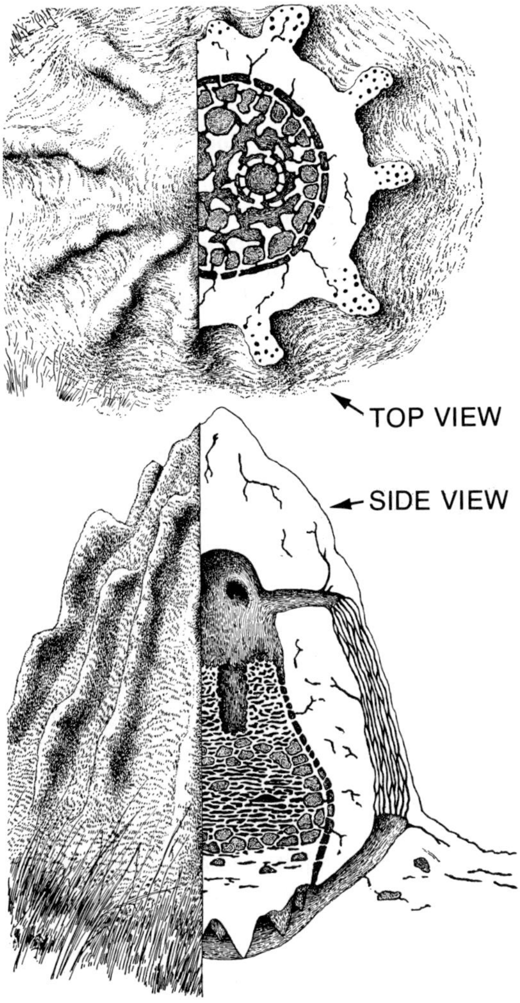

FIG. 13.2. Great mound of the termite *Macrotermes* may be several meters tall. In the conspicuous ridges of its exterior surface are many small air channels which connect to the inner chamber. Air in the chamber, warmed by the termites' metabolic activities, rises and is forced laterally into the air channels where it cools and sinks, entering the chamber for another circuit. Within the ridges diffusional exchange with the outer atmosphere removes CO_2 from the circulating air and replenishes it with O_2. The ridges provide a large surface area for exchange. (From M. Lüscher: *Scientific American*, July, 1961. Reproduced with permission from Scientific American, Inc. All rights reserved.)

the other hand, more roles may be defined in complex vertebrate societies (and advanced human societies seem by far the most complex) than among insect castes. Even where role differentiation is highly restricted, however, perhaps involving only the establishment of certain members as sentinels or chief defenders, a social group certainly benefits.

But social differentiation is more than apportionment of different tasks; it is also the establishment of recognized rights and restrictions among individuals. The differentiation of a society into a social hierarchy or a set of territorial property owners involves intraspecific competition that may be fierce while the process is occurring; but once individuals find a slot, recognized both by them and by others, competition is far lower than in the undifferentiated society. Competition is not reduced entirely, because most individuals retain upward aspirations in a hierarchial system or in a system of unequal territories; however, for the most part the *status quo* is accepted. Dominant individuals benefit at the expense of weaker ones from such systems of differentiation, but for the group as a whole this may be advantageous,* and even for the weaker members it may be advantageous much of the time.

To understand these benefits, consider the two forms that intrapopulation competition can take. Let us suppose this competition is for some resource such as food or nest sites. On the one hand there can be **scramble competition,** in which no individual claims any proprietary rights, but each obtains what it can from a pool available to all. If the resource is abundant relative to the demand, all members will be satisfied and competition will, in fact, be minimal. But when each individual has equal access to a *scarce* resource, disaster may ensue from a scramble. If the scarce resource is food no member of the population may get enough to live; if it is nesting space, suitable areas may become so crowded by breeders that survival of offspring is negligible in the whole population. Scramble competition is the sort one would expect in an undifferentiated aggregation.

On the other hand there is the possibility of **contest competition,** the sort more typical of well differentiated societies. Here competition is not so much for resources *per se* as for the

* Controversies have brewed in recent years over whether it is possible for natural selection to favor group advantages as opposed to individual advantages, *e. g.* whether it can favor forms of individual behavior that sacrifice the individual for the good of the group. We will return to this question later; here the suggestion is simply that the group benefits because most of its individuals benefit.

FIG. 13.3. Prairie dog at its burrow in the Wichita Mountains Wildlife Refuge in Oklahoma. Note the distinctive herbaceous vegetation ringing the mound. Several types of prairie dog predator are less common here than formerly, and this overpopulated town contains as much bare soil as edible plant matter. (Photograph by author.)

black-tailed prairie dog so unattractive—and vulnerable—to its most dangerous modern predator, the cattleman.

A number of years ago, I watched the demise of a prairie-dog town on the plains northeast of Greeley, Colorado, without realizing at the time the importance of social structure and population abundance in rodent survival. This town was complete with prairie dogs, burrowing owls, rattlesnakes, and cottontails. The "poison experts" got in their licks and killed all but about a dozen prairie dogs.

Now, I realize that the coterie system had been destroyed, that the widely distributed remaining prairie dogs were probably unfamiliar with each other, and that they were uncooperative because of the long distances between them. Also, they were too few in numbers—the thousand eyes that formerly located enemies were reduced below the critical minimum for warning against enemies.

In the second summer after the poisoning, wheatgrasses, pigweeds, and other tall plants grew faster than the prairie dogs could clip them down for good visibility. At last, only one prairie dog remained. I saw it sitting lonesomely on its mound in the late summer of 1949. I suspect that a coyote or a golden eagle ended its existence.

D. F. Costello (1969), *The Prairie World* (p. 96)

Comparisons within the Genus Marmota

Cousins of prairie dogs, the large ground squirrels known as marmots have been investigated thoroughly in recent years by David Barash (1974 and references cited therein). Barash began his work under John Emlen, who has catalyzed an unparalleled diversity of socioecological investigations at the University of Wisconsin. Barash first pursued an intensive study of the Olympic marmot which inhabits alpine meadows in the coastal mountains of Washington, then made comparative observations on several other species, including the common woodchuck (groundhog). Barash's findings are summarized in Table 13.2. The correlations he has found between social characteristics and environment are so excitingly consistent that after investigating two or three species Barash found he could predict very

TABLE 13.2

Socioecological Correlations in the Genus Marmota

Data are from Barash (1974).

Species	Where Studied	Length of Growing Season	Growth Rate of Young Marmots	Dispersal Age of Young	Breeding Frequency of adult ♀	Social Attributes
Olympic marmot (*M. olympus*)	Washington	40 to 70 days	Slow, matures in 3 years	2 years	Biennial	Highly social, living in distinct (though small) colonies. Greeting ritual performed frequently; free visitation among burrows. 2-year-olds are driven from colony only if they wander and return too infrequently to be recognized as natives.
Hoary marmot (*M. caligata*)	Montana	Short	Slow, matures in 3 years	2 years	Probably biennial	Similar to Olympic marmot. Colony consists of a dominant ♂ with a few adult ♀s, 2-year-olds, yearlings and pups. Closely integrated social structure. No territories or unique home ranges.
Alpine marmot (*M. marmota*)	Switzerland	Short	?	?	Yearly	Similar to Olympic marmot; greeting ritual less frequent, but other indices of strong sociality exist.
Yellow-bellied marmot (*M. flaviventris* (high altitude populations)	Colorado	Short (*c*40 to 70 days)	Slow?	?	?	Similar to Olympic marmot
Yellow-bellied marmot (*M. flaviventris*) (mid-altitude populations)	Wyoming, Colorado	70 to 100 days	Moderate, matures in 2 years	1 year	Yearly, though sometimes biennial	Adults moderately social: live in large adult colonies, but individuals have unique (though overlapping) home ranges. Social interactions infrequent, but greeting ritual occurs one to two times/day.
Woodchuck (*M. monax*)	Pennsylvania	>150 days	Rapid, often matures in 1 year	<1 year	Yearly	Adults solitary and aggressive; social displays and greeting rituals are minimal or lacking. Mothers drive young from nest at early áge.

accurately the degree of sociality he would find in species yet to be studied. His explanation for the social differences among marmots resembles Nelson's explanation for the distinctive social characteristics of frigate birds. Barash believes length of growing season largely determines the social differences in marmots by dictating the length of the juvenile dependency period. The eastern woodchuck encounters the mildest climate and longest growing season. The young grow rapidly, become independent at an early age, and are driven off by the mother at weaning; among adults there is virtually no social interaction. The Olympic marmot experiences a harsh climate with a short growing season; the young mature gradually over 2 years, establishing intimate social relationships among themselves and with their mother. Perhaps because of this long period of early socialization, the adults themselves live in colonies. Greeting rituals exist similar to the identification kiss of the prairie dog, and the colony is a tightly knit, highly interactive unit with a communal burrow system. Dispersal is not automatic, as in the case of the woodchuck, but occurs only if a colony member wanders away and fails to return often enough for ritual greetings and visitations; if it is absent too long, it is treated as a foreigner and rebuffed aggressively. Wandering of newly adult 2-year-olds is much greater at times of high colony density and is reduced when the population is low. This is a usefully flexible population-regulating device in the erratic environment of the Olympic Mountains, where winter mortalities vary sharply from year to year. After a severe winter there may be no spring dispersal of the few 2-year-olds that have survived. After a winter of low mortality and high adult reproduction, all 2-year-olds may disperse.

Marmot species living in similarly harsh environments are socially similar to the Olympic marmot and unlike the woodchuck. Barash was fortunate in being able to study one species, the yellow-bellied marmot, which occupies a range of elevations in the Rocky Mountains that defines a considerable range of growing seasons. Not only did he find a population that nicely displayed intermediate levels of sociality in an environment with an intermediate growing season, but he found a high altitude population of yellow-bellied marmots which was socially similar to the Olympic marmot and the other species experiencing short growing seasons. The *intraspecific* social variation in this species strongly implies that the correlations Barash found represent true environmental causation. When dealing with different species one can always argue that they have different genetic predispositions toward a particular social system. Different populations of the *same* species are genetically much more similar, and arguments about genetic predisposition ring slightly hollow when the social differences in these populations are as pronounced as in populations of different species. To cement his case, Barash hopes to study woodchucks in Alaska, predicting that populations of this species in a harsh northern environment will be much more sociable than populations in Pennsylvania.

House Mice and Population Density

A study of house mice by P. K. Anderson (1961) provides additional evidence of environment-controlled intraspecific social variation. This work has been overshadowed by the more dramatic demonstrations by Calhoun 1962 of social "pathology" in artificially crowded rat populations, but Anderson's paper is more significant because it describes natural populations as well as crowded laboratory populations. House mice "in the wild" exhibit a wide range of population densities reflecting habitats as diverse as meadows, stacks of corn stalks, and well stocked cellars (Fig. 13.4).

As shown in Figure 13.4, highly interesting social transformations accompany changes in population density. The best known type of social organization in house mice is the nonlinear dominance hierarchy, which is seen in very crowded populations. Here dominance is established among the males by violent combat, with one individual exerting despotic sway over other males and the latter assuming roughly equal subordinate rank. Females are nonagressive and nonhierarchical, although their activities are influenced by the male hierarchy.

At intermediate densities house mouse populations may be characterized by nonaggressive, relatively nonsocial behavior of the members ("togetherness"), while at low densities the characteristic social system is one of individual, defended foraging territories maintained by both males and females.

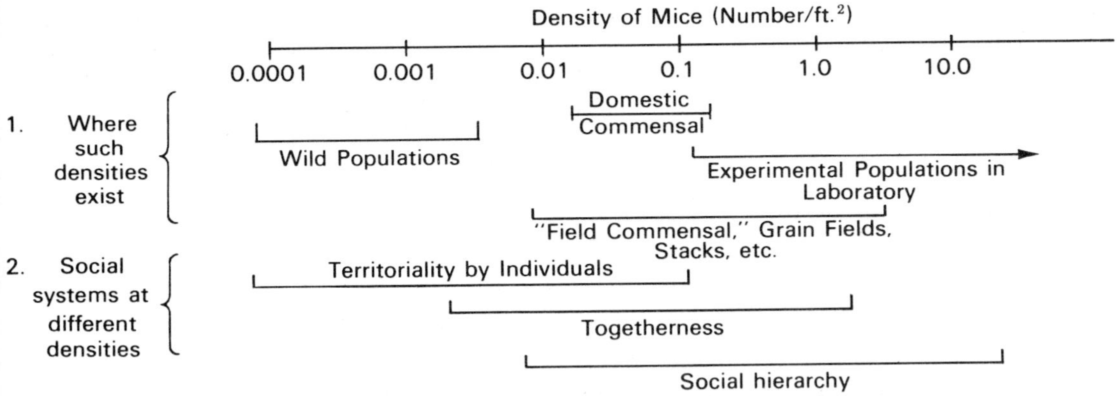

FIG. 13.4. Relationship of population density to social system in house mice. The "commensal" populations are those that benefit from close association with man. (Figure much modified from P. K. Anderson: *Transactions of the New York Academy of Science*, series 2, *23*:447–451, 1961.)

The correlation of social structure with population density in this example is suggestive, but other density-associated variables may also play a determining role. For example, in the dense, hierarchial populations there usually is lack of opportunity to emigrate; the habitat may be a cage, or may be physically removed from other suitable habitats (as are houses or stacks of corn shocks). On the other hand, territorial populations typically *are* able to emigrate; in fact, the young are expelled from maternal territories at weaning. It would be interesting to know, too, how important the abundance and patchiness of food is to the social structure of house mouse populations. In analogous studies it has been found that nectar-feeding African sunbirds set up defended feeding territories only when nectar-containing flowers are at an intermediate density. The territorial system disappears when flowers are so sparse that extreme effort and energy expenditure would be required to defend a territory large enough to feed an owner, or when flowers (or the nectar levels in each flower) are so abundant that the energy expenditure of territorial defense would yield no real benefit (Gill and Wolf, 1975; Kolata, 1975). Quite possibly the social organization of house mice responds to the same sorts of resource differences, rather than to population density *per se*.

PRIMATE SOCIOECOLOGY

Centrality of Social Life to Primates

In some ways man's nearest relatives, the apes and monkeys, are very unlike him. In technological ability, for example, termites or squirrels are considerably more specialized:

> Most primates do not shape their environment in any adaptive way. The sleeping nests of the large apes are poor, roofless constructions built for only one night. Monkeys use their sleeping lairs exactly as they find them, without making any artificial alterations. No higher primates are known to gather or store food, and therefore they have no stable home base which could accommodate a sick or wounded group member for a day or two . . . . The hand-to-mouth economy of primates forces every one of them to seek food and water every day. When natural resources fail, they starve.
>
> Hans Kummer (1971), *Primate Societies* (pp. 36 and 37)

Given these limitations, it is not surprising that nonhuman primates are restricted to relatively warm parts of the world. Man, in sharp contrast, has spread more widely than any other animal species.

Socially, too, most other primates are rather unspecialized:

> Among primates, every group member has to feed, drink, flee, and climb a tree for himself. There is no sharing or passing of gathered food or prepared shelter. Mutual assistance is negligible or absent. Only infants are carried and nursed. In such an economy, the group seems materially useless.
>
> Hans Kummer (1971), *Primate Societies* (p. 42)

Yet many primates *do* live in groups, and in a few cases there *is* a measure of cooperative behavior, though far less than, say, in a hunting pack of wolves. In fact, despite the limitations Kummer catalogues, it is possible to suggest that group living *per se* is a primary specialization of primates, for long association between young and old facilitates the learning upon which primates rely so much for survival. As with the prairie dog, sociability is as basic to the survival of many primates as is a properly functioning heart or digestive system. This has been demonstrated in the laboratory studies of socially deprived monkeys conducted by Harlow and his associates (*cf.* Harlow *et al.*, 1971).

Social Variety in Different Habitats

The twin sparks kindling my own interest in primate socioecology were a visit to the Washington Zoo and the reading of Crook and Gartlan's brief, seminal synthesis, *Evolution of primate societies* (1966). The zoo experience came as I walked up the hill to the entrance: before my eyes or nose told me I had arrived my ears already were filled with the raucous, echoing whoop of gibbons in the far-off monkey house; and all that day, as we made our way around the zoo, the gibbon's noise pollution followed us. Gradually a question formed in my mind. Why should gibbons dominate the eardrum? Why not gorillas? lions? hippopotami? And gradually an answer suggested itself: in the rain forests through which these apes swing so airily, vocal communication is an ideal mode. Visual signals are unsuitable, given the density of treetop foliage and the rapidity of gibbon movement; if members of a group are to keep in social contact with one another it may best be by sound. For the first time I really appreciated how closely a species' mode of communication is tied both to its environment and to the average distance between individuals. Yet one can see these associations everywhere: in the brilliant color patches and limited vocalization of baboons in their closely spaced troops on an open grassland; in the drab but highly vocal forest thrushes when compared to the colorful but more silent thrushes (*e. g.* bluebirds) of more open habitats; and in the intimate identification kiss of prairie dogs in their small-scale, spatially fixed "towns." "The Case of the Howling Gibbons" made me begin thinking of animal behavior in new, more habitat-oriented ways.

Crook and Gartlan's paper (1966) fueled this new perspective. The ideas they presented remain stimulating, although they require a degree of modification resulting from the avalanche of primate studies that has appeared since the mid-60s. Later reviews by Crook himself (1970) and by Eisenberg *et al.* (1972) have attempted an updating. Some of the major socioecological speculations of these workers are merged in the pages that follow.

Crook and Gartlan brought a phylogenetic perspective to their comparison of primate societies, but infused it with a consideration of habitat and dietary influences. The earliest primates lived in forests, and even today forests are the habitat of most primates. Forest foods include insects, leaves and fruits, and dietary specialization for one of these three food types seems an important social determinant. The earliest primates were insectivorous forest species, and the primitive species retaining these habits today are also the most primitive socially. Most such species are small, and live solitary lives except for brief consort pairings during the breeding season and relatively short-lived mother-young units. But in what way is their social simplicity related to their lowly evolutionary status? Perhaps one should think less of an inherent inability to be social than of a social conservatism related to diet. Insect food is scattered and not as abundant in a forest as plant food; moreover, much of it can be obtained by slow, careful probing under bark and through leaves at night, when many arboreal predators are inactive. The nocturnal habits of these small primates (which include prosimians such as the tarsier and galago and also certain lemurs) probably serve as a discouragement to sociality, but grouping into societies might serve little useful purpose for these organisms even if they were diurnal.

Adaptive radiation from this primitive primate stock has involved the evolution of leaf-eating habits by some forest monkeys, of fruit-eating habits by others, and—perhaps most recently—of more varied diets by still others, coupled with emancipation from forest habitats. These evolutionary changes have been accompanied by a general increase in size, which might be related either to more abundant food supplies (particularly for leaf eaters and omnivores) or to the adoption of diurnal habits with greater consequent exposure to predators. Ground-dwelling primates are particularly exposed to predators, and in these species considerable sexual dimorphism has evolved with males tending to be larger than females and specialized for defense.

We might expect the impetus to sociality and

the form such sociality takes to differ in these major adaptive lines. Consider the following contrasts. Leaf-eating species of evergreen tropical forests have an abundant year-round food supply. Preference for particular tree species confers spatial patchiness to this food supply, but once the location of favored trees is known they may be visited with assurance that food will be there. On the other hand, the generally omnivorous primates of open country depend heavily on roots and young shoots. If the climate is not too dry their food is more evenly distributed than that of specialist forest leaf eaters, but is equally abundant; in very dry habitats, however, the food of open country species tends to occur in predictable local patches but is not always abundant in these patches. Our third adaptive group, forest primates that specialize in fruit, utilize a food supply that is *both* patchy and unpredictable: in moist tropical forests there are always some trees in fruit, but the tree that yields a bounty one month will be barren the next.

These differences suggest social consequences. In the case of leaf-eating species a stable group life might be advantageous, younger members learning from their elders where to find favored trees and the whole group moving from tree to tree within a familiar **home range** (foraging area). This pattern in fact characterizes many leaf-eating species of both the Old World and the New, including the African colobus monkeys, the Asian langurs and the American howler monkeys. The home range of leaf-eating species need not be large to support a good-sized troop, but other troops would overexploit it if allowed to forage there also. The findings of Eisenberg *et al.* (1972) therefore seem reasonable: "We can correlate an arboreal, diurnal, leaf-eating niche with a species having a social structure that tends toward a uni-male system with a small home range and the employment of chorusing behavior to effect spacing." There is considerable variability about this pattern, partly because some leaf-eating species eat considerable quantities of fruit as well, and partly because different species are characterized by differing degrees of male-male tolerance. Struhsaker (1974), for example, describes the leaf-eating red colobus monkey of Uganda as living in heterosexual groups of about 50 members, including several adult males, but the related black-and-white colobus generally lives in smaller groups, several adult females and immatures being associated with a single male.

To species that depend heavily on fruits, it seems adaptive to live in highly fluid troops that break up into smaller groups to search for food trees; when food is found, the rest of the troop is called in by excited vocalization. Such a pattern has evolved independently in the Old World chimpanzee and New World spider monkeys. Chimpanzees, however, live today in rather varied habitats and their social organization seems to differ somewhat depending on local circumstances. Troops inhabiting extensive, undisturbed forest habitats seem to show particular fluidity; intratroop bonds may be relatively weak and individuals sometimes transfer allegiance to other troops. That this looseness is a corollary of the frequent splitting of such troops to forage is suggested by the more cohesive troop pattern described for chimpanzee populations exploiting more predictable food supplies (such as banana plantations).

The greatest social complexity and largest troop size to be found among primates occurs in the semi-terrestrial baboon societies of open country. Particularly in relatively moist, food-rich savanna habitats, baboons have evolved an unusual degree of male-male tolerance, which can be associated with a habit of active troop defense against predators. In drier habitats there tend to be smaller, uni-male foraging systems; here predators may be equally threatening, but the safety of large-troop living has yielded to the reality of a smaller, more patchy food supply.

We thus see among the primates many relatively clear-cut instances of convergent social evolution in similar ecological circumstances, and divergent social evolution in different ecological circumstances. Habitat differences clearly are part of the story, but habitat *per se* seems less determining of a social system than total way of life. For as we see in the next section, interspecific social differences may characterize primates coexisting in the *same* habitat, if they rely on different foods and different ways of dealing with predators.

Social Variety in a Single Habitat

Let us consider three species, the yellow baboon and the patas and vervet monkeys (Fig.

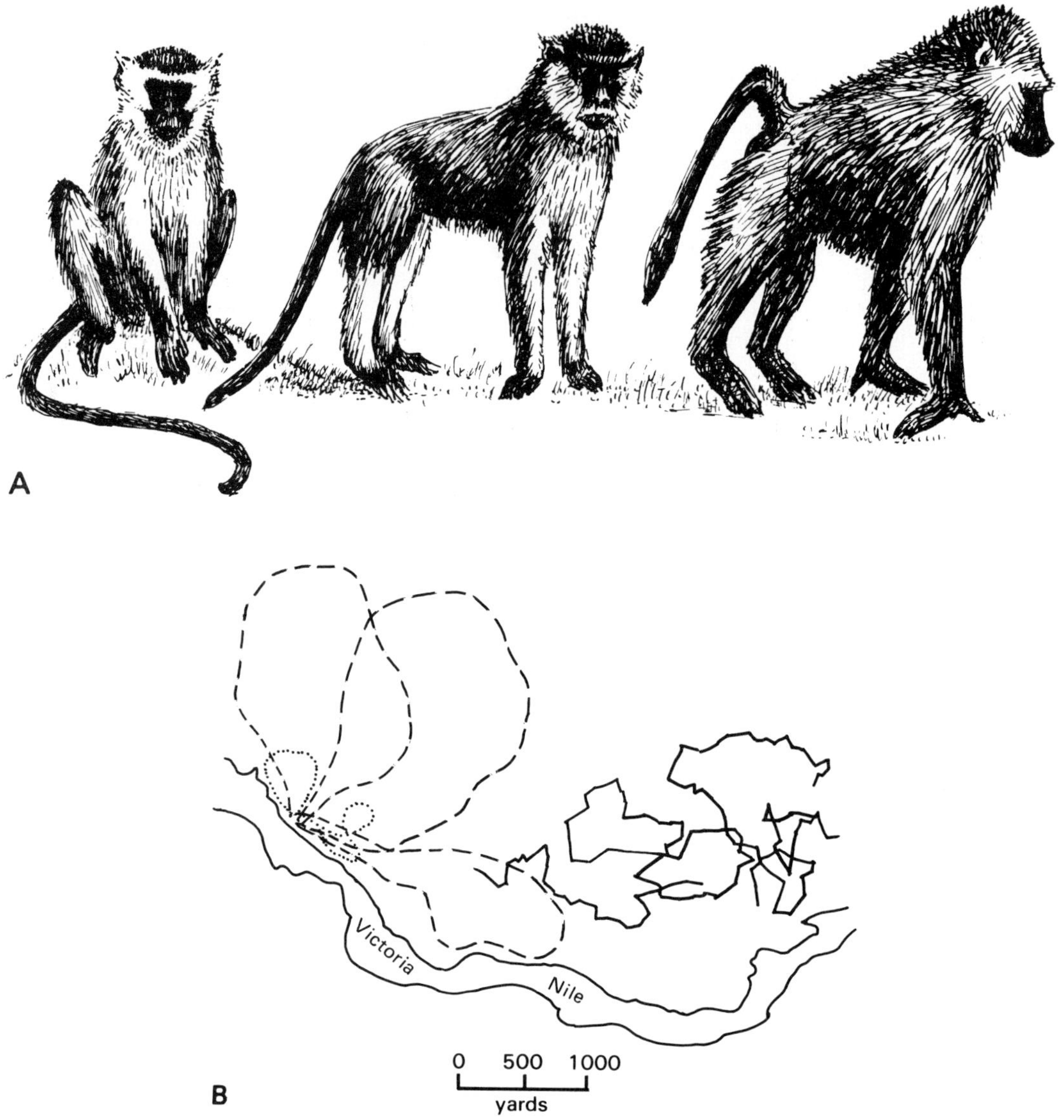

FIG. 13.5. (*A*) *Left to right:* vervet and patas monkeys and yellow baboon (male), showing differences in size and physique. (Drawing by Sarah Allan.) (*B*) As described in the text, the daily travels of each species differ in character. Vervets (*dotted line*) stay in group territories adjacent to the river. Baboons (*dashed lines*) venture well into the savanna but return to riverine forest trees to sleep. Patas monkeys (*solid line*) remain in the savanna day and night. (From K. R. L. Hall: *Journal of Zoology, London 148:* 15–87, 1965. Reproduced with permission from the Zoological Society of London.)

13.5*A*), which often live in neighboring troops on the savannas of East Africa. The term "single habitat" must be used loosely, because these savannas really are a mosaic of grassland and scattered trees, with more continuous tree cover ("gallery forest") along watercourses. It is a single macrohabitat, within which the three species typically distribute themselves somewhat differently.

Social contrasts between these species are pronounced (Table 13.3). Patas monkeys live in small, one-male groups, vervets in larger multi-male groups, and yellow baboons in still larger, highly cohesive and complex multi-male groups characterized by considerable role differentiation. Of the three, only vervets forage within territorial boundaries which are defended against other troops. Why do these differences

TABLE 13.3

Comparisons among Vervet Monkeys, Patas Monkeys and Yellow Baboons Living in a Savanna-Forest Mosaic

Data are from Struhsaker (1967a and 1967b); Hall (1965); Washburn and Devore (1961)

	Vervet (*Cercopithecus aethiops*)	Patas (*Erythrocebus patas*)	Baboon (*Papio cynocephalus*)
Life style	Semi-terrestrial, but spends considerable time in trees of gallery forest	Highly terrestrial; usually found in open savannah	Intermediate between patas and vervet
Food	Beans of acacia trees, various ground foods	Varied foods near ground: grasses, seeds, insects	Varied; roots important
Size of individuals	Small	Fairly small	Large and formidable, especially males
Sexual dimorphism	Slight	Marked; males larger, conspicuously colored	Very marked; males much larger, with large canine teeth
Group size and number of adult males	24 (average); several males	Smaller than vervet (9 to 30); only one adult male	Large: up to 185 members. Numerous males
Group sub-divisions	Slight	None	Main troop contains many subgroups differing in age, sex and role
Hierarchy within group	Single ± linear hierarchy	Perhaps weakly developed in adult females; the single male is usually the leader	Marked and complex; subgroups may have their own hierarchy
Defended group territories?	Yes	No	No

exist? Let us examine each species and its habits in more detail.

Baboon troops are the most socially complex and highly integrated of any primate except man. Since these troops often are very large one might explain their complex organization simply as a function of size. But this explanation is insufficient: equally large, *unorganized* aggregations are found in many animal species. Rather, both the size and the organization of baboon troops seem related to their aggressive response to potential predators. The troop actively defends itself when attacked, but intimidates most predators from even initiating attacks simply by its size and cohesion:

> As the troop moves the less dominant adult males and perhaps a larger juvenile or two occupy the van. Females and more of the older juveniles follow, and in the center of the troop are the females with infants, the young juveniles and the most dominant males. The back of the troop is a mirror image of its front, with less dominant males at the rear. Thus, without any fixed or formal order, the arrangement of the troop is such that the females and young are protected at the center. No matter from what direction a predator approaches the troop, it must first encounter the adult males.
>
> S. L. Washburn and Irven DeVore (1961)

Individuals stay close to each other; baboons cannot run far without becoming winded, and a straggler in the open is highly vulnerable.

It thus appears that the size, cohesion and differentiation of baboon troops, as well as the size and weaponry of individual males, all are linked to the threat of predators in an open habitat, and to a particular strategy—aggressive defense—of dealing with predators. This strategy, in turn, may be related both to the feeding habits of baboons—digging roots takes time, and the group moves slowly—and to their limited running ability. What is cause and what is effect is difficult to state, but clearly what exists is an adaptive complex.

Patas monkeys must cope with predator threats even more severe than those encountered by baboons. They move far into the open grassland and remain there even at night, when baboons and vervets retire to tall trees in the gallery forest (Fig. 13.5*B*). Unlike baboons, patas groups deal with predators by concealment and distraction: typically small and usually rather widely dispersed, a group moves quite rapidly through an area, feeding on easily gathered, above-ground food. The single adult male usually scouts ahead, and if a predator appears he typically carries out a diversionary display, leaping and bouncing conspicuously while the females and young crouch motionless in their dispersed positions. If this tactic fails the group takes flight: patas are slim, long-legged, and probably the fastest runners of any primates.

Patas males are highly intolerant of one another, ensuring that the groups remain small. Maturing males split off and form their own troops. Given the small stature of the patas, a strategy of concealment and/or flight from predators is not surprising, and a small, dispersed group is less conspicuous than a large, cohesive one.

Vervets stay closest to the tall trees of the gallery forest, and scramble up these for safety. Smallest in size, they are less able to outrun a predator than the lanky patas, and unsuited to the group defense tactics used against predators by baboons. But vervets *do* defend their foraging territories against other vervet troops, perhaps because these foraging areas are relatively small and the number of trees bearing acacia beans (a favorite food) is limited.

Thus we see three quite different primate societies operating within the same locality, each being highly adapted to a particular way of life. Each social system becomes comprehensible and even reasonable when viewed as a trait that has evolved by natural selection in company with the other traits of each species.

BREEDING SYSTEMS OF BIRDS

Major Types of Mating Systems

Studies of the diverse types of avian breeding unit reveal ecological correlations similar to those characterizing the more permanent social systems we have discussed for mammals. Nelson's study of the great frigate bird has introduced us to this topic already, but to illustrate further the types of questions presently being raised about avian breeding systems, we will describe studies of the polygynous system of the bobolink and the lek of the prairie chicken. First, however, let us survey the full spectrum of avian mating systems (Table 13.4). Monogamous systems are far more common among birds than multiple-mate systems, and can be considered "conservative" mating systems in that two parents tend to lavish attention on relatively few young, *i. e.* those produced from one or two clutches of eggs in a given breeding season. The young of such monogamous matings typically are altricial (helpless at birth), and paternal assistance doubtless helps to promote their survival. The most conservative monogamous systems among birds are communal enterprises: each nest is tended not only by the parents but by nonbreeding adults belonging to the parents' "clan;" in this system very few young are raised in proportion to the number of adults present. Monogamy in general, and particularly communal monogamy, often seems associated with relatively mature, stable habitats having fairly stable bird populations. Birds of such habitats are likely to be equilibrium species, and their monogamous system seems geared to persistence *in situ;* relatively few young are produced, but these are carefully nurtured.

Polygyny and polyandry, on the other hand, are opportunistic breeding systems involving less parental care but more offspring. To be successful, these systems require habitats relatively rich in food, where the young do not require the attentions of both parents. In addition, it obviously is advantageous if the young are precocial (able to forage on their own soon after hatching, like the domestic chicken), and the bird species employing these mating patterns typically are precocial. In polygynous systems relatively few males monopolize the breeding activity. In polyandrous systems the males, in a fascinating role reversal, assume the

TABLE 13.4

Major Types of Avian Mating System

Type	Description	Other Aspects	Frequency of Occurrence	Examples
A. Monogamy (only one mate)				
1. Solitary	A single pair nests by itself	Solitary nature usually enforced by territorial defense	Very common	Most songbirds; hawks
2. Colonial	Dense associations of monogamous pairs	Usually there are small defended territories (Fig. 12.2)	Very common, especially where suitable breeding sites are limited	Many sea and shore birds: gulls, herons, etc.
3. Communal	One breeding pair is assisted by other nonbreeding individuals in raising young	Nesting territory is defended by the group, and ownership is probably passed from generation to generation within the group	Rare	Many species of jay (Brown, 1974)
B. Polygyny (>1 "wife")				
1. Simultaneous	Male has several nesting mates simultaneously, maintaining a bond with each	Male may defend a territory including several nesting females, or may simply maintain a harem without a real territory	Not common; usually in open, patchy habitats	Bobolink, red-winged blackbird (Verner and Willson, 1966)
2. Promiscuous	Male may mate with many females, but otherwise is not associated with them; no real pair bond formed	Males set up small territories in a mating arena (lek) and compete directly for females. Often the males are very showy; young are typically precocious	Not common; typcially in grassland habitats	Prairie chicken and prairie race of wild turkey; sage grouse
C. Polyandry (>1 "husband")				
1. Sequential	Female lays a clutch of eggs for male to incubate, then moves to another nest; may incubate last clutch herself	Young are precocious	Rare	Spotted sandpiper (sometimes); northern phalarope; rhea sanderling (Jenni, 1974)
2. Simultaneous	Female maintains pair bonds with all mates incubating her eggs.	Female may defend a large territory in which several mates are nesting; males defend own nest areas; young are precocious	Very rare	American jacana (Jenni and Collier, 1972; Jenni, 1974)

function of incubating eggs and tending young. This allows specialization of the female for egg production; sometimes much larger than her several mates, she produces many offspring in a single breeding season. The opportunism of these systems, particularly polyandry, lies in a maximization of reproductive output that results, in favorable years, in rapid growth of the population.

It is tempting to suggest that polyandrous behavior, in which a female deposits eggs in several nests for her mates to incubate, is the evolutionary antecedent of the "nest parasitism" of cuckoos and cowbirds, whose females lay eggs in the nests of other species.

Ecological understanding of avian mating systems has been especially advanced through comparative studies of closely related species. Investigations involving sandpipers and tropical weaver birds exemplify this approach. Certain species of sandpiper practice solitary monogamy, others sequential polyandry, still others polygyny of one or another type; and this variety has made possible an especially revealing comparative study by Pitelka *et al.* (1974) which supports the socioecological correlations of Table 13.4. Spotted sandpipers even exhibit distinct *intra*specific variations in mating system, populations having an excess of males apparently being polyandrous but populations with a more even sex ratio being monogamous (Jenni, 1974). Among weaver birds Crook (1964) found the following correlations: species inhabiting poorly lit forest and eating insects typically are monogamous and territorial. These species build solitary arboreal nests and engage in active pursuit during courtship. Weaver species of well lit savannas eat seeds and grain and nest in conspicuous arboreal colonies (Fig. 13.6); their courtship is relatively static (within the colony) and the breeding systems are polygynous. Still other species, of purer grassland areas, breed monogamously on the ground in widely spaced territories even though they feed together in flocks. Crook thinks that territoriality here may provide protection against predation, by keeping the vulnerable nests few and far between. Alone among weavers, these species engage in conspicuous individual courtship acrobatics, displaying in the air well away from the nest. The need both to attract a mate and to keep the nest site hidden may explain why aerial courtship evolved in these species.

FIG. 13.6. Nests of a weaver bird colony in Uganda. (Photograph by author.)

Simultaneous Polygyny in Bobolinks

A study by Stephen Martin (1974) illustrates well the sorts of ecological questions that can be raised about a social system, and the way such questions can be investigated. Martin studied a nesting population of bobolinks—small, insect- and seed-eating birds of the American Midwest—in a Wisconsin meadow. He knew at the start of his study that bobolinks practice simultaneous polygyny, with one male defending a territory occupied by two or more nesting mates. Polygyny in the bobolink is somewhat surprising, since the young are altricial and require substantial parental feeding; furthermore, the harem tactics of dominant male bobolinks altogether exclude many other males from breeding and from helping the females raise young. Bobolink habitats, however, seem to be the type in which simultaneous polygyny arises most often: they typically vary locally in their suitability for nesting, but provide sufficient food in favorable patches for several broods of offspring. If a male can establish territorial rights to a favored area several females may be induced to nest there as his mates. But such an arrangement can only be adaptive, Martin reasoned, if the nesting success of these females is greater than it would be if they paired monogamously with all the males, utilizing less favorable habitats when necessary but providing their offspring with the full attentions of two parents. Martin could not test this hypothesis completely, because there was no monogamous control population; but the behavior of polygynous male bobolinks does provide a control of sorts in that they typically lavish attention on the first mate and her brood while neglecting the nests of later mates. By comparing the success of primary nests with those of later ones, Martin could therefore learn something of the reproductive cost of polygyny. Not surprisingly there *is* a cost, but Martin discovered a number of secondary adaptations that keep it relatively low, perhaps less than the cost of monogamous nesting in poorer habitats.

Martin's methods typify the sorts of procedure used by socioecologists: long, careful observations of natural populations which are disturbed as little as possible. He was interested in comparing egg loss and hatchling mortality in primary nests *versus* other nests, but he also attempted finer comparisons: of the rate of food delivery to nestlings, of the kinds of food delivered, and of nestling growth rates. Growth measurements required handling of the nestlings, but they were carefully returned to the nests after each weighing. Food delivery rates were determined by observation with binoculars from strategically placed towers, and food types were recorded (in 4000 photos) by cameras mounted several meters from the nest and triggered from the towers when parents returned with food items. Martin's 1974 paper did not include any data on food differences, but did contain records of food delivery rates and hatchling growth rates (Fig. 13.7). Martin found that unassisted secondary females foraged very hard within short distances from the nest, and made almost as many food deliveries as did the paired male and female attending the primary nest. But the primary pair foraged much farther afield, and perhaps obtained better food items: as can be seen, the weights of primary nestlings were consistently higher than those of secondary nestlings. Some of the latter starved, in fact, and survival to fledging was only 49% in secondary nests compared to 64% in primary nests. Both these percentages are high, however, compared to average nestling survival among comparable monogamous species. Thus the polygynous system seems to pass the "success test," even though secondary females are somewhat less

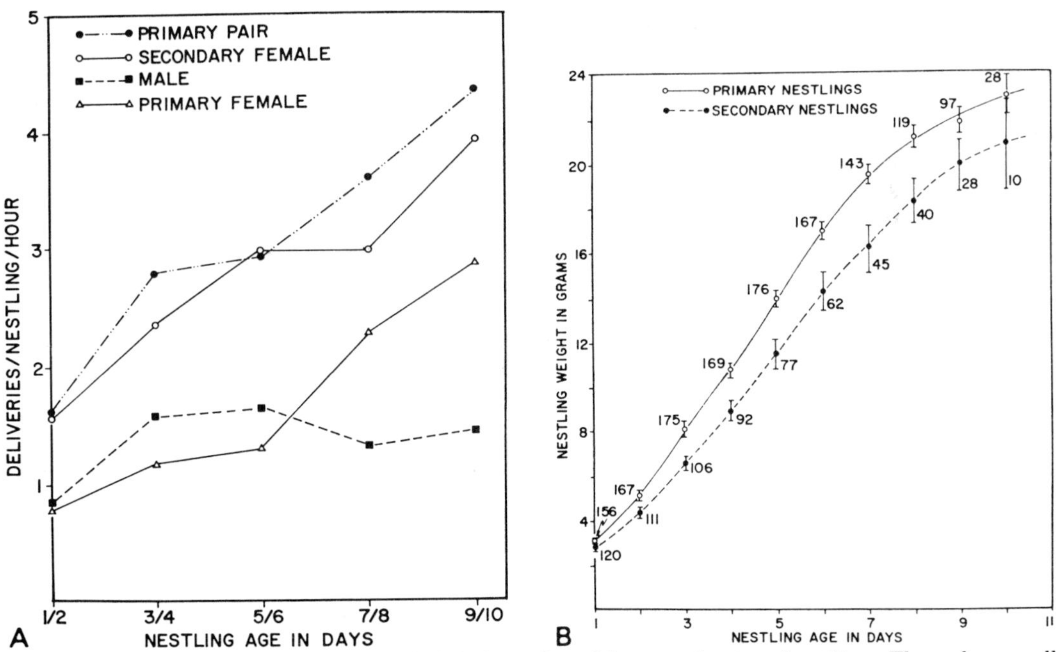

FIG. 13.7. Foraging activities of nesting bobolinks and resulting growth rates of nestlings. The male normally spends all his time assisting the primary female, while the secondary female must feed her nestlings alone. Despite her great individual effort (*A*), her nestlings are less well fed than those of the primary nest (*B*). The numbers beside each point in *graph B* are the number of nestlings weighed at that age. (From S. G. Martin: *American Zoologist, 14:* 109–119, 1974. Reproduced with permission.)

successful than primary pairs in raising young.

What explains the relatively great success of even the secondary females? For one thing, they average one less egg per clutch than primary females (4.8 *versus* 5.5), so ordinarily have one less mouth to feed. For another, bobolinks are characterized by an asynchrous hatching schedule which seems to function as a hedged bet against hard times. The last egg is laid a day after incubation begins, the result being that one nestling is smaller and younger than the others. If food is abundant the runt survives; if not, it soon starves and there is more for the others. This can be viewed as a form of unequal scramble competition that reduces the danger of the whole clutch succumbing. Finally, Martin discovered that while males *ordinarily* feed only their primary brood, they are alert to crises among the other broods; for example, if a secondary female hatches an unusually large number of nestlings and all are in danger of starving despite her attentions, the father will lend assistance at the secondary nest. Thus four notable features—industrious foraging and lower egg production by secondary females, asynchronous hatching of offspring, and a degree of flexibility of male parental behavior—can be viewed as adaptive correlates of bobolink polygyny, the polygyny itself being an adaptive correlate of a habitat that is quite patchy in its concentration of food resources.

Lek Mating System of the Prairie Chicken

In the highly promiscuous polygyny of lek systems, males gather at a tight, territorially subdivided breeding "arena" and try to entice females into breeding with them by elaborate displays. This is a mating system in which no lasting pair bonds are formed; the female departs the arena to nest and raise her precocial young alone, or with a small group of other females.

Lek systems seem particularly characteristic of ground birds inhabiting open grasslands. In America, several species of prairie grouse breed in leks, whereas the ruffed grouse of eastern forests does not (Wiley, 1974). A similar but *intra*specific contrast exists between prairie and forest populations of the wild turkey (Watts and Stokes, 1971). These authors believe that lek breeding benefits prairie turkeys in two ways: nonbreeding males on the fringes of the arena provide warning against predators, and the excitement associated with mass courtship helps to telescope the population's breeding activity into the brief, late winter rainy season. The chicks are born during the subsequent short-lived flush of vegetable and insect food. Interestingly, a few bird species of tropical woodland (*e. g.* birds of paradise) also breed in leks, and these typically are fruit eaters whose period of abundant food, like the prairie turkey's, is seasonal. A mammalian instance of lek breeding occurs in the case of Uganda kob, an antelope species of open savannas which is subject to considerable predator pressure (Buechner and Roth, 1974).

The lek system of the greater prairie chicken (Fig. 13.8) serves as an interesting case study. The locations of this species' mating arenas (leks) are as traditional as small town lovers' lanes: year after year, successive generations return to court on the same turf. Robert Robel and his associates at Kansas State University have studied mating activities at three arenas since 1964 (Robel, 1972; Robel and Ballard, 1974), and describe the system as follows. Males arrive at the breeding ground early in March and set up a social hierarchy which results in a division of the arena into territories, inner territories being claimed by the most dominant individuals and territories near the arena's fringe by less dominant males. Still other males are excluded from the arena altogether: the three leks studied by Robel *et al.* accommodated only about half of the adult males in their study area.

Females appear at the lek in flocks of 6 to 12 during late March and early April, but little mating occurs until the middle of April. When ready to mate, females come to the lek in smaller flocks (1 to 6 birds) possessing their own hierarchy, the dominant hen aggressively preventing other females from mating until she has mated herself. This is possible because the two dominant males (α and β) are highly favored as mates and the females generally wait until one of them is available. On the three leks studied by Robel's group, 6 males (the α and β individuals of each lek) accounted for 92% of all the observed copulations. When, in an experiment, the 3 most dominant males of one lek were removed, 3 nonbreeding males quickly entered the arena and a redistribution of territories

FIG. 13.8. Mating arena (lek) of the prairie chicken. Courting or fighting males inflate conspicuous bare throat sacs, raise their prominent head feathers, and strut with ruffled feathers and spread wings. The overall effect is to make them appear much larger. After mating, the females leave to raise their broods alone. The males remain on the lek until the breeding season ends. (Drawing by Sarah Allan, based on an exhibit at the American Museum of Natural History. Courtesy of the American Museum of Natural History.)

occurred. Much confusion resulted, however: there was a great deal of fighting among the males throughout the remaining 6 weeks of the breeding season, and no really stable hierarchy was established. Matings were greatly inhibited during this period because other males on the lek harrassed courting pairs, and the investigators concluded that a stable hierachical system of males is very important to successful reproduction. Among females, reproductive success proved to be strongly correlated to earliness of mating, which in turn may be partly a function of the female's dominance position. Forty-four per cent of nestings begun before May 5 were successfully completed, while only 10% of nestings begun after May 5 were successful. Increased predator pressure on late nestings is believed to be part of the reason for this trend. Moreover, clutch sizes (number of eggs per nest) were found to decrease as the season progressed. Thus, in this lek system, recruitment to the population was largely the work of early breeding females and of a very select group of males. The theoretical genetic advantages of such restriction have been suggested already. Studies by other investigators suggest, however, that breeding is a less exclusive affair in some prairie chicken populations than in the one Robel studied. But whatever the advantages of lek systems in terms of increased genetic fitness, we still must ask why lek systems occur only rarely, and in particular ecological circumstances.

In the case of the paririe chicken, Robel's records of low and highly seasonal reproductive success suggest a need for efficient, early mating if the population is to succeed. Undoubtedly significant, too, is the high mortality of prairie chickens later in life: as noted by Robel and Ballard, their mean longevity in Kansas is only 1.6 years. Low survivorship suggests an obvious advantage to early maturation of this species, an advantage that is particularly great for females. Since a few males can service many females, and since males are not needed in rearing the young, there should be less selective pressure for early maturation of males, and indeed male prairie chickens (like males of most members of the grouse family) mature at a later age than do females (Wiley, 1974). Given a high mortality rate, this means that the breeding population normally has an overabundance of females. Another advantage of the lek system is thereby suggested: it is a system that permits all females to mate, even when relatively few breeding males are present.

In sum, the adaptive significance of the prairie chicken lek probably lies in the way this mating system helps the population to offest those aspects of the environment which limit the suitable breeding season and sharply curtail both nesting success and later survival. The prairie environment is drought-prone, and hazardous for young and adults alike, but for a brief season each year conditions are less inimical to reproductive success. The lek system of the prairie chicken seems to facilitate maximal production of young during this time.

SOCIAL COOPERATION AND ALTRUISM

There is an alternative to the scramble competition of nonsocial populations and the contest competition of socially organized ones. It is *sharing*—social cooperation rather than competition. Sharing implies an apportionment of resources according to need; at least ideally, it minimizes the wastefulness of dominance systems, where during stressful times some individuals continue to luxuriate while others starve. Sharing requires cooperation—organisms acting together for mutual benefit rather than individually for individual benefit—and cooperation often involves deferred rather than immediate rewards to the individual. In one sense highly cooperative social systems represent the pinnacle of sociality, yet so rare are such systems that one suspects they are favored by natural selection only in special circumstances.

It is important to distinguish forms of cooperative behavior which are mutually beneficial to the participants from the type of behavior known as **altruism** in which individuals seem to willingly sacrifice their welfare or even their lives for the good of others. Evolution of the former is easily explained: any trait that benefits the individual is selected because individuals possessing that trait leave more offspring, as a rule, than those not possessing the trait. Individuals with genes for cooperative traits therefore increase proportionally in successive generations if cooperation meant greater reproductive output. But could true altruistic behavior evolve in this way? Some biologists do not believe that self-sacrificing behavior ever could be favored by selection at the level of the individual, and have suggested that altruism could only have arisen through selection favoring particular group characters **(group selection)** rather than operating only through the success or failure of individuals. These workers seem to believe that altruistic individuals in a population would leave fewer offspring than selfish ones, and hence selfish traits would always be favored by selection over self-sacrificing ones. Let us consider this problem briefly before considering the more straightforward subject of cooperation for mutual benefit.

The writings of Wynne-Edwards (1962 and 1964), the chief modern champion of group selection, suggest one approach to the question of social altruism. His ideas have stimulated vigorous dissent by Lack (1966), Wiens (1966), and many others. Much of the controversy revolves not about altruism at all, but about social traits that seem very useful to most of the individuals practicing them. For example, the lek mating system serves a prairie chicken population well, but if it excludes most males of a population from breeding, it represents a *disservice* to these males at the individual level. If so, how could lek behavior have evolved through individual selection? The obvious answer is that it could *easily* evolve through individual selection, since some individuals *do* benefit albeit at the expense of others. In "traditional" natural selection, *i. e.* selection at the individual level, one does not expect all members to benefit. Intraspecific competition is one of the agencies by which unequal reproductive success, the basis of evolutionary change, is accomplished. Operating through its benefit *to some individuals*, lek behavior incidentally benefits the group as a whole, just as different mating behaviors result in overall group advantages in other environments. The group's adaptive features are molded through individual traits that benefit some, not necessarily all, members of the group. In a lek system unsuccessful males are not sacrificing themselves; they are being *unwillingly* sacrificed in a competitive struggle that benefits other individuals. One of the criticisms of Wynne-Edwards is that he tends to view social systems like the lek as products of group selection, when the logically more attractive mechanism of individual selection is a perfectly adequate explanation of the phenomenon.

But can group selection be denied when we are asked to explain true self-sacrifice, altruism in its real sense? The answer frequently is yes. Consider the most widespread form of social altruism, namely the self-sacrificing behavior of parents with regard to their offspring. Parents may contribute their lives in fruitless defense of their young, when they themselves could easily escape. This form of altruism is quite easy to explain by individual selection if the behavior, *on the whole*, results in more of an individual's young surviving to the next generation; the young will carry in their genes the same altruistic impulse, and it will spread through the population via their preferential survival over the offspring of parents which were *not* altruistic. Altruistic parental behavior obviously is more likely to evolve in species whose members raise relatively few young in a lifetime, and is less likely to evolve in species whose members raise many broods. For species of the first type, one set of offspring represents a considerable part of the parent's total reproductive contribution, *i.e.* an investment worth defending. But in species whose adults breed often and prolifically parents would, *on the whole*, leave more successful offspring through a policy of abandoning a threatened brood and trying again. Individual selection might favor self-sacrificing parental behavior among animals of the first type, but not those of the second.

Altruism that extends beyond the parent-offspring relationship is more difficult to explain without recourse to group selection ideas. For example, how does one account for cases in which adult individuals do not breed themselves, but assist other adults to raise their young? The helpers do not *seem* to be acting in a way that maximizes their own genetic contribution to the population. Nevertheless, currently favored arguments suggest that they are. Selfless helpers in many cases may be young adults, whose child-rearing success may be relatively low at this time, and whose help may represent a net individual benefit in that their experience results in their being more successful when they themselves reproduce later. Or the helpers may be old, postreproductive relatives of the breeding pair, maximizing their own genetic contribution to the population by assisting their descendents—not their own offspring, but their children's children. In fact, communal altruism of this sort seems in most cases to be explicable by an extension of our explanation of parental altruism. This extension has been termed **kin selection;** it is perhaps the only type of true group selection that need be postulated for any of the known examples of social altruism. Kin selection is not a whole new form of natural selection (as is the Wynne-Edwards version of group selection); to the idea of selection acting at the level of the individual it simply adds the idea of selection acting at the level of genetic lineages.

To illustrate this idea, and also to indicate the sorts of ecologic situations that might favor kin selection, we may consider Brown's ideas (1974) regarding the evolution of communal breeding in jays. Populations of the Mexican jay, for example, are characterized by relatively few breeding pairs but each pair is assisted by nonbreeding adults in raising their brood. Not all species of jay exhibit this form of social altruism; it tends to characterize those species occupying stable, climax vegetation types rather than transient or early successional habitats. The Mexican jay, one of the most altruistic species, is found in mature pine/oak woodland, and seems to be an equilibrium species adapted for persistence in this stable habitat rather than for colonizing ability or simple fecundity. A strategy of fewer offspring and better adult care would probably serve the species better than would the opportunistic strategy of many offspring and minimal care. Moreover, the equilibrium habitat of the Mexican jay is patchy: it contains favorable and not so favorable areas in terms of suitability for jays. Hence, there could be an advantage in members of the population laying claim to particular parcels of favored territory. Kin selection might ensure that this claim is permanent: "the most productive territories of an area are handed down from generation to generation within the ownership of the same genetic lineages or clans" (Brown, 1974). But the communal altruism of kin selection would only arise, Brown reasons, if individuals of the clan could, on the average, maximize *their own* reproductive success by remaining within the favored territory "in a subordinate, helping condition but ready to breed should the parent disappear." Once they became breeders, of course, the altruistic tradition in the clan would benefit *them.*

But Brown visualizes a further evolutionary step, in which benefit to the individual might no longer be selected for, but rather benefit to the clan. A well established, inbred clan would represent a collection of similar genetic traits in genetically similar individuals, whose shared traits would be successfully passed on *whether or not* all clan members were reproductively successful as individuals. Whatever traits then favored the perpetuation of the clan would be those that were selected, and these could include traits of communal, not just parental, altruism.

Social altruism, then, resolves into a special form of selfishness in which benefits to the individual may be deferred, or may be realized (in the case of parental sacrifice) only in the evolutionary sense of the word "benefit:" maximization of the individual's contribution, or in the case of advanced kin selection, of a clan's collective contribution, to the population's gene pool.

Now let us consider cooperation that provides *mutual benefits* to members of a society. Such cooperative behavior needs no intricate evolutionary argument to explain its existence. Obviously, selection for individual benefit is sufficient explanation. Here, then, we can proceed directly to the question of the ecology of cooperative behavior. Clearly, cooperation for mutual benefit is most likely to arise in societies where the potential benefits of cooperation are great. What may not be immediately clear is that sometimes the potential benefits are minimal. Consider first the case of group-hunting mammals. Among vertebrates, carnivores such as wolves, lions, Cape hunting dogs, and spotted hyenas display unequalled levels of social cooperation despite their fierce way of life and fearful weaponry. Cooperation is especially useful to these species because of where they live and what they feed on. Their habitat, for the most part, is open country and their prey consists of organisms which frequently are as large or larger than themselves. To capture a caribou on the tundra wolves must be able not only to run it down but to kill it. By taking turns in the chase, and by assaulting the quarry together, both needs are achieved and food is obtained. Likewise lions, heavy-bodied and relatively clumsy, are able to set ambushes in long grass and capture prey by the cooperation of "drivers" and "lurkers." On the other hand, carnivores which hunt smaller prey, either by speed (cheetahs) or in forest habitats where individual stealth is more possible (foxes) possess only minor social tendencies, let alone a well-developed cooperative system. It is not difficult to visualize strong selective pressure toward cooperation among wolves or lions, but weak or even negative selection for cooperation in cheetahs or foxes.

Compared to group-hunting carnivores, societies of nonhuman primates are quite uncooperative. Foraging is an individual matter, or one controlled by social dominance. Mutual grooming *is* cooperative, of course, and there seems to be a limited amount of kin selection altruism within the societies of many primate species, with older individuals preferentially caring for their own younger siblings. In baboon societies, additionally, there is cooperation in troop defense and limited cooperation between dominant males, who maintain their favored social position by group repression of ambitious subordinate males. But when we consider its advanced state in most human societies, the minimal development of cooperative behavior in the social systems of other primates is surprising. The cooperation that ties together modern human societies may very well have its roots in the predatory habits and open country habitat of early man. In such a habitat, human cooperation in hunting must have been strongly favored, just as it has been in the wolves of open tundras. Cooperation in defense also was useful, of course, as it is in the case of open country baboons; but cooperation in hunting probably was the greater spur (Crook, 1966).

SUGGESTED READING

BANKS, E. M., AND M. F. WILLSON, eds. Ecology and evolution of social organization. *American Zoologist* 14 (1): 7–264, 1974. A collection of most interesting symposium papers that serves as a fine entry to the whole subject of this chapter and provides many references.

ELLIS, P. E., ed. Social Organization of Animal Communities (Symposia of the Zoological Society of London No. 14). London, Academic Press, 1965. Another fine collection of symposium papers.

HAMILTON, W. D. The genetical evolution of social behavior.

Journal of Theoretical Biology 7: 1–52 (1964). A difficult but seminal paper on kin selection and altruism.

KING, J. A. The social behavior of prairie dogs. *Scientific American* 201: 128–140 (Oct., 1959).

KUMMER, H. Primate Societies: Group Techniques of Ecological Adaptation. Chicago, Aldine-Atherton, Inc., 1971. An excellent paperback introduction to the voluminous primate literature.

WATTS, C. R., AND A. W. STOKES. The social order of turkeys. *Scientific American* 224: 112–118 (June, 1971).

WIENS, J. A. Group selection and Wynne-Edwards' hypothesis. *American Scientist* 54: 273–287 (1966). A good analysis of the group selection controversy.

WILSON, E. O. The Insect Societies. Cambridge, Harvard University Press, 1971. An extremely thorough and authoritative work.

WILSON, E. O. Sociobiology: The New Synthesis. Cambridge, Harvard University Press, 1975. Mind-stretching, voluminous, and somewhat controversial.

WYNNE-EDWARDS, V. C. Animal Dispersion in Relation to Social Behavior. Edinburgh, Oliver and Boyd, 1962. This book sets forth Wynne-Edwards' theories of social function and group selection, and has aroused more controversy than ten ordinary books.

chapter 14

Numerical Fluctuation of Populations

Population, when unchecked, increases in a geometrical ratio. Subsistence increases only in an arithmetical ratio . . . . Through the animal and vegetable kingdoms, nature has scattered the seeds of life abroad with the most profuse and liberal hand. She has been comparatively sparing in the room and the nourishment necessary to rear them.

Thomas Malthus, *An Essay on the Principle of Populations* (*1798*)

BY WAY OF REVIEW

None but the short-sighted doubt Malthus' thesis: the reproductive potential of any population, if long unchecked, far outstrips the capacity of the earth to support that population. If we believe Malthus we must deduce that populations ordinarily are restrained by various forces, which by increasing mortality or retarding the natural rate of increase bring deaths and births into approximate balance.

What is the character of these forces? Why do populations fluctuate over time instead of establishing and maintaining stable levels? Why are the population fluctuations of some species more violent than those of others?

Previous chapters have laid the groundwork for discussion of such questions. For example, we know already that intrinsic reproductive rate varies among species. Given optimal growth conditions, populations of some species are capable of extremely rapid multiplication; maximal growth rates are much slower in other species. Differences of generation time partly explain such differences of intrinsic reproductive rate: organisms that mature in a week can produce 52 generations, each geometrically larger than the preceding one, in a year; other organisms take 5, 10 or 20 years to mature and produce a single new generation.

But intrinsic reproductive rate depends on more than generation time. Two species of tree, for instance, may have similar generation times but differ enormously in seed production. Consider the effect on intrinsic reproductive rate of the equilibrium strategy of self-perpetuation (as in white oak, which produces relatively few seeds) and the opportunistic strategy (as in sycamore, which produces millions yearly). Similar contrasts characterize equilibrium and opportunistic species of animal.

In Chapter 3 we raised the possibility that population fluctuations of river-running salmon serve an adaptive function, by preventing the build-up of large predator populations or the evolution of predators which feed only on salmon. Thus we already have one possible answer to the question of why populations fluctuate: it may be an adaptive strategy. Periodical cicadas (Fig. 14.1) provide another instance, and one that is particularly convincing, of the apparent evolution of population variability as a predator-avoidance strategy. But there is little reason to believe that *all* population fluctuations are the adaptive product of natural selection, and in this chapter we must explore other explanations.

To the question of why some populations fluctuate more than others, we have a suggested answer from Chapter 13: among species that

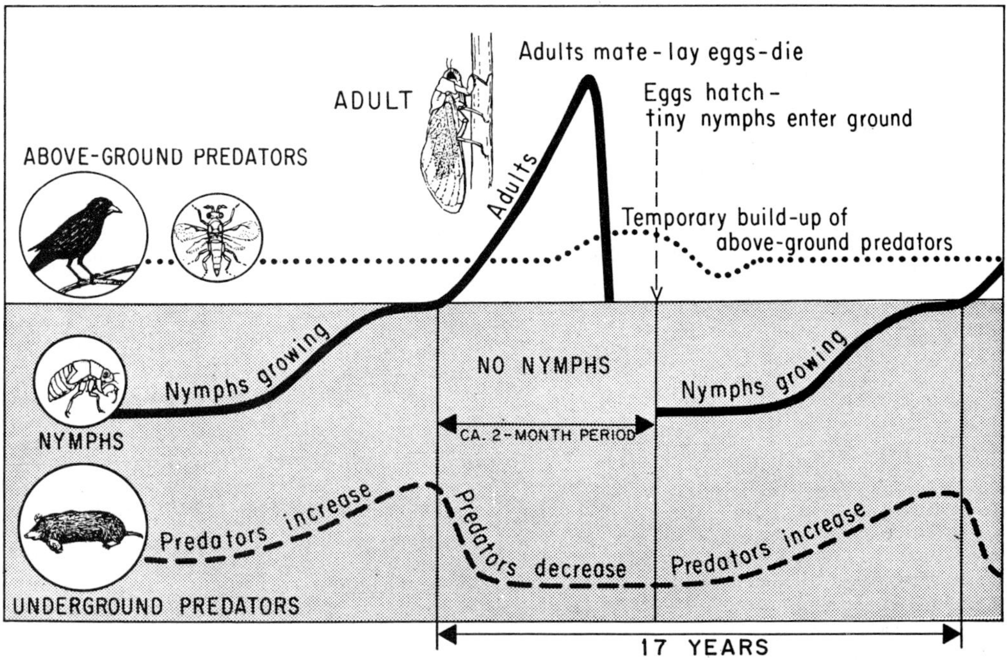

FIG. 14.1. Life cycle of the 17-year cicada, with suggested abundance patterns of some of its predators. Above-ground predators either have alternate prey or maintain very low populations based on out-of-phase emergence of small numbers of adults. In either case, predators are unlikely to be present in matching numbers when the cicadas emerge en masse every 17 years. Underground predator populations possibly build slowly as the nymphs slowly mature, but emergence leads to a drastic cutback in food supply. Probably, therefore, no predator species can *specialize* on cicada nymphs. (From M. Lloyd and H. S. Dybas: *Evolution 20:* 133–149, 1966.)

engage in contest competition, the populations may exert some degree of self-control on numbers, whereas populations of many species characterized by scramble competition lack self-control and fluctuate more wildly as resources vary. Also, however, the habitat of a population and the structure of the community to which it belongs seem related to its degree of numerical stability. For example, Bourliere and Hadley (1970) suggest that dry tropical savannas and cold tundras are particularly conducive to **irruptive population outbreaks,** enormous increases followed by mass emigration from a locality, as in the case of swarming locust species and Arctic lemmings. On a broader scale, there is evidence that populations in communities containing few species tend to fluctuate more markedly than do populations belonging to more complex communities. In simple communities like that of the tundra even equilibrium species may be characterized by great numerical instability (Fig. 3.23), whereas in species-rich communities like those of mature rain forests population fluctuations typically are subdued.

Additional, *intra*specific considerations regarding population fluctuation were suggested in Chapter 12. For example, within a single species those populations occupying marginally habitable areas tend to be less stable than populations occupying more favorable habitats. Naturally, "marginally habitable" in this context refers to the requirements of the species in question, not of all species; thus rain forest populations of a species not really suited to rain forest may conceivably be quite unstable despite the complexity of the community they belong to.

Given such a multiplicity of causes—and we have not yet mentioned them all—we should not wonder that populations of different species, or even of the same species in different circumstances, should differ widely in their patterns of fluctuation over time. Figures 14.2 through 14.5 provide examples of this variety for preliminary comparison. We will be referring back to these figures from time to time.

In the literature of population dynamics, commercially important species of insect pest, food fish and fur-bearing mammal loom very

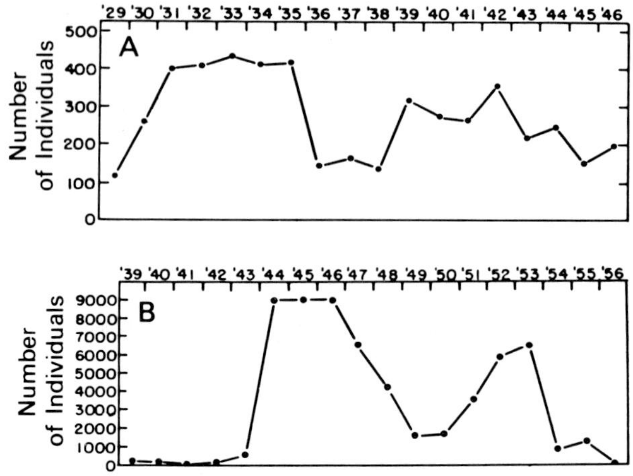

FIG. 14.2. (*A*) Autumn population levels of bobwhite quail on a 4500-acre tract of partly agricultural, partly wooded land in southern Wisconsin. Spring censuses also were taken, and in many years there was an inverse relationship between the number of adults in spring and the subsequent rate of population gain. In other words, reproductive recruitment to the population was better in years of low spring density, poorer in years of high spring density. Numbers fluctuated by a factor of less than 4 during this 17-year period. (*B*) Autumn population levels of muskrat on 1000 acres of marsh in north-central Iowa. Habitat changes greatly affected reproductive activity and behavior, resulting in enormous disparities between spring-to-fall recruitment in different years: +2018% in 1944, the second year of a brief spell when the marsh was fully flooded and maximally habitable, and −78% in 1956, a drought year when much emigration occurred. (Both figures are from P. L. Errington: *Cold Spring Harbor Symposia on Quantitative Biology 22:* 287–300, 1957.)

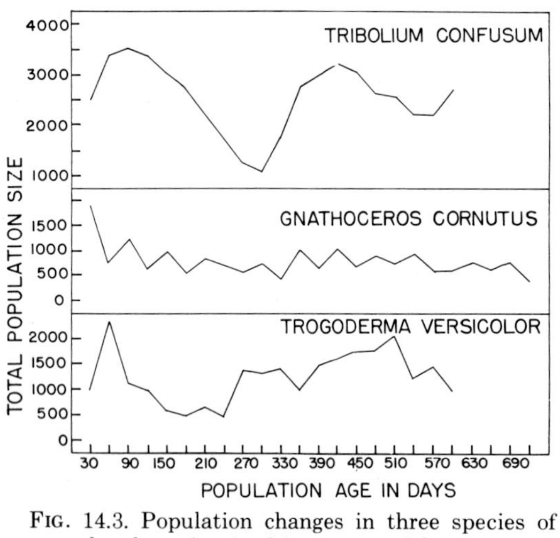

FIG. 14.3. Population changes in three species of granary beetle maintained in separate laboratory cultures under identical conditions. Each species has its own distinct pattern of fluctuation. The laboratory conditions probably were not dissimilar from the natural conditions these species encounter in stores of grain. (From T. Park, E. Gregg and C. Lutherman: *Physiological Zoology 14:* 395–430, 1941.)

large. Because fluctuations of such populations have had major economic impact, useful long term records sometimes were started long before specialized population ecologists appeared on the scene (Fig. 3.23). Even today, the most complete records and many of the most talented students of population change reside in government fisheries laboratories, departments of economic entomology, and similar institutions having an applied bent.

At the end of Chapter 12 will be found a series of references describing elementary techniques of censusing. Populations obviously must be enumerated if we are to follow their fluctuations, but we will not describe such methods here.

PARAMETERS OF NUMERICAL CHANGE

Recruitment and Loss

If a population is to be numerically stable there must be a balance between the processes of recruitment (birth and immigration) and the processes of removal (death and emigration). If a population's size is changing, the rate of change is the overall difference between these additive and subtractive processes. Population changes thus can be expressed as

$$\frac{\Delta N}{\Delta t} = \frac{B + I}{\Delta t} - \frac{D + E}{\Delta t}$$

where ΔN is the numerical change over a given time period Δt, and where B, I, D and E are births, immigrants, deaths and emigrants over the same time span.

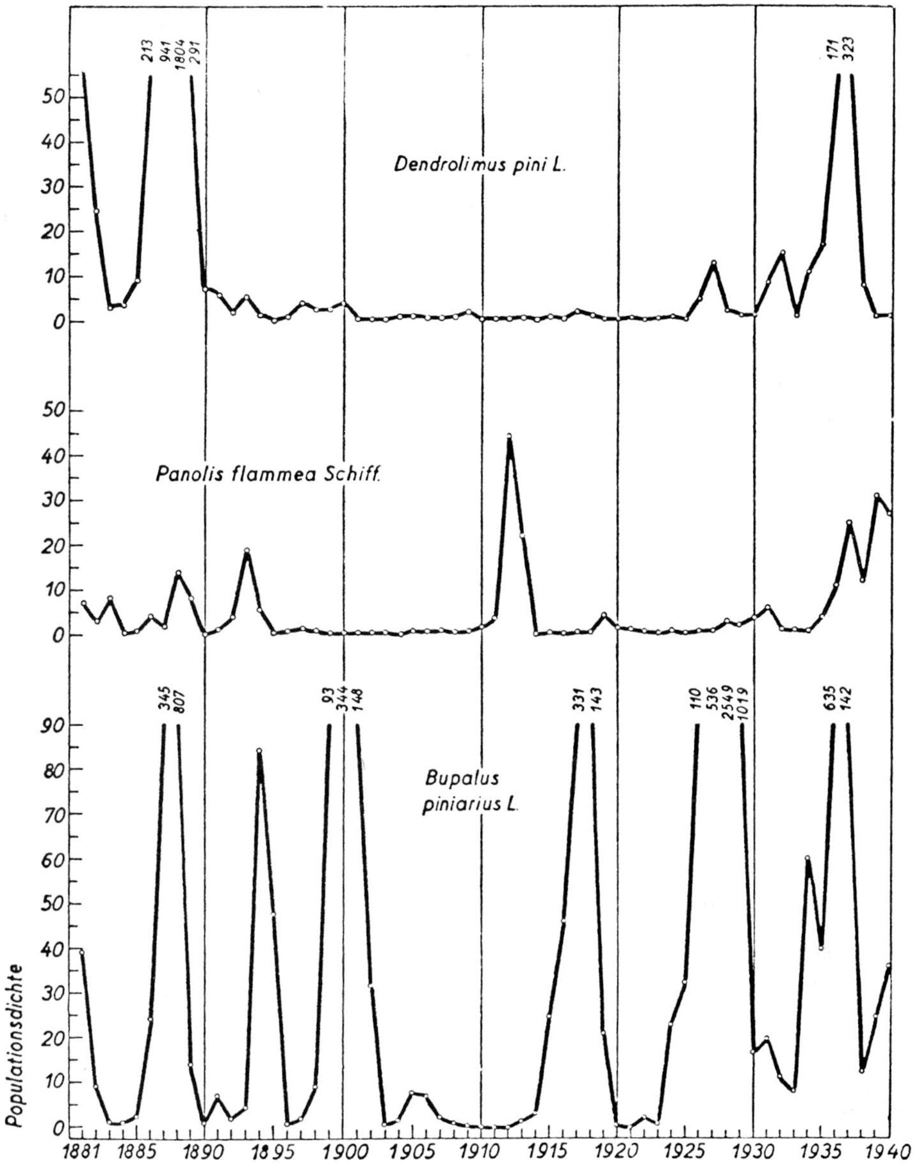

FIG. 14.4. Long term population records of three forest insect pests in Germany. Irregular, explosive irruptions from very low density characterize two of the species; the less dramatic fluctuations of the third still are on the order of 50-fold. (From F. Schwerdtfeger: *Zeitschrift fur angewandte Entomologie, 28:* 254–303, 1941. Reproduced with permission.)

All population change is mediated through these agencies, but their relative importances are not always the same. Sometimes immigration or emigration is an extremely powerful influence: think of the Arctic tundra, with its irruptions of lemmings and its seasonal changes in migratory bird and caribou numbers. In other situations immigrations and emigrations are appreciable, but essentially balance each other. An example might be the grasshopper populations of two meadows separated by a patch of forest. There is exchange between these populations—each gains immigrants and loses emigrants—but there may be near-equivalence of migrational gains and losses. If we judge this to be the case, we can posit births and deaths as the overriding parameters of each population's fluctuations. It is often dangerous to simply make

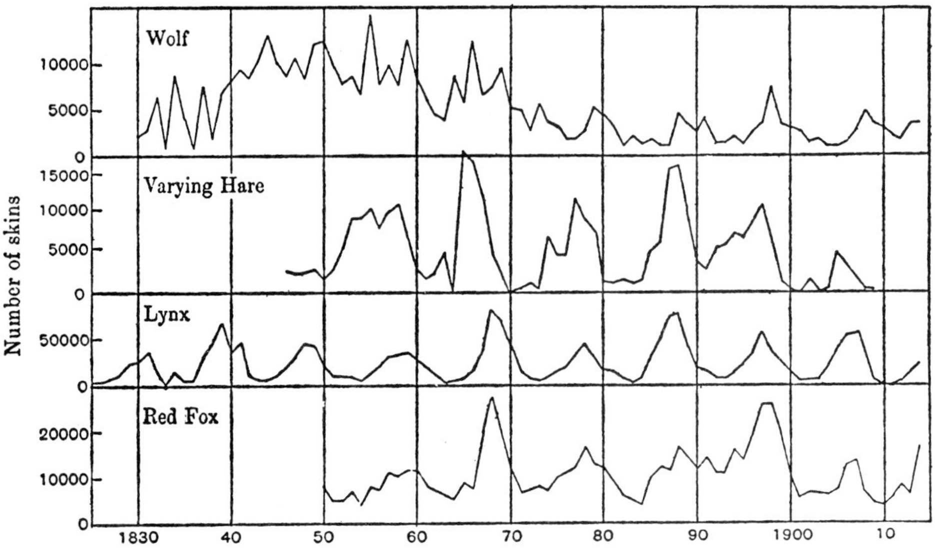

FIG. 14.5. Population fluctuations of four Canadian mammals, as deduced from pelt records of the Hudson Bay Company. Several of these patterns illustrate the tendency of certain species toward quite regular cycles of abundance. After C. Hewitt, from C. Elton: *Animal Ecology*. New York: The Macmillan Co., 1927.)

an assumption that migration is balanced, but even if we are not sure of this we may be able to determine that migrants are few relative to the numbers that remain, and hence conclude that migrations are a minor influence on a population's fluctuations.

In the case of still other populations, changes due to migration are nearly or totally impossible. Think of mice on an island, beetles in a bag of flour, snails in an aquarium or bacteria in a culture flask. Here numerical changes *must* be due to unequal rates of birth and death. Analyses of population change become much simpler in such circumstances. Therefore, theoretical treatments like the one that follows often start with the assumption that migrations are nil or that their positive and negative aspects are balanced.

Theoretical Growth of a Population Introduced to a Favorable Habitat

The most familiar model of population growth is the S-shaped model shown in Figure 14.6*A*. This is the **logistic curve,** first applied as a population model by Verhulst in 1838. It is based on the propositions of Malthus (*a*) that unrestrained populations grow in a geometrical progression and (*b*) that this growth does not remain unrestrained, but eventually is stopped by the inability of the environment to support more individuals.

Implicit in the second proposition is the notion of an **environmental carrying capacity:** for any population, this is the maximum number of individuals that the environment can sustain. The gradual levelling of a sigmoid logistic curve as it approaches carrying capacity conveys the message that birth and death rates gradually come to equal each other as the limits of the environment are reached, resulting in a population that is stable over time.

S-shaped growth patterns like that of Figure 14.6*A* are explicable in terms of positive and negative feedback. Positive feedback is an accelerating phenomenon that underlies all situations—in populations or elsewhere—where growth is geometric or exponential. The nature of positive feedback is illustrated with beautiful clarity in the growth pattern of a single plant (Fig. 14.7), when one plots the growth of individual leaves in addition to total growth. In populations, positive feedback occurs when restraints on growth are small. By Malthus' law, reproduction under such conditions far outstrips mortality. Each generation therefore is larger than the previous one, and *simply because it is larger* it produces more offspring. Because positive feedback is a self-multiplying process, the logistic curve initially becomes steeper and steeper.

Eventually, however, the curve reaches an inflection point and begins to flatten. Now

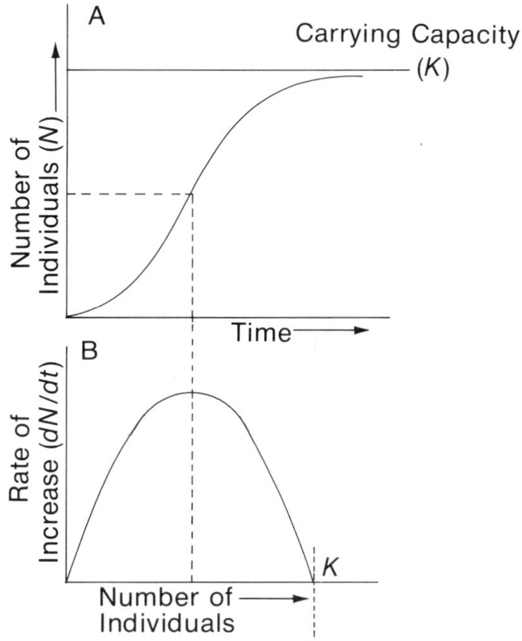

FIG. 14.6. Two graphical versions of logistic population growth. In (*A*), a plot of numbers against time, the curve is sigmoid. This is the familiar graph of logistic growth. But if, instead, the rate of growth of the population is plotted against numbers (*B*), the curve is parabolic: when numbers reach carrying capacity, growth rate is zero. In terms of the logistic model the two curves are equivalent. The points of inflection of the two curves are equivalent, too, in that each represents the density at which the population is growing most rapidly. Note that this density is half the carrying capacity.

positive feedback is losing out to negative feedback, which is not a self-multiplying process but a self-damping one. At this stage of population growth we can imagine the restraints of a finite environment aggressively asserting themselves; the death rate begins to gain on the birth rate, and when the two finally are equal growth ceases. Without specifying the nature of this restraint we can call it the **environmental resistance** to growth. During the positive feedback phase it is relatively small, but during the negative feedback phase it is large. This is the essence of negative feedback: the larger the population becomes, the greater is the resistance to further growth.

The complete sigmoid growth pattern is described by a simple differential equation:

$$\frac{dN}{dt} = rN\frac{(K - N)}{K}$$

where dN/dt is the change in population size over a short time t; r is the intrinsic reproductive rate of the species (known also as the reproductive or biotic potential); N is the population size at the beginning of the given time period; and K is the carrying capacity of the environment. $(K-N)/K$ is the negative feedback term: the value of this fraction ranges from nearly 1, when N is very small relative to K, to zero when $N = K$. At these extremes the equation reduces to $dN/dt = rN$ (when population growth is not being restrained at all) and $dN/dt = 0$ (when the population has reached carrying capacity and is not growing at all).

r, the intrinsic reproductive rate, is defined as (births − deaths)/time per individual of the population under optimal, nonrestraining conditions. r is constant for a given species according to the strict definition, but sometimes in growth experiments r may be defined as optimum growth under a particular set of abiotic constraints, *e.g.*, at a particular temperature. In either case, the steepness of the growth curve during the positive feedback phase is determined by the value of r. As we know, r differs widely among species; that is, the rates of maximum growth are different for different species.

The classic logistic equation describes the growth patterns of some real populations very well (Figs. 14.8 and 14.9), but it describes others quite poorly. There are at least three possible causes for a poor fit. The first difficulty lies in the negative feedback term $(K-N)/K$. According to the equation this term varies linearly between 1 and 0 as the population increases. But there is no reason to believe that a steady, linear increase of environmental resistance actually occurs. Rather, environmental resistance may be applied irregularly. During early growth stages the reproductive success of a population actually may *increase* with increasing numbers, rather than being negatively affected from the start. A generalization termed **Allee's Principle** recognizes this fact: it states that a population may be too sparse for maximal reproductive success per individual, and that optimal growth conditions actually are found at densities that are low but not too low. Toward the upper end of the curve one also can doubt the linearity of environmental resistance. When a population becomes too dense for maximal growth the restraints need *not* increase gradually and

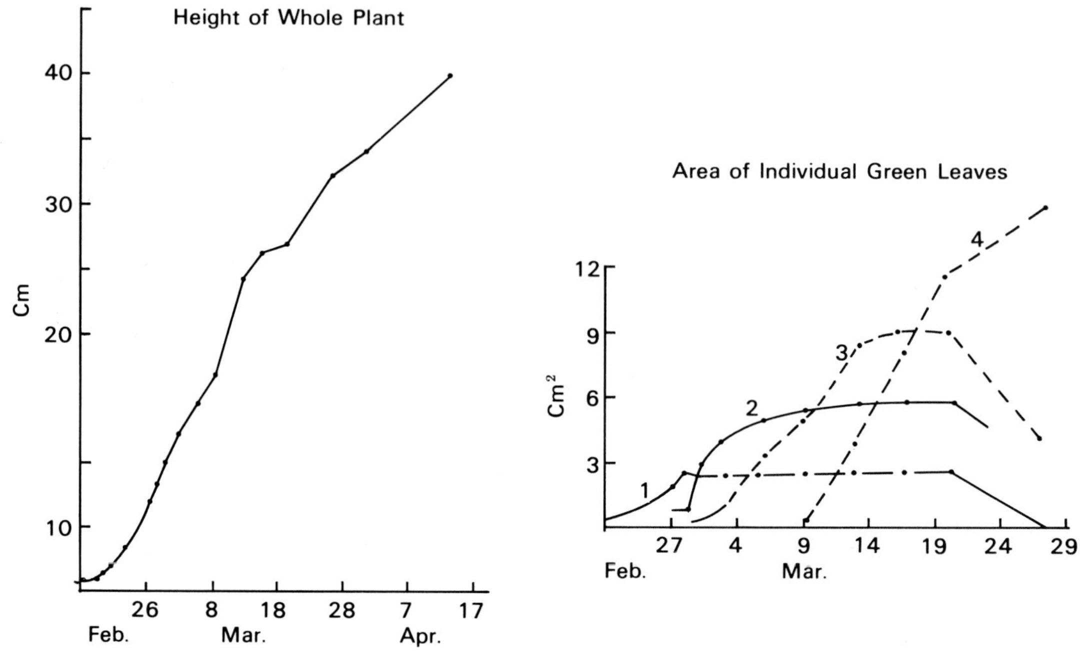

FIG. 14.7. Individual organisms as well as populations experience logistic growth. In the graph at *left*, growth in height of a corn plant is plotted against time. At *right* is plotted the growth of leaves successively produced by this plant. Each leaf grows to a certain size, then stops growing but remains green for a time. During this time it is channeling its photosynthetic product into the rest of the plant as is evident from the fact that each successive leaf grows larger (and, generally, faster) before it levels off. This is the essence of the growth-acceleration phenomenon of positive feedback. (Data of author.)

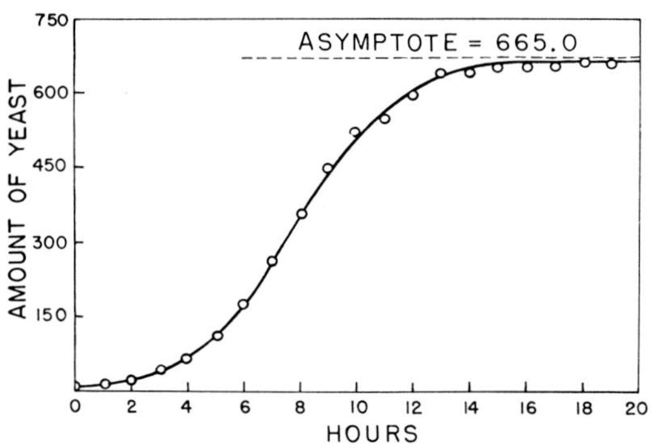

FIG. 14.8. Growth curve of yeast is one of the most perfect approximations of the logistic model. Growth easily can be followed by periodically measuring the turbidity of a culture with a spectrophotometer. (From W. C. Allee, *et al.*: *Principles of Animal Ecology*. Philadelphia, W. B. Saunders and Co., 1949; after Pearl.)

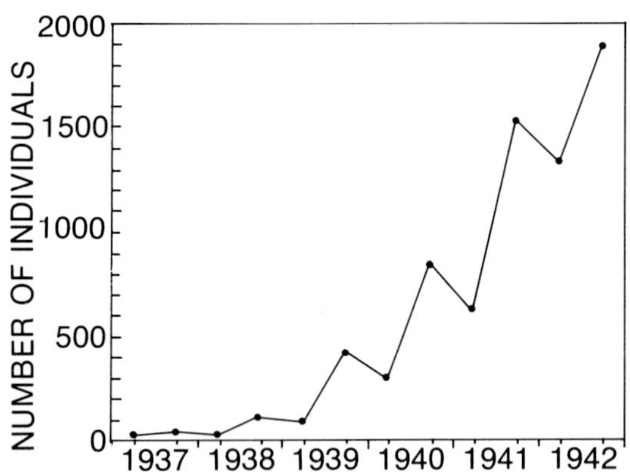

FIG. 14.9. Growth of a population of pheasants introduced to an island in Puget Sound, Washington, is saw-toothed because of winter mortality but otherwise corresponds to the logistic model. Positive feedback can be seen in the successively greater numerical increases in each breeding season—with the exception of the last, when the population shows signs of starting to level. (Modified from E. P. Odum: *Fundamentals of Ecology*, Ed. 2. Philadelphia: W. B. Saunders Co., 1959; original data from A. S. Einarson: ***Murrelet, 26:*** 39–44, 1945.)

steadily, as the equation states. Rather, restraints may be imposed rather suddenly, in the manner of threshold phenomena, and a population may suffer real setbacks in numbers rather than converging smoothly with the carrying capacity.

As a second difficulty, we should consider the possibility that negative feedback is relatively unimportant in some real populations. If somehow a population is maintained nearly always well below the carrying capacity, then the logistic equation with its negative feedback term will provide a poor, perhaps even misleading, description of its numerical behavior.

Finally, even if negative feedback *is* important in controlling a population, the logistic equation is inappropriate if the population is characterized by distinct, nonoverlapping generations (May, 1974). Many temperate zone insects, for example, exist at any one time as single-aged populations, born nearly simultaneously and destined to breed, lay eggs and die before the new generation of hatchlings appears on the scene. A different equation must be formulated to describe the dynamics of populations such as these; the classic logistic equation theoretically is applicable only to populations like man's, in which the generations overlap.

Actual Population Patterns

Figure 14.10 illustrates some of the imperfect logistic possibilities just described. When compared to real patterns (Figs. 14.2 to 14.5), one or another of these modified logistics frequently resembles the real pattern more closely than does the classic logistic curve. Except at very low density (when Allee's Principle may be operating), nearly all populations experience positive feedback when well below the carrying capacity, so the lower limb of a sigmoid curve ordinarily is recognizable early in the growth process. It is the upper limb which frequently loses its classic character, because populations differ markedly in the fidelity with which they stabilize at carrying capacity, and in how (or how often) they approach it. In Figure 14.10, *curve A* might characterize a population that is controlled by finely tuned negative feedback mechanisms. Population B behaves like A except that initially it overshoots the carrying capacity considerably; then the population adjusts to the carrying capacity in dying oscillations, *i.e.*, it becomes finely tuned over time. Population C is controlled by more coarsely tuned negative feedback mechanisms that allow it repeatedly to overshoot the carrying capacity, but environmental resistance clamps down suddenly after each overshoot so that the population sags abruptly. Carrying capacity in these diagrams is shown as a constant, but in the case of populations like C we should think realistically of a varying carrying capacity that periodically is reduced by the depredations of an oversized population (like cattle overgrazing a pasture). The carrying capacity then recovers, *i.e.* rises, after the population has fallen. Populations controlled by extremely coarse, abruptly imposed negative feedback mechanisms often take on a J-shaped or multiple-J pattern of fluctuation like that of population C.

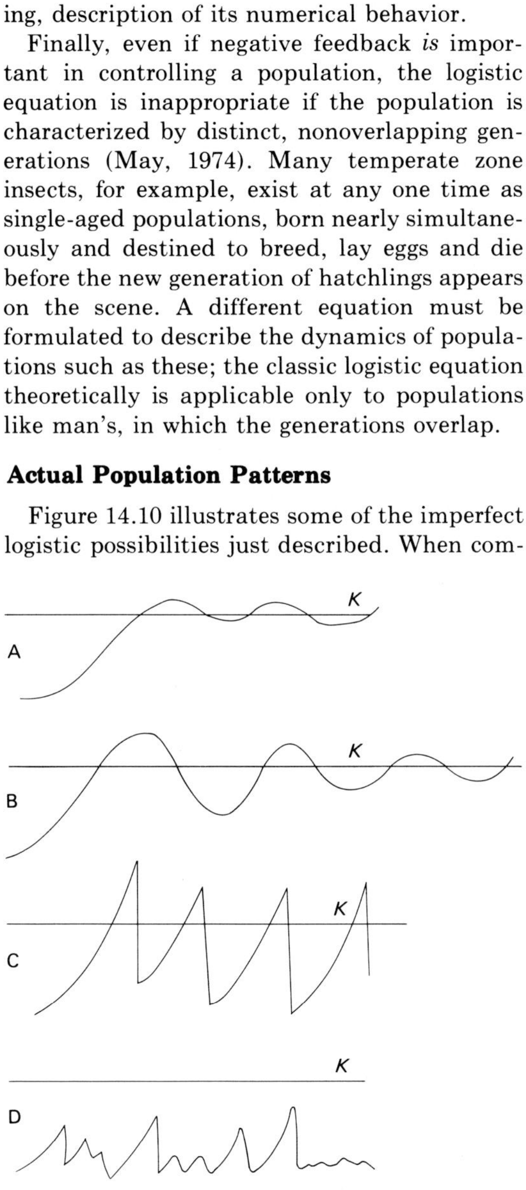

FIG. 14.10. Contrasting types of population curve described in text. *A*, *B*, and *C* are "real" variants of the logistic model. *D* is not, because negative feedback cannot be responsible for the population downturns.

Curve D in Figure 14.10 represents the possibility of a population that very seldom approaches carrying capacity, being controlled by repeated adverse events unrelated to negative feedback controls. These events depress the population, but are independent of population size in the severity of their effect. Hence they are not acting in negative feedback fashion. The simple logistic equation does not recognize this possibility, hence cannot account for such density-independent fluctuations. We will return to all these curves when we consider the kinds of mechanisms that can regulate populations. First, however, we should deal with two topics directly related to the logistic equation and its sigmoid curve.

r- and K-adapted Species

In earlier chapters we spoke of successional (opportunistic) species and of equilibrium species. The former have evolved strategies of high intrinsic reproductive rate and great dispersal ability, suitable for the colonization of temporary habitats. The latter have evolved strategies of persistence in a persistent habitat: lesser reproduction but greater provision for the young, often coupled with greater competitive resilience. These contrasting adaptive modes sometimes are termed *r* and *K* strategies, after the terms of the logistic equation that designate reproductive potential and carrying capacity. Species of the opportunistic type are termed *r* strategists and species of the equilibrium type *K* strategists. Opportunistic species have evolved toward a maximization of reproductive potential (*r*), thus allowing them to increase very rapidly when a favorable habitat is encountered. But this trait of rapid reproduction almost ensures that populations of opportunistic species, if given a chance, approach the carrying capacity of a habitat too rapidly to avoid overshoot and consequent die-back. If only for this reason, species that are *r* strategists are likely to be characterized by unstable populations even if they manage to persist for a time in a community. Equilibrium *K* strategists, on the other hand, approach the carrying capacity relatively slowly because of their limited reproductive potential. They are less likely to seriously overshoot it, and indeed their whole strategy seems suited to maintaining a relatively stable population near the carrying capacity. Pure *K* strategists and pure *r* strategists represent extremely divergent evolutionary solutions to the problem of self-perpetuation. Many species, as we have seen, are compromises combining some elements of each strategy. If *curve D* in Figure 14.10 is not purely hypothetical, *i.e.* if species really exist whose populations nearly always are far below carrying capacity, then such species might be expected to be the purest *r* strategists of all, because they almost never reach densities at which selection might favor an equilibrium *K* strategy.

Exploitation of a Population: the Maximum Sustainable Yield

A healthy population of organisms is a renewable resource: if part of the population is harvested, the losses theoretically are made up by reproduction. Whenever man uses another species' population as a resource, two practical questions are (*a*) what is the maximum harvest that can be taken on a sustained basis? and (*b*) how does one manage the population to obtain this maximum? From the theoretical logistic curve (Fig. 14.6) one can obtain theoretical answers. The maximum sustainable yield occurs at the point of inflection of the sigmoid curve of numbers against time, *i.e.* at half the carrying capacity. At this stage, equivalent to the top of the parabola of Figure 14.6*B*, the growth rate of the population is at a maximum. Below the point of inflection the stock is reproducing rapidly as individuals, but the number of reproducing individuals is too low for maximum output on a population basis. Above the point of inflection the population no longer is growing as rapidly; negative feedback processes are slowing the rate of recruitment. Hence the management technique that promotes a maximum sustainable yield is one that harvests just enough to keep the population size at the point of inflection of the growth curve. For most commercially exploited populations, *e.g.* of fish or timber, the measurement of interest is not number of individuals but some measure of biomass, for example tons or board feet. But in theory, biomass also should follow a sigmoid growth curve, so we may simply substitute biomass units for number of individuals in the equation. Maximum sustainable yield still will be at the point of inflection of the curve.

When we move from theory to reality, achievement of maximum sustainable yield loses its simplicity. Now one is dealing not with a population of homogeneous units but (if the generations overlap) with a population of organisms differing in age, hence in reproductive status and individual growth rate. We may recall Slobodkin's laboratory studies of ecological efficiency (Chapter 10); he had great difficulty actually *achieving* a maximum sustained harvest of *Daphnia*, even though he had nearly complete control of the system. To develop a complete maximum yield strategy for an exploited population one needs to know a great deal about the biology of the average individual in that population and about overall mortality from causes other than exploitation. If one is a forester, for example, one needs to measure growth rates of individuals at different ages and stocking densities, and, if relying on natural reproduction, the age at which seed is first borne and the average quantity of seed produced per tree.

The fishing industry is one in which sustainable yield problems constantly arise. For an exploitable population of fishes one wants to know the degree of fishing pressure the stock will take and the effect on the population of different harvesting methods. For instance, using gill nets of smaller or larger size may be the difference between sound management and decimation of the stock. In ocean fisheries one does not even know the population size to start with, and management programs must begin with the simple gathering of statistics. If one keeps records of the catch per unit effort (*e.g.* the weight caught per boat per day) over the years spanning the development of a one-species fishery, these records will provide an index of how much pressure the population is feeling (Fig. 14.11). One then can devise management regulations ensuring a sustainable yield. However, one still may have no clue as to whether this is the *maximum* sustainable yield. If the fishery is already established when the gathering of statistics begins, the situation is even harder to diagnose: men who fish for a livelihood will not easily be persuaded to relax their efforts so that the analyst can study the population under different fishing pressures. But long term records of catch per unit effort, age structure of the catch, and average size at different ages will yield considerable management information; and if for some reason fishing effort *does* change, real insight into maximum sustainable yield may be possible. In the North Sea, where plaice stocks had long been exploited, such a change in effort occurred during World War II, and it became evident that prior to this time overfishing had kept the population too young for maximum sustainable yield. Relaxation of fishing pressure during the war allowed stocks to build up, and in the postwar years large plaice were sufficiently abundant so that effort could be concentrated on them. Yields rose considerably above prewar yields even with somewhat less fishing effort, and they remained high even after the immediate postwar plaice population had been pruned back by half (Gulland, 1972). Apparently what happened is that during the war the population reached an age structure more conducive to high recruitment, and was maintained there by the concentration of postwar fishing effort on larger individuals.

Interestingly, in the case of the plaice fishery Beverton and Holt (1957) proved that a change of fishing effort really is not necessary to the determination of maximum sustainable yield if sufficient records of other population parameters are available. Simply by analyzing prewar fishery statistics Beverton and Holt learned enough about the plaice population to predict mathematically what the yield would be under various exploitation schemes (Fig. 14.12). The agreement of the postwar catch with their predictions showed that it is quite possible, in the case of exploited fish populations, to develop mathematical models sufficiently close to real-world processes to be practical tools. Beverton and Holt's approach to fisheries problems probably is the major work to date, on any group of organisms, that treats both the theory and the actuality of population dynamics.

It is obvious from the foregoing that a sustainable yield may be less than maximal because of underharvesting. It should be obvious also that a less than maximal, but still sustainable, yield may be obtained by *overharvesting*, if by this term we mean exploiting the population in a way that creates an age structure inappropriate for maximum recruitment (either of individuals or of biomass). Harvesting organisms that are still small, growing rapidly, and perhaps have barely reached reproducing age, exemplifies an exploitation policy that adversely affects age structure and results in lower than maximal yields. Such yields may or may not be sustainable, depending on the seriousness of the pressure. The policy that brings *highest* sustainable yields of many fishes, and presumably many other organisms too, is one where individuals are not harvested at all until they reach a certain age or weight and then are harvested as quickly and completely as possible. In social populations whose reproductive success is dependent on a stable, age-related organization, this exploitation policy might *not* be optimal because the population would be socially disrupted. But where social complications can be discounted, the appropriate age for harvest is determined by biological characteristics of the average individual—when its own growth rate begins to slow, when it reaches reproductive age, and how many offspring it produces in a given breeding season—as well as by the rate of natural, preharvest mortality in the population. These factors also help to determine the percentage of the total population which should be

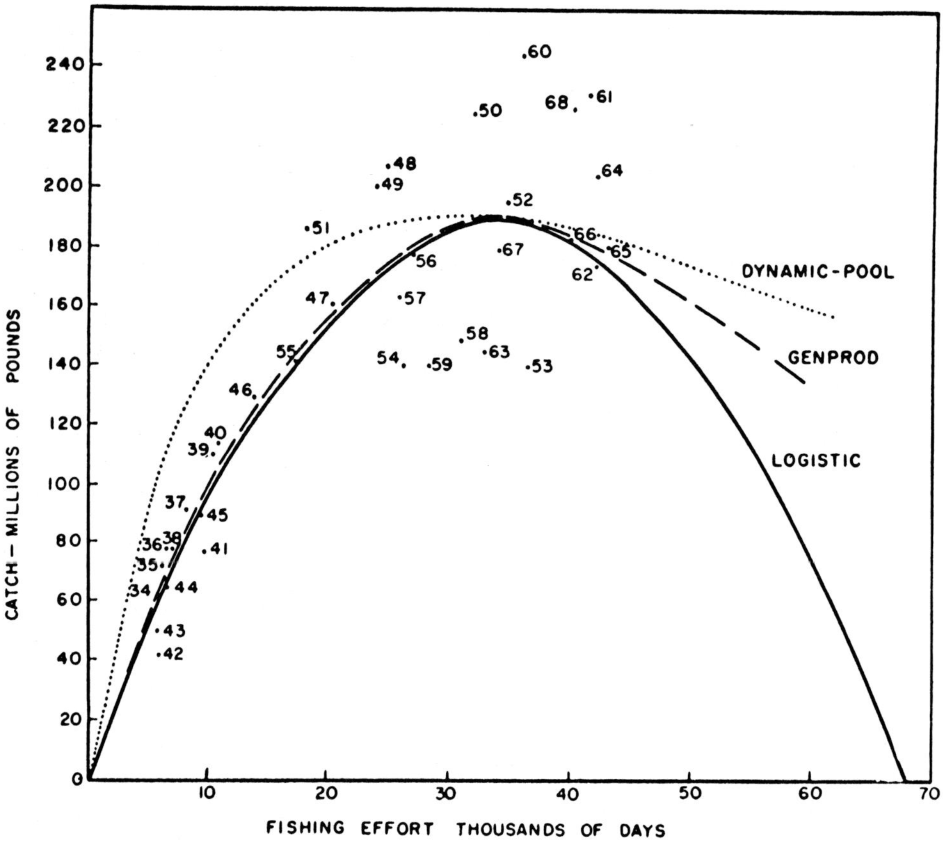

FIG. 14.11. Plot of catch versus fishing effort allows assessment of the state of yellowfin tuna stocks in the eastern Pacific. The numbers represent the years of record, 1934 to 1968. Up to 1950 the catch increased steeply as fishing effort increased. Thereafter the relationship became uncertain from year to year, but never after 1950 was the catch per unit effort as high as in early years. The diagnosis is that the fishery had caught up to the available stock, and the prognosis is that the total catch will suffer an absolute decrease if fishing effort continues to expand. The three curves are predictive models based on slightly different assumptions regarding the biological basis of the decline in yield per unit effort. (From the 1968 Inter-American Tropical Tuna Commission's Annual Report, as reproduced in J. A. Gulland: *Population Dynamics of World Fisheries*. Seattle: Division of Marine Resources, University of Washington, 1972.)

harvested for maximum sustainable yield. This figure differs considerably among species (Table 14.1). Much basic information about a species is necessary before its populations can be effectively managed for maximum sustainable yield. One begins with simple theory (the logistic growth curve) but must go far beyond.

FACTORS REGULATING POPULATION SIZE

Fueling a Controversy

Population ecologists conduct painstaking censuses, sometimes for decades, to follow the dynamics of particular populations. They may conduct experiments, too, if the populations are amenable to manipulation. In a single, carefully studied population the causal factors of numerical change sometimes can be pinpointed quite clearly, but population ecologists are cautious about extrapolating their conclusions to other populations even of the same species. If they try to do so, often they are accused of unwarranted generalization. If population ecologists cannot even agree on population control processes at the level of a species, imagine the reaction to a paper appearing in 1960 which, in three pages of deduction from casual observation, made sweeping generalizations regarding the regulatory factors not of particular populations nor even of particular species, but of whole trophic levels. To population ecologists the message

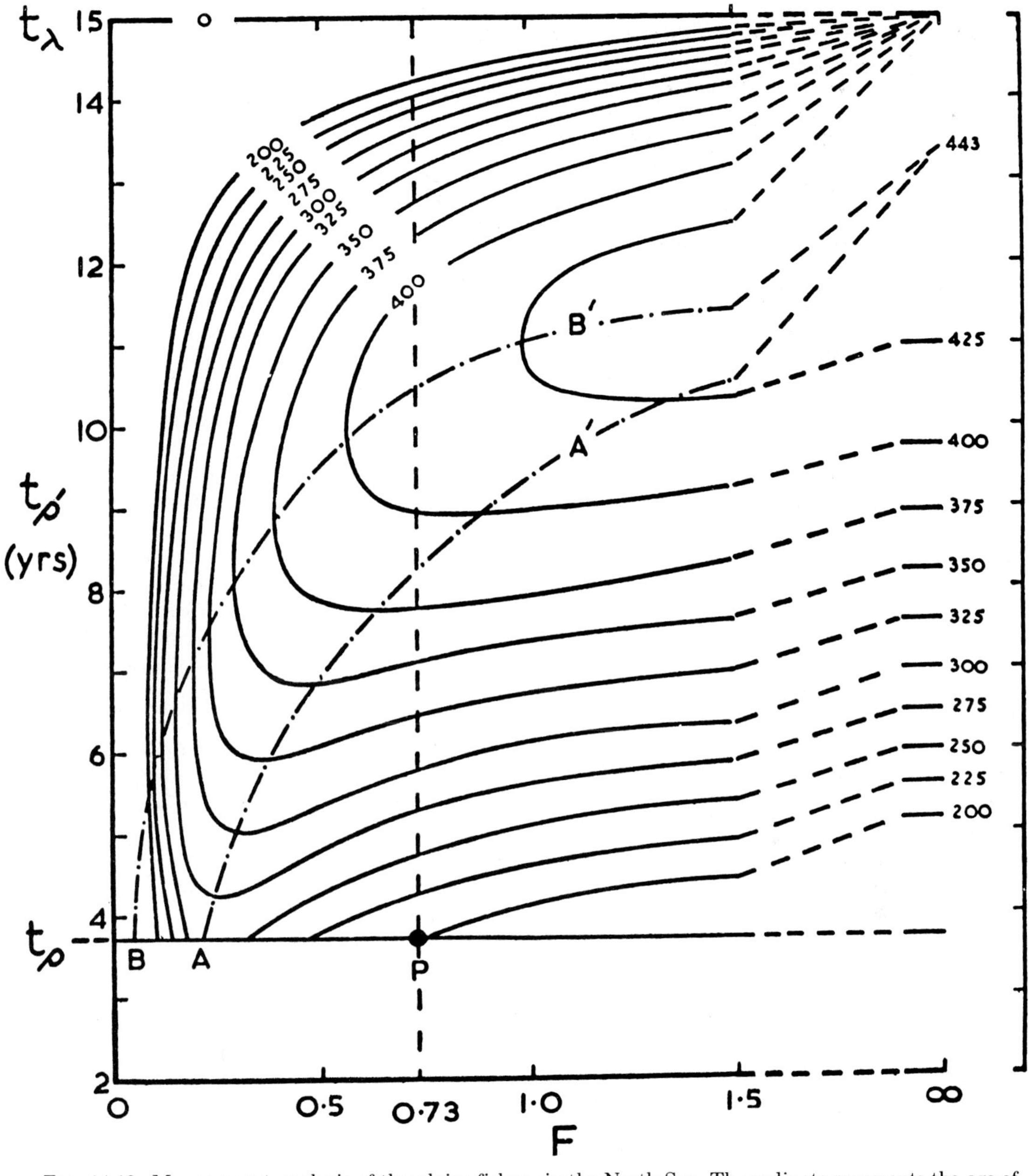

FIG. 14.12. Management analysis of the plaice fishery in the North Sea. The ordinate represents the age of the fish at first capture, while the abscissa represents the fishing intensity. There is a maximum possible age of first capture (t_λ) represented by the natural lifespan of the fish, also a minimum possible age of first capture (t_ρ) below which the fish are too small to use. The curved contour lines are lines of equal yield: they indicate that the same yield can be achieved by various combinations of fishing intensity and fishing method. The highest yields (400-plus units) are predicted when high fishing rate is combined with high age at first capture. The same fishing intensity that represented overfishing before World War II (indicated here by *P*) could achieve close to the maximum yield if gear were used that did not capture the fish until they were 9 years old. *Curves A—A′* and *B—B′* outline hypothetically successful management strategies involving temporary reduction of fishing intensity from the prewar point *P*, followed by a simultaneous buildup of intensity and age at first capture until maximum sustaining yields are reached. (From Beverton and Holt, 1957; as reproduced in J. A. Gulland: *Population Dynamics of World Fisheries*, Seattle: Division of Marine Resources, University of Washington, 1972.)

TABLE 14.1

Maximum Sustainable Yields for Different Species

Unless otherwise indicated, the figures are given as the harvestable percentage of the total population, and assume that the population has an optimum age structure. Most data are from Watt (1968), who compiled them from various sources. Data for seals are from McLaren (1962).

Organism	Maximum Sustainable Yield	Comments
Blowflies	99% of emerging adults per day	A few adults produce enough larvae to overpopulate a decaying carcass. When overpopulated, most larvae succumb to intraspecific competition.
Flour beetles	3% per day	Laboratory population
Daphnia	23% per day	Laboratory population
Guppies	2% per day	Laboratory population
Unicellular algae	13% per day	Mass culture experiment
Ringed seal	7 to 10% per year	Baffin Island population

seemed to be, "You can't see the forest for the trees."

Three eminent ecologists had collaborated on this paper: Nelson Hairston, Frederick Smith and Lawrence Slobodkin, all of the University of Michigan. Partly on its own and partly through the many reactive papers it has stimulated (*e.g.* Murdoch, 1966; Ehrlich and Birch, 1967; Slobodkin *et al.*, 1967), this paper has become so celebrated that now it frequently is referred to simply by the initials of its authors (HSS) instead of its full title ("Community Structure, Population Control and Competition"). The claims of HSS chiefly apply to terrestrial ecosystems, and the article's basic points can be made quite quickly.

1. In most habitats of today's world dead organic matter decays relatively quickly, dissipation of this material seemingly being in balance with the rate at which it is formed. No longer, as in past ages like the coal-forming Carboniferous, do large amounts of incompletely decomposed material accumulate over time; rather, decaying matter seems to be utilized more or less completely. This observation suggests that the abundance of *decomposers* is regulated, in general, by the rate at which their supply of dead organic matter is renewed; *i.e.* (in the words of HSS), decomposers are "resource-limited."

2. Except in relatively rare cases, the abundance of *terrestrial plant life* does not seem limited by the organisms that feed on them. The landscape normally remains green and "the plants are abundant and largely intact." Rather, the abundance of terrestrial plant life seems limited in most cases by insufficient light and/or moisture. Thus, according to HSS, terrestrial plants also are resource-limited.

3. *Herbivores*, if they do not ordinarily limit the abundance of plants, are not likely themselves to be limited by their food supply. Moreover, although adverse or beneficent weather often strongly affects the size of herbivore populations, the frequently enormous success of herbivores introduced to new lands and unfamiliar climates suggests that weather is not normally the *primary* control of herbivore numbers. What possibilities remain? Since plant food usually is present in abundance, competition with other herbivores for food is not a likely control, but limitation by predators remains a candidate. Pointing again to examples of herbivores that have multiplied to pest proportions when introduced to new habitats free of predators (or when predators have been removed from an herbivore's native habitat), HSS conclude that predators usually *do* limit populations of herbivores.

4. But if *predators* limit the abundance of herbivores, it seems logical that the numbers of predators are controlled by the resource whose abundance they limit, *i.e.* their food supply. Hence, say HSS, predators dependent on herbivores are food-limited. Predators higher up in the food chain, *i.e.* secondary or tertiary carnivores, presumably also are limited by food supply, although an argument invoking decreasing energy supply with each step in the food chain might be the best one here. (HSS do not really discuss higher level predators.)

The conclusions of HSS seem backed by powerful logic, but they have been attacked on at least two grounds. First, the statements are so sweeping and over-arching that even if generally true, they may accommodate many exceptions. Population ecologists therefore cannot really assume that the conclusions of HSS apply to the species or populations they are studying. Turning the forest/trees argument around, population ecologists would say that the trees are important and different even if the forest as a whole obeys HSS.

Second, the limiting mechanisms proposed by HSS all are *direct density-dependent mechanisms, i.e.* control mechanisms that conform to negative feedback models of population control. By implication, HSS deny the existence or at least the importance of regulating mechanisms that are *density-independent;* their analysis leaves no room for populations which, like that of Figure 14.10*D*, might seldom be seriously subject to density-dependent controls. Many population ecologists would agree with HSS in minimizing the significance of density-independent regulation. But many others would disagree: the controversy over the relative importance of density-dependent and density-independent population regulation is one of the most persistent in ecology. Before we look at this controversy, let us make sure we know what sorts of regulatory mechanisms we are talking about when we use the adjectives density-dependent and density-independent.

Inverse Density Dependence

Allee's Principle implies that there are times in the life of many populations when a decrease in numbers stimulates a further decrease, the magnitude of this effect being inversely proportional to population size. This phenomenon is opposite to the "direct" density-dependence that is usually stressed by population ecologists, and we can deal with it relatively briefly. In fact, inverse density-dependence is not a regulatory or stabilizing process at all, but is *destabilizing*. Consider the difficulty an animal encounters in finding a mate when population density is very low. The few widely spaced individuals in such a population often may go unmated, and the problem obviously becomes more serious as numbers decline, rather than correcting itself as a negative feedback process would do. To cite a real example, the suggestion has been made that the rare Indian rhinoceros presently is dwindling in numbers partly because of its already low numbers. The rhino is a solitary animal, seeking out other rhinos only at mating time, so individuals in the sparse population become widely spaced in the nonbreeding season and ultimately isolated. Small populations of species more social than the Indian rhino may not suffer spatial fragmentation, but may be adversely affected by inbreeding, which of course also becomes more extreme as the population dwindles. In each of these cases reduction of population size leads to further reduction, not to recovery. There is a positive feedback toward extinction.

Less obviously and in a different way, predation and parasitism also can act on populations as inverse density-dependent factors. Consider the case of a predator which feeds on several types of prey, one of which is increasing in abundance. As this prey population grows, the predator population probably will concentrate more and more heavily on it—hence behave in the direct density-dependent fashion of logistic theory—but if the predator is an insufficient curb the prey population may progressively "escape" its influence. Prey numbers will continue to increase and the predator population, saturated, will take a constant quantity (but increasingly small proportion) of the available prey. This is inverse density-dependence again, but with numerical consequences opposite those felt by a population "suffering" from Allee's Principle. In the first case the destabilizing trend was toward extinction; in the present one, the trend is toward destabilizing release from a previously important curb to growth. We will return to predator-prey relationships later, but should note from this example that predators or parasites may act as *direct* density-dependent factors when prey populations are relatively small, but as *inverse* density-dependent factors when the same populations are large. The inverseness in this instance derives from the decreasing effectiveness of the predator as a controlling influence on its prey population.

Of course, if the predator is able to multiply as rapidly as its prey, inverse density dependence need not arise. Usually it is when the reproductive potential of the predator is much less than that of the prey that we see the inverse effect. In such cases, if other controls do not seriously intrude, a growing prey population may quite literally escape from effective regulation by a predator population which may have been a very effective regulator when the prey density was low.

Direct Density Dependence

It already may be clear that this term refers to environmental resistance as classically defined in the logistic equation. When an environmental factor acts in direct density-dependent fashion it is a true regulator in the sense of opposing the trend (either up or down) that is in progress. It is a negative feedback brake that is applied with increasing strength as a population increases and is progressively relaxed as density falls. Whatever this brake, and however gradually or abruptly it operates, it is a direct density-dependent factor. By pressing down when the population grows and easing when it declines, direct density-dependent factors are stabilizing influences—albeit, at times, very ineffective ones—because they oppose both growth and decline. In this respect the action of direct density-dependent influences is opposite that of inverse density-dependent influences.

Table 14.2 lists a number of regulatory mechanisms that commonly act in direct density-dependent fashion. These mechanisms fall easily into two groups: **external environmental agencies** which impose themselves directly upon a population, and **self-adjusting mechanisms** *within* populations that sense the level of environmental resistance and react accordingly. Starvation due to food exhaustion exemplifies the first type of regulation; physiologically or socially mediated birth limitation that forestalls starvation exemplifies the second. Self-adjusting mechanisms sometimes are called **internal mechanisms.**

Let us consider the external mechanisms first. We know already that predators and parasites do not always act in direct density-dependent fashion. According to simple models, however, rapidly reproducing parasites and pathogens *should* do so always, because their capacity for multiplication is at least as rapid as that of the host. Theoretically, only slower-growing populations of macropredators should lag in response to an increase in their prey and perhaps lose control. In the real world, however, one seldom is dealing with simple, two-species predator-prey or parasite-host systems. In nature, for example, fast-growing parasite populations may themselves fall victims to hyperparasites, and thereby lose control of their own prey. An example is provided by Clark (1964) for a multi-population complex of Australian insects. The species involved are an herbivorous plant louse (*Cardiaspina albitextura*), its chief parasites (wasp-like species of the genus *Psyllaephagus*), and *their* parasites (other species of *Psyllaephagus*, together with flies of the genus *Syrphus*). An increase in density of *Cardiaspina* initiates a corresponding increase in its parasites, but an increase in their hyperparasites soon decimates the parasite population, and *Cardiaspina* is effectively released from its parasitic controls. Parasites may not be the chief brake even on low density *Cardiaspina* populations, but at such times the parasites at least are acting in direct density-dependent fashion. However, their influence becomes *inversely* density-dependent when the numbers of *Cardiaspina* exceed a certain level; because of the hyperparasites, *Cardiaspina* "escapes" from its own parasites even though the latter are capable of reproducing just as rapidly as *Cardiaspina*.

Despite such ambiguities and exceptions, predators and parasites probably act for the most part in direct density-dependent fashion, *i.e.* their numbers vary directly with the density of their prey and so help to stabilize the numbers of the latter. Therefore predators and parasites find a place in Table 14.2. In this table the other mechanisms listed as external all represent resources, in the broad sense. They are aspects of an environment's carrying capacity that may be overtaxed if a population becomes too large. Resource depletion of one type or another is an especially common external mechanism of population control among plants and lower animals. Here shortage of some necessity typically accentuates intraspecific scramble competition, scarcity of the resource being the direct cause of increased mortality.

Populations of social animals may be directly resource-limited too, but here the response to a dwindling supply often is an accentuation of **contest competition,** and the resource may be conserved by internal population control mechanisms rather than becoming directly limiting. The internal mechanisms of Table 14.2 can be considered indirect adjustments to resource stress. In fact, when HSS (page 349) use the term "resource-limited" they clearly mean it to include cases where species limit themselves in response to resource shortages.

In actuality there is no species whose populations are regulated completely by internal mechanisms. But if sensitive, self-adjusting mechanisms of density regulation have evolved in a species, its populations can be expected to fluctuate less wildly than do coexisting populations of species that lack self-adjusting mechanisms. This perhaps is a dangerous generaliza-

TABLE 14.2

Classification of Direct Density-dependent Regulatory Mechanisms

Mechanism	Explanation of Action	Notes or Examples
I. External Mechanisms		
Nutrient supply	Obviously becomes limiting only at high population density	Many animal populations overeat their food supply; plant populations may exhaust their water or nutrient supply
Availability of space	Obviously becomes limiting only at high population density	Often seen in animals having specific needs as to nesting site, shelter, etc.
Availability of light	Plants may shade out their own offspring	Involved in self-thinning of many plant populations
Predators and parasites	Often, but not always, increase as density of affected prey population increases	May alternate between direct density-dependent and inverse density-dependent action; see text
Epidemic disease	Incidence usually increases as density increases	Ease of transmission is related to density of the population
II. Internal Mechanisms		
Density-related resistance to disease	Physiological changes resulting from stress cause lower disease resistance at high densities	Noted in captive populations of rats, deer, etc.
Density-related levels of aggressiveness	Warfare, individual strife, etc. often increase with density and many affect both mortality and birth rate	Probably operates in many animal societies
Territoriality	An established feeding or breeding territory is kept and defended, preventing overpopulation	Territories may be somewhat elastic, shrinking under population pressure and expanding under population relaxation, but limits to elasticity exist, discouraging growth at high density and encouraging it at low density
Density-responsive practices of infanticide or geronticide	Excess offspring (or aged) are sacrificed or neglected and allowed to die	Common in human societies, both past and present; many bird clutches include a runt which is nourished or neglected depending on the available food supply
Density-responsive birth rate/individual	Density-induced physiological changes in female result in adjustment of litter size or fertility	Birds' clutch size often seems related to fluctuations in available food; crowding in rats reduces fertility and frequency of successful mating; special case: use of contraceptive methods may be density- or resource-related in man
Density-responsive growth rate/individual	Retarded growth rates under crowded conditions are sometimes due not to lack of food, but to "psychological" mechanisms	Tadpoles and many fishes, even when amply fed, grow more slowly under crowded conditions than at low density; this reduces the food demand of large populations and forestalls the external mechanism of starvation

tion, but what we are saying is that the negative feedback mechanisms of internally regulated populations tend to be more finely tuned than those of externally regulated populations. Of course, it is easy to think of exceptions: the self-adjusting mass migrations of the collared lemming, for example, are a case of very coarse internal tuning that results in great fluctuations of population size. Moreover, we must repeat what has just been said: no populations are regulated completely by internal mechanisms, and some internal mechanisms probably are quite ineffective.

Many internal mechanisms are socially mediated, and a few ecologists take the extreme view that social behavior evolved quite simply as an internal mechanism of density control. Moreover, they suggest that social behavior, in all its variety, *still* can be explained largely in these terms. Wynne-Edwards (1962, 1965), is the chief proponent of this view. He regards, for example, the familiar "dawn chorus" of bird populations, and also their frequent propensity to mass in flocks, as density-sensing **(epideictic)** devices. These forms of behavior provide individual birds with information which elicits from them appropriate intensities of flexible, population-adjusting behavior such as territoriality. If the voices of many robins are raised in early morning song, each male presumably is "instructed" to defend his nesting territory more vigorously against competitors. Perhaps he also will expand or contract the area he is prepared to defend.

Using many such examples, Wynne-Edwards has developed an essentially unitary hypothesis: population self-regulation is the underlying stimulus to almost all social behavior and organization. Not surprisingly, this concept has encountered the same sort of resistance as did the generalizations of HSS. As is clear from Chapter 13, ecologically oriented behaviorists such as John Crook (1965) find far too much environmentally induced variety in social organization to be satisfied with a unitary hypothesis. Wynne-Edwards' reply to these critics might resemble the reply of HSS to theirs: "You can't see the forest for the trees." But Wynne-Edwards is criticized from another side by population ecologists like the late David Lack (1966), who claim that social mechanisms of population control usually are secondary to nonsocial mechanisms, even among the higher vertebrates. Exhaustive studies of bird populations made Lack sceptical regarding the real role in population control of epideictic displays and subsequent social mechanisms like territoriality. Rather, Lack concluded, external mechanisms together with physiological adjustments of fecundity (Table 14.2) are the prime regulators of bird populations and probably of animal populations generally. Certainly the adjustment of fecundity, be it the clutch size of birds or snails (Eisenberg, 1966) or the number of seeds produced by a plant, is a widely documented internal mechanism of direct density-dependent regulation. Adjustments of fecundity often are *not* socially induced; in plants obviously they cannot be. But few ecologists would argue as vigorously as did Lack that social behavior seldom or never plays much role in adjustments of fecundity. In many social species the two seem closely related, social behavior triggering physiological changes that affect the birth rate.

Related to fecundity adjustment, and also mediated physiologically, are density-related adjustments of individual growth rates (Table 14.2). These are found in populations of tadpoles (*cf.* John and Fenster, 1975) and some fishes (Weatherly, 1966). In nature the mechanism of reduced individual growth rate clearly functions to forestall direct food limitation: members of the population grow more slowly when crowded, and may reach adulthood in stunted condition. But experiments show that growth retardation is not directly due to inadequate nutrition (as is the case, for example, with crowded plants), but is due to psychological or social mechanisms acting in direct density-dependent fashion. If they are crowded, tadpoles in the lab or bluegill sunfish in a farm pond will experience retarded growth even if provided with plenty of food. Quite probably the stunted adults also will produce fewer eggs per individual than would well grown adults.

Ecologists like Lack and Wynne-Edwards stand at opposite poles regarding the significance of social mechanisms of population control. Other ecologists fall somewhere in between. Many, for example, feel that some social mechanisms are important but question the importance of a hormonally based social mechanism that has been termed the **adrenal stress syndrome.** This syndrome involves density-related changes in aggressiveness and in resistance to disease (Table 14.2). As a self-adjusting mechanism of population regulation the syndrome has been studied especially by John Christian (*cf.* Christian and Davis, 1964) and John Calhoun. In Calhoun's famous study of crowded rats (1962), density-related physiological stress, mediated through overstimulation of the adrenal glands, seems to underlie a complex of density-reducing agencies. In Calhoun's rats this syndrome included heightened aggressiveness of males, poor maternal care, decreased fertility and lactation among females, and lowered resistance to disease, the result being proportionately greater mortality (particularly of young) in dense populations than in sparse ones. The centrality of aggressiveness to this whole complex was convincingly demonstrated by Vessey (1967), who fed aggression-muting tranquilizers to captive mouse populations (Fig. 14.13). The irascible undoctored populations stabilized at lower densities than the same populations on tranquilizers.

Argument about the adrenal stress syndrome centers on the fact that it has been proven important only in artifically crowded populations restrained from emigration. Outside the laboratory there is only one well documented case of this syndrome having a density-adjusting effect. On James Island in Chesapeake Bay, a population of introduced Sika deer, protected from predators and denied the chance to emigrate, grew to high density. Eventually, during two successive winters, there was a progressive and spectacular die-off amounting to 78% of the original herd. Most of the die-off occurred in the first winter, and autopsies indicated physiological disorders associated with adrenal overstimulation (Christian *et al.*, 1960). After the die-off the surviving population, presumably under less physiological stress, began to increase again.

But if not an actual laboratory situation, the Sika deer situation was akin to one. The adrenal stress syndrome in its totality has yet to be convincingly documented in truly wild, unconfined populations. Of course, fully natural habitats exist very like that of James Island, at least in terms of forced confinement; and particularly if predator pressure is low, the adrenal stress syndrome should be looked for in such circumstances. It may be, too, that in populations having the opportunity to emigrate lesser de-

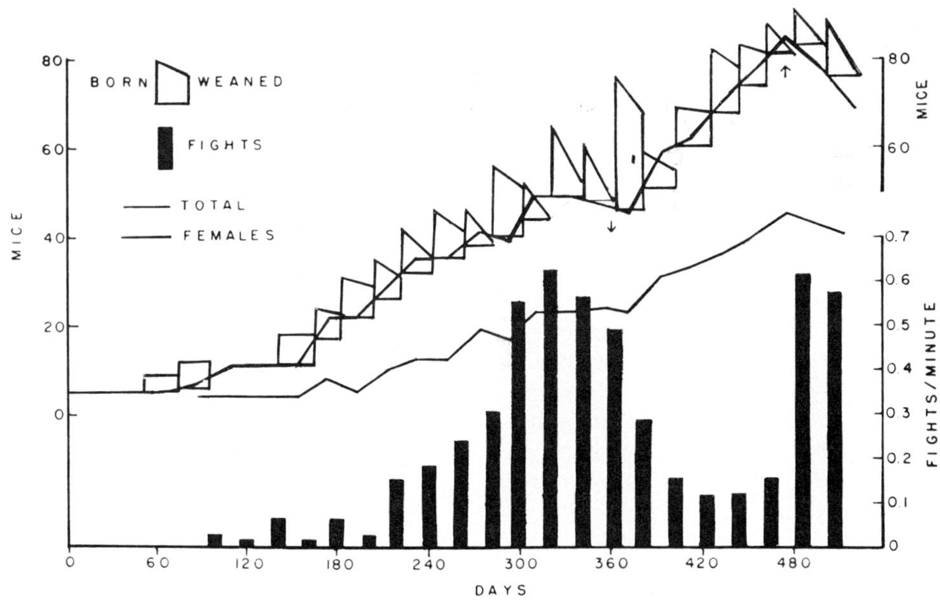

FIG. 14.13. Tranquilizer experiment with house mice. Within 330 days after introduction of a pair of wild mice to a cage of 1.6 m^2 floor area, the population apparently had reached carrying capacity of ±45 mice. Aggression, mostly initiated by males, had increased as the population grew. The tranquilizer chlorpromazine then was added to the food (*first arrow*): aggression declined and the population grew rapidly to 80. When the tranquilizer was discontinued (*second arrow*), fighting abruptly increased, and the population started to fall. Changes in infant survival were the chief vehicle of these population changes. (From S. Vessey: *Ecology* 48: 367–376, 1967.)

grees of adrenal stress play a role in differential emigration. In vertebrates, adrenal stress may always be an associate of high levels of aggression, and if we turn from the syndrome *in toto* to aggression alone, we find many instances of its operation in other ways as an internal, density-dependent regulator. Changing levels of aggression control the size of breeding or feeding territories in populations of many species, and in others clearly are related to the amount of emigration. Marmots (Barash, 1974), deer mice (Healey, 1967), red grouse (Watson, 1972) and lemmings are among the organisms in which density-related aggression seems associated with emigration rates. Among all these species, forced emigration from a familiar and favorable location is accompanied by increased mortality, so emigration is not just a mechanism of population spacing but also usually is one of population reduction.

Where one finds evidence for a regulatory role of variable aggression, one should not assume that it always is the only or even the most important mechanism of self-adjustment in the population. For example, aggression is but one of several agencies that have been invoked to explain the sometimes spectacular mass emigrations of lemmings.

The complexities of density-dependent population regulation are well illustrated by studies of red grouse on the heather moors of Scotland, and we may conclude our discussion of density-dependent regulation with this case history. The red grouse is not closely related to American species of grouse, but rather belongs to the group containing the Arctic ptarmigan. Red grouse populations have been studied for many years by Adam Watson, David Jenkins, and numerous associates. Their ideas have changed over the years regarding population regulatory mechanisms in this species, but in a paper which incorporates several of the ideas we have discussed, Watson (1972) summarizes recent thinking on the subject.

Changing degrees of territorial aggressiveness seem central to the population dynamics of red grouse. The males establish territories in autumn which are larger and, strangely, more vigorougly defended when the population is declining than when it is increasing. This is because males raised in years of poor breeding success are more aggressive than males raised in more favorable years. Most winter mortality is socially induced, being very high among birds which do not manage to gain territories (or, if females, which do not succeed in obtaining a mate). Hence aggressive, territory-expanding behavior following years of poor reproductive success initially serves to reduce the population still further, although it probably promotes greater breeding success among survivors the following spring. The offspring of this small but relatively successful breeding stock are comparatively unaggressive and establish quite small, contiguous territories. Hence there are relatively many breeders when the offspring reach maturity, but this stock generally has poor breeding success. The chicks that are reared successfully are consequently aggressive, and the population is back to the starting point in the cycle (Fig. 14.14).

This account twice contains examples of a complicating phenomenon: **delayed density dependence,** a form of direct density dependence in which a population trend initially is reinforced rather than immediately encountering negative feedback. Here, low breeding success is followed by territorial aggressiveness which serves to further reduce the population, and high breeding success is followed by relaxed territorial aggressiveness that promotes survival the following winter. But each of these trends is reversed by differential breeding success at a later time. Delayed density dependence obviously promotes a fluctuating rather than stable population, as will become clearer when we discuss population cycles of predators and prey. In the case of red grouse, however, the fluctuations are muted by the relative longevity of these birds, which retain throughout life their aggressive or nonaggressive tendencies. At any one time the population will contain some highly aggressive males (born in years of low breeding success) and some less aggressive males (born in years of high breeding success), rather than birds of only one type.

Thus far the red grouse story seems a refutation of David Lack's thesis that social mechanisms do not really regulate bird populations. But Watson at least bows in Lack's direction. Note from Figure 14.14 that the breeding success of hens is related to their nutritional status. Breeding success in this species is mediated largely through differential mortality of chicks, not adjustment of clutch size; but Watson suggests that the mortality of chicks is strongly

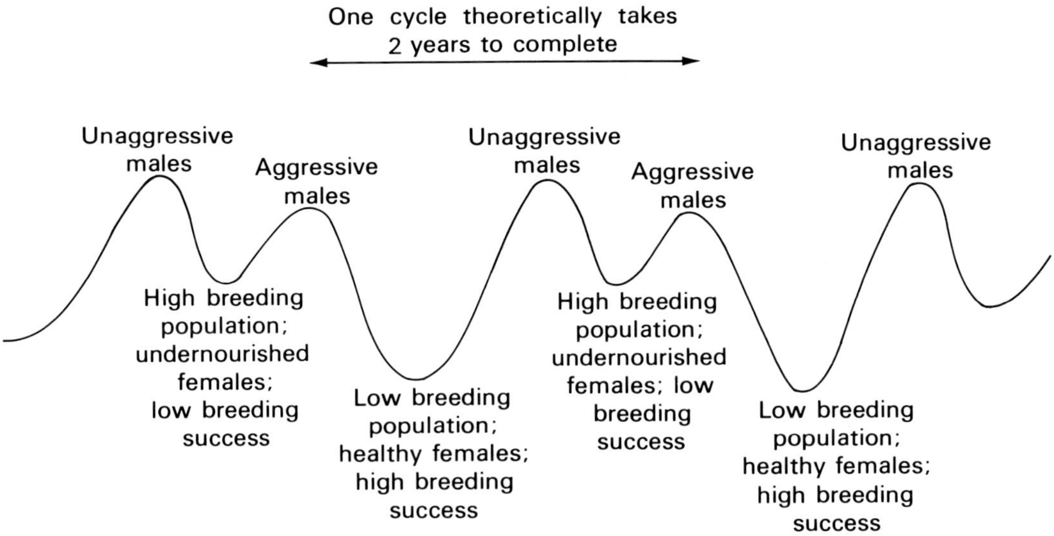

FIG. 14.14. Apparent elements in the self-regulation of red grouse populations, as described in the text. The curve in this diagram is highly schematic, meant only to suggest logical consequences, not real population changes. Because some birds are relatively long-lived, the real population contains a mixture of aggressive and nonaggressive males and the changing proportions of these influence the population in less extreme fashion than this simplified diagram might indicate. For the same reasons, the idea of a cycle completing itself every 2 years probably is over-simplified. These complexities do not make the suggested regulatory mechanism any less valid.

influenced by a pre-existing "egg factor"—better or poorer nourishment of the embryo within the egg, as determined by the foods available to the hen. Thus, breeding success in spring is determined by the nutritional status of the hens, which in turn is related to the number of birds that must share the available food (the breeding stock). Since breeding stock size is determined by the degree of territorial aggressiveness of the males, *i.e.* the size of their territories, we see here an interplay of both external and social mechanisms in regulating the population.

Watson remains uncertain about what causes male survivors of relatively unsuccessful broods to be more aggressive than males from successful large broods. Heightened aggressiveness, he suggests, may be a component of the "egg factor" derived from poorly nourished mothers, or it may be inherent in the posthatch physical robustness developed by male chicks which need not compete with many other chicks for food. Watson essentially rules out the adrenal stress syndrome as a regulator of red grouse populations, noting that in grouse the relationship to population density of aggressiveness—a key element in the adrenal stress complex—is the opposite of that found in Calhoun's rats and Vessey's mice. In the delayed density dependence of red grouse, aggressiveness is greater at low densities than at high. Hence the adrenal stress syndrome is not likely to be functioning at high densities.

Density-independent Factors

The differing viewpoints of Wynne-Edwards, Lack, Christian and others represent disagreements about which density-*dependent* mechanisms really are important in population regulation. These arguments are recent and relatively minor compared to arguments about the significance of density-*independent* factors in population regulation. Here controversies have swirled for years, and we must plunge in once again.

What are density-independent factors? They are adverse agencies whose effect on a population's numbers is proportionally the same at any density. Whether the population is large or small such factors will cause death or retard birth with the same intensity. There is no variable braking pressure that depends on density, *i.e.* no negative feedback.

Weather events—floods, storms, droughts, temperature changes—normally act in density-independent fashion. A late frost will kill the same fraction of a large tomato crop as of a small one—not the same number of plants, but the same fraction. Similarly, most forms of pollution have density-independent effects. In a stream, it does not matter whether a pollution-sensitive stonefly is abundant or rare prior to introduction of a pollutant; most or all of the population will be killed by it regardless of initial density.

Some factors may be density-independent for certain species and density-dependent for others. Consider fire in a forest community like the New Jersey Pine Barrens (Fig. 4.15). The intensity of any fire will be related to the density of dominant species like pitch pine and scrub oak, since they are major items of fuel. Fire damage to populations of these species thus will be density-dependent. But differences in density of a minor species like pitcher plant will have no real influence on the severity of a fire. Even if abundant, pitcher plants provide very little fuel, so to populations of pitcher plant fire is a density-independent factor. One can think of similar exceptions with regard to the supposed density-independence of weather: for example, limited amounts of shelter might be sufficient for a small population and allow it to survive a winter storm nearly unscathed, but would be insufficient to fully protect a larger population. Under these special circumstances the effect of the storm would be density-dependent.

Weather, pollutants, and fire: despite exceptions, these are the agencies most often mentioned when density-independent factors are discussed.

What is the density independence controversy about? It is obvious that density-independent influences can and do greatly affect the numbers of many populations. This is not the issue. But many ecologists claim that only direct density-dependent factors, with their negative feedback effect at both high and low densities, ever can truly *regulate* a population, and that no population can persist very long without such regulation. Their arguments go back to Malthus: any population, if unchecked, will grow exponentially until resource limitation—which is density-dependent—sets in. The question then becomes: can density-independent factors be the *primary* controls of any plant or animal population?

The logical answer—no—is bedeviled in the minds of many ecologists by the behavior of real populations they have studied. The fortunes of insect populations, particularly, often seem tied so closely to weather fluctuations that weather seemingly must be given primacy. The numerical changes of such populations seem explicable only occasionally in terms of density-dependent control, but *often* seem explicable in terms of density-independent control (see, *e.g.* Fig. 14.4). Is it possible that such populations are kept so far below carrying capacity by density-independent factors that density-dependent influences are superfluous (as in Fig. 14.10*D*)?

The conflict between theory and observation is expressed well in a review of insect population dynamics by Varley and Gradwell (1970). On the one hand they say:

> In nature, density-independent mortalities are often the key factor causing population change. Their action has to be mimicked in population models which seek to explain and predict population change.

But they also say:

> . . . to date, mathematical population models which do not include density-dependent mortalities are unstable. They neither allow the continued existence of a species, nor have they been able to mimic or explain population level. Such models predict that population densities should, in the long term, either increase or decrease indefinitely.

Some of the theoretical objections to density-independent control could be met if a density-independent setback to growth were to be applied to a population repeatedly and more or less regularly. Regular repetition of this growth-curbing factor is *essential* if the population is to be kept below levels where it is subject to resource limitation. Only thus can Malthus be confounded. Are there any predictably recurring, density-independent agencies? Storms, floods and fires must be ruled out as too erratic. But seasonal weather changes are very regular and repetitive, and might well serve as major controls of species having an appropriate natural life span. Hutchinson (1965), pursuing this point, defines what is meant here by "an appropriate life span." It is a natural life span that is no longer than the cyclically occurring event, but is not too much shorter either. This definition fits the life span of most temperate zone insects, which grow to maturity (or even produce a couple of generations) during one warm season, then lay eggs and die before winter comes. Why must the life span be relatively short? because if the time to maturity is longer than the intervals between the control event, then individuals in a viable population obviously must survive the control event in appreciable numbers. In fact, temperate species having long life spans have evolved adaptations to seasonal weather changes that minimize mortality from this cause. But if many individuals persist from year to year, they will serve as an enlarging base for geometric growth and the population will progressively expand. If it is to be a long term controlling factor of population size, seasonal weather change therefore must kill virtually the whole pre-existing population (except its overwintering eggs).

Just as generation time cannot be too long, so it cannot be too short. Populations of warm weather bacteria cannot be regulated primarily by seasonal weather change because they multiply so rapidly that they would become huge, and resource-limited, long before winter arrived.

Let us look again at warm weather insects in the temperate zone. Their life span is appropriate because the environmental carrying capacity can remain far above the one, two or four generations they might produce in a summer. But if the eggs overwinter, is it not likely that generations in successive summers will be increasingly large (simply because each generation produces more than the replacement number of eggs)? The answer would be "yes" if the parenthetical statement were true, but often it may be false. Although seasonal change is inevitable, growing seasons are not the same length from year to year. If a multiple-generation insect produces three generations during one favorable summer, the number of eggs hatching the following spring ordinarily will be large. But frosts may occur earlier the succeeding fall, limiting that year's crop to two generations—hence less overwintering eggs. Or unpredictable summer storms or droughts might decimate the large spring generation even before it reached reproductive age. Given the appropriate biological prerequisites—sensitivity to weather and life cycles of the proper length—a density-independent combination of regular and irregular weather events could effectively limit a population nearly all of the time.

But if chance weather events are part of the formula, then chance would sometimes favor continued growth—say, in a consecutive series of favorable years—and the population might expand to levels where density-dependent factors came into play. If we look again at Figure 14.4, an explanation of the sort just given seems quite reasonable: density-independent controls could be predominating 90% of the time, but they occasionally relax and allow the population to skyrocket. The contrast between the huge occasional density and the normal low density of this population suggests that the normal density is far below the environment's carrying capacity in terms of resources. Of course, a pattern like those of Figure 14.4 might be explained also by density dependence: it might be related to a predator which normally kept the population in check but occasionally lost control. But many entomologists believe that the populations they study are controlled primarily by weather, and the evidence they cite is convincing. Just because a population pattern theoretically can be explained by density dependence doesn't mean that this is the best explanation.

Are insect populations the only ones sensitive to control by seasonal events? No, but they have been so intensively studied that we should not wonder that most of the firmest believers in density-independent regulation are entomologists. If they look, however, limnologists may well find examples of density-independent regulation in the plankton of a lake. Populations of some diatoms, for example, quite possibly are restrained each summer not primarily by nutrient limitation but simply because they consist of heavy cells which sink rapidly out of the photosynthetic zone when the water is warm and its density and viscosity are low. Finally, we can think of populations in marginal environments, where the possibility exists even of long-lived species being limited primarily by density-independent events that occur erratically but predictably in a statistical sense. Consider timberline populations of bristlecone pine in the Western United States. These trees grow and reproduce slowly, and often are sparsely distributed. Are bristlecone populations limited by resources, predators or pathogens?—or by the occasional ultra-severe winter, windstorm or lightning storm? Not simply mortality, but recruitment problems also, must be considered. Perhaps most of the time the harsh timberline climate hinders reproduction, and only occasionally does a mild season allow seedlings to become established. To learn the relative importance of density dependent and density-independent influences in keeping bristlecone populations low would require a careful research program. It might, in fact, require a research program involving many generations of human observers, since a successful bristlecone pine may live thousands of years.

Thus proponents of density-dependent control and proponents of density-independent control each seem to have considerable maneuvering room. We may hope that the polarization between them is nearly past. Indeed, many arguments on this matter have been prolonged by semantic differences over what really can be regarded as density-independent and what as density-dependent factors. (See, for example, the exchange between Ehrlich and Birch, 1967, and Slobodkin *et al.*, 1967.) A more constructive, middle-of-the-road approach is that of Horn (1968*b*), whose very simple mathematical treatment (Fig. 14.15) goes far toward explaining, and thereby resolving, the polarity of views that has tended to characterize the question of density-independent control.

PREDATORS AND PREY: ADDITIONAL CONSIDERATIONS

The Sustained Yield Question

This chapter's survey of population dynamics has skimmed many topics, but none more inadequately than predator-prey relationships. Many population ecologists have focused their major efforts in this area. Food webs, after all, are the glue of communities, and what are food webs but interacting sets of predator-prey relationships?

In the discussion that follows we will enlarge the usual definition of "predator" to include herbivores and parasites, and the term "prey" to include plants and the hosts of parasites. A predator population then becomes any animal population that consumes the substance of other living organisms, and a prey population any that is the target of consumers.

In Chapter 7 we touched on the concept of optimization at the community level. Evolutionary change is a process of optimization at the species level, and in the case of predators and prey the expectation might be that over time predators would evolve habits that provide a maximum sustainable yield of their prey. To

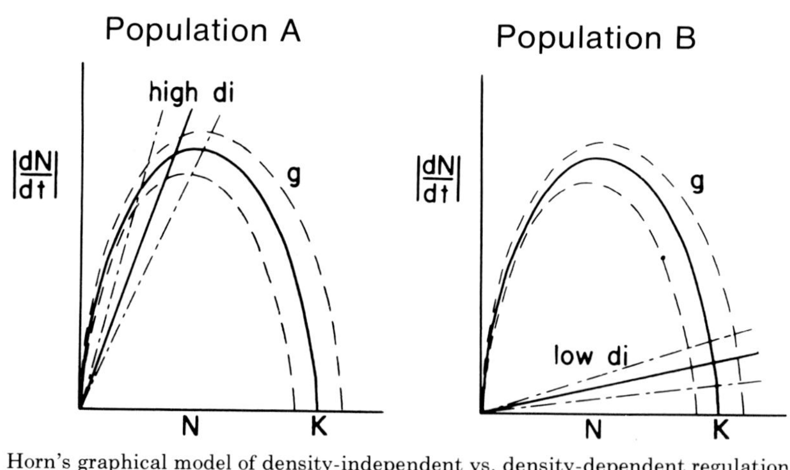

FIG. 14.15. Horn's graphical model of density-independent vs. density-dependent regulation. The parabolic *curve g* is the logistic curve, *i.e.* the theoretical pattern of population growth and stabilization under density-dependent influences. The straight line d_i is the death rate due to density-independent influence—always linearly related to *N*, but very high in the case of population *A* and very low in the case of population *B*. By plotting *g* and d_i against *N* on the same set of axes, it is seen that a high d_i may intersect *g* before negative feedback has seriously slowed population growth. At this point of intersection, *g* and d_i theoretically cancel each other out, stabilizing *N* at a density far below (population *A*) or very close to (population *B*), the carrying capacity (*K*). Of course d_i is not always constant, nor is *K*. Deviations from constancy are indicated by dashed lines in the case of each population. By projecting downward from points of intersection, however, it can be seen that for population *A*, fluctuations in density-independent death rate (d_i) affect *N* more than do fluctuations in density-dependent growth rate (*g*), whereas the opposite is true in the case of population *B*. Hence *A* is susceptible to density-independent control but *B* is not. (From H. S. Horn: *Ecology 49:* 776–778.)

some degree this has happened: Connell (1961b) found that a barnacle-eating snail, *Thais lapillus*, selectively chooses larger (hence older and slower-growing) barnacles as food. By keeping its prey population in a state of relative youth and rapid growth, the *Thais* population ensures more food for itself over time. But the snail population almost certainly does not adjust its rate of predation to achieve a *maximum* sustainable yield. Probably it could not do so even if snails were infinitely wise. There are too many unbalancing extras. Other species populations in the community affect the numbers of snails and barnacles with varying severity, as do the continual abiotic changes in this intertidal environment. It is unreasonable to expect that a natural predator-prey relationship could operate according to the principles of maximum sustainable yield, and we might conclude that evolutionary selection cannot even work strongly in this direction. Even in environments more stable than the intertidal zone, community processes seem far too complicated and population sizes subject to too many influences.

One class of predator-prey relationship, that of internal parasites and their hosts, may be a partial exception to this statement. An internal parasite inhabits an *unusually* stable environment, and one in which the complication of other species may be reduced (though probably no host contains only one kind of parasite). Given these conditions, a parasite species really may evolve toward a maximum, or at least highly efficient, sustaining yield relationship with its host species. The principle of parasite and host adapting to each other is well established: in a highly evolved relationship the parasite satisfies its needs and the host is not seriously debilitated. Charles Elton has pointed out that the essence of successful parasitism, as opposed to other forms of predation, is that the parasite manages to live off income while the typical predator consumes capital. If it really is possible for internal parasites to adapt their demands closely to the sustained supply abilites of the host, the reason certainly is that other influences intrude relatively little on the relationship. For free-living predators there really is no evidence that maximum sustaining yield relationships are being optimized by evolution.

Models of Predator-Prey Dynamics

The gist of the foregoing is that in a complex community subject to fluctuating abiotic influences it is unrealistic to think of a predator population maintaining its prey population in some kind of optimal, stable age structure and density. The predator population is but one of many influences affecting its prey, and even by itself the predator population is not sufficiently flexible always to apply the appropriate pressure for optimal "management." In reality, the relationship of predator to prey often is one of short term opportunism: when other factors do not impede, predators frequently overexploit the prey population and then either suffer from reduction of their own food source or turn to other sources of sustenance while the prey population recovers. If other influences prevent overexploitation, they often do so by diminishing the numbers of predators to levels that are insufficient to be major controls of the prey population.

Mathematically minded ecologists have long been interested in the intertwined fortunes of predator and prey, but the problems of modeling their population phenomena are very great if one tries to incorporate all the complexities of the real world. Perhaps, however, even simple models can sharpen our insight.

The earliest attempt to analyze predator-prey dynamics mathematically was the model of Lotka and Volterra in the 1920's. This model simply ignores all extraneous influences, treating the predator and prey populations as two parts of a closed interacting system. Figure 14.16 illustrates its prediction, namely that the abundances of predator and prey will each fluctuate over time in regular, cyclic fashion. In addition to the premise of no outside influence, the Lotka-Volterra model assumes that the functional relationships between predator and prey do not vary at all as density varies. Direct density-dependence is assumed at all times between the two species, and, in addition, there is no provision for the operation of Allee's Principle whenever either population drops very low. More recent mathematical treatments of predator-prey interaction make allowance for certain of these possibilities, but remain very general. The most influential of recent models are those of Rosenzweig and MacArthur (1963;

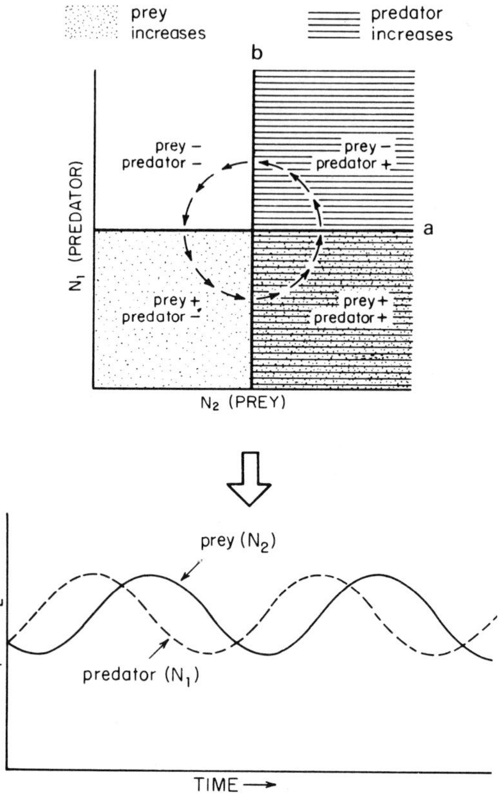

FIG. 14.16. Graphical representation of the Lotka-Volterra predator-prey model, as developed by Wilson and Bossert (1971). In the *upper graph* the numbers of predator and prey form the two axes. The *horizontal line* extending to the ordinate from point *a* represents a critical density of predators. If there are more predators than this, their grazing pressure will cause the prey population to decline. If there are less predators, the prey population will be able to increase. The *vertical line* extending from point *b* to the abscissa represents a similar critical density of prey. To the right of this line the prey population is large enough to allow growth of the predator population; to the left it is too small. In each of the quadrants so defined the trend of predator relative to prey is indicated by the arrows. Arrows pointing both *up and out* indicate that both populations increase. Arrows pointing *down and in* indicate declining trends in both populations. At times, however, the predator is increasing while the prey is decreasing, and vice versa. This pattern is clear in the *lower graph* which shows oscillations of the two populations over time and which also makes clear that the critical densities *a* and *b* of each species in the *upper graph* are predicated on the premise that both species are at particular points in their coupled cycles. The assumption is that each species is the only factor influencing the numbers of the other. (From E. O. Wilson and W. H. Bossert: *A Primer of Population Biology*, Stamford, Conn.: Sinauer Associates, 1971.)

see also Rosenzweig, 1969 or Wilson and Bossert, 1971). Each of the newer models predicts, like that of Lotka and Volterra, cyclic oscillations of interacting predator and prey populations over time; but depending on the premises the oscillations may remain stable in amplitude, may gradually decline in amplitude, or may increase until one or both populations becomes extinct as it plunges from an ever higher peak to an ever deeper trough. With caution, these models can be applied to the real world. They suggest what might be the effect of predator populations and prey populations on one another under different sets of circumstances, given the assumption (not always true) that a particular predator population and a particular prey population do influence each other's numbers significantly. Perhaps the most realistic of Rosenzweig and MacArthur's models is one which postulates a refuge for the prey such that when their numbers drop below a certain level they are essentially inaccessible to predation: all prey individuals have secure hiding places, or their numbers are so low that the predator goes elsewhere to hunt. This model, like the simpler one of Lotka and Volterra, predicts cyclic oscillations of constant magnitude, *i.e.* a see-saw pattern over time. In view of the existence of cyclic population phenomena in nature (*e.g.* Figs. 3.23 and 14.5) it is interesting how many of these predator-prey models predict regular oscillations of predator and prey densities.

How predators really behave

If a predator population is heavily dependent on a single type of prey, then changes in the population size of the prey normally should have a major influence on the population size of the predator. Barring other controls, the predator population will be high when food is abundant; if the prey population falls, the predator population also should fall. But fluctuations of the two populations should not be completely in phase, as is predicted in Figure 14.16 and evident in Figure 3.23. Changes in the predator population mostly lag behind those of the prey. This phase shift simply may be the lag that occurs between any stimulus (here the upturn or downturn of prey density) and any response. But the response of vertebrate predators to a prey increase involves a more pronounced lag, because the prey normally are capable of more rapid reproduction than the predator. If the prey population is the main food source of the predator, however, the predator curve tends to catch up with that of the prey during periods of downturn, because numerical losses by starvation or emigration are comparatively rapid predator responses to food shortages, more rapid than is reproductive increase during periods of food abundance.

One of the destabilizing influences in predator-prey population dynamics is *delayed density dependence* of predator pressure. In the example of red grouse self-regulation we saw that delayed density dependence tends to perpetuate oscillations in population density, and actually to increase the amplitude of oscillation. In the present case, because predator increases lag behind those of the prey, the predator population tends to reach greatest abundance when the prey population already is falling, resulting in extremely high predator pressure at this time. If the system were operating as a simple, direct density-dependent system, the predator pressure would lessen as prey density declined. Here, at least initially, the reverse is true. Due to very heavy predator pressure the prey population plunges precipitously, being released and starting recovery only when the predator population collapses and/or turns to other food sources. If the latter occurs, delayed density-dependence may also act as the prey population begins to recover: preoccupied with other foods, the predator initially neglects the increasing prey population, allowing it to get a strong start because of abnormally weak predation. Delayed density-dependence, of course, is a delay in the operation of *direct* density-dependence; during the delay inverse density-dependence is operating temporarily and exerting its destabilizing effect. Delayed density-dependence should be a particularly common aspect of predator-prey relations when the predator is a long-lived, slow-reproducing vertebrate; but it also is common to situations involving relatively short-lived predators such as insects. We see again here how delayed density independence fosters population fluctuations, not numerical constancy.

Crawford Holling (1959 and 1965) has provided considerable insight into the way predators actually respond to changes in prey density. In a pine plantation in Ontario Holling studied predation on the cocoons of a dominant insect, the pine sawfly, by three small mammals: the masked shrew, the short-tailed shrew and the deer mouse. He censused the mammals by trapping, and censused cocoon density by counts in sample quadrats. The cocoons were spun on the forest floor and were the only stage of the life cycle on which the mammals preyed. Hence, unlike some of the predators we have been discussing, these mammals must have relied on alternate prey in some seasons. When they preyed on sawfly cocoons they ate only the contents, and Holling could count the empty cocoons and identify the specific predator by the way each cocoon had been opened.

In analyzing this system, Holling realized that he was dealing with two different aspects of predator response to changes in prey density. The first he called the **functional component:** how did the predators respond *individually* to an increase in cocoon density? Did each animal's rate of intake increase linearly with the supply, or did the predators shift from their alternate diets to sawflies in some nonlinear way? Figure 14.17 indicates that for each species the shift was nonlinear, but to a different degree. The contrasts perhaps reflect differing preferences for the sawfly diet or differing abilities to find cocoons when scarce. On an individual basis, of course, each animal reached a saturation level, so these functional curves assumed a sigmoid shape.

The second aspect of predator response is the one we have been concerned with up to now: the **numerical component.** How do the densities of the three predator species change with a change in prey numbers? Obviously this response is related to the functional response of each predator as an individual. Does it really make use of the extra prey available? How completely and how rapidly does it turn from other, perhaps preferable, foods to the burgeoning supply of this particular prey? Figure 14.18 shows the numerical response of the predator populations. Two of the predator species actually increased in numbers but one (the short-tailed shrew) did not; apparently its numbers were being controlled by something else.

In terms of effect on the *prey* population, what counts is the predators' functional and numerical responses combined. This **total response** can be calculated from the functional

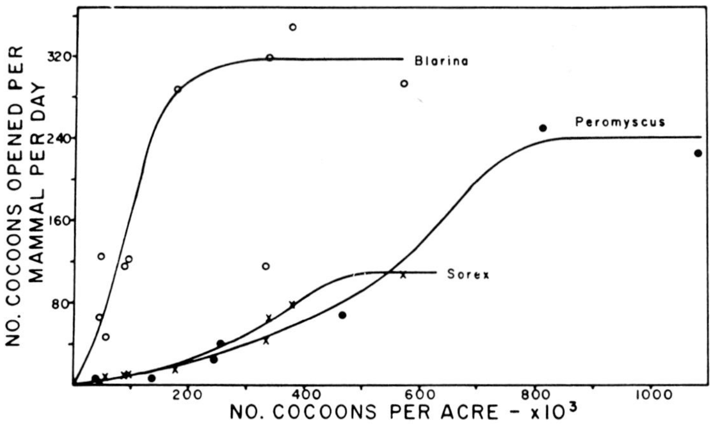

FIG. 14.17. Functional responses of the predators in Holling's study to changes in prey density. (From C. S. Holling: *The Canadian Entomologist, 91:* 293–320, 1959. Reproduced with permission.)

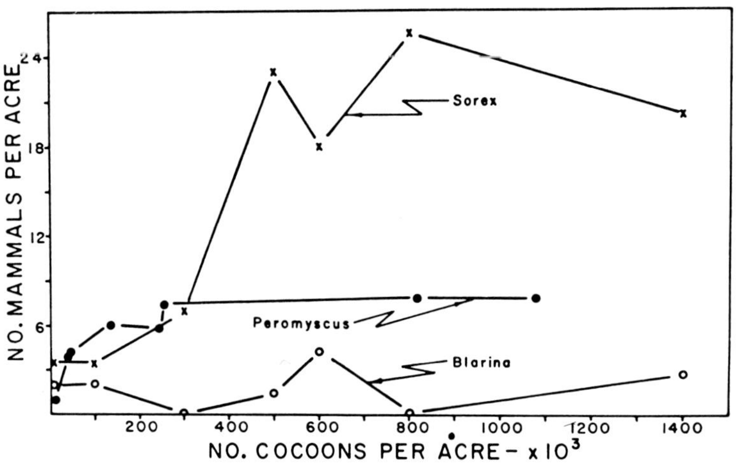

FIG. 14.18. Numerical responses of Holling's predators to changes in prey density. (From C. S. Holling: *The Canadian Entomologist, 91:* 293–320, 1959. Reproduced with permission.)

and numerical data for different prey densities, and is shown in Figure 14.19. It can be seen that for all predators combined, predation intensity rises to a peak at an intermediate density of prey, then falls off. Beyond this peak the sawfly population progressively escapes control by its predators in inverse density-dependent fashion. Functionally the existing predator individuals are sated; numerically, predator reproduction cannot keep up with the prey. Holling makes an interesting point here: the total response of each predator species peaks at a different prey density, thereby expanding the range of prey density which is subject to direct and relatively effective density-dependent predator influence. *Blarina* is most effective at low densities, *Sorex* at intermediate densities, and *Peromyscus* at relatively high densities, though at still higher densities the prey population escapes even from *Peromyscus*. From a community standpoint Holling has uncovered a subtle niche difference among these species in their relation to a particular prey. From a population standpoint he arrives at a much broader generalization: the more kinds of predators that make use of a given prey, the more effectively will the prey be controlled by their combined response. The converse obviously is true as well: the more types of prey a given predator makes use of, the less control a particular prey population will have over the numbers of the predator. These are simply new derivations of ideas already familiar to us: the observation that populations tend to be more stable in a complex community than in a simple one, and the notion that a generalized diet buffers, for a species, the effects of changes in its resource base. It is easy to understand why populations of mammals in the Arctic, where there are relatively few species of predator and prey, often fluctuate so strongly, and it also is easy to guess which of the Arctic species in Figure 14.5 is the greatest food generalist.

Conclusions

Despite their simplicity, mathematical models of predator-prey dynamics develop predictions which seem consonant with many real predator-prey population patterns. Smith (1974) summarizes these predictions as follows.

1. If the predator is limited by the supply of prey, the interaction may lead to regular fluctuations in numbers of both species.
2. "If the prey is resource-limited, and not limited by the predator, this will tend to damp out the oscillations."
3. "If the predator is limited by some factor other than the prey, this will tend to damp out the oscillations."
4. "If there is cover available which makes a constant number of prey unavailable to the predators, this will tend to damp out the oscillations," (but will not eliminate them, according to Rosenzweig and MacArthur).
5. Fluctuations may become greater and greater, perhaps culminating in extinction of the prey population, if the predator is able to maintain pressure at low prey density (by hunting the prey very efficiently and perhaps making use of other prey).

Predators and prey, then, can affect each other's numbers in various ways and with varying degrees of severity. A combination of predator species probably can control a prey species better than can a single predator species, because as Holling showed, the predators' effectiveness as control agents may be maximal at different points in the prey population's growth curve. But some prey populations seem to escape regularly from control by their predators, resulting in cyclic peaks of abundance which exceed the prey's resource base and result in die-backs initially unrelated to predators. This is a frequent interpretation (though not the only possible one) of the population die-backs of Arctic hare in Figures 3.23 and 14.5; die-backs may be initiated by the hare population exceeding the carrying capacity, then accentuated by overbearing predator pressure on the dwindling prey (delayed density-dependence), resulting in a deep trough in the prey curve. If the carrying capacity for the predator is determined essentially by the abundance of one prey species, as it may be in the case of lynx and hare, then the predator population obviously must fall too, when its carrying capacity collapses.

A last example illustrates the enormously wide range of pressure that some predator populations exert on their prey at different

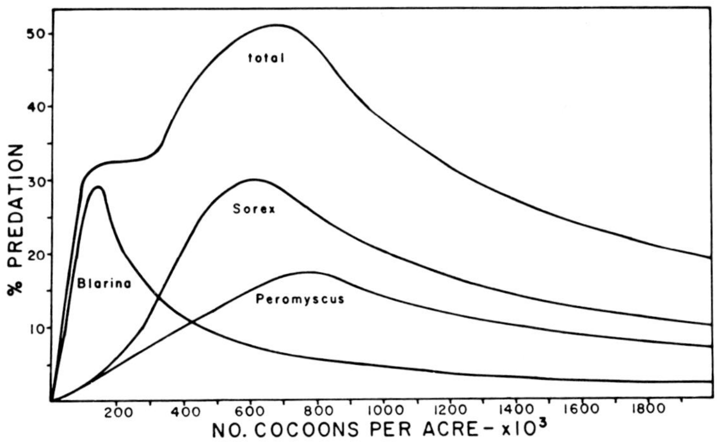

FIG. 14.19. Total responses of Holling's predators to changes in prey density. (From C. S. Holling: *The Canadian Entomologist, 91:* 293–320, 1959. Reproduced with permission.)

times. Over a 3-year period in a Berkeley, California park, Oliver Pearson (1966) recorded a 47-fold variation in the number of carnivores (the numerical component) and a 5-fold variation in the number of meadow mice eaten by the average carnivore per month (the functional component). Overall predation on meadow mice (the total response) varied 214-fold during this period. In 1961–62, when a large meadow mouse population was crashing, predators (mostly cats) helped the crash along by consuming over 90% of the available meadow mice in each of 4 consecutive months. Less than half a year later, as the meadow mouse population began to recover, less than 5% of the available meadow mice were being consumed per month: most predators had sought better hunting grounds or other types of prey after the crash. Pearson concluded that in this case predators, "instead of stabilizing the mouse cycle tended to accentuate it by lowering the troughs and allowing the peaks to grow with relatively little control."

If we think back to the conclusion of HSS that herbivores are predator-limited, and if we realize also that most of the prey species we have been discussing are herbivores, it is obvious that contradictions exist between the findings of men like Pearson and the generalizations of HSS. This does not mean that either is wrong. It does mean that generalizations have exceptions, sometimes many of them. This being the case, population dynamics remain a fertile field for study and further discovery.

SUGGESTED READINGS

Allee, , W. C. *et al.* Principles of Animal Ecology. Philadelphia, W. B. Saunders Co., 1949. A major reference work with a good section on populations, written earlier than many of the controversies aired in the present chapter.

Burnet, F. M., and D. White. The Natural History of Infectious Disease. Cambridge University Press, 1972 (4th edition). A highly readable introduction to the populational and evolutionaly aspects of disease.

Cushing, D. H. Fisheries Biology: A Study in Population Dynamics. Madison, University of Wisconsin Press, 1968. A relatively brief text on a complex subject.

Gulland, J. A. Population Dynamics of World Fisheries. Seattle, Division of Marine Resources, University of Washington, 1972. Useful transcript of a series of lectures by a world expert on this subject.

Krebs, C. J. Ecology: The Experimental Analysis of Distribution and Abundance. New York, Harper and Row, 1972. One of the best treatments of population ecology in any textbook.

Lack, D. The Natural Regulation of Animal Numbers. Oxford University Press, 1954. This and the next book set forth Lack's ideas, which are very different from those of Wynne-Edwards.

Lack, D. Population Studies of Birds. London, Oxford University Press, 1966. Interesting case studies and a long appendix dealing with population regulation controversies.

McLaren, I., ed. Natural Regulation of Animal Populations. New York, Atherton Press paperback edition, 1971. An excellent collection of papers that presents the various sides of the controversies over population regulatory mechanisms.

Murdoch, W. and A. Oaten. Predation and population stability. In A. Macfadyen, ed. Advances in Ecological Research, Vol. 9. New York, Academic Press, Inc., 1975. A long and quite technical review.

Varley, G., and G. Gradwell. Recent advances in insect population dynamics. *Annual Review of Entomology* 15: 1–24, (1970). A good review and introduction to the insect literature.

Warren, K. B., ed. Population Studies: Animal Ecology and Demography (Cold Spring Harbor Symposia on Quantitative Biology, Vol. 27). Cold Spring Harbor, New York, The Biological Laboratory, 1957. A landmark symposium.

Watt, K. E. F. Ecology and Resource Management. New York, McGraw Hill, 1968. Deals with problems of sustained yield, pest management, and other practical aspects of population ecology.

Wilson, E., and W. Bossert. A Primer of Population Biology. Stamford, Conn., Sinauer Associates paperback edition, 1971. A brief introduction to the mathematics and models of population ecology. Relatively elementary, but still not easy reading.

Wynne-Edwards, V. C. Animal Dispersion in Relation to Social Behavior. Edinburgh, Oliver and Boyd, 1962. Wynne-Edwards' big book. Shorter expositions of his arguments may be found in *Science* 147: 1543–1548 (1965) and *Scientific American* 211:68–74 (Aug., 1964).

chapter 15

Populations: Adaptation and Extinction

. . . there is relative constancy in the abundance of species populations. The word "relative" must be emphasized because changes in numbers must be related to a species' real potential for fluctuations; to paraphrase Lack, the changes observed are mere "ripples" compared to the possible "waves".

David Pimentel (1968), *Population Regulation and Genetic Feedback*

. . . to the heath hen something more than death has happened, or, rather, a different kind of death. There is no survivor, there is no future, there is no life to be recreated in this form again. We are looking upon the uttermost finality which can be written, glimpsing the darkness which will not know another ray of light. We are in touch with the reality of extinction.

Henry Beetle Hough (1933), *A Bird That Man Could Kill*

INTRODUCTION

The quotation from Pimentel serves equally well as an epilogue to Chapter 14 and a prologue to this one. It is a useful bridge, for we are not yet finished with population dynamics. In this chapter we will consider, among other things, the role genetic change may play in muting the fluctuations of populations. We also will consider more long term aspects of population adaptation. By this we mean, essentially, Darwinian evolution: for it is at the level of the population that natural selection operates to bring about genetic adaptation. Finally, we will look at the process of extinction, and some of the factors that bring it about.

Pimentel's statement reminds us that although populations indeed fluctuate, sometimes markedly, they generally change in numbers far less than they actually could, given the exponential nature of unrestricted population increase. MacArthur and Connell (1966) have calculated that "if bacteria reproduce every 20 minutes, exponential growth for 36 hours would produce a layer of bacteria a foot deep over the entire earth, which within the next hour would be over our heads." These authors also recall Darwin's calculation that a pair of elephants would give rise to a population of 19 million after only 750 years, if their reproduction and that of their descendents was environmentally unimpeded—and this despite the constraints of elephant biology, namely a generation time of 30 years and a biological limit of six young per female. That mankind has never suffocated in a soup of bacteria nor perished beneath thundering herds of elephants is eloquent tribute to the forces that restrain all populations from unlimited growth. Nor, as species, have elephants or many bacteria succumbed to the forces of species extinction. The fortunes of these species have waxed and waned, but over time the waxings and wanings have been far more conservative than is theoretically possible. This conservatism of population numbers, with respect both to maxima and minima, is perhaps the strongest card in the hand of ecologists who insist that direct density-dependent mechanisms are primary regulatory agents of *all* populations.

Yet extinction is not infrequent, both of populations and less often of whole species or subspecies, like the heath hen. Even the most enduring of "living fossils" is not immortal. A common estimate is that 99% of all species that have appeared on Earth are now extinct; the diversity of our present biosphere is but a tiny fraction of the cumulative diversity of life. Hence we cannot ignore extinction, although early in this chapter we will be seeing how populations are genetically buffered against it.

GENETIC BUFFERING OF POPULATION DYNAMICS

Genetic Feedback in the Simple Sense

The idea that natural selection is significantly involved in short term population fluctuations, *i.e.* that such fluctuations are explicable partly in terms of genetic change, came first from E. B. Ford. Beginning more than 50 years ago, Ford's careful studies of wild populations of butterflies first demonstrated the occurrence of rapid genetic change in fluctuating populations. This led him to believe that natural selection can act with surprising speed and that genetic adaptation might be important in relatively short term processes of population regulation. Ford's ideas, recounted in his book *Ecological Genetics* (1971), have been seized and elaborated upon by numerous ecologists in recent years.

Let us first consider how genetic "buffering" might operate during a period of declining numbers. At such times, mortality within the population should not be occurring randomly. Rather, it should be occurring by survival of the fittest and elimination of the less fit, resulting in a natural selection of particular genotypes and an "adaptive" honing of the population as it declines. (A **genotype** is an individual or group of individuals having a distinctive combination of genes.) Regardless of what agency is doing the selecting, the population as it dwindles comes increasingly to consist of individuals unusually resistant to the mortality factor. This is true even if no new individuals are born during this period: the population acquires increasing resistance to further mortality because only relatively resistant individuals remain. We are not suggesting acquisition of greater resistance by *individuals*, but rather a progressive slowing of the rate of population decline once the less resistant members are eliminated. Hence the population decrease tends to "bottom out" unless the environmental change is so pronounced that no members of the population can resist it.

If the environment remains harsh the remnant population, relatively slowly, may breed genotypes that are even more resistant, and the population may slowly recover numerically. Ordinarily this is not what happens. Rather, the environmental stringency relaxes and the population is able to grow without further genetic adaptation. In fact, as Ford first discovered in studies of the butterfly *Euphydryas aurinia*, a population growing rapidly after a period of respression usually becomes highly variable. Natural selection is relaxed during this period of low environmental resistance, and many unusual genetic combinations may survive from matings which in harsh times would leave no viable offspring. Thus, there is a weakening of the average fitness of the population which automatically makes the population more vulnerable to any new environmental stringency. When such a stringency comes, then, the population is bound to stop growing and go into a decline. Thus, natural selection acts to make population peaks less high, as well as population troughs less deep.

Ford believes there is a reciprocal phenomenon as well: a speeding up of evolutionary adaptation made possible by population fluctuations. As a population grows under relaxed selection, most of the new genotypes that appear may be expected to be ecologically weak, but by chance one or two may be more fit in some respect than any pre-existing genotypes. These new genetic combinations will spread through the population during periods of population downturn, if the downturn is caused by agencies to which the new genotypes are especially resistant. Over time, then, the population becomes better adapted to these agencies. Evolution in this manner can occur rapidly as a result of population fluctuation. In Ford's study of *Euphydryas aurinia* the average individual born after a population surge and subsequent decline was quite different in appearance from the average individual born just 5 years previously, when the population was at a low ebb. Almost certainly the change in appearance was due to selection, though its adaptive significance is unknown.

To summarize, genetic pruning during episodes of population decline normally builds a progressive resistance to further population reduction, and relaxed selection during periods of population increase normally allows unfit individuals to proliferate, almost ensuring an eventual population downturn. These changes may be termed **genetic feedback**. Genetic feedback is a negative feedback process of direct density-dependent control: it helps reduce the amplitude of population fluctuations even while ensuring that some fluctuation occurs. Over longer time periods, however, these fluctuations can produce overall adaptation by introducing and spreading successful new genotypes in the population, and this adaptation dampens the amplitude of fluctuation still further. At least in part, this progressive adaptation lies behind the progressive stabilization of many populations introduced into new environments (*cf.* Fig. 14.10*B*), or exposed to new types of mortality such as introduced diseases. Pimentel (1961) seems to have had only this progressive, longer term type of genetic adjustment in mind when he coined the term "genetic feedback." Here we use it in a short term sense as well.

Let us consider two examples of genetic feedback, first the short term aspect and then Pimentel's longer term aspect. In British Columbia, W. D. Wellington (1964) found he could visually characterize populations of tent caterpillars as genetically "healthy" or "unhealthy" simply by comparing the sizes and shapes of a sample of tents. Tent characteristics proved indicative of the degree of activity of their builders: large, elongate tents being built by highly active caterpillars and small, compact tents by sluggish caterpillars (Fig. 4.17). Adult moths of each caterpillar type retain the juvenile temperament. There is an advantage to having both types of individual in the population, since highly active adults travel far and enlarge the area of habitation, whereas the sluggish members remain where they were born and exploit the habitat very effectively if it is a favorable one. But over time, a local population tends to become dominated by sluggish individuals, since it continually is losing active members (and their genotype) by emigration. The vitality of the local population declines: sluggishness in a favorable environment may not be harmful, but a population dominated by sluggards is ripe for demise if the environment turns sour. This periodically happens in British Columbia in the form of weather shifts. Sluggish populations even in normally optimal habitats frequently are decimated at such times. So are many vigorous populations, if they have colonized marginal habitats during periods of favorable weather. Ordinarily, however, at least some vigorous individuals have recolonized optimal areas that are dominated by sluggish populations. It is these vigorous individuals or their offspring that survive to initiate a new outbreak when better tenting weather arrives.

A likely interpretation of this story is that periods of climatic favorability allow large but relatively unhealthy populations to develop, and periods of climatic stringency prune the population back to a core of vigorous, fit individuals. Here genetic feedback, a direct density-dependent process, combines with the density-independent mechanism of weather to influence population size over time.

Human medical history contains many examples of the longer term, adaptive process of genetic feedback. Populations exposed to an initially devastating disease gradually accommodate so that it becomes a less serious mortality factor. In an ecologically oriented treatment of infectious disease, Burnet and White (1972) discuss the case of yellow fever:

> For two and a half centuries, yellow fever held possession of the Spanish Main, slaying English, Spanish and Carib, soldier, buccaneer and merchant alike, but strangely sparing the negro slaves. In time there grew up a local population in the West Indies which seemed to be immune to its attacks but almost every newcomer had to face its peril.

Why were the negro slaves resistant to yellow fever from the beginning? Because it was a disease endemic to West Africa, and transplanted black populations had already gone through the process of evolving resistance through natural selection. The rest of the Caribbean community went through the adaptive

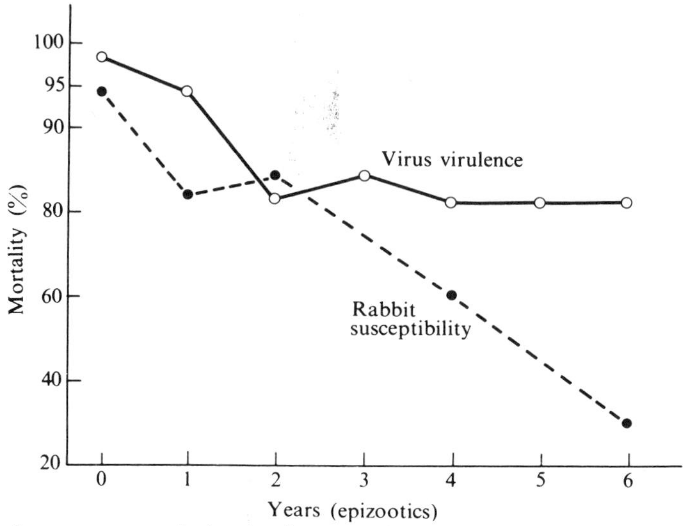

FIG. 15.1. Host:pathogen accommodation in the case of myxomatosis. Introduced to Australia in 1859, rabbits quickly spread into all but extreme desert and rain forest regions of the continent and became a major pest. A rabbit-specific myxoma virus, deliberately introduced as a control measure, became established in wild populations in 1950–1951, spreading extremely rapidly and causing enormous mortality. But mortality dropped in ensuing years. Virus strains taken in later years from infected rabbits in the field proved less virulent than the initial strain, when tested on standard laboratory rabbits. Likewise, rabbits born into remaining wild populations proved, in successive generations, to be progressively more resistant when experimentally infected with a standard, initially highly virulent virus strain. This is a case of extremely rapid natural selection for resistance to a pathogen. The rabbit now is a pest again in parts of Australia. (From F. M. Burnet and D. O. White: *Natural History of Infectious Disease*, 4th Ed., Cambridge, Cambridge University Press, 1972).

process there, since yellow fever had not been endemic to the lands of their ancestors.

In the United States tuberculosis furnishes a similar example. It was introduced to American Indians by contact with Europeans, apparently not being known to the continent previously. In the first three or four generations after being introduced to the disease, American Indians were inordinately susceptible to contracting it, and more often died from it than did infected nonIndians. According to Burnet and White, this genetic susceptibility now seems to have vanished. Tuberculosis remains a major disease among American Indians, but its importance now is due chiefly to environmental influences (such as poor living conditions) rather than genetic hypersusceptibility.

The acquisition of disease resistance by a population has been documented especially well in studies of the interaction of rabbits and the myxoma virus introduced to control them in Australia (Fig. 15.1). Here there is evidence of the virus adapting to its host as well, by becoming less virulent. Pimentel (1968) makes clear that this sort of mutual evolutionary adjustment should have a progressively stabilizing effect on populations of any two interacting species, whether the interaction be parasite-host, predator-prey or competitive in character.* But stabilizing genetic adjustment need not involve another species; another example is the acquisition of resistance to DDT by populations of many mosquito species.

"Active" Genetic Feedback

The types of genetic feedback we have described so far are fundamental to the theories of natural selection and evolution, and the ideas may be quite familiar. But in some situations, genetic feedback may be integrated into population dynamics more intimately or "actively," different genotypes with different behavioral attributes actually serving to initiate the upswings and downturns of a population curve (Chitty, 1967).

The work of Charles Krebs and his associates

* There obviously are exceptions to this statement, such as the cicada and salmon populations discussed at the beginning of Chapter 14, but they probably are not very common.

serves as the strongest basis of this hypothesis. These investigators have concentrated their studies on small rodents (Krebs *et al.*, 1973; Krebs and Myers, 1974), particularly lemmings and meadow mice. Both types of rodent tend to follow quite pronounced numerical cycles, and for two species of meadow mouse there now is evidence both of pronounced genetic change during a population decline (meaning stronger selection against some genotypes than others) and of recurring changes during successive cycles in the propensity of the population to disperse. Moreover, the differences in dispersal propensity have been linked to these different genotypes. The suggestion Krebs *et al.* offer is that in populations of meadow mice, and probably lemmings, dispersal during periods of increase is a major mechanism in minimizing serious overshoot of the carrying capacity, and that genotypes with a propensity for dispersal are selected for during periods of population increase. These genotypes seem to carry aggressive qualities making their possessors relatively intolerant to crowding. In a real sense, Krebs *et al.* suggest, there may *within* these populations be subgroups corresponding to *r* strategists and *K* strategists, and selection at different stages in the cycle favors first one subgroup and then the other. Dispersal-prone, *r* strategist genotypes reproduce disproportionately during the period of population growth, but their dispersal eventually is sufficient to end population growth, since what is left behind are slower-reproducing *K* strategist genotypes. These individuals may remain at high density for some time, but the population eventually declines. Although the mechanisms of this decline are still obscure, Krebs and his co-workers seem to believe that both the increasing and decreasing phases of these rodent cycles are partially programmed through selection pressures that favor one genotype during periods of growth and another during periods of decline.

We may note that in this example one of the ideas basic to "simple" genetic feedback—that selection is relaxed during periods of population growth—is contradicted. Obviously during periods of population increase selection *can* be intense for traits associated with social dominance or tolerance to crowding, though probably not for other kinds of traits. One of the chief differences between "simple" and "active" genetic feedback is that selection pressures are maintained at all times in the latter.

An hypothesis of cyclic genetic feedback in prairie chicken populations can conclude this discussion. Robel (1972) has proposed that genetically based changes in level of aggression

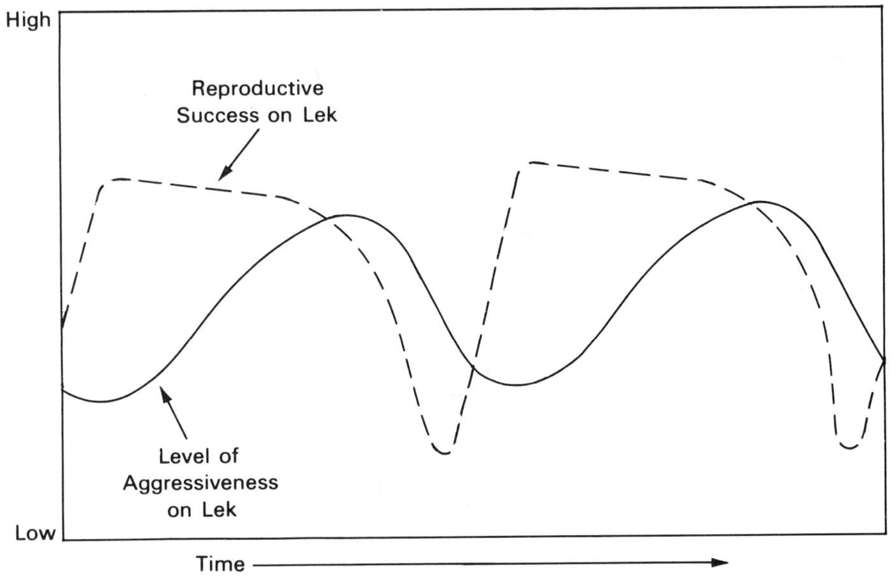

FIG. 15.2. Robel's hypothesis of cyclic changes in aggressiveness as a regulator of prairie chicken populations. See text for explanation.

help to regulate numbers in these populations also. From Chapter 13 we recall that prairie chickens breed on a lek, where a few males from the total pool dominate the breeding. Competition among males for these breeding rights is intense, and Robel believes that the most aggressive do most of the breeding. Since aggressive males pour their genes into the population pool, and since in each generation only the most aggressive males get to breed, selection brings about a progressive and rapid increase in population aggressiveness. According to the hypothesis, aggressive interactions on the lek reach such high levels every 5 to 10 years that reproductive activity is greatly impaired (Fig. 15.2). Males are constantly fighting; females, too, may devote more energy to pecking order squabbles then to actual breeding. A population crash ensues; of the small amount of successful breeding that occurs, most is "offstage" between subordinate males and females unable to penetrate the lek. Subordinates, of course, have relatively low levels of aggression; their offspring, too, should be unaggressive. This period of population collapse, then, represents a brief phase of selection away from aggressiveness. Subsequently there is a resumption of successful lek breeding by the now less combative population. Competition on the lek renews selection for aggressive traits and the cycle starts another turn.

Robel acknowledges that this scheme is unproven, and it has been criticized by workers who doubt that the lek system is as effective as he believes in selecting for aggressiveness. But Robel's hypothesis, as well as the ideas of Krebs, show how genetic change in a population might be implicated not just in preparing the ground for numerical reversals but actually in starting those reversals.

GENETIC POLYMORPHISM WITHIN POPULATIONS

Genetic variety is any population's chief buffer against extinction. Every individual, usually in very small ways, differs from other individuals in its capacity to survive the varied stresses of the environment. The mating of different individuals constantly reshuffles the variety in a population, producing new genetic combinations that may be adaptive in new ways. When we speak of natural selection we mean enhanced survival and reproductive output of individuals having successful gene combinations, and the proliferation of these gene combinations in the population. If the population reproduces sexually there is a constantly self-renewing range of variation that is subject to evolutionary adaptation via selection.

Genetic polymorphism is a special aspect of this intrapopulation variety. Genetic polymorphism is the occurrence in the same population of two or more distinct forms of a species, each form being maintained by natural selection. Each form typically is well adapted to some aspect of the environment but more vulnerable than coexisting forms to some other aspect. A balance of selective forces exists, such that each polymorphic type is maintained in the population at a frequency greater than could be expected simply by chance. This is because each polymorphic type has some special advantage and is being favored by selection. Of course, if an environmental change causes one form to lose its selective advantage, it may rapidly become very rare in the population: **balanced polymorphism** will give way to **transient polymorphism** as the population is remolded by selection.

Besides being unduly common in a population (when they exist at all), polymorphic types differ from each other sharply. As the word implies, "polymorphic" initially referred to abruptly discontinuous *visible* differences among members of a population. We realize now, however, that many genetically based biochemical or physiological discontinuities in populations are completely invisible (though detectable by other means), yet represent the same evolutionary phenomenon as visible polymorphisms. Today, therefore, all distinct, selection-maintained types within a population are considered examples of genetic polymorphism, whether or not these types actually *look* different. Hence we would include Krebs' *r* and *K* genotypes in meadow mouse populations as cases of polymorphism.

The meadow mouse example illustrates the fundamental way in which genetic polymorphism is maintained: each type has the selective edge at some time or in some circumstances. Two further examples of balanced polymorphism may be useful, however. In England, populations of the land snail *Cepaea nemoralis* are polymorphic with regard to shell color, some being brown, others pink or greenish, and others boldly banded with black. In a famous study, Cain and Sheppard (1950) discovered the apparent reason: selective predation on these color phases by thrushes. Each type was protected in certain habitats by its color and here the others were more vulnerable. In meadow habitats green individuals blended inconspicously with the grass; in forest habitats brown individuals were overlooked because they were difficult to distinguish from leaf litter; and in mixed weedy habitats the banded form was especially well camouflaged.

This is a nice story, but it sounds anthropomorphic. How do we know the camouflage worked against thrushes, not just against curious humans? Cain and Sheppard were ingenious naturalists as well as students of evolution, and they made use of the habits of the thrush. On discovering a snail, a thrush carries it to a rock which is used as an anvil on which to break the shell. It was relatively simple to demonstrate strong selective predation by comparing the frequency of different shell types strewn near anvil rocks with the frequency of each shell type in the live population. Most populations of *Cepaea* live in habitats that rather strongly favor one of the polymorphic types, but several other types usually persist in the population also. Presumably this is because the habitats are sufficiently diverse to favor the rarer types in certain microhabitats or at certain times of year. In a field, for example, green snails might be better hidden than brown ones during most of the active season, but the reverse might be true in autumn when the grass and weeds turn brown.

In America the tiger swallowtail butterfly is polymorphic (Fig. 15.3). J. M. Burns (1966) elucidated the selective aspects of this story, which proves to be a classic tale of love and war. Only females of the tiger swallowtail are polymorphic, but the two color phases of the female are so different that most observers never suspect they are the same species. Male tiger

FIG. 15.3. Yellow and brown color phases of the female tiger swallowtail (*Papilio glaucus*), an example of genetic polymorphism. (Drawing by Sarah Allan.)

swallowtails (which are always yellow) may share this recognition problem; at any rate, they select yellow females preferentially and tend to ignore the brown ones. What then preserves the brown form in the population? If reproductively disadvantaged it soon should become vanishingly rare.

As a matter of fact, brown females *are* vanishingly rare in certain parts of the species' range. But some populations of tiger swallowtail dwell within the range of another butterfly, the pipevine swallowtail, that is unpalatable to birds. In such regions, adverse sexual selection against the brown tiger swallowtail is offset by the brown form's protective resemblance to the pipevine species. Compared to their yellow sisters, brown females in these populations are relatively immune to bird predation, so the brown form thrives despite its reduced attractiveness to yellow males. As Table 15.1 shows, the polymorphic balance varies greatly among different populations of tiger swallowtails, the brown form being commonest in localities where the pipevine swallowtail is especially abundant.

Among plants many instances of polymorphism have been recognized. One example, found in a number of species, is heterostyly, in which different plants in the same population bear flowers differing in the relative positions of anthers and stigma. Heterostyly appears to favor cross-pollination: it facilitates successful transfer of pollen by insects between different flower types (i.e. different plants), while making pollen transfer between the identical flowers of a single plant more difficult. Probable examples of cryptic (invisible) polymorphism have been recognized also in plant populations, one being the production of seeds having quite different germination requirements. Cavers and Harper (1966) suggest that germination polymorphism probably is especially common and adaptive in weed species, noting that it expands the variety of sites suitable for colonization and favors intermittent germination over a period of years, "with each seed germinating only when its unique requirements of moisture, light and/or temperature" are met. This may be one reason why species like dandelion and ragweed are so ubiquitous.

Chromatography, electrophoresis and serological typing now are used widely to identify cryptic genetic polymorphisms in animals. Electrophoresis of blood proteins was the method used by Krebs and his co-workers to

TABLE 15.1

Frequencies of Brown and Yellow Females in Different Parts of the Tiger Swallowtail's Range

Different frequencies in different regions represent local balances, determined primarily by the abundance of the pipevine swallowtail, a species which is mimicked by brown females of the tiger swallowtail. The pipevine swallowtail is essentially a southern species, but is rare in peninsular Florida. For explanation see text. Adapted from J. M. Burns (1966).

Locality	Number of Females Captured	Yellow Females	Brown Females
Northern New England	—	*c.100%*	*c.0%*
Baltimore, Md.	29	55	45
Mt. Lake, Va./W. Va.	84	14	86
Great Smoky Mts.	15	7	93
Northern Georgia	33	3	97
Central Florida	1018	93	7

TABLE 15.2

Some Assets and Liabilities of Different Blood Types

Blood Group	Effect on Possessor (Average)
A	1. Heightened susceptibility to cancer of the stomach 2. Heightened susceptibility to pernicious anemia 3. More susceptible to infantile bronchial pneumonia than O individuals (?)
O	1. 35% more likely to develop duodenal ulcers than possessors of other blood types 2. Less susceptible to rheumatic fever 3. Longer expectation of life (in male Caucasians) than possessors of other blood types 4. Shorter expectation of life (in female Caucasians) than possessors of other blood types
B	1. Shorter expectation of life (in male Caucasians) than possessors of other blood types 2. Longer expectation of life (in female Caucasians) than possessors of other blood types

differentiate the polymorphic genotypes of meadow mouse that selection favors at different phases of the population cycle. The cryptic polymorphisms that earliest were recognized as such were in man: the blood groups of the O, A, B series. In the 1940s E. B. Ford realized that these genetic types all were too abundant in human populations to be of neutral or purely negative selective value, and predicted that each would be found to be associated with greater or lesser susceptibility to different diseases. This prediction is being borne out, although (as might be expected from medical studies) it is the *disadvantages* of particular blood types that are being recognized first, rather than the advantages (Table 15.2). Interestingly, the frequency of the different blood types is not the same in different human races, suggesting that historically the risk of stomach cancer, duodenal ulcer or rheumatic fever has differed for populations living in different parts of the world.

The sickle cell trait provides another example of cryptic polymorphism in man. This is a polymorphism between dominant and recessive forms of a single gene that affects the oxygen-carrying abilities of hemoglobin. Individuals wholly recessive for this trait have seriously defective hemoglobin, and usually die before reaching reproductive age. The recessive gene therefore should be subject to strong negative selection, and over a few generations should become very rare in the population. This has not happened, however; in many parts of Africa the recessive gene is common, as is the sickle cell disability it causes. What balancing factor favors possessors of the recessive gene? The unraveller of the mystery, Anthony Allison, has told the story well (1956). In areas where the sickle cell trait is common, malaria also is common and a major mortality factor. Surprisingly, however, sickle cell heterozygotes (individuals that possess one recessive sickling gene and one normal gene) are relatively immune to malaria. Moreover, in these individuals the normal gene largely overrides the sickling gene as a determinant of oxygen-carrying ability. As a result, heterozygotic individuals actually are more fit than individuals possessing two normal genes. Since selection favors these heterozygotes it favors the retention of the sickling gene in the population, even though individuals unlucky enough to inherit *two* recessive genes normally die, and always are seriously impaired. Selection for the heterozygote, *i.e.* for protection against malaria, must be very strong for the sickling gene to remain as common in some African populations as it does, since there is steady removal of this gene from the population in the form of double-recessives who do not live to reproduce. When Allison did his studies this polymorphism seemed to have achieved an approximate balance in most populations he looked at.

One would predict that in nonmalarial regions the selective balance for and against the sickling gene would be replaced by unbalanced selection against it. This is happening among American Blacks: the incidence of the sickling gene, originally high in American Blacks because of their African origin, now is declingly rapidly. Because malaria is not a problem in this country, sickle cell polymorphism has become a transient polymorphism trending toward monomorphic nonsickling.

ADAPTIVE DIFFERENCES BETWEEN POPULATIONS

Ecological Races (Ecotypes)

We turn now from adaptive genetic processes *within* populations to an examination of differences between *separate* populations of the same species. Many species consist of a number of discrete populations scattered across a range that is discontinuously habitable. If migratory interchange is low between such populations, each will tend over time to become genetically differentiated from the others. This sometimes is a matter of chance: if a population was founded by a very few individuals, their particular genetic complement may cast its stamp strongly on the population as a whole and alone be sufficient to make it detectably different from other populations. But ordinarily, differences between isolated populations of the same species are the result of different selective pressures on the different populations. No two localities, even in the same biome, are environmentally identical, and separated populations therefore will not remain identical either: they will become adaptively molded to their local environment. Genetically distinct populations of this type are termed **ecological races, eco-**

types, or **subspecies.** (The terms are essentially synonyms.) If ecological races are separated long enough their genetic divergence may become very marked. Now if they are somehow brought together, individuals of the two ecotypes may be unable to interbreed successfully. At this point the process of speciation, begun by the formation of ecological races, is complete. We no longer can treat the two races as variants of one species; they now are two species. If an ecological race is not too isolated from others—if a degree of interpollination or intermigration remains possible—its genetic exchange with populations of other ecotypes will limit its divergence and it will remain simply an ecotype itself. Reproductive isolation, normally achieved through stringent *geographical* isolation, is necessary for the step from ecotype to new species to be accomplished.

Because individual organisms have abilities of **acclimation** (individual, nonhereditary adjustment to differing environments), it sometimes is difficult to identify true ecotypes simply by field observation of different populations. Is the stunted plant population on a mountaintop a true dwarf ecotype, or is the stunting simply a matter of poor nutrition and slow growth? Is the dark-skinned lizard population on a dark lava flow really different genetically from the light-skinned population on nearby sand dunes? To resolve such questions one obviously should rear members of the two contrasting populations under the same conditions. If sharp differences remain, they must be genetic and we are dealing with true ecotypes. If the differences disappear, what we witnessed in the wild must have been nongenetic individual acclimation.

Let us look at examples. Thirty-two races of song sparrow have been identified in North America through differences in song or plumage (Fig. 15.4). Most of these probably are true ecotypes, although experiments of the type suggested above have not been performed to prove it, nor have adaptive elements of these differences been identified. Note that some song sparrow ecotypes are extremely localized. Between these ecotypes and others there usually are real geographical barriers—uninhabitable territory—that limit range extension and genetic exchange. The wide-ranging ecotypes, on the other hand, typically have a common border with other wide-ranging ecotypes. These ecotypes occur in areas that are generally not discontinuously suitable for song sparrow; here, then, there is little likelihood of localized populations becoming genetically isolated from each other, and this explains the wide range of each ecotype. Also, these wide-ranging ecotypes typically merge their distinctive characters with those of neighboring ecotypes across broad zones of transition where the populations are intermediate in character. Here the "barrier" between ecotypes simply is distance, not any real geographical discontinuity.

Among plants it is common to find ecotypic variation in germination requirement and photoperiodic response, especially in wide-ranging species which are exposed to quite different influences in different parts of their range (Table 15.3). But great geographical range is not always a prerequisite for the production of ecotypes. In one part of Wales, for example, Cavers and Harper (1966) found that maritime and inland populations of the weed *Rumex crispus* (dock) differed in their germination requirements for light. Seeds from the maritime population germinated well in darkness, a response the authors regard as adaptive to this population's habitat on open shingle beaches, where seeds often slip down between loose stones into a dark resting place. On the other hand, seeds from the inland population did *not* germinate well in darkness. At inland sites a dark seedbed usually means one that is shaded and crowded with competing plants. Because they require light to germinate, seeds of this population ordinarily will begin growth only on well lit, cleared ground where their chances of survival are high. Presumably dock, like many weeds, has seeds capable of long dormancy when immediate growth conditions are not favorable. Seeds which do not germinate immediately have not given up the fight; they just are waiting for better odds.

Clinal Variation

Even if the distribution of a wide-ranging species is continuous, so that it comprises but one extensive population, natural selection in different parts of the range may well bring about adaptive differentiation related to local circumstances. The result is frequently a **cline**, a genetically based, geographically progressive

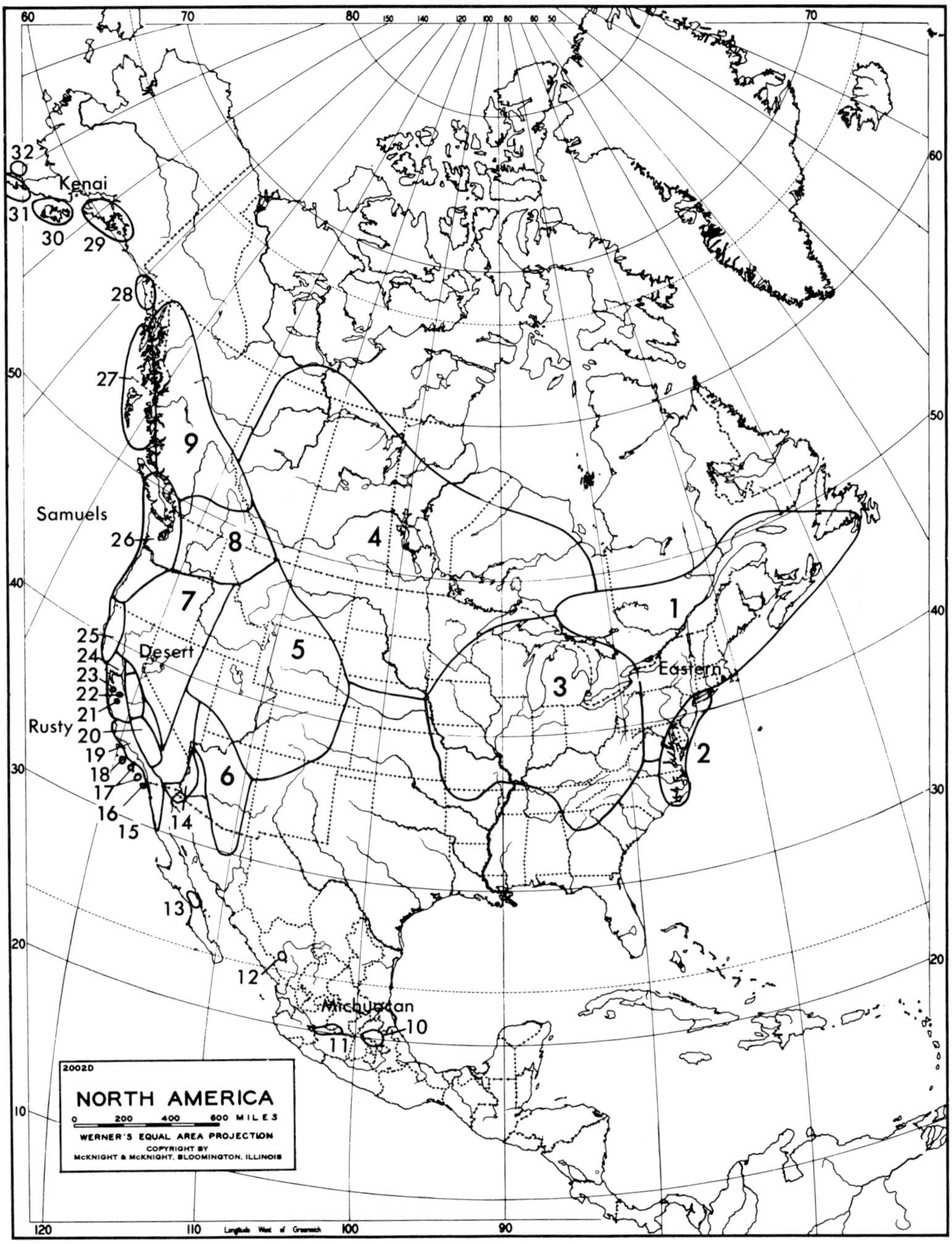

FIG. 15.4. Distribution map of the recognized races of song sparrow in North America. Six of the races are illustrated (next page) and their ranges identified on the map. (From G. L. Stebbins: *Processes of Organic Evolution*, Ed. 2, pp. 90–91, Englewood Cliffs, N. J.: Prentice-Hall, Inc., 1971. Reproduced with permission.)

change in the character of the population. Clines generally are made possible by distance, genetic differentiation being possible because widely spaced portions of the population do not exchange genetic material directly. Of course indirect exchange via members of the population occupying intermediate territory would occur to some extent, but it is insufficient to

prevent specific adaptation to local circumstances. Clines may also characterize populations existing across very sharp environmental gradients, *e.g.* up a mountainside. Clines as well as ecotypes are represented by the plants in Table 15.3 and by the contiguous races of song sparrow in eastern North America. Often, as in the case of the song sparrow, gradual change over a broad geographic area is recognized somewhat arbitrarily by scientists: they simply chop the cline into pieces and call each segment an ecological race. Sometimes one need not be so arbitrary: while progressive, the cline may be a **stepped cline** with clear-cut discontinuities where, because of some partially effective local barrier to gene flow, variation between segments of the population is quite sharp. McNaughton (1966) has identified north-south clines in a number of characteristics of the common cattail (*Typha latifolia*), and we may use two of these to illustrate the phenomenon (Fig. 15.5). Cattails, of course, grow only in wet places, and we *could* regard these changes as clinal variation between populations of different marshes rather than within a single population. The term cline can be used in either case, so long as there is a geographically progressive trend of variation. If the cline involves a series of sharply differentiated ecotypes, implying real problems of gene exchange over a species' whole range, one sometimes encounters situations in which adjacent ecotypes are interfertile but ecotypes at the extremes of the cline are not (if their members are brought together). This poses an interesting problem of classification—should one call the whole group a single species or several?—but this is a problem for the taxonomist, not the ecologist.

EXTINCTION: FOUR CASE HISTORIES

Despite negative feedback mechanisms that oppose it, and despite the resilience of individual acclimation and population adaptation via evolution, extinction occurs. It occurs because some agents of population change are not negative feedback mechanisms and because environmental shifts may be so sudden or so extreme as to overpower adaptive capacities. The extinction of a particular population is not really a rare event in nature; such extinction often goes unnoticed because the population is re-established by immigration from elsewhere. In our day, however, the extinction of *all* populations of a species is dishearteningly common. Let us use four examples among North American birds to gain a flavor of the agencies that promote extinction.

Health Hen

By chance there is a great plain on Martha's Vineyard, and despite the fact that the island is relatively small, it has never been considered amiss to speak of the vastness of this great plain of scrub oak, sweet fern, alder, blasted pine—and of the heath hen. Here, then, in sound of the roaring surf, amid

TABLE 15.3

Ecotypic Differences in Alpine Sorrel (Oxyria digyna)

This small perennial plant is widespread in the Arctic, and in the United States it occurs on many of the high mountains in the West. Arctic and American alpine populations of *Oxyria digyna* have been found to differ in many ways (*cf.* Mooney and Billings, 1961; Billings and Mooney, 1968).

Arctic	U. S. A. (Sierras and Rockies)	Suggested Significance
Vegetative reproduction predominant: extensive rhizomatous root system	Reproduction by seed; less extensive root system	Shorter growing season in Arctic may favor the production of offspring nourished by parent plant: advantage of wider dispersal is enjoyed by U. S. plants
Flowers only when exposed to a very long (or continuous)photoperiod	Flowering induced by shorter photoperiod (15 hr)	Days never get as long in the U. S.; different photoperiods ensure summer flowering of the populations in their respective regions
Forms winter buds at 15 to 13-hr photoperiod	Forms winter buds at 12-hr photoperiod	Day length of autumn's onset is different in the two regions
Seeds germinate less well at low temperatures than do those of U. S. alpine populations	Seeds germinate better at low temperatures than do those of Arctic populations	Summer night-time temperatures are lower on high U. S. peaks than in Arctic
Leaves have higher chlorophyll content, and relatively low light saturation level	Leaves have lower chlorophyll content and relatively high light saturation level	More intense sunlight in U. S. alpine habitats
Maximum photysynthesis occurs at relatively low temperature	Maxmum photosynthesis occurs at higher temperature	More intense sunlight and longer growing season in U. S. alpine habitats: summer days often are hotter
	High altitude Wyoming population is genetically dwarfed	Possibly a relict survivor of very harsh conditions here in last Ice Age

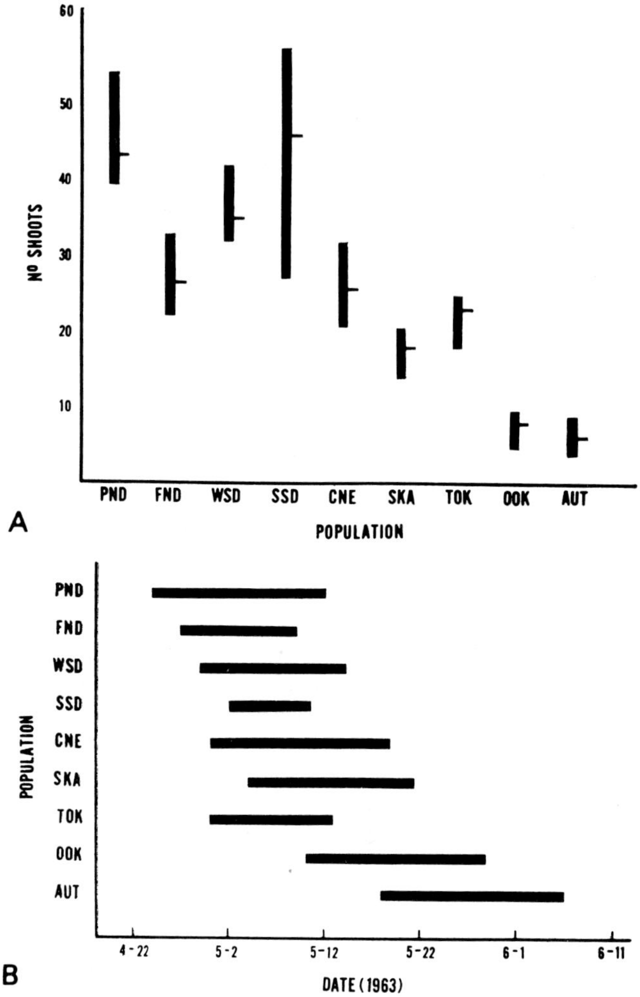

FIG. 15.5. Clinal variations in the common cattail, *Typha latifolia.* Representatives were gathered from marshes along a north-south transect from North Dakota to Texas, then grown under uniform conditions at Austin, Texas. (*A*) Vegetative reproduction from rhizomes. A cline exists in terms of vegetative reproduction from rhizomes. The collection sites are ranked from most northerly to most southerly in this graph. The thin *horizontal lines* represent means and the *vertical bars* represent ranges. Rhizomes of the most northerly cattails produced 9 times as many shoots as did those of the most southerly cattails. (Table 15.3 illustrates the same reliance on vegetative reproduction in northerly populations of another plant.) (*B*) Dates of flowering period. Cattails also vary clinally in terms of flowering period. When grown together there is a progressive tendency for plants from more northerly populations to flower earlier. (The staggering of flowering data in the Austin experiment might not occur in the plants' native habitats, because longer days come earlier in the spring in southern marshes than in northern ones. The plants probably are adapted to appropriate day lengths in their own latitude as the trigger for flowering.) The abbreviated locality names are as follows: PND, Pembina, North Dakota; FND, Fargo, North Dakota; WSD, Watertown, South Dakota; SSD, Salem, South Dakota; CNE, Columbus, Nebraska; SKA, Salina, Kansas; TOK, Tonkawa, Oklahoma; OOK, Oklahoma City, Oklahoma; AUT, Austin, Texas. (From S. J. McNaughton: *Ecological Monographs 36:* 297–325, 1966.)

such great and monotonous distance all encompassed in small space, the prairie chicken of the east lived a century beyond its time and then died . . . .

H. B. Hough (1933)

Part of Hough's eloquent obituary to the heath hen appeared at the beginning of this chapter. In this further passage the author, who was editor of Martha's Vineyard's local paper,

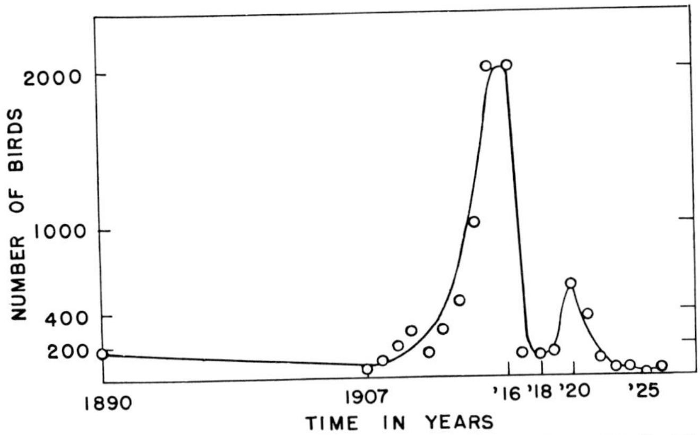

FIG. 15.6. Last years of the Martha's Vineyard heath hen population. (From W. C. Allee, *et al.*: *Principles of Annual Ecology*, Philadelphia: W. B. Saunders and Company, 1949; after Gross.)

provides a capsule image of the heath hen's history and habitat. We already know something of the ways of the prairie chicken (*cf.* Fig. 13.8). The heath hen was its eastern relative, considered by some a separate species and by others an ecological race. It was not a true forest bird; although it dwelt in the well forested states of the Atlantic seaboard it frequented flood plain meadows and scrubby sand barrens, usually near the sea. Originally very common in such habitats, it was easily hunted and for a time was even marketed commercially. By 1791 seasonal hunting restrictions were deemed necessary for the heath hen on Long Island, but here and in most other parts of its range protection did not long deter the extinction of local populations. On the mainland it persisted longest in the largest and wildest of suitable habitats, the New Jersey Pine Barrens; but after 1869 only one island population remained, off the coast of Massachusetts. This population was followed closely for many years, and provides one of the most detailed records of extinction that is available (Fig. 15.6).

A relatively reliable count of this population puts its size at about 200 in 1890. For the next 16 years, the numbers seem to have dwindled progressively until, after a destructive fire in 1906, fewer than 50 individuals remained. In 1907 stringent protection measures were initiated: a heath hen reservation was set aside and attempts were made to control predators such as wild domestic cats. As can be seen, the population responded dramatically. But in 1916 there were drastic setbacks: a devastating spring fire interrupted the nesting season, and cold temperatures and gales the following winter took both a direct and an indirect toll. The weather's indirect effects seem to have been especially serious: driven southward by the severe winter, goshawks converged on the island in unusual numbers and preyed heavily on the heath hens. The population of breeding pairs fell from 1000 to less than 100. It rebounded briefly, but about 1920 went into a final decline. The causes of this critical downturn are not clearly understood, but a skewed sex ratio (many fewer females than males), inbreeding, and poultry disease all are thought to have been important in the final stages. The numbers fell to 13 in 1927, and to 1 by December 1928. The mateless survivor held out, alone, until the spring of 1932.

What emerges from this history? First, the role of predation: reduction of the population by man, by cats and hawks, finally by disease. Second, the effect of fire and weather, both of which also may have affected predation rates. Third, the operation of Allee's Principle as the population became very small: reproduction and survival were adversely affected by influences that would not affect a large population. Finally, the vulnerability of this last population. Its island habitat at first afforded a measure of protection from the forces that decimated mainland populations, but in the end the island prairie was a prison where unusual pressures, such as the goshawk invasion, could not be evaded, and where losses probably would not have been restored by immigration even if populations had existed elsewhere. The adverse role of fire, which swept over most of the habitat at least twice in this period, is ironic, as Matthiessen (1959) notes: for it was fire-molded landscapes that the heath hen originally found most to its liking. But fire in this last habitat posed special risks to the population, because Martha's Vineyard was an island from which escape was impossible.

Passenger Pigeon

We have written in Chapter 4 of the passenger pigeon, a bird formerly incredibly abundant in the deciduous forest region of North America. We have no good numerical tally of its decline, and unlike the heath hen no single, sedentary population to follow to extinction. For the passenger pigeon was almost constantly on the move. In Aldo Leopold's words (1949), the species was a biological storm:

> Yearly the feathered tempest roared up, down, and across the continent, sucking up the laden fruits of forest and prairie, burning them in a travelling blast of life. Like any other chain-reaction, the pigeon could survive no diminution of his own furious intensity. Once the pigeoners had subtracted from his numbers, and once the settlers had chopped gaps in the continuity of his fuel, his flame guttered out with hardly a sputter or a wisp of smoke.

Can we extract insight from legend in the case of this bird? For legend exists in abundance, and much disagreement remains regarding the causes of the pigeon's demise.

Leopold's account provides one clue that often is overlooked. The passenger pigeon seemingly was a fugitive species, *i.e.* an extreme opportunist. It moved here and there, responding to the erratic production of food in different segments of the deciduous forest region from year to year. Probably its total numbers fluctuated greatly even before the white man reached America. Leopold suggests, as have others, that forest clearance reduced the natural food supply of the pigeon, but it was still present in huge numbers in the 1870s, only 30 years before it disappeared from the wild. It is not unlikely that the great Wisconsin nesting of 1871 represented a natural overshoot of the carrying capacity, a bloated population that would have declined even without the help of man. (Matthiessen (1959) estimates that this nesting comprised 136 million breeding birds in an area of 750 square miles—180,000 per square mile over a breeding ground 75 miles long and 10 miles wide.) Predation by man, however, seems ultimately responsible for the failure of the pigeon to recover from its population downturn. This is the opinion of Schorger (1973), the chief authority on the passenger pigeon, and he documents a supporting history of steady human hunting pressure during the waning years of the 1800s.

The communal habits of the pigeon made it a particularly easy prey, even as its numbers declined; not only the great nestings, but even much smaller ones, provided a sufficient concentration of birds to lure the commercial hunter. Young birds (squabs) were favored for the table and especially easy to collect. Although laws regulating the exploitation of nesting aggregations began appearing as early as 1862, they were widely ignored. Of the Wisconsin nesting of 1871, Matthiessen (1959) writes that "hundreds of thousands, indeed millions, of dead birds were shipped out at a wholesale price of fifteen to twenty-five cents a dozen, on the cars of the same railroads which, by opening the great eastern markets, were accommodating the exit of the bison." Transportation improvements in this period obviously were greatly increasing human access to previously remote nesting grounds. In 1878 a famous nesting near Petoskey, in north-central Michigan, was decimated by man in the same manner as that in Wisconsin, and substantial (though smaller) nestings were similarly exploited in Wisconsin again (1882), Missouri (1883), Pennsylvania (1886), and elsewhere. Schorger notes that there was enormous wastage in the harvesting: for every bird actually taken to market, several others usually perished. Particularly vulnerable were squabs too young to be taken. The pigeons' breeding season was staggered, but hunters tended to plunder the nesting grounds as soon as the first squabs were old enough for market. Even though they were not collected, younger squabs perished at this time in great quantities because of nest destruction or because their parents had been killed or driven away.

This sort of commercial pressure had been building rapidly in the 1850s and 1860s, but judging from the great Wisconsin nesting of 1871 it was insufficient to curb the pigeon population. Thereafter, however, particularly after

1878, the continued persecution of nesting flocks seems to have resulted in a progressive inability of the pigeons to replace their losses by reproduction. The species was rare everywhere by the late 1880s, but small flocks still were exploited for market as late as 1892. The last specimen shot in the wild was taken in Ohio in 1900.

What happened in those last pitiful years? Did the species, reduced to very small flocks, fail to reproduce efficiently because the social stimulus of communal breeding was absent? This is a possibility. Certainly, at the very end, man's predatory pressure must have eased somewhat, but perhaps not enough. We can imagine a new hunting stimulus, acquisition of a rare trophy, replacing the commerical stimulus and perhaps administering the *coup de grâce*. One can argue over whether man ***started*** the pigeon's final decline, and no one knows whether he finished it, but the responsibility for this extinction clearly is his. As we saw in the last chapter, a predator that maintains strong and efficient pressure on a prey population that sinks to a low level can drive it to extinction. This seems to be what happened here.

Bermuda Petrel

In this example and the next, we are dealing with species not yet extinct but under siege. The Bermuda petrel, or cahow, is a small oceanic bird that feeds mainly on squid and visits land only to breed. The breeding ground is restricted to the small Bermuda archipelago, and here the cahow suffered near-extinction shortly after the arrival of man in the early 1600s. In this respect the cahow was not atypical: many island species, unused to man and his domestic carnivores, have been rapidly exterminated when these invaders arrived. The cahow in fact *was* exterminated from the main island of Bermuda, and for 300 years was considered extinct. But after 1900, specimens began to be sighted sporadically, and in 1951 a small breeding colony was discovered on some rocky, uninhabited islets. Under stringent protective measures this lone population increased slightly during the following 15 years, but comprised only about 100 birds in 1968, when a new threat of extinction was recognized. In that year Charles Wurster and David Wingate reported that reproductive success of the petrels was declining, and suggested as probable cause an accumulation of damaging amounts of DDT in the eggs and chicks. A number of potentially toxic substances, including DDT, mercury, and strontium 90, tend to become concentrated in the issues of carnivores situated at the end of long food chains (Table 15.4), since these substances are resistant to decay and tend to be stored rather than excreted by animals lower in the food chain. Carnivorous sea birds such as the cahow are quite prone to such poisoning, since they form the terminus of unusually long food chains. In these birds. DDT disrupts reproductive hormonal balances and causes the production of eggs with thin shells that often break prematurely.

DDT was found to be concentrated in eggs and dead chicks of the cahows. Furthermore,

TABLE 15.4

Concentration of DDT Residues in an Estuarine Food Chain

If toxic substances such as DDT, methyl mercury or strontium-90 are ingested more rapidly than they are excreted or broken down metabolically, they tend to accumulate in the tissues. This process can be greatly magnified through food chain concentration, since each individual at a given trophic level eats far more than its own weight of organisms from the next lower trophic level. In the example below, organisms roughly corresponding to four successive levels in a food chain were analyzed for DDT residues. Enormous concentration is evident, with potentially toxic levels being reached at the last step. The study site was the Carmans River estuary on Long Island. The data are from Woodwell *et al.* (1967).

	DDT Residues	Average Concentration Factor	Trophic Level
	ppm		
Water	0.00005		
Zooplankton	0.04	800×	2
Shrimp	0.16		
Silverside	0.23	10,800×	3
Fundulus	1.24[a]		
Needlefish	2.07	60,400×	4
Common tern	3.91		
Cormorant	26.4	528,000×	5

[a] Perhaps operated partly at trophic level 4.

the data of Wurster and Wingate for the years 1958 to 1967 revealed a declining rate of successful reproduction that, if extrapolated, would reach zero in 1978. If reproduction really fails completely beyond that date, man will have dealt the cahow a "second extinction," this one more real than the first. Some hope remains, however. DDT use in the United States has dropped sharply in recent years, and several threatened mainland species are on the rebound. Many other countries do not have the stringent DDT regulations that now exist in the United States; but if the DDT residues reported in the cahow were mostly of American origin, contamination should be declining as the insecticide slowly degrades and is not replaced in the Cahow's diet.

Eastern Bluebird

What has happened to the bluebird? In parts of the American West, the western bluebird is still common; but the eastern bluebird, native to all states east of the Rockies, now is a very rare sight in most of the orchards, pastures, fields, and suburbs it once frequented. It is not yet on the endangered species list, but almost certainly many *local* populations have become extinct. The decline of the bluebird in the past 25 years seems attributable to several causes, but the principal one seems to be competition with introduced species for nesting sites (Zeleny, 1972). The bluebird almost certainly was adversely affected by the surge of insecticide use in the 1950s and 1960s; it is an insect eater, and probably suffers directly from accumulation of insecticides in its tissues, as well as indirectly through the elimination of some of its preferred foods. Furthermore, bluebirds wintering in the southern states are susceptible to severe mortality when their winter food supply (mostly berries) becomes snow- or ice-covered. Zeleny believes that heavy storm losses in the winter of 1958 still have not been made up. But the chief present stumbling blocks to bluebird populations are the house sparrow and the starling, both introduced from England and now extremely abundant in most parts of the eastern bluebird's range. These species begin nesting earlier in the spring than does the bluebird, and claim all or nearly all of the tree-hole sites that the bluebird favors. In the Middle-Atlantic states, Zeleny recently has initiated an Audubon Society nest box project that has boosted bluebird populations in several localities. The boxes are provided with holes too small to be used by starlings, and house sparrow nests are regularly removed to allow free space for bluebirds. With human help, the eastern bluebird may yet return to abundance. As the starling spreads westward, however, the western bluebird may suffer a comparable decline unless it receives similar encouragement.

CAUSES OF EXTINCTION: AN OVERVIEW

In the foregoing case histories many of the common agents of extinction are represented. Man was mentioned often: presently he is a major actor in this story all over the world. Islands also figured prominently in these case histories. In fact, a disproportionate number of species extinctions occur on islands, a phenomenon we will consider in the next section.

Table 15.5 attempts, for birds and mammals, to evaluate the principal causes of extinction or increasing rarity worldwide since 1600 A.D. Man is implicated, either directly or indirectly, in approximately three-quarters of all cases. Hunting and habitat destruction are the chief modes of human pressure on other species, but the species he introduces to alien habitats also are a significant extinction factor.

TABLE 15.5

Pressures for Extinction of Mammals and Birds Since 1600 A.D.

Data on extinct and beleagured species have here been evaluated in terms of the major causes of extinction or threatened extinction. Adapted from ***Wildlife in Danger***, by James Fisher, Noel Simon and Jack Vincent. Copyright © 1969 by the International Union for Conservation of Nature and Natural Resources. Reprinted by permission of The Viking Press.

	Mammals		Birds	
	Extinct	Threatened	Extinct	Threatened
	%			
All nonhuman causes	25	14	24	32
Human agencies				
1. Hunting	33	43	42	24
2. Predators introduced by man	17	8	15	11
3. Nonpredators introduced by man	6	6	4	3
4. Habitat destruction	19	29	15	30

From Table 15.5 we also can gain an idea of whether a particular extinction agency is becoming more or less important, by comparing its "extinct" and "rare" percentages. The former can be considered a measure of the relative importance of each agency in the past, the latter a measure of their present relative importances. New pressures for extinction, such as pesticides, are not included in this accounting.

As a hunter, man might be expected to act in direct density-dependent fashion when he is exploiting a species commercially. As a population or entire species becomes rare, its harvesting should become economically unprofitable, and hunting pressure should decline. But judging from Table 15.5, man's hunting pressures often do not work this way; if they did, there would be no extinctions from hunting.

From the standpoint of other species, the problem is twofold. First, man sometimes acts as a generalized predator, harvesting a rare species as part of a larger, still profitable harvest of other organisms. The endangered blue whale is an excellent example (Fig. 15.7). Because other whales remained relatively abundant in the Antartic during the 1960s, whaling fleets continued to operate there, killing blue whales whenever they could because of their great size. Specific protection for the blue whale has now been enacted into law, but it may be too late.

Second, individuals of some species become more valuable to man, either economically or in terms of prestige, as the species becomes rarer. Hunting pressure upon these species therefore does not decline in direct density-dependent fashion. Many wild species hunted for their fur are still pursued vigorously even when rare: vicuña, wolverine, and snow-leopard coats simply become more expensive.

In the case of trophy hunting we think mostly of large mammals. A sordid story of 1971 illustrates the pressures toward extinction of this practice. In America there is an exclusive group of big-game hunters calling themselves the Grand Slam Club; these are people who have shot all four types of North American bighorn sheep: the Dall sheep of Alaska and the Yukon, the Stone bighorn of British Columbia, the Rocky Mountain bighorn of the United States, and the desert bighorn of Mexico and California. Certain of these varieties, like the Rocky Mountain bighorn, are in little present danger of extinction; but others are quite rare. It is possible to obtain permits to shoot them all, but fortunately the number of permits is controlled; for the rare desert bighorn, the protagonist of this story, there is complete legal protection in California and a limit of 30 permits per year in Mexico. These regulations, then, are designed to make the hunting of bighorn sheep density-dependent: fewer of the rare varieties are taken. But the existence of the Grand Slam Club provides a strong tacit incentive to circumvent the law: there is in fact a long waiting list for desert bighorn permits, and hunters who have shot the other three must be sorely tempted to try something illegal. Several years ago a California taxidermist won cheers from conservationists for leading a campaign to protect and restore the desert bighorn in that state. Later, the sorry truth emerged that his campaign was a smoke screen for the lucrative poaching service he actually was running; trophy hunters were paying big fees to be taken by this man on secret hunts for desert bighorn.

It is evident from Table 15.5 that habitat destruction is a serious and growing vehicle of extinction. This is hardly surprising, given man's increasing population and increasing technological capacity to alter the planet, and given the habitat specificity of most types of organism. Burt (1973) has drawn attention to the clear correlation that exists in the United States between human population density and the extinction of wild species (Fig. 15.8). He suggests that this correlation is mainly due to pervasive habitat modification of populous areas.

Human pressures promoting extinction of other species give little promise of relaxing. We may hope that laws and conservation awareness will stimulate growing counter-pressures for preservation, and that certain pressures of the past will ease. But habitat destruction and new threats, such as pesticides, polychlorinated biphenyls, and other industrial pollutants, will not disappear. These pressures are byproducts of man's other activities. Their repercussions for other species seldom are considered until too late, or, if considered beforehand, they often are given low priority. In the 1960s, for example, a group of ecologists was assigned by the government to study the potential biological effects of weather modification by man. Large scale weather modification is ever nearer to reality,

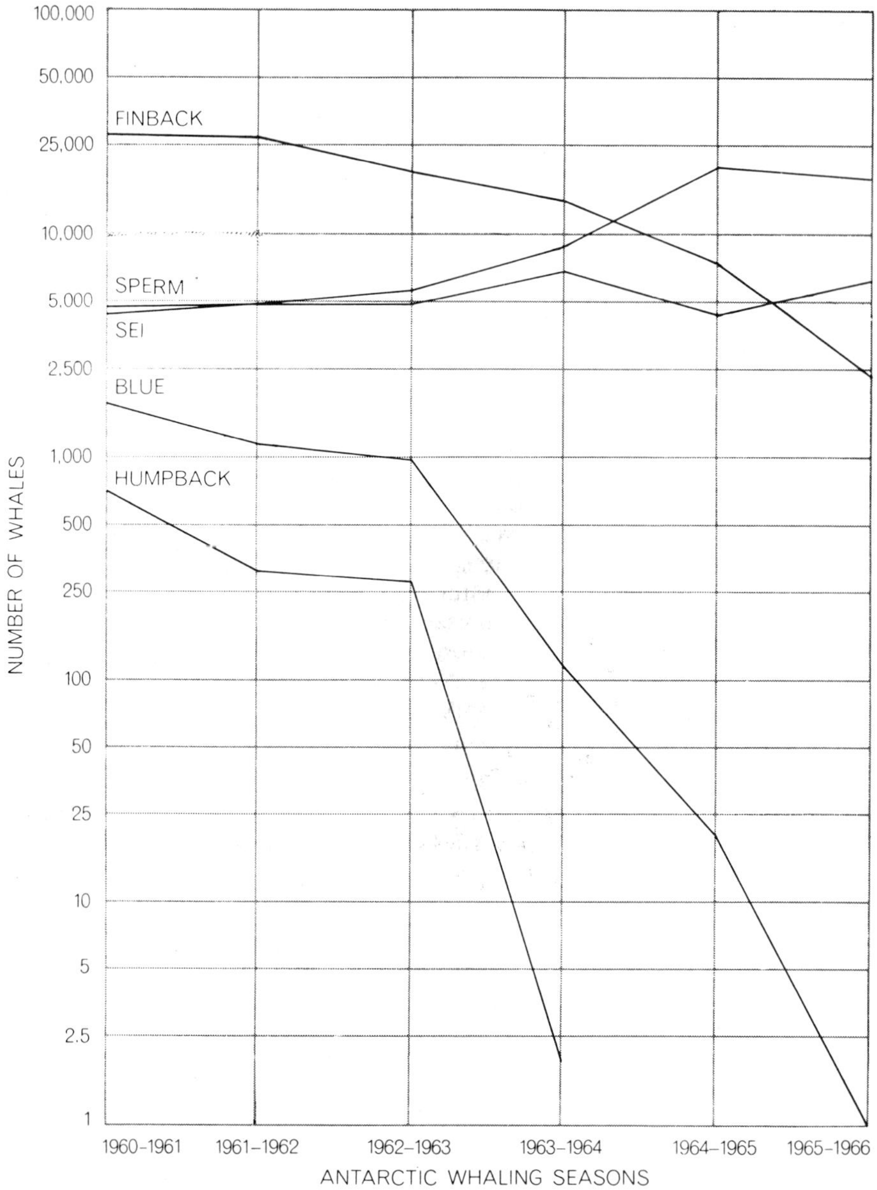

FIG. 15.7. Decline of blue and humpback whales in antarctic waters. Finback, sperm, and sei whales kept the fishery viable despite the increasing rarity of the first two species. (From S. McVay: *Scientific American*, August 1966. Reproduced with permission from Scientific American, Inc. All rights reserved.)

yet the predictions of the study group—massive disruption of natural communities, and probable extinction of many species—have not been widely publicized and seemingly have not influenced the planners.

ISLANDS AND SPECIES EXTINCTION

Of the 94 bird species suffering extinction since 1600, 85 have lived on relatively small islands; only 9 have been continental species. Statistics for other animal and plant groups probably would be similarly weighted. Island species, especially those of small islands, are terribly vulnerable.

Why should this be? Two competing theories of biogeography provide explanations, and each may carry considerable truth. The older theory, espoused particularly by P. J. Darlington

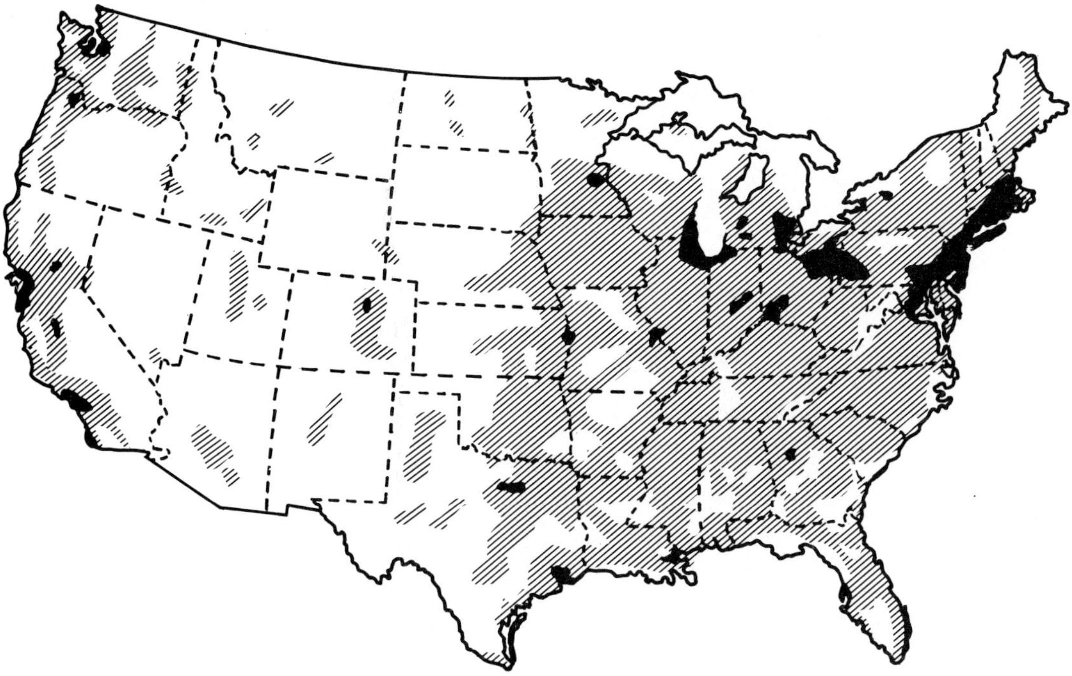

TABLE I. THE GEOGRAPHY OF EXTINCTION IN THE UNITED STATES**

BIRDS	MAMMALS
I. Coastal Zones and Offshore Islands	
*Great auk—1844	*Gull Island mole—1898
*Labrador duck—1875	*Amargosa meadow vole—1917
*Heath hen—1932	*Sea mink—1890
Dusky seaside sparrow	*Steller's sea cow—1768
Light-footed clapper rail	Delmarva Peninsula fox squirrel
Cape sable sparrow	Marrow Bay kangaroo rat
Southern bald eagle	Salt marsh harvest mouse
Brown pelican	Red wolf
Everglade kite	Florida panther
Attwaters greater prairie chicken	Florida manatee
California least tern	Key deer
Whooping crane	Columbian white-tailed deer
California clapper rail	
II. Eastern Deciduous Forest	
*Passenger pigeon—1914	*Eastern bison—1832
*Carolina parakeet—1914	*Eastern elk—1880
American peregrine falcon	*Eastern cougar—1899
Ivory-billed woodpecker	Indiana bat
Red-cockaded woodpecker	Eastern timber wolf
Bachman's warbler	
Kirtland's warbler	
III. Prairie	
	*Badlands bighorn sheep—1910
	*Plains wolf—1926
	Black-footed ferret
IV. Mountain, Desert and Arid Lands	
California condor	*Merriam elk—1900
Masked bobwhite	San Joaquin kit fox
Yuma clapper rail	Sonoran pronghorn
Mexican duck	Tule elk

*Extinct species. All others listed are on the endangered species list.

**From the official Bureau of Sports & Wildlife register.

TABLE 2. THE GEOGRAPHY OF THE AUDUBON SOCIETY'S BLUE LIST OF BIRDS*

I. Coastal Zones and Offshore Islands	
Great white heron	Black vulture
Reddish egret	Swallow-tailed kite
White pelican	Osprey
Black-crowned night heron	Audubon's Caracara
White-faced glossy ibis	Sparrow hawk
Wood ibis	Limpkin
White ibis	Snowy plover
Fulvous tree-duck	American oyster-catcher
Turkey vulture	Least tern
II. Eastern Deciduous Forest	
Black-crowned night heron	Red-shouldered hawk
Turkey vulture	Loggerhead shrike
Black vulture	Barn owl
Eastern bluebird	Red-headed woodpecker
Sharp-shinned hawk	Bewick's wren
Cooper's hawk	Bachman's Pine Woods sparrow
III. Prairie	
White pelican	Turkey vulture
Double-crested cormorant	Black vulture
Black-crowned night heron	Ferruginous rough-legged hawk
Sharp-tailed grouse	Marsh hawk
Least tern	Prairie falcon
Loggerhead shrike	Burrowing owl
IV. Mountain, Desert, and Arid Lands	
Mexican duck	Gila woodpecker
Turkey vulture	Bewick's wren
Black vulture	Loggerhead shrike
Harris's hawk	Gray vireo
Sage grouse	Western bluebird
Mountain bluebird	

* Birds on the Blue List ("early warning system") are usually more common and more widespread than endangered species.

FIG. 15.8. Human population density and the geography of extinction in the United States. The lengths of the species lists, particularly in Table 1, correlate discouragingly well with average human densities in the indicated habitat regions. Densities greater than 260 humans/mi.^{2} are shown on the map in black; densities between 260 and 25 are indicated by hatching; and densities of less than 25 are shown in white. (From D. E. Burt: *The Explorer, 15:* 4–10, 1973. Map modified from the Hammond Medallion World Atlas, 1971, p. 189. Reproduced with permission.)

(1957), suggests that the larger the land mass, the more "dominant" its flora and fauna. Large land masses, says Darlington, are centers of evolutionary diversification where species become honed to fitness through competitive and predatory interaction with many other species. Such floras and faunas, on the whole, have expansionist tendencies and are relatively successful in displacing the biotas of smaller land masses to which they gain access. Smaller land masses, on the other hand, particularly those that are isolated, tend (says Darlington) to be evolutionary backwaters where selection for maximum fitness has not been so stringent because competitive and predatory pressures have not been so diverse. Islands, being particularly small and often isolated, therefore have faunas and floras that often are unable to cope with animal or plant invaders from larger land masses. Darlington documents this theory with much historical and paleontological evidence, showing that natural expansions of floras and faunas have largely involved the invasion of smaller land masses from larger ones rather than vice versa. Islands that are quite isolated seem particularly susceptible to extinction because their faunas and floras often are depauperate and "unbalanced." The Galapagos Islands, for example, have no true woodpeckers, only a finch that rather imperfectly fills the woodpecker niche, grasping a thorn in its beak to root under bark for grubs. If a South American woodpecker managed to colonize the Galapagos, the competition might be very unequal.

The second theory of island vulnerability derives from the work of MacArthur and Wilson (1967), and rests quite simply on their hypothesis that extinction of local populations occurs quite regularly on all land masses, large and small (see p. 222). But on larger land masses there usually are several local populations, and if one is exterminated, recolonization of the vacated habitat often occurs quite quickly. Therefore species of large land masses not only have a buffer against total species extinction, but local population extinctions may go completely unnoticed.

Many island species are very restricted geographically, often to a single archipelago or a single small island. In many cases, like the cahow, the species consists of a single population. Thus, if that population disappears, so does the whole species.

To the basic ideas of MacArthur and Wilson can be added additional considerations that help to explain island extinctions: the populations involved often are small; geographic restriction makes whole species vulnerable to localized events like hurricanes or fires; and so forth. The reader doubtless can think of other reasons. In the case of epidemic diseases, the smallness and isolation of island populations may have special repercussions for both host and pathogen as populations. An introduced pathogen often has devastating results if the host population on an island is small but dense, and has not previously been exposed to it. In 1875, 20% of the human population of Fiji died from newly introduced measles. On the other hand, a disease like measles needs a relatively large and spaced-out host population if it is to persist indefinitely. Otherwise, a major epidemic will immunize nearly the entire population and the number of individuals remaining susceptible will be too low to maintain the disease in the host population. In any isolated host population "of less than a few hundred thousand people measles will die out after an epidemic and will not reappear unless it is imported from somewhere else" (Burnet and White, 1972). Thus, on islands, pathogens like measles tend to sweep through a host population, then become extinct.

Obviously the term "island" should take on a larger meaning in discussions such as this, for what we are talking about really applies to any isolated patch of habitat and its local populations, be it a real island in an ocean, a lake in the desert, or the bit of tundra on a Mexican volcano.

WHY WORRY ABOUT EXTINCTION?

The pace of species extinction in our day apparently is far greater than during most of Earth's history, and man seems the main cause. But need we be concerned? We select those species which are of use to us and encourage them by domestication; why should we care about whooping cranes, Indian rhinos, and vanishing plant species?

Opinions *do* vary among ecologists about the seriousness of the extinction problem. Viewed dispassionately, the claims of some conservationists seem overdrawn. But extinction seems not just to be rapid today, but to be accelerating, and that is cause for real concern. Why?

For one thing, we find new and practical uses for many wild creatures. The manatee, for example, outperforms most machines as a clearer of weeds from clogged waterways. Already rare in most parts of its range, the manatee is on the endangered species list in the United States. Presumably it can be nutured back to abundance; but what if it, like the related Steller's sea cow, had been hunted to extinction a century ago? In some parts of Africa, farming of wild herbivores now appears more efficient and profitable than cattle raising; and where still abundant, wild game of all sorts attracts tourist revenue to countries that depend on it. In America the bison, once nearly extinct, shows economic promise. These developments are recent; extinction nearly made some of them impossible.

Other species, many of them endangered, are useful to man either as "mirrors" or as early warning systems. We learn about ourselves from gorillas, orang-utans, and other primates; we would know less if the opportunity to study them was gone. We learn, too, when birds are poisoned by DDT or leopards in the Staten Island zoo are poisoned by aerosols from leaded gas. These species have served, in Roger Tory Peterson's words, as "environmental litmus paper," alerting us to potential human dangers we would otherwise have overlooked. When such species are gone, so also will be the warning system.

To an ecologist, the central question is what extinctions may do to the stability of natural communities. It is important to note that many man-related agents of extinction are very selective. Poisons acting by food chain concentration, for example, act selectively against top carnivores. One wonders what repercussions there might be in croplands if rodent-eating birds of prey were eliminated. Several species of hawk and eagle already have been severely stressed by DDT, and although the chief danger from this insecticide may be past, many other substances act in similar ways. Unfortunately, we frequently do not know enough about the role a wild species plays in its ecosystem to visualize clearly what will happen if it is removed. Any time an important member of a food chain is eliminated or its numbers seriously altered, the repercussions are partially or largely unpredictable, and often are very damaging to human interests. In a more general sense, too, extinction reduces the diversity of a community, always raising the possibility, as we have seen in earlier chapters, of reduced stability, *i.e.* of more widely fluctuating population levels among the remaining species.

Not all extinctions are grieved over like the heath hen's. Presumably no one bewails the imminent worldwide extinction of the smallpox bacillus in a blitz mounted by the World Health Organization. Many would cheer to see the same fate befall cornworms and potato bettles, if it could be done without causing worse problems. But many species that man considers enemies are only local problems, and local control should be considered a preferable alternative to global exterextermination. Many other species that man eliminates are not pests at all but are victims of his greed as a predator or are simply innocent bystanders to "progress." We may feel anger or remorse at such extinctions; what we might also feel is loss.

SUGGESTED READING

Adaptation

BILLINGS, W. D., AND H. A. MOONEY. The ecology of Artic and alpine plants. *Biological Reviews* 43: 481–529 (1968). Interesting examples of ecotypic adaptation.

CONNELL, J. H. *et al.*, Readings in Ecology and Ecological Genetics. New York, Harper and Row paperback, 1970. A useful collection of readings, some of which relate to matters in this chapter.

ELTON, C. S. The Ecology of Invasions by Animals and Plants. New York, John Wiley and Sons, 1958. Deals as much with extinction as with adaptation.

FORD, E. B. Ecological Genetics. New York, John Wiley and Sons, 1964. A basic reference by the pioneer student of polymorphism and natural selection operating in wild populations.

KETTLEWELL, H. B. D. Darwin's missing evidence. *Scientific American* (Mar., 1959). The famous story of transient polymorphism and rapid selection in the peppered moth.

MCLAREN, I. A., ed. Natural Regulation of Animal Populations. New York, Atherton Press, 1971. In addition to many good papers on population dynamics in the broader sense, this collection contains papers by Pimentel and Chitty on genetic feedback.

STEBBINS, G. L. Processes of Organic Evolution. Englewood Cliffs, New Jersey, Prentice Hall, Inc., 1966. A fine basic paperback text.

Extinction and conservation

BERRY, J., AND OSGOOD. Chemical Villains. St. Louis, C. V. Mosby Co., 1974. A catalogue of the new breed of threats to other species.

COX, G. W., ed. Readings in Conservation Ecology. New York, Appleton Century Crofts, 1974 (2nd edition). A good paperback collection.

EHRENFELD, D. Conserving Life on Earth. New York, Oxford University Press, 1972. Interesting case histories.

FISHER, J., N. SIMON AND J. VINCENT. Wildlife in Danger. New York, Viking Press, 1969. A descriptive catalogue of endangered species throughout the world, beautifully produced and authoritative.

MATTHIESSEN, P. Wildlife in America. New York, Viking Compass paperback, 1964. A compelling historical account of the effects of European settlement on American wildlife, from colonial days to the present.

Threatened Wildlife of the United States. Washington, Department of the Interior, Bureau of Sport Fisheries and Wildlife Resource Publication No. 34, 1973. A very valuable compendium.

chapter 16

Epilogue: The Human Avalanche

If the system of life is such that a harmonious balance of conflicting interests has been reached where every element is either hostile or indifferent to every other, may we not trust much to the outcome where, as in human affairs, the spontaneous adjustments of nature are aided by intelligent effort, by sympathy, and by self-sacrifice?

S. A. Forbes (1887), *The Lake as a Microcosm*

These lines close a now classic paper on lake communities by one of America's pioneer ecologists. Forbes' words carry brave optimism, but are framed as a question with a hint of wistfulness. Already in Forbes' day America's eastern forests had been ravaged by the lumber boom. Already bison were nearly gone from the prairie. The heath hen had been squeezed back to its last refuge on Martha's Vineyard, and the last great nesting of passenger pigeons was 9 years into history.

But in Forbes' day the specter of atomic annihilation was unknown. America's population was 60 million, not today's 210 million; world population was 1.3 billion and growing at half the rate of today's, which threatens to double (to 8 billion) in the next 35 years. In Forbes' day the crisis of vanishing resources was not yet upon us. The frontier was still there. Air was still clean outside the cities, and when Forbes described his beloved lakes he wrote of natural communities, not the communities of pollution. Modern ecologists have inherited Forbes' wistfulness; many have not inherited his optimism.

If an ecologist were to write a history of mankind, he might well adopt D. D. Kosambi's definition (Fig. 16.1), for it is man's technological revolutions, piling on one another with accelerative speed ever since the first pebble tools were invented, that have turned him from an equilibrium species to a boom-and-burst opportunist. By standard ecological criteria man is, indeed, a *K* adapted species. Only a few carefully nurtured offspring are produced per individual, and these mature slowly. Hence man's intrinsic rate of increase (r) is relatively low, like that of an equilibrium species. Moreover, he does not seem biologically specialized for rapid dispersal, as are classic opportunistic species.

Why, then, is man behaving like a locust plague unleashed upon the globe, multiplying and spreading to its farthest corners? He is doing so because technological advance, mediated through man's adaptable behavior and communicative skills, has supplied the means of dispersal and has raised the environment's carrying capacity for our species. If the carrying capacity rises steeply and continuously, as it has for man during the past 400 years, even a slow-reproducing *K* adapted species can build up mighty momentum.

We know, though, that Malthus cannot be denied forever; Earth's carrying capacity for man is finite, and the strain already is showing. Some experts feel the human population already has burst through Earth's long term carrying capacity, and must fall back. The question is whether we can guide civilization into some new equilibrium with nature, or whether that equilibrium will be imposed by catastrophes of our own making.

A look at history is not reassuring. Although

FIG. 16.1. "History is the presentation in chronological order of successive changes in the means and relations of production." In these words D. D. Kosambi captures the essence of man's changing relationship to his environment, and ecologists should find his definition of history most agreeable. Almost of necessity, Kosambi took an ecological approach when writing of his native India: "India has virtually no historical records worth the name. . . . But what is history? If history means only the succession of outstanding megalomaniac names and imposing battles, Indian history would be difficult to write. If, however, it is more important to know whether a given people had the plough or not than to know the name of their king, then India has a history." (Reprinted with permission from *The Culture and Civilization of Ancient India in Historical Outline*, by D. D. Kosambi (1965). London, Routledge and Kegan Paul, Ltd.; New York, Pantheon Books.)

carrying capacity has risen, the rise has not been smooth, and increasing populations seemingly have been insensitive to environmental limits time and again. Despite his wisdom, man has been as self-centered as any other species, and has paid for it dearly. Currently it is speculated that the Pleistocene hunting populations of North America were boom-and-burst phenomena: they decimated their own food supply and collapsed, only later developing demographic balance between their resources and their technology. Likewise, overpopulation and exhaustion of soil fertility almost certainly contributed to the collapse of great early civilizations like those of Mesopotamia and Yucatan.

Given these lessons from the past, we clearly cannot lay all the blame for our "ecological crisis" on a Judeo-Christian ethic that instructs man to subjugate the earth, although such an argument has been powerfully stated (White, 1967). Man's intense sociality is perhaps more basically to blame: both within religion and without it, we perhaps are so preoccupied with our own kind that we fail to be very far-sighted about nature until she forces us to be. History tells us that our sensitivity to nature is largely a function of natural scarcity. Only when resources are overtaxed does man seem to think of living within a carrying capacity. Equilibrium populations have indeed been established in the past, if Deevey (1960) is correct (Fig. 16.2). Doubtless these equilibria always have been initiated the hard way; although once the lesson of adversity was learned, it often was retained, at least till some technological innovation upset the balance.

Can we halt the human juggernaut? It seems

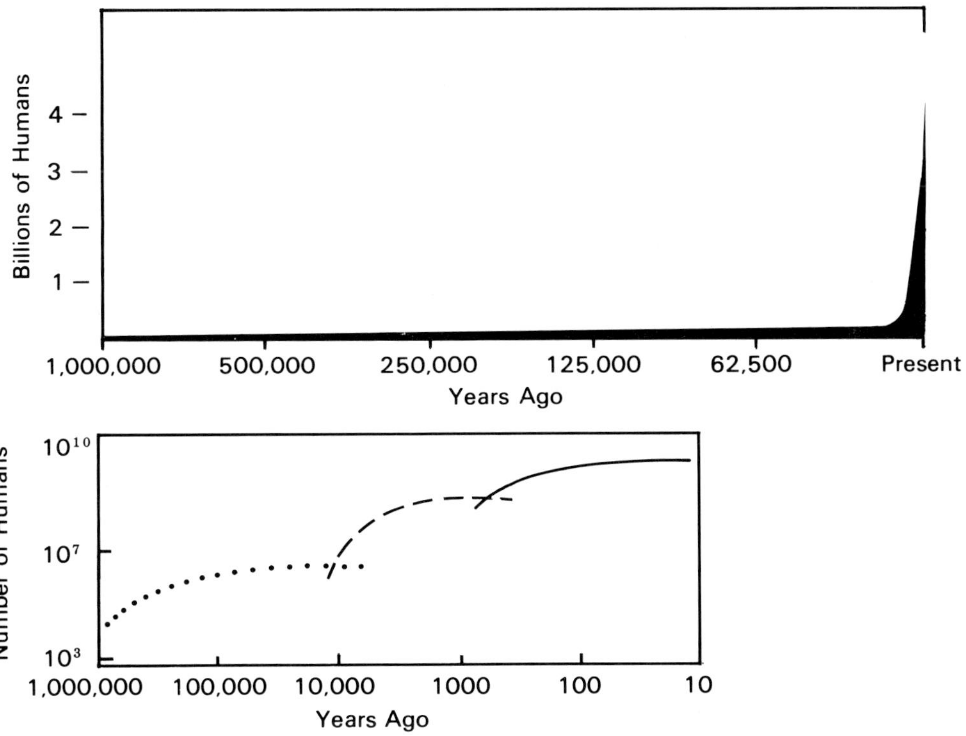

FIG. 16.2. Growth of human population is usually depicted on an arithmetic scale as a long, steepening curve (*top*), but this obscures what may have been long periods of relative population stability in the past. Human numbers probably rose significantly with the tool-making revolution, and certainly did so during the agricultural revolution: there was an estimated 16-fold increase in human population between 10,000 and 6,000 years ago, as animal domestication and farming became widespread. Plotting human population growth on a logarithmic scale (*bottom*) dramatizes these early population spurts and emphasizes the subsequent stagnation of growth as ancient populations approached carrying capacities lower than today's. (From E. S. Deevey, Jr.: *Scientific American*, September 1960. Reproduced with permission from Scientific American, Inc. All rights reserved.)

to have achieved a momentum all its own; burgeoning populations demand more and more and, while they can, continue to grow. Room for maneuver is slight; nations are impelled onward, as is China today. Traditional Chinese philosophy, some believe, instilled in that people a sense of man's oneness with nature; but if such a sense really exists, it is as tenuous as the "land wisdom" of other peoples, and seems similarly related to past experiences with overexploitation. In most parts of China forests disappeared long ago; floods and declining soil fertility were the result, as the population grew burdensome on the land. Today's Chinese government is pushing reforestation in some areas, but only recently has it begun pursuing a policy of population limitation. The lessons of the past seem only partly learned in China, and the immense short term strain of feeding its huge population leaves little time or inclination for thoughts about a more enduring integration with nature. Visiting China in 1972 for the Associated Press, Barbara Tuchman wrote as follows:

> Insecticide is so important in today's China that in one ballet we saw in Peking the girl dancers appeared with hand sprayers as part of their costumes. Except for locusts, insects seem extraordinarily absent, and birds, too, in consequence. For the sake of agricultural yield China has taken a long step toward silent spring.

But we know that man must come to terms with nature. Some new and enduring equilibrium must be striven for. Achieving it without catastrophe will be extraordinarily difficult: quite probably we will be required to retreat voluntarily from our present pinnacle of global dominance to some more modest pinnacle. No other species ever has willfully done this, nor has man without hard lessons first. But if we are to recover the optimism of Forbes in 1887, we must first act on the words of Sir Peter Medawar in *The Future of Man* (Basic Books, Inc., New York, 1959):

> The inference we can draw from an analytical study of the differences between ourselves and other animals is surely this: that the bells which toll for mankind are—most of them anyway—like the bells on Alpine cattle; they are attached to our own necks, and it must be *our* fault if they do not make a cheerful and harmonious sound.

SUGGESTED READING

(This is just a sampling of the huge, somewhat redundant literature on man's population and environmental crisis.)

CLOUD, P. E., JR., ed. Resources and Man. San Francisco, W. H. Freeman and Co., 1969. An excellent series of articles by experts.

EHRLICH, P. R., AND A. H. EHRLICH. Population, Resources, Environment. 1972 (2nd edition). A fact-filled text and reference.

The Human Population. A special issue of *Scientific American* (Sept., 1974).

JOHNSON, W. H., AND W. C. STEERE. The environmental Challenge. New York, Holt, Rinehart and Winston, 1974. A very good paperback.

MURDOCH, W. W., ed. Environment. Stamford, Conn., Sinauer Associates, 1975 (2nd edition). Another excellent book composed of chapters by a series of experts.

STEBBINS, G. L. The natural history and evolutionary future of mankind. *American Naturalist* 104: 111–126 (1970). By a well known student of evolutionary adaptation.

WAGNER, R. H. Environment and Man. New York, W. W. Norton Co., 1974 (2nd edition) A popular, wide-ranging textbook.

APPENDIX

Indices for Community Comparison: Similarity, Diversity and Equitability*

A SIMPLE INDEX OF SIMILARITY

This index is discussed by Sanders (1960), who used it to compare the invertebrate communities inhabiting substrates of different texture on the ocean floor. The index can be used in almost any situation where one wishes a measure of the degree of similarity (or difference) between communities in terms of species composition and dominance. It is of little use in comparing communities of completely different habitats, because it relies on some species, at least, being possessed in common by the communities undergoing comparison. My students have used this index, for example, to compare the associations of epiphytic algae growing on different types of rock or on different species of aquatic plant; and also to compare forest compositions on distinctive soil types (Table A.1). As is clear from Table A.1, the index measures the compositional affinity between pairs of communities. If one wishes to compare more than two communities, separate two-way indices should be calculated between each community and every other one in the series. Then the separate indices can be compared. Indices are calculated very simply, as shown in Table A.1. For those species possessed in common by two communities, percentage composition in each community is calculated and expressed as a decimal. The lower of the two decimal figures for each species is recorded in a separate column, and these lower percentages are added to derive the index. The range of the index is from 0 (no species in common) to 1 (identical communities). An index of 0.7 or above indicates a very high degree of similarity.

INDICES OF DIVERSITY

Diversity indices allow either similar or completely different community types to be compared. They usually are a combined measure of (*a*) the number of species in a community and (*b*) their relative abundance, *i.e.* the degree of dominance that prevails. The simplest measure of diversity ignores dominance and is simply a tabulation of the number of species in a community or component group. Pianka (1966b) suggests that this is the most useful measure in some cases, and has used it, *e.g.* to compare the lizard faunas of different desert habitats. Species richness indices that standardize for sample size also are common, and very simple. For example, the species richness of a sample may be expressed as S/N, $S/(\log N)$, or $(S - 1)/(\log N)$, where S represents the total number of kinds and N the total number of individuals.

The measure of diversity used most widely by ecologists is the Shannon-Weaver index, a measure that is sensitive to *both* species richness and relative species abundance. The Shannon-Weaver index is determined by the formula:

$$H = -\Sigma p_i \ln p_i$$

where H = diversity and P_i = the frequency of species i. (Each species, in turn, is processed as species i and the sum of all values is the diversity index for the community.) Lloyd *et al.* (1968) have published an easy-to-calculate

* These indices almost always are calculated not for whole communities, but for component groups within communities: *e.g.* trees, birds, phytoplankton, fishes, soil invertebrates. These are censused by species (or other taxonomic category) and the indices are then calculated.

TABLE A.1

An Example of Similarity Analysis

Within the oak-hickory region of southeastern Pennsylvania, serpentine soils bear a distinctive vegetation dominated by pitch pine. One finds both types of forest adjacent to one another, merging through a rather abrupt transition zone. How different are these forests? Sanders' index provides a quantitative measure. (Data were collected by Suzanne Bailey.)

Species	Relative abundance in:			Decimal figures used for:		
	Oak-hickory forest	Serpentine forest	Transition forest	Oak-hickory/ serpentine comparison	Oak-hickory/ transition comparison	Serpentine/ transition comparison
White oak	0.302	0.222	0.221	0.222	0.221	0.221
Red oak	0.209			0	0	0
Scrub oak		0.125		0	0	0
Blackjack oak		0.083		0	0	0
Black oak		0.028		0	0	0
Post oak		0.014		0	0	0
Pitch pine		0.444	0.426	0	0	0.426
Black cherry	0.070	0.042	0.118	0.042	0.07	0.042
Bigtooth poplar		0.042	0.073	0	0	0.042
Red maple	0.186		0.088	0	0.088	0
Dogwood	0.186		0.044	0	0.044	0
Hickory spp.	0.042			0	0	0
American Beech		0.029		0	0	0
Sum (= index of similarity)				0.264	0.423	0.731

TABLE A.2

Sample Calculation of Shannon-Weaver Index for the Breeding Bird Community of Quaker Run Valley

(Calculation from Lloyd *et al.* (1968); original bird data from Saunders (1936).)

Species	Individuals (n_i)	$n_i \log_{10} n_i$
Yellow warbler	48	80.6996
Yellowthroat	36	56.0269
Song sparrow	22	29.5333
Catbird	10	10.0000
Woodcock	4	2.4082
Veery	2	0.6021
Sums	122 (= N)	179.2701 ($\Sigma n_i \log_{10} n_i$)

Also, $N \log_{10} N$ for $N = 122$ is 254.5359, and c, when calculating diversity to the base e, is 2.302585. Thus:

$H = (2.302585/122)\ (254.5359 - 179.2701) = 1.4206$

transposition of this formula (together with tables of values ready to plug into it), as

$$H = c/N\,(N \log_{10} N - \Sigma n_i \log_{10} n_i)$$

where c is a scale factor for conversion of logs from base 10 to any other base ($c = 1$ for base 10, 2.302585 for base e), N = the total count of all species, and n the count of species i. As an example, the species diversity of a group of breeding birds along a stream in New York State is calculated in Table A.2.

The maximum possible value of the Shannon-Weaver diversity index is dependent on sample size: for example, it is 4.6 for a sample of 100 individuals in which every individual is of a different species, and 6.9 for a sample of 1000 individuals in which this holds true. The minimum possible value is zero. In the real world, Shannon-Weaver values <1 represent very low diversity, and values above 3 represent relatively high diversity. The former might characterize, for example, the invertebrates of a polluted stream or the plants of a salt marsh, while the latter might characterize the invertebrates of an unpolluted stream or the trees of a mixed mesophytic deciduous forest.

Other diversity indices exist, but are used less commonly. Some are essentially visual, being dependent on graphical comparison of species abundance patterns (see Chapter 7; also Odum, 1971, p. 149; Patrick, 1963 and 1968).

EQUITABILITY

As just stated, the maximum possible Shannon-Weaver value is a function of sample size. Indices of equitability are independent of differences in sample size, and so are often used as a further step beyond the calculation of Shannon-Weaver indices. Equitability indices are used to compare *relative abundance patterns* in different communities or samples, since they are independent of the actual number of species involved. The simplest measure of equitability is simply the ratio

$$H/H_{max}$$

where H is the Shannon-Weaver diversity of the sample and H_{max} would be the diversity obtaining when every individual in the sample was of a different species. H_{max}, like H, is easily calculated from the tables of Lloyd *et al.* (1968). The ratio $H{:}H_{max}$ often is called Margalef's equitability index, or Margalef's measure of redundancy. A second index of equitability, proposed by Lloyd and Ghelardi (1964) uses as a standard not the maximum possible Shannon-Weaver diversity but the diversity that a sample of given size would have if it conformed to MacArthur's broken stick model of relative abundance (see Chapter 7). Lloyd and Ghelardi provide tables that allow their index to be easily calculated.

Bibliography

ADDICOTT, J. F. (1974) Predation and prey community structure: an experimental study of the effect of mosquito larvae on the protozoan communities of pitcher plants. *Ecology* 55: 475–492

AHEARN, G. A. (1970) The control of water loss in desert tenebrionid beetles. *J. Exp. Biol.* 53: 573–595

ALLEE, W. C. *et al.* (1949) Principles of Animal Ecology. Philadelphia, W. B. Saunders and Co.

ALLISON, A. C. (1956) Sickle cells and evolution. *Sci. Amer.* 195(2): 87–94

AMOS, W. H. (1959) The life of a sand dune. *Sci. Amer.* 201(1): 91–98

ANDERSON, A. J., AND E. J. UNDERWOOD (1959) Trace-element deserts. *Sci. Amer.* 200(1): 97–106

ANDERSON, E. (1967) Plants, Man and Life. Berkeley, California, University of California Press

ANDERSON, P. K. (1961) Density, social structure, and nonsocial environment in house-mouse populations and the implications for regulation of numbers. *Trans. N. Y. Acad. Sci. Ser. II* 23(5): 447–451

ANDRESEN, J. W. (1973) Forestry thrives in Chicago. *Amer. Forests* 79(4): 12–15 and 58–59

ANDREWARTHA, H. G. (1961) Introduction to the Study of Animal Populations. Chicago, University of Chicago Press

ANTHONY, A. (1955) Behavior patterns in a laboratory population of prairie dogs. *J. Mammal.* 36: 69–78

ARON, W. I., AND S. H. SMITH (1971) Ship canals and aquatic ecosystems. *Science* 174: 13–21

ART, H. W., F. H. BORMANN, G. K. VOIGT, AND G. M. WOODWELL (1974) Barrier island forest ecosystem: role of meteorologic nutrient inputs. *Science* 184: 60–62.

AUCLAIR, A. N., AND G. COTTAM (1971) Dynamics of black cherry (*Prunus serotina* Erhr.) in southern Wisconsin oak forests. *Ecol. Monogr.* 41: 153–177

AXELROD, D., AND H. P. BAILEY (1969) Paleotemperature analysis of tertiary floras. *Palaeogeogr. Palaeoclimatol. Palaeoecol.* 6: 163–195

AYALA, F. J. (1969) Experimental invalidation of the principle of competitive exclusion. *Nature* 224: 1076–1079

AYALA, F. J. (1970) Invalidation of the principle of competitive exclusion defended. *Nature* 227: 89–90

BAKELESS, J. (1950) The Eyes of Discovery. Philadelphia, J. B. Lippincott Co.

BARASH, D. P. (1974) The evolution of marmot societies: a general theory. *Science* 185: 415–420.

BARBOUR, C. D., AND J. H. BROWN (1974) Fish species diversity in lakes. *Am. Naturalist* 108(962): 473–489

BARBOUR, M. G. (1968) Germination requirements of the desert shrub *Larrea divaricata*. *Ecology* 49: 915–923

BARBOUR, M. G. (1970a) Germination and early growth of the strand plant *Cakile maritima*. *Bull. Torrey Bot. Club* 97: 13–22

BARBOUR, M. G. (1970b) Seedling ecology of *Cakile maritima* along the California coast. *Bull. Torrey Bot. Club* 97: 280–289

BARBOUR, M. G. (1972) Seedling establishment of *Cakile maritima* at Bodega Head, California. *Bull. Torrey Bot. Club* 99: 11–16

BARBOUR, M. G. (1973) Ecologist and consultant: do we have to be both? *Ecology* 54(1): 1–2

BARBOUR, M. G., AND D. V. DIAZ (1973) *Larrea* plant communities on bajada and moisture gradients in the United States and Argentina. *Vegetatio* 28: 335–352

BARD, G. E. (1952) Secondary succession on the piedmont of New Jersey. *Ecol. Monogr.* 22: 195–215

BARTHOLOMEW, G. A. (1964) The roles of physiology and behavior in the maintenance of homeostasis in the desert environment. In G. M. Hughes, ed., Homeostasis and Feedback Mechanisms (Symposium 18 of the Society for Experimental Biology), pp. 7–30, Cambridge, Cambridge University Press

BARTHOLOMEW, G. A., AND J. W. HUDSON (1961) Desert ground squirrels. *Sci. Amer.* 205: 107–116

BARTHOLOMEW, G. A., J. W. HUDSON, AND T. R. HOWELL (1962) Body temperature, oxygen consumption, evaporative water loss, and heart rate in the poor-will. *Condor* 64: 117–125

BASKIN, J. M., AND C. C. BASKIN (1972) Germination characteristics of *Diamorpha cymosa* seeds and an ecological interpretation. *Oecologia* 10: 17–28

BATES, M. (1960) The Forest and the Sea. New York, Random House, Inc.

BATES, M. (1963) Animal Worlds. New York, Random House, Inc.

BAZZAZ, F. A. (1968) Succession on abandoned fields in the Shawnee Hills, Southern Illinois. *Ecology* 49(5): 924–936

BEAUCHAMP, R. S. A. (1964) The Rift Valley lakes of Africa. *Verh. Internat. Verein. Limnol.* 15: 91–99

BEETON, A. M. (1965) Eutrophication of the St. Lawrence Great Lakes. *Limnol. Oceanogr.* 10: 240–254

BENTLEY, P. J. (1966) Adaptations of amphibia to arid environments. *Science* 152(3722): 619–623

BENTLEY, P. J., AND W. F. C. BLUMER (1962) Uptake of water by the lizard *Moloch horridus*. *Nature* 194: 699–700

BERKLAND, J. O., AND L. A. RAYMOND (1973) Pleistocene glaciation in the Blue Ridge Province, southern Appalachian Mountains, North Carolina. *Science* 181: 651–653

BEVERTON, R. J. H., AND S. J. HOLT (1957) On the dynamics of exploited fish populations. *Fishery Invest.*, Ser. 2, No. 19. London, H. M. Stationery Office.

BILLINGS, W. D. (1938) The structure and development of old-field pine stands and certain associated physical properties of the soil. *Ecol. Monogr.* 8: 437–499

BILLINGS, W. D. (1964) Plants and the Ecosystem. Belmont, California, Wadsworth Publishing Co.

BILLINGS, W. D., AND H. A. MOONEY (1968) The ecology of arctic and alpine plants. *Biol. Rev.* 43: 481–529

BJORKMAN, O. *et al.* (1972) Photosynthetic adaptation to high temperatures: a field study in Death Valley, California. *Science* 175(4023): 786–788

BJORKMAN, O., AND J. BERRY (1973) High-efficiency photosynthesis. *Sci. Amer.* 229(4): 80–93

BOGERT, C. M. (1959) How reptiles regulate their body temperature. *Sci. Amer.* 200(4): 105–120

BORCHERT, J. R. (1950) Climate of the central North American grassland. *Assoc. Amer. Geogr. Ann.* 40: 1–39

BORMANN, F. H. (1953) Factors determining the role of loblolly pine and sweet gum in early old-field succession in the Piedmont of North Carolina. *Ecol. Monogr.* 23: 339–358

BORMANN, F. H. (1972) Unlimited growth: growing, growing, gone? *Bioscience* 22(12): 706–709

BORMANN, F. H., AND G. E. LIKENS (1970) The nutrient cycles of an ecosystem. *Sci. Amer.* 223(4): 92–101

BOTKIN, D. B., J. F. JANAK, AND J. R. WALLIS (1973) Estimating the effects of carbon fertilization on forest composition by ecosystem simulation. In G. M. Woodwell and E. V. Pecan. eds. Carbon and the Biosphere (Brookhaven Symposia in Biology No. 24), pp. 328–344, Washington, D.C., U.S.A.E.C. Technical Information Center

BOURLIERE, F. (1963) Observations on the ecology of some large African mammals. In F. C. Howell and F. Bourlière, eds., African Ecology and Human Evolution, pp. 43–54, Chicago, Aldine Publishing Co.

BOURLIÈRE, F. (1973) Comparative ecology of rain forest mammals. In B. J. Meggers and E. S. Ayensu, eds., Tropical Forest Ecosystems in Africa and South America, pp. 279–292, Washington, D.C., Smithsonian Institution Press

BOURLIÈRE, F., AND M. HADLEY (1970) The ecology of tropical savannas. *Annu. Rev. Ecol. Systematics* 1: 125–152

BRADLEY, W. G. (1968) Food habits of the antelope ground squirrel in southern Nevada. *J. Mammal.* 49(1): 14–21

BRADLEY, W. G., AND R. A. MAUER (1971) Reproduction and food habits of Merriam's kangaroo rat, *Dipodomys merriami. J. Mammal.* 52: 497–507

BRADSHAW, W. E. (1972) Action spectra for photoperiodic response in a diapausing mosquito. *Science* 172: 1361–1362

BRAY, J. R. (1964) Primary consumption in three forest canopies. *Ecology* 45: 165–167

BRAY, J. R., AND J. T. CURTIS (1957) An ordination of the upland forest communities of southern Wisconsin. *Ecol. Monogr.* 22: 217–234

BRAY, J. R. AND E. GORHAM (1964) Litter production in forests of the world. *Adv. Ecol. Res.* 2: 101–157

BRIGGS, J. C. (1969) The sea-level Panama Canal: potential biological catastrophe. *Bioscience* 19(1): 44–47

BROCK, T. D. (1967) The ecosystem and the steady state. *Bioscience* 17: 166–169

BROECKER, W. (1970) Man's oxygen reserves. *Science* 168: 1537–1538

BROWER, L. P. (1969) Ecological chemistry. *Sci. Amer.* 220(2): 22–29

BROWN, J. H., JR. (1960) The role of fire in altering the species composition of forests in Rhode Island. *Ecology* 41(2): 310–316

BROWN, J. L. (1974) Alternate routes to sociality in jays—with a theory for the evolution of altruism and communal breeding. *Amer. Zool.* 14:63–80

BRUNSKILL, G. J., AND S. D. LUDLAM (1969) Fayetteville Green Lake, New York. I. Physical and chemical limnology. *Limnol. Oceanogr.* 14: 817–829

BRYLINSKY, M., AND K. H. MANN (1973) An analysis of factors governing productivity in lakes and reservoirs. *Limnol. Oceanogr.* 18(1): 1–14

BRYSON, R. A. (1966) Air masses, streamlines and the boreal forest. *Geograph. Bull.* 8: 228–269

BUECHNER, H. K., AND H. D. ROTH (1974) The lek system in Uganda kob antelope. *Amer. Zool.* 14: 145–162

BUELL, M. F., H. F. BUELL, J. A. SMALL, AND T. G. SICCAMA (1971) Invasion of trees in secondary succession on the New Jersey piedmont. *Bull. Torrey Bot. Club* 98(2): 67–74

BURGESS, E. W. (1925) The growth of the city: an introduction to a research project. In R. E. Park *et al.*, eds., The City, pp. 47–62, Chicago, University of Chicago Press

BURNET, F. M., AND D. O. WHITE (1972) Natural History of Infectious Disease, 4th Ed., Cambridge, Cambridge University Press

BURNS, G. P. (1923) Measurement of solar radiant energy in plant habitats. *Ecology* 4: 189–195

BURNS, J. M. (1966) Preferential mating *versus* mimicry: disruptive selection and sex-limited dimorphism in *Papilio glaucus. Science* 153(3735): 551–553.

BURT, D. E. (1973) The geography of extinct and endangered species in the United States. *Explorer* 15(3): 4–10

CAIN, A. J., AND P. M. SHEPPARD (1950) Selection in the polymorphic land snail *Cepaea nemoralis* (L.). *Heredity* 4: 275–294

CALDER, W. A., AND K. SCHMIDT-NIELSEN (1967) Temperature regulation and evaporation in the pigeon and the roadrunner. *Am. J. Physiol.* 213: 883–889

CALHOUN, J. B. (1962) Population density and social pathology. *Sci. Amer.* 206(2): 139–148

CANTER, H. M., AND J. W. G. LUND (1951) Studies on plankton parasites. 3. Examples of the interaction between parasitism and other factors determining the growth of diatoms. *Ann. Bot. London N. S.* 15: 359–371

CANTLON, J. E. (1953) Vegetation and microclimates on north and south slopes of Cushetunk Mountain, New Jersey. *Ecol. Monogr.* 23(3): 241–270

CARSON, R. (1951) The Sea Around Us. New York, Oxford University Press

CAULEY D. L., AND J. R. SHINNER (1973) The Cincinnati raccoons. *Nat. Hist.* 82(9): 58–60

CAVERS, P. B., AND HARPER, J. L. (1966) Germination polymorphism in *Rumex crispus* and *Rumex obtusifolius. J. Ecol.* 54: 367–382

CHEW, R. M., AND A. E. CHEW (1965) The primary productivity of a desert shrub (*Larrea tridentata*) community. *Ecol. Monogr.* 35: 355–375

CHITTY, D. (1967) The natural selection of self-regulatory behavior in animal populations. *Proc. Ecol. Soc. Australia* 2: 51–78

CHRISTIAN, J. J. (1970) Social subordination, population density, and mammalian evolution. *Science* 168(3927): 84–90

CHRISTIAN, J. J., AND D. E. DAVIS (1964) Endocrines, behavior and population. *Science* 146: 1550–1560

CHRISTIAN, J. J., V. FLYGER, AND D. E. DAVIS (1960) Factors in the mass mortality of a herd of Sika deer, *Cervus nippon. Chesapeake Sci.* 1(2): 79–95

CLAPHAM, A. R. (1936) Overdispersion in grassland communities and the use of statistical methods in plant ecology. *J. Ecol.* 24: 232–251

CLARK, L. R. (1964) The intensity of parasite attack in relation to the abundance of *Cardiaspina albitextura* (Psyllidae). *Aust. J. Zool.* 12: 349–361

CLOUD, P. E., JR. (1968) Atmospheric and hydrospheric evolution on the primitive earth. *Science* 160: 729–736

CLOUD, P. E., JR. (1974) Evolution of ecosystems. *Amer. Sci.* 62(1): 54–66

CODY, M. L. (1971) Finch flocks in the Mohave Desert. *Theoret. Population Biol.* 2(2): 142–158

CODY, M. L. (1974) Optimization in ecology. *Science* 183: 1156–1164.

COLE, L. C. (1946) A theory for analyzing contagiously distributed populations. *Ecology* 27: 329–341

CONNELL, J. H. (1961a) The influence of interspecific competition and other factors on the distribution of the barnacle *Chthamalus stellatus. Ecology* 42(4); 710–723

CONNELL, J. H. (1961b) Effects of competition, predation by *Thais lapillus* and other factors on natural populations of the barnacle *Balanus balanoides. Ecol. Monogr.* 31: 61–104

COOK, R. C. (1962) The story of Mauritius: from the dodo to the stork. *Population Bull.* 18(5): 93–115

COOPER, C. F. (1960) Changes in vegetation structure and growth of southwestern pine forests since white settlement. *Ecol. Monogr.* 30(2): 129–164

COOPER, C. F. (1961) The ecology of fire. *Sci. Amer.* 204(4): 150–160

COPELAND, B. J. (1965) Evidence for regulation of community metabolism in a marine ecosystem. *Ecology* 46(4): 563–564

COSTELLO, D. F. (1969) The Prairie World. New York, Thomas Y. Crowell Co.

COULTER, G. W. (1963) Hydrological changes in relation to biological production in southern Lake Tanganyika. *Limnol. Oceanogr.* 8(4): 463–477

COX, G. W. (1976) Laboratory Manual of General Ecology. Dubuque, Iowa, William C. Brown Co.

CROOK, J. H. (1964) The evolution of social organization and visual communication in the weaver birds (Ploceinae). *Behavior* Suppl. 10: 1–178

CROOK, J. H. (1965) The adaptive significance of avian social organizations. In P. V. Ellis, ed., Social Organization of Animal Communities (Symp. Zool. Soc. London 14), pp. 181–218

CROOK, J. H. (1966) Co-operation in primates. *Eugenics Rev.* 58(2): 63–70

CROOK, J. H. (1970) The socio-ecology of primates. In J. H. Crook, ed., Social Behavior in Birds and Mammals, pp. 103–166, New York, Academic Press

CROOK, J. H., AND J. S. GARTLAN (1966) Evolution of primate societies. *Nature* 210(5042): 1200–1203

CROSSLEY, D. A., JR, AND M. P. HOGLAND (1962) A litter bag method for the study of microarthropods inhabiting leaf litter. *Ecology* 43: 571–573

CUMMINS, K. W. (1974) Structure and function of stream ecosystems. *Bioscience* 24(11): 631–641

CURTIS, J. T. (1959) The Vegetation of Wisconsin. Madison, University of Wisconsin Press

CURTIS, J. T., AND R. P. MCINTOSH (1951) An upland forest continuum in the prairie-forest border region of Wisconsin. *Ecology* 32: 476–496

DANSEREAU, P. (1957) Biogeography: An Ecological Perspective. New York, The Ronald Press Co.

DARLINGTON, P. J., JR. (1957) Zoogeography. New York, John Wiley and Sons, Inc.

DASMANN, R. F. (1968) Environmental Conservation, 2nd Ed., New York, John Wiley and Sons, Inc.

DASMANN, R. F. (1976) Environmental Conservation, 4th Ed., New York, John Wiley and Sons, Inc.

DAUBENMIRE, R. F. (1959) Plants and Environment: A Textbook of Plant Autecology, 2nd Ed., New York, John Wiley and Sons, Inc.

DAVIS, M. (1969) Palynology and environmental history during the Quaternary period. *Amer. Sci.* 57(3): 317–332

DEEVEY, E. S., JR. (1947) Life tables for natural populations of animals. *Quart. Rev. Biol.* 22: 283–314

DEEVEY, E. S., JR. (1956) The human crop. *Sci. Amer.* 194(4): 105–112

DEEVEY, E. S., JR., (1960) The human population. *Sci. Amer.* 203(3): 195–204

DEEVEY, E. S., JR., (1969) Coaxing history to conduct experiments. *Bioscience* 19(1): 40–43

DEEVEY, E. S., JR. (1970) Mineral cycles. *Sci. Amer.* 223(3): 148–158

DEEVEY, E. S., JR. (1973) Sulfur, nitrogen and carbon in the biosphere. In G. M. Woodwell and E. V. Pecan, eds., Carbon and the Biosphere (Brookhaven Symposia in Biology No. 24), pp. 182–190, Washington, D.C., U.S. A.E.C., Technical Information Center

DELWICHE, C. C. (1970) The nitrogen cycle. *Sci. Amer.* 223(3): 136–146

DETTWILLER, J. (1970) Deep soil temperature trends and urban effects at Paris. *J. Appl. Meteorol.* 9: 178–180

DINESEN, I. (1937) Out of Africa. New York, Random House, Inc.

DOBSON, H. F. H. *et al.* (1974) A summary and comparison of nutrients and related water quality in Lakes Erie, Ontario, Huron and Superior. *J. Fish Res. Board Canada* 31(5): 731–738

DOEKSEN, J., AND J. VANDER DRIFT, eds. (1963) Soil Organisms. Amsterdam, North-Holland Publishing Co.

DORF, E. (1969) Paleobotanical evidence of Mesozoic and Cenozoic climatic changes. *Proc. N. Amer. Paleontol. Conv.* Part D: 323–346

DOWNS, A. A. AND W. E. MCQUILKIN (1944) Seed production of southern Appalachian oaks. *J. Forestry* 42: 913–920

DUMONT, R. (1969) False Start in Africa, 2nd Ed., New York, Praeger

EDMONDSON, W. T. (1969) Eutrophication in North America. In Eutrophication: Causes, Consequences, Correctives. pp. 129–149, Nat. Acad. Sci./Nat. Res. Council Pub. 1700

EDMONDSON, W. T. (1972) Nutrients and phytoplankton in Lake Washington. In G. E. Likens, ed., Nutrients and Eutrophication: The Limiting Nutrient Controversy (Special Symposia I, Am. Soc. Limnol. Oceanogr.), pp. 172–193, Lawrence, Kansas, Allen Press

EDNEY, E. B. (1967) Water balance in desert arthropods. *Science* 156(3778): 1059–1066

EDWARDS, C. A., AND G. W. HEATH (1963) The role of soil animals in breakdown of leaf material. In J. Doeksen and J. Vander Drift, eds., Soil Organisms, pp. 76–84, Amsterdam, North-Holland Publishing Co.

EDWARDS, C. A., D. E. REICHLE, AND D. A. CROSSLEY, JR. (1970) The role of soil invertebrates in turnover of organic matter and nutrients. In D. E. Reichle, ed., Analysis of Temperate Forest Ecosystems, pp. 147 172, Berlin, Springer-Verlag

EDWARDS, R. Y. (1954) Fire and the decline of a mountain caribou herd. *J. Wildlife Management* 18: 521–526

EGLER, F. (1971) The romance of roadside ragweed. *Conservationist* 26(1): 27

EHRLICH, P. R., AND L. C. BIRCH (1967) The "balance of nature" and "population control." *Am. Naturalist* 101: 97–107

EHRLICH, P. R., AND A. H. EHRLICH (1973) Coevolution: heterotypic schooling in Caribbean reef fishes. *Am. Naturalist* 107(953): 157–160

EHRLICH, P. R., AND P. H. RAVEN (1967) Butterflies and plants. *Sci. Amer.* 216(6): 104–113

EISENBERG, J. F. *et al.* (1972) The relation between ecology and social structure in primates. *Science* 176(4037): 863–874

EISENBERG, R. M. (1966) The regulation of density in a natural population of the pond snail, *Lymnaea elodes*. *Ecology* 47(6): 889–906

EKMAN, S. P. (1953) Zoogeography of the Sea. London, Sidgwick and Jackson

ELTON, C. S. (1958) The Ecology of Invasions by Animals and Plants. New York, John Wiley and Sons, Inc.

ERIKSSON, E. (1963) The yearly circulation of sulfur in nature. *J. Geophys. Res.* 68(13): 4001–4008

ERRINGTON, P. L. (1957) Of population cycles and unknowns. *Cold Spring Harbor Symp. Quant. Biol.* 22: 287–300

ESTES, R. D. (1967) Predators and scavengers. *Nat. Hist.* 76(2): 20–29 and 76(3): 38–47

FARNER, D. S., AND J. R. KING, eds. (1971) Avian Biology, Vol. 1. New York, Academic Press

FEDOROV, A. A. (1966) The structure of the tropical rain forest and speciation in the humid tropics. *J. Ecol.* 54: 1–11

FELGER, R. S., AND C. H. LOWE (1967) Clinal variation in the surface-volume relationships of the columnar cactus *Lophocereus schottii* in northwestern Mexico. *Ecology* 48(4): 530–536

FENSTER, D. M. *et al.* (1972) A biological survey of the Conestoga Creek Basin. Unpublished report, Biology Department, Franklin and Marshall College.

FERTIG, D. S., AND V. W. EDMONDS (1969) The physiology of the house mouse. *Sci. Amer.* 221(4): 103–110

FINCH, V. C., G. T. TREWARTHA, A. H. ROBINSON, AND E. H. HAMMOND (1957) Elements of Geography. New York. Collier-Macmillan Co.

Fisher, J., N. Simon, and J. Vincent (1969) Wildlife in Danger. New York, Viking Press, Inc.

Flint, R. F. (1971) Glacial and Quaternary Geology. New York, John Wiley and Sons, Inc.

Florer, L. E. (1972) Palynology of a postglacial bog in the New Jersey Pine Barrens. *Bull. Torrey Bot. Club* 99(3): 135–138

Forbes, S. A. (1887) The lake as a microcosm. *Bull Peoria Sci. Assoc.* 1887: 77–87 (Reprinted in *Ill. Nat. Hist. Surv. Bull.* 15: 537–550 (1925).)

Forbes, S. E. *et al.* (1971) The White-tailed Deer in Pennsylvania. Harrisburg, Pennsylvania, Pennsylvania Game Commission. *Research Bulletin* 170

Ford, E. B. (1971) Ecological Genetics. New York, John Wiley and Sons, Inc.

Fowells, H. A. (1965) Silvics of Forest Trees of the United States. United States Department of Agriculture Handbook No. 271

Fowke, E. (1970) Lumbering songs from the Northern Woods. American Folklore Society Memoir 55, Austin, Texas, University of Texas Press

Frieden, E. (1972) The chemical elements of life. *Sci. Amer.* 227(1): 52–60

Froment, A. (1972) Soil respiration in a mixed oak forest. *Oikos* 23:273–277

Gates D. M. (1963) The energy environment in which we live. *Am. Sci.* 51: 327–348

Gates, D. M. (1971) The flow of energy in the biosphere. *Sci. Amer.* 224(3): 88–100

Gates, D. M. (1972) Man and His Environment: Climate. New York, Harper and Row

Gates, D. M., R. Alderfer, and E. Taylor (1968) Leaf temperatures of desert plants. *Science* 159(3818): 994–995

Gause, G. F. (1970) Criticism of invalidation of principle of competitive exclusion. *Nature* 227:89

Geiger, R. (1959) The Climate Near the Ground, 3rd Ed., Cambridge, Harvard University Press

Geiger, R. (1965) The Climate Near the Ground, 4th Ed., Cambridge, Harvard University Press

Geis, A. E. (1974) The new town bird quadrille. *Nat. Hist.* 83(6): 54–61

George, C. J., and D. McKinley (1974) Urban Ecology: In Search of an Asphalt Rose. New York, McGraw-Hill Co.

Gibbs, R. J. (1970) Mechanisms controlling world water chemistry. *Science* 170: 1088–1090

Gill, F. B., and L. L. Wolf (1975) Economics of feeding territorality in the golden-winged sunbird. *Ecology* 56: 333–345

Golley, F. B. (1972) Energy flux in ecosystems. In J. A. Wiens, ed., Ecosystem Structure and Function, pp. 69–90, Corvallis, Oregon State University Press

Gomez-Pompa, A. *et al.* (1972) The tropical rain forest: a nonrenewable resource. *Science* 177: 762–765

Goulden, C. (1969) Temporal changes in diversity. In G. M. Woodwell and H. H. Smith, eds., Diversity and Stability in Ecological Systems (Brookhaven Symposia in Biology No. 22), pp. 96–102, Upton, N. Y., Brookhaven National Laboratory

Graenicher, S. (1906) Some notes on the pollination of flowers. *Bull. Wisc. Nat. Hist. Soc.* 4: 16–18

Graham, E. H. (1947) The Land and Wildlife. New York, Oxford University Press

Graham, E. H. (1969) Natural Principles of Land Use. New York, Greenwood Press

Grant, V. (1951) The fertilization of flowers. *Sci. Amer.* 184(6): 52–56

Graybill, D. L. (1971) Mast production in northwestern Pennsylvania forests. *Bull. Ecol. Soc. Amer.* 52(4): 52

Grime, J. P. (1966) Shade avoidance and shade tolerance in flowering plants. In R. Bainbridge *et al.*, eds., Light as an Ecological Factor, pp. 187–208, New York, John Wiley and Sons, Inc.

Gulland, J. A. (1972) Population Dynamics of World Fisheries. Seattle, Division of Marine Resources, University of Washington

Gwynne, M. D., and R. H. V. Bell (1968) Selection of vegetation components by grazing ungulates in the Serengeti National Park. *Nature* 220: 390–393

Hadley, N. F. (1972) Desert species and adaptation. *Amer. Sci.* 60(3): 338–347

Hairston, N. G., F. E. Smith, and L. B. Slobodkin (1960) Community structure, population control, and competition. *Amer. Nat.* 94: 421–425

Hall, K. R. L. (1965) Behavior and ecology of the wild patas monkey, *Erythrocebus patas*, in Uganda. *J. Zool.* 148: 15–87

Hardin, G. (1969) Not peace, but ecology. In G. M. Woodwell and H. H. Smith, eds., Diversity and Stability in Ecological Systems (Brookhaven Symposia in Biology No. 22), pp. 151–161, Upton, N. Y., Brookhaven National Laboratory

Hare, F. K. (1963) The Restless Atmosphere, 3rd Ed. New York, Harper and Row (Harper Torchbooks)

Harlow, H. F. *et al.* (1971) From thought to therapy: lessons from a primate laboratory. *Amer. Sci.* 59(5): 538–549

Harper, J. L., P. H. Lovell, and K. G. Moore (1970) The shapes and size of seeds. *Annu. Rev. Ecol. Systematics*, 1: 327–356

Harris, C. D., and E. L. Ullman (1945) The nature of cities. *Ann. Amer. Acad. Pol. Soc. Sci.* 242: 7–17

Harriss, R. C. (1966) Biological buffering of oceanic silica. *Nature* 212(5059): 275–276

Hartmann, F., ed. (1973) The metro forest. *Nat. Hist.* 82(9): 45–84

Hayes, F. R., J. A. McCarter, M. L. Cameron, and D. A. Livingstone (1952) On the kinetics of phosphorus exchange in lakes. *J. Ecol.* 40: 202–216

Heady, H. F. (1966) Influence of grazing on the composition of *Themeda triandra* grassland, East Africa. *J. Ecol.* 54: 704–727

Healey, M. C. (1967) Aggression and self-regulation of population size in deermice. *Ecology* 48: 377–392

Heezen, B. C., and C. D. Hollister (1971) The Face of the Deep. New York, Oxford University Press

Heinselman, M. L. Fire in the virgin forests of the Boundary Waters Canoe Area, Minnesota. *Quat. Res.* 3: 329–382

Hesse, R., W. C. Allee, and K. P. Schmidt (1951) Ecological Animal Geography, 2nd ed. New York. John Wiley and Sons, Inc.

Hett, J. M., and O. L. Loucks (1971) Sugar maple (*Acer saccharum* Marsh.) seedling mortality. *J. Ecol.* 59: 507–520

Hickey, R. J. (1971) Air pollution. In W. W. Murdoch, ed., Environment, pp. 189–212. Stamford, Conn., Sinauer Associates

Hock, R. J. (1970) The physiology of high altitude. *Sci. Amer.* 222(2): 55–62

Hodges, J. D., and D. R. M. Scott (1968) Photosynthesis in seedlings of six conifer species under natural environmental conditions. *Ecology* 49(5): 973–998

Holland, H. D. (1972) The geologic history of sea water—an attempt to solve the problem. *Geochim. Cosmochim. Acta* 36: 637–651

Holling, C. S. (1959) The components of predation as revealed by a study of small mammal predation of the European pine sawfly. *Can. Entomol.* 91: 293–320

Holling, C. S. (1965) The functional response of predators to prey density and its role in mimicry and population regulation. *Mem. Entomol. Soc. Can.* 45: 1–60

Horn, H. S. (1968a) The adaptive significance of colonial

nesting in Brewer's blackbird (*Euphagus cynocephalus*). *Ecology* 49: 682–694

HORN, H. S. (1968b) Regulation of animal numbers: a model counter-example. *Ecology* 49: 776–778

HOUGH, H. B. (1933) A bird that man could kill. *Vineyard Gazette* (Martha's Vineyard, Mass.), April 21, 1933

HOUGHTON, H. G. (1954) On the annual heat balance of the Northern Hemisphere. *J. Meteorol.* 11: 1–9

HURLEY, A. C. (1973) Fecundity of the acorn barnacle *Balanus pacificus* Pilsbry: a fugitive species. *Limnol. Oceanogr.* 18: 386–393

HUTCHINSON, G. E. (1943) The biogeochemistry of aluminum and of certain related elements (concluded). *Quart. Rev. Biol.* 18: 331–363

HUTCHINSON, G. E. (1951) Copepodology for the ornithologist. *Ecology* 32: 571–577

HUTCHINSON, G. E. (1957a) Concluding remarks. *Cold Spring Harbor Symp. Quant. Biol.* 22: 415–427

HUTCHINSON, G. E. (1957b) *A Treatise on Limnology*, Vol. 1. New York, John Wiley and Sons, Inc.

HUTCHINSON, G. E. (1961) The paradox of the plankton. *Am. Naturalist* 95: 137–145

HUTCHINSON, G. E. (1964) The lacustrine microcosm reconsidered. *Amer. Sci.* 52: 334–341

HUTCHINSON, G. E. (1965) The Ecological Theater and the Evolutionary Play. New Haven, Yale University Press

HUTCHINSON, G. E. (1967) A Treatise on Limnology, Vol. 2. New York, John Wiley and Sons, Inc.

HUTCHINSON, G. E. (1970) The biosphere. *Sci. Amer.* 223(3): 44–53

HUTCHISON, J. H., AND J. L. LARIMER (1960) Reflectivity of the integuments of some lizards from different habitats. *Ecology* 41: 199–209

HYNES, H. B. N. (1960) The Biology of Polluted Waters. Liverpool, Liverpool University Press

IRVING, L. (1966) Adaptations to cold. *Sci. Amer.* 214(1): 94–101

JACKSON, D. C., AND K. SCHMIDT-NIELSEN (1964) Countercurrent heat exchange in the respiratory passages. *Proc. Natl. Acad. Sci. U.S.A.* 51: 1192–1197

JACKSON, R. M., AND F. RAW (1966) Life in the Soil. New York, St. Martin's Press

JANZEN, D. H. (1967a) Fire, vegetation structure and the ant × acacia interaction in Central America. *Ecology* 48(1): 26–35

JANZEN, D. H. (1967b) Interaction of the bulls-horn acacia (*Acacia cornigera* L.) with an ant inhabitant (*Pseudomyrmex ferruginea* F. Smith) in eastern Mexico. *U. Kans. Sci. Bull.* 47(6): 315–558

JANZEN, D. H. (1970) Herbivores and the number of tree species in tropical forests. *Am. Naturalist* 104: 501–528

JENNI, D. A. (1974) Evolution of polyandry in birds. *Amer. Zool.* 14: 129–144

JENNI, D. A., AND G. COLLIER (1972) Polyandry in the American jacana (*Jacana spinosa*). *Auk* 89(4): 743–765

JOHN, K. R., AND D. FENSTER (1975) The effects of partitions on the growth rates of crowded *Rana pipiens* tadpoles. *Amer. Midland Nat.* 93: 123–130

JOHNSON, R. E. (1968) Temperature regulation in the white-tailed ptarmigan, *Lagopus leucurus*. *Comp. Biochem. Physiol.* 24: 1003–1014

KEAST, A., AND D. WEBB (1966) Mouth and body form relative to feeding ecology in the fish fauna of a small lake, Lake Opinicon, Ontario. *J. Fish, Res. Bd. Can.* 23(12): 1845–1874

KEETON, W. C. (1972) Biological Science, 2nd Ed. New York, W. W. Norton and Co., Inc.

KEEVER, C. (1950) Causes of succession on old fields of the Piedmont, North Carolina. *Ecol. Monogr.* 20: 229–250

KEEVER, C. (1973) Distribution of major forest species in southeastern Pennsylvania. *Ecol. Monogr.* 43: 303–327

KEITH, L. B. (1963) Wildlife's Ten-year Cycle. Madison, Wisconsin University Press

KELLOGG, W. W. *et al.* (1972) The sulfur cycle. *Science* 175(4022): 587–596

KENAGY, G. J. (1972) Saltbush leaves: excision of hypersaline tissue by a kangaroo rat. *Science* 178(4064): 1094–1096

KENNEDY, V. C. (1963) Sediment transported by Georgia streams. *U.S. Geol. Surv. Water Supply Paper No.* 1668

KEVAN, D. K. M. (1962) Soil Animals. London, H. F. Witherby and Co., Ltd.

KEYFITZ, N. (1966) Population density and the style of social life. *Bioscience* 16(12): 868–873

KILHAM P. (1971) A hypothesis concerning silica and the freshwater planktonic diatoms. *Limnol. Oceanogr.* 16(1): 10–18

KING, J. A. (1955) Social behavior, social organization and population dynamics in a black-tailed prairie dog town in the Black Hills of South Dakota. *Contr. Lab. Vert. Biol.* (*U. Mich.*) 67: 1–123

KING, J. A. (1959) The social behavior of prairie dogs. *Sci. Amer.* 201(4): 128–140

KIRCHER, H. W., W. B. HEED, J. S. RUSSELL, AND J. GROVE (1967) Senita cactus alkaloids: their significance to Sonoran Desert *Drosophila* ecology. *J. Insect Physiol.* 13: 1869–1874

KLIKOFF, L. G. (1967) Moisture stress in a vegetational continum in the Sonoran Desert. *Amer. Midland Nat.* 77(1): 128–137

KOCHEL, R. C., AND R. A. DORSEY (1972) Rates of denudation in the Conestoga River Basin, 1972, Lancaster County, Pennsylvania. Unpublished Report, Geology Department, Franklin and Marshall College

KOLATA, G. B. (1975) Sociobiology (II): the evolution of social systems. *Science* 187(4172): 156–157

KORSTIAN, C. F., AND T. S. COILE (1938) Plant competition in forest stands. *Duke Univ. School For. Bull.* 3

KOZLOVSKY, D. G. (1968) A critical evaluation of the trophic level concept. I. Ecological efficiencies. *Ecology* 40: 48–60

KREBS, C. J. *et al.* (1973) Population cycles in small rodents. *Science* 179: 35–41

KREBS, C. J., AND J. H. MYERS (1974) Population cycles in small mammals. *Adv. Ecol. Res.* 8: 268–399

KRUTCH, J. W. (1951) The Desert Year. New York, William Sloan Associates

KUENZLER, E. J. (1961a) Structure and energy flow of a mussel population in a Georgia salt marsh. *Limnol. Oceanogr.* 6: 191–204

KUENZLER, E. J. (1961b) Phosphorus budget of a mussel population. *Limnol. Oceanogr.* 6: 400–415

KULLBERG, R. G. (1971) Algal distribution in six thermal spring effluents. *Trans. Amer. Micros. Soc.* 90: 412–434

KUMMER, H. (1971) Primate Societies: Group Techniques of Ecological Adaptation. Chicago, Aldine-Atherton, Inc.

LACK, D. (1966) Population Studies of Birds. Oxford, Clarendon Press

LANDSBERG, H. E. (1970) Man-made climatic changes. *Science* 170(3964): 1265–1274

LANGBEIN, W. B., AND S. A. SCHUMM (1958) Yield of sediment in relation to mean annual precipitation. *Trans. Amer. Geophys. Union* 39: 1076–1084

LARSON, D. W. (1972) Temperature, transparency and phytoplankton productivity in Crater Lake, Oregon. *Limnol. Oceanogr.* 17(3): 410–417

LAVE, L. B., AND E. P. SESKIN (1970) Air pollution and human health. *Science* 169:(3947): 723–733

LEHNER, P. N. (1964) The mourning dove—its status in New York. *Conservationist* June–July: 6–8

LEMON, P. C. (1967) Effects of fire on herbs of the southeastern United States and Central Africa. *Proc. Tall Timbers Fire Ecol. Conference* 6: 113–128

LEOPOLD, A. (1949) A Sand County Almanac and Sketches Here and There. New York, Oxford University Press

LEVINS, R. (1968) Evolution in Changing Environments: Some Theoretical Explorations. Princeton, N. J., Princeton University Press

LEWIS, O. (1966) The culture of poverty. *Sci. Amer.* 215(4): 19–25

LI, H. (1969) Urban botany: need for a new science. *Bioscience* 19(10): 882–884

LIETH, H., AND J. S. RADFORD (1971) Phenology, resource management, and synagraphic computer mapping. *Bioscience* 21(2): 62–70

LIKENS, G. E., AND F. H. BORMANN (1974) Acid rain: a serious regional environmental problem. *Science* 184: 1176–1179

LIKENS, G. E., F. H. BORMANN, N. M. JOHNSON, D. W. FISHER, AND R. S. PIERCE (1970) Effects of forest cutting and herbicide treatment on nutrient budgets in the Hubbard Brook watershed-ecosystem. *Ecol. Monogr.* 40: 23–47

LINDSEY, A. A. *et al.* (1961) Vegetation and environment along the Wabash and Tippecanoe Rivers. *Ecol. Monogr.* 31: 105–156

LINDSEY, C. C. (1966) Body sizes of poikilotherm vertebrates at different latitudes. *Evolution* 20: 456–465

LIVINGSTONE, D. A. (1963) The sodium cycle and the age of the ocean. *Geochim. Cosmochim. Acta* 27: 1055–1069

LIVINGSTONE, D. A. (1971) Biogeochemical cycles. In V. G. Dethier, R. C. Lewontin, and M. Lloyd, eds., Topics in Animal Behaviour. Ecology and Evolution, pp. 103–107, New York, Harper and Row

LIVINGSTONE, D. A. (1975) Late Quaternary climatic change in Africa. *Annu. Rev. Ecol. Systematics* 6: 249–280

LLOYD, M., AND H. S. DYBAS (1966) The periodical cicada problem. I. Population ecology. *Evolution* 20(2): 133–149

LLOYD, M., AND R. J. GHELARDI (1964) A table for calculating the "equitability" component of species diversity. *J. Anim. Ecol.* 33: 217–225

LLOYD, M., J. H. ZAR, AND J. R. KARR (1968) On the calculation of information—theoretical measures of diversity. *Amer. Midland Nat.* 79(2): 257–272

LODHI, M. A. K., AND E. L. RICE (1971) Allelopathic effects of *Celtis laevigata*. *Bull. Torrey Bot. Club* 98(2): 83–89

LOTKA, A. J. (1956) Elements of Mathematical Biology. New York, Dover

LUND, J. W. G. (1971) An artificial alteration of the seasonal cycle of the plankton diatom *Melosira islandica* subsp. *subarctica* in an English lake. *J Ecol.* 59(2): 521–534

LÜSCHER, M. (1961) Air-conditioned termite nests. *Sci. Amer.* 205(2): 138–145

LUTZ, H. J., AND R. F. CHANDLER, JR. (1946) Forest Soils. New York, John Wiley and Sons, Inc.

MACAN, T. T. (1970) Biological Studies of the English Lakes. New York, American Elsevier Publishing Co., Inc.

MACARTHUR, R. H. (1957) On the relative abundance of bird species. *Proc. Natl. Acad. Sci. U.S.A.* 45: 293–295

MACARTHUR, R. H. (1958) Population ecology of some warblers of northeastern coniferous forests. *Ecology* 39: 599–619

MACARTHUR, R. H., AND J. CONNELL (1966) The Biology of Populations. New York, John Wiley and Sons, Inc.

MACARTHUR, R. H., AND R. LEVINS (1967) The limiting similarity, convergence and divergence of coexisting species. *Amer. Nat.* 101: 377–385

MACARTHUR, R. H., AND J. W. MACARTHUR (1961) On bird species diversity. *Ecology* 42(3): 594–598

MACARTHUR, R. H., AND E. O. WILSON (1967) The Theory of Island Biogeography. Princeton, N. J., Princeton University Press

MACFADYEN, A. (1963) The contribution of the microfauna to total soil metabolism. In J. Doeksen and J. Van der Drift, eds., Soil Organisms, pp. 3–17, Amsterdam, North-Holland Publishing Co.

MACHTA, L. (1973) Prediction of CO_2 in the atmosphere. In G. M. Woodwell and E. V. Pecan, eds., Carbon and the Biosphere, pp. 21–31, (Brookhaven Symposia In Biology No. 24) Washington, D.C., U.S. A.E.C., Technical Information Center.

MACLULICH, D. A. (1937) Fluctuations in the numbers of the varying hare (*Lepus americanus*). *Univ. Toronto Studies Biol. Ser. No. 43*

MAGUIRE B., JR. (1971) Phytotelmata: biota and community structure determination in plant-held waters. *Annu. Rev. Ecol. Systematics* 2: 439–464

MARKS, P. L. (1974) The role of pin cherry (*Prunus pensylvanica* L) in the maintenance of stability in northern hardwood ecosystems. *Ecol. Monogr.* 44(1): 73–88

MARKS, P. L., AND F. H. BORMANN (1972) Revegetation following forest cutting: mechanisms for return to steady-state nutrient cycling. *Science* 176: 914–915

MARTIN, P. S. (1973) The discovery of America. *Science* 179(4077): 969–974

MARTIN, P. S., AND H. E. WRIGHT, JR. (1967) Pleistocene Extinctions: The Search for a Cause. New Haven, Yale University Press

MARTIN, S. G. (1974) Adaptations for polygynous breeding in the bobolink, *Dolichonyx oryzivorus*. *Amer. Zool.* 14: 109–119

MARTIN, W. E. (1959) The vegetation of Island Beach State Park, New Jersey. *Ecol. Monogr.* 29: 1–46

MATTHIESSEN, P. (1959) Wildlife in America. New York, The Viking Press, Inc.

MAY, R. M. (1974) Biological populations with nonoverlapping generations: stable points, stable cycles, and chaos. *Science* 186(4164): 645–647

MCDIFFETT, W. F., A. E. CARR, AND D. L. YOUNG (1972) An estimate of primary productivity in a Pennsylvania trout stream using a diurnal oxygen curve technique. *Amer. Midland Nat.* 87(2): 564–570

MCDOUGALL, W. B. (1958) Botany of the museum and Colton Ranch area. II. Plant competition. *Plateau* 30: 79–82

MCDOUGALL, W. B. (1966) Botany of the museum and Colton Ranch area. IV. Vegetation changes in Field Five. *Plateau* 39: 134–142

MCLAREN, I. A. (1962) Population dynamics and exploitation of seals in the eastern Canadian Arctic. In E. D. Le Cren and M. W. Holdgate, eds., The Exploitation of Natural Animal Populations, pp. 168–183, New York, John Wiley and Sons, Inc.

MCNAB, B. (1971) On the ecological significance of Bergmann's Rule. *Ecology* 52: 845–854

MCNAUGHTON, S. J. (1966) Ecotype function in the *Typha* community-type. *Ecol. Monogr.* 36: 297–325

MCNAUGHTON, S. J., AND L. L. WOLF (1970) Dominance and the niche in ecological systems. *Science* 167(3915): 131–139

MCPHEE, J. (1968) The Pine Barrens. New York, Farrar, Straus and Giroux

MCVAY, S. (1966) The last of the great whales. *Sci. Amer.* 215(2): 13–21

MEADOWS, P. S., AND J. G. ANDERSON (1966) Micro-organisms attached to marine and freshwater sand grains. *Nature* 212(5066): 1059–1060

MEDAWAR, P. B. (1959) The Future of Man. New York, Basic Books, Inc.

MILGRAM, S. (1970) The experience of living in cities. *Science* 167: 1461–1468

MINSHALL, W. (1969) Community structure in natural stream systems. In K. W. Cummins, ed., The Stream Ecosystem, pp. 2–4, Michigan State University, Institute of Water Research, Technical Report No. 7

MOONEY, H. A., AND W. D. BILLINGS (1961) Comparative physiological ecology of arctic and alpine populations of *Oxyria digyna*. *Ecol. Monogr.* 31: 1–29

MOONEY, H. A., AND E. L. DUNN (1970) Photosynthetic systems of mediterranean climate shrubs of California and Chile. *Amer. Nat.* 104: 447–453

MORGAN, A. H. (1930) Field Book of Ponds and Streams. New York, G. P. Putnam's Sons

MORIBER, G. (1974) Environmental Science. Boston, Allyn and Bacon, Inc.

MORSE, D. H. (1970) Ecological aspects of some mixed-species foraging flocks of birds. *Ecol. Monogr.* 40(1): 119–168

MORTIMER, C. H. (1952) Water movements in lakes during summer stratification; evidence from the distribution of temperature in Windermere. *Phil. Trans. Royal Soc. Lond. (B)* 236: 355–404

MOTT, J. J. (1972) Germination studies on some annual species from an arid region of western Australia. *J. Ecol.* 60: 293–304

MOTT, J. J. (1973) Temporal and spatial distribution of an annual flora in an arid region of Western Australia. *Tropical Grasslands* 7(1): 89–97

MOYNIHAN, M. (1962) The organization and probable evolution of some mixed species flocks of neotropical birds. *Smithsonian Miscell. Coll.* 143(7): 1–140

MUENSCHER, W. C. L. (1935) Weeds. New York. The Macmillan Co.

MURDOCH, W. W. (1966) "Community structure, population control, and competition"—a critique. *Amer. Nat.* 100: 219–226

NAGY, J. G., AND TENGERDY, R. P. (1968) Antibacterial action of essential oils of *Artemisia* as an ecological factor. II. Antibacterial action of the volatile oils of *Artemisia tridentata* (big sagebrush) on bacteria from the rumen of mule deer. *Appl. Microbiol.* 16: 441–444

NEAL, E. (1965) Woodland Ecology. Cambridge, Mass., Harvard University Press

NEALES, T. F., A. A. PATTERSON, AND V. J. HARTNEY (1968) Physiological adaptation to drought in the carbon assimilation and water loss of xerophytes. *Nature* 219: 469–472

NELSON, D. J. (1969) The stream ecosystem: terrestrial-lotic community interactions. In K. W. Cummins, ed., The Stream Ecosystem, pp. 14–19, Michigan State University, Institute of Water Research, Technical Report No. 7

NELSON, J. B. (1967) Etho-ecological adaptations in the great frigate-bird. *Nature* 214: 318

NEWELL, N. D. (1963) Crises in the history of life. *Sci. Amer.* 208(2): 76–92

NIERING, W. A., AND R. H. GOODWIN (1974) Creation of relatively stable shrublands with herbicides: arresting "succession" on rights-of-way and pasturelands. *Ecology* 55: 784–795

NIERING, W. A., R. H. GOODWIN, AND S. TAYLOR (1970) Prescribed burning in southern New England: introduction to long-range studies. *Proc. Annu. Tall Timbers Fire Ecol. Conference* 267–286.

NIERING, W. A., R. H. WHITTAKER, AND C. H. LOWE (1963) The saguaro: a population in relation to environment. *Science* 142(3588): 15–23

NIXON, C. M. (1970) Deer populations in the Midwest. *In* White-tailed Deer in the Midwest, United States Department of Agriculture Forest Service Research Paper NC-39, pp. 11–18

ODUM, E. P. (1950) Bird populations of the Highlands (North Carolina) Plateau in relation to plant succession and avian invasion. *Ecology* 31: 587–605

ODUM, E. P. (1962) Relationships between structure and function in the ecosystem. *Jap. J. Ecol.* 12: 108–118

ODUM, E. P. (1969) The strategy of ecosystem development. *Science* 164: 262–270

ODUM, E. P. (1971) Fundamentals of Ecology, 3rd Ed. Philadelphia, W. B. Saunders Co.

ODUM, H. T. (1956) Primary production in flowing waters. *Limnol Oceanogr.* 1: 102–117

ODUM, H. T. (1957) Trophic structure and productivity of Silver Springs, Florida. *Ecol. Monogr.* 27: 55–112

ODUM, H. T. (1971) Environment, Power, and Society. New York, Wiley-Interscience

ODUM, H. T., AND R. C. PINKERTON (1955) Time's speed regulator: the optimum efficiency for maximum power output in physical and biological systems. *Amer. Sci.* 43: 331–343

OECHEL, W. C., B. R. STRAIN, AND W. R. ODENING (1972a) Photosynthetic rates of a desert shrub, *Larrea divaricata* Cav. under field conditions. *Photosynthetica* 6(2): 183–188

OECHEL, W. C., B. R. STRAIN, AND W. R. ODENING (1972b) Tissue water potential, photosynthesis, ^{14}C-labeled photosynthate utilization, and growth in the desert shrub *Larrea divaricata* Cav. *Ecol. Monogr.* 42: 127–141

OHMART, R. D. (1973) Observations on the breeding adaptations of the roadrunner. *Condor* 75: 140–149

OOSTING, H. J. (1942) An ecological analysis of the plant communities of Piedmont, North Carolina. *Amer. Midland Nat.* 28: 1–126

OVERLEASE, W. R. (1969) Early succession on abandoned cornfields in southern Chester County, Pennsylvania. *Proc. Pa. Acad. Sci.* 43: 101–105

PAINE, R. T. (1966) Food web complexity and species diversity. *Amer. Nat.* 100: 65–73

PARK, T. (1962) Beetles, competition and populations. *Science* 138: 1369–1375

PARK, T., E. V. GREGG, AND C. Z. LUTHERMAN (1941) Studies in population physiology. X. Interspecific competition in populations of granary beetles. *Physiol. Zool.* 14: 395–430

PATRICK, R. (1963) The structure of diatom communities under varying ecological conditions. *Ann. N. Y. Acad. Sci.* 108: 359–365

PATRICK, R. (1968) The structure of diatom communities in similar ecological conditions. *Amer. Nat.* 102(924): 173–184

PATTERSON, C. C. (1965) Contaminated and natural lead environments of man. *Arch. Environ. Health* 11: 344–360

PEARSON, O. P. (1948) Metabolism and bioenergetics. *Sci. Monthly* 46: 131–134

PEARSON, O. P. (1954) The daily energy requirements of a wild anna hummingbird. *Condor* 56: 317–322

PEARSON, O. P. (1966) The prey of carnivores during one cycle of mouse abundance. *J. Animal Ecol.* 35: 217–233

PERINO, J. V., AND P. G. RISSER (1972) Some aspects of structure and function in Oklahoma old-field succession. *Bull. Torrey Bot. Club* 99: 233–239

PERRY, T. O. (1971) Dormancy of trees in winter. *Science* 171: 29–36

PETERSON, R. T., ed. (1967) The Bird Watcher's Anthology. New York, Harcourt, Brace and Co.

PHARIS, R. P. (1966) Comparative drought resistance of five conifers and foliage moisture content as a viability index. *Ecology* 47(2): 211–221

PHELPS, R. H. (1974) The sonorous spring. *New York Times Magazine*, April 14, 1974: 14–26

PHILLIPSON, J. (1966) Ecological Energetics. New York, St. Martin's Press

PIANKA, E. R. (1966a) Convexity, desert lizards and spatial heterogeneity. *Ecology* 47(6): 1055–1059

PIANKA, E. R. (1966b) Latitudinal gradients in species diversity: a review of concepts. *Amer. Nat.* 100: 33–46

PIELOU, E. C. (1969) An Introduction to Mathematical Ecology. New York, Wiley-Interscience

PIMENTEL, D. (1961) Animal population regulation by the genetic feedback mechanism. *Amer. Nat.* 95: 65–79

PIMENTEL, D. (1968) Population regulation and genetic feedback. *Science* 159: 1432–1437

PIRIE, N. (1959) Chemical diversity and the origins of life. In F. Clark and R. C. M. Syuge, eds., The Origin of Life on the Earth, pp. 76–83, Oxford, Pergamon Press

PITELKA, F. A. *et al.* (1974) Ecology and evolution of social

organization in artic sandpipers. *Amer. Zool.* 14: 185–204
PLATT, R. (1953) A Pocket Guide to the Trees. (Cardinal Edition) New York, Pocket Books, Inc.
PRESTON, F. W. (1948) The commonness, and rarity, of species. *Ecology* 29: 254–283
PRESTON, F. W. (1969) Diversity and stability in the biological world. In G. M. Woodwell and H. H. Smith, eds., Diversity and Stability in Ecological Systems (Brookhaven Symposia in Biology No. 22), pp. 1–12, Upton, N. Y., Brookhaven National Laboratory

RACINE, C. H. (1971) Reproduction of three species of oak in relation to vegetational and environmental gradients in the southern Blue Ridge. *Bull. Torrey Bot. Club.* 98(6): 297–310
RAGOTZKIE, R. (1974) The Great Lakes rediscovered. *Amer. Sci.* 62: 454–464
REDFIELD, A. C. (1958) The biological control of chemical factors in the environment. *Amer. Sci.* 46: 204–221
REICHLE, D. E., ed. (1970) Analysis of Temperate Forest Ecosystems. Berlin, Springer-Verlag
REINERS, W. A. (1968) Carbon dioxide evolution from the floor of three Minnesota forests. *Ecology* 49: 471–483
RENSCH, B. (1936) Die Gesichte des Sundabogens: eine tiergeographische Untersuchung. Berlin, Gebrüder Borntraeger
REX, M. A. (1973) Deep-sea species diversity: decreased gastropod diversity at abyssal depths. *Science* 181(4104): 1051–1053
RICE, E. L. (1971) Inhibition of nodulation of inoculated legumes by leaf leachates from pioneer plant species from abandoned fields. *Amer. J. Bot.* 58: 368–371
RICE, E. L., AND S. K. PANCHOLY (1972) Inhibition of nitrification by climax ecosystems. *Amer. J. Bot.* 59: 1033–1040
RICE, E. L., AND S. K. PANCHOLY (1973) Inhibition of nitrification by climax ecosystems. II. Additional evidence and possible role of tannins. *Amer. J. Bot.* 60: 691–702
RICHARDS, P. W. (1973) The tropical rain forest. *Sci. Amer.* 229(6): 58–67
RISSER, P. G., AND G. COTTAM (1968) Carbohydrate cycles in the bulbs of some spring ephemerals. *Bull. Torrey Bot. Club* 95: 359–369
ROBEL, R. J. (1972) Possible function of the lek in regulating tetraonid populations. *Proc. 15th Internatl. Ornithol. Congress*, pp. 121–133
ROBEL, R. J., AND W. B. BALLARD, JR. (1974) Lek social organization and reproductive success in the greater prairie chicken. *Amer. Zool.* 14: 121–128
ROOT, R. B. (1967) The niche exploitation pattern of the blue-gray gnatcatcher. *Ecol. Monogr.* 37: 317–350
ROOT, R. B. (1973) The organization of a plant-arthropod association in simple and diverse habitats: The fauna of collards, *Brassica oleracea*. *Ecol. Monogr.* 43: 95–124
ROSENZWEIG, M. L. (1968) Net primary productivity of terrestrial communities: prediction from climatological data. *Amer. Nat.* 102(923) 67–74
ROSENZWEIG, M. L. (1969) Why the prey curve has a hump. *Amer. Nat.* 103: 81–87
ROSENZWEIG, M. L., AND R. H. MACARTHUR (1966) Graphical representation and stability conditions of predator-prey interactions. *Amer. Nat.* 97: 209–223
ROSENZWEIG, M. L., AND J. WINAKUR (1969) Population ecology of desert rodent communities: habitats and environmental complexity. *Ecology* 50(4): 558–572
ROWE, G. T. (1971) Benthic biomass and surface productivity. In J. T. Costlow, ed., Fertility of the Sea, Vol. 2, pp. 441–454, New York, Gordon and Breach
RUMNEY, G. R. (1968) Climatology and the World's Climates. New York. The Macmillan Co.
RUTTNER, F. (1963) Fundamentals of Limnology, 3rd Ed. Toronto, Toronto University Press
RYTHER, J. H., AND W. M. DUNSTAN (1971) Nitrogen, phosphorus and eutrophication in the coastal marine environment. *Science* 171: 1008–1013

SAGAN, C. (1970) Life. Encyclopaedia Britannica, 14th Ed., Chicago, Encyclopaedia Britannica, Inc.
SAGAN, C. *et al.* (1973) Climatic change on Mars. *Science* 181: 1045–1049
SALISBURY, E. J. (1925) The structure of woodlands. *Veröff. Geobot. Inst. Rübel* 3: 334–355
SALTER, P. S. (1974) Toward an ecology of the urban environment. In W. H. Johnson and W. C. Steere, eds., The Environmental Challenge, pp. 238–266, New York, Holt, Rinehart and Winston, Inc.
SANCHEZ, R. (1971) Phytochrome involvement in the control of leaf shape of *Taraxacum officinale* L. *Experientia* 27(10): 1234–1237
SANDERS, H. L. (1960) Benthic studies in Buzzards Bay. III. The structure of the soft bottom community. *Limnol. Oceanogr.* 5: 138–153
SANDERS, H. L. (1968) Marine benthic diversity: a comparative study. *Amer. Nat.* 102: 243–282
SANDERS, H. L. (1969) Benthic marine diversity and the stability-time hypothesis. In G. M. Woodwell and H. H. Smith, eds., Diversity and Stability in Ecological Systems (Brookhaven Symposia in Biology, No. 22), pp. 71–81, Upton, N. Y., Brookhaven National Laboratory
SANDERS, H. L., AND R. R. HESSLER (1969) Ecology of the deep-sea benthos. *Science* 163: 1419–1424
SAUNDERS, A. A. (1936) Ecology of the Birds of Quaker Run Valley, Allegany State Park, New York. Albany, New York State Museum Handbook 16
SCHINDLER, D. W. (1974) Eutrophication and recovery in experimental lakes: implications for lake management. *Science* 184: 897–899
SCHMID, W. D. (1969) Physiological specializations of amphibians to habitats of varying aridity. In C. C. Hoff and M. L. Riedesel, eds., Physiological systems in Semiarid Environments, pp. 135–142, Albuquerque, University of New Mexico Press
SCHMIDT-NIELSEN, K. (1959) The physiology of the camel. *Sci. Amer.* 201: 140–151
SCHMIDT-NIELSEN, K. (1964) Desert Animals. Oxford, Oxford University Press
SCHMIDT-NIELSEN, K. (1972) How Animals work. Cambridge, Cambridge University Press
SCHMIDT-NIELSEN, K., AND B. SCHMIDT-NIELSEN (1953) The desert rat. *Sci. Amer.* 189(1): 73–78
SCHOLANDER, P. F. (1955) Evolution of climatic adaptation in homeotherms. *Evolution* 9: 15–26
SCHORGER, A. W. (1973) The Passenger Pigeon. Norman, Okla., University of Oklahoma Press
SCHREIDER, E. (1964) Ecological rules, body-heat regulation and human evolution. *Evolution* 18: 1–9
SELF, P. (1973) The city and its environment. In J. Benthall, ed., Ecology in Theory and Practice, pp. 81–100, New York, The Viking Press, Inc.
SERRUYA, C. (1969) Les Dépots du Lac Léman en Relation avec L'Évolution du Bassin Sédimentaire et les Caractères du Milieu Lacustre. Thesis, University of Geneva, Geneva, Imprimerie Kundig
SHANKS, R. E., AND F. H. NORRIS (1950) Microclimatic variation in a small valley in eastern Tennessee. *Ecology* 31(4): 532–539
SHORT, L. L. *et al.* (1973) Tropical rain forests. *Science* 181: 893–895
SIGAFOOS, R. S. (1964) Botanical evidence of floods and flood plain deposition. Washington, D.C., United States Geological Survey Professional Paper 485-A
SLACK, A. V. (1970) Defense against Famine: The Role of the Fertilizer Industry. Garden City, N. Y., Doubleday and Co., Inc.
SLOBODKIN, L. B. (1959) Energetics in *Daphnia pulex* populations. *Ecology* 40: 232–243

SLOBODKIN, L. B., F. E. SMITH, AND N. G. HAIRSTON (1967) Regulation in terrestrial ecosystems, and the implied balance of nature. *Amer. Nat.* 101: 109–124

SMITH, J. M. (1974) Models in Ecology. Cambridge, Cambridge University Press

SMITH, K. L., JR. AND J. M. TEAL (1973) Deep-sea benthic community respiration: an in situ study at 1850 meters. *Science* 179(4070): 282–283

SOHOLT, L. F. (1973) Consumption of primary production by a population of kangaroo rats (*Dipodomys merriami*) in the Mojave Desert. *Ecol. Mongr.* 43(3): 357–376

SPARLING, J. H. (1967) Assimilation rates of some woodland herbs in Ontario. *Bot. Gazette* 128: 160–168

SPIGHT, T. M. (1967) Species diversity: a comment on the role of the predator. *Amer. Nat.* 101: 467–474

SPRUGEL, D. G. (1974) Natural Disturbance and Ecosystem Responses in Wave-regenerated *Abies balsamea* Forests. Ph.D. Dissertation, Yale University School of Forestry

SPURR, S. H. (1956) Forest association in the Harvard Forest. *Ecol. Monogr.* 26(3): 245–262

SPURR, S. H. (1964) Forest Ecology. New York, The Ronald Press Co.

STACKHOUSE, D. (1973) Secondary Succession in Lancaster County Forests with Special Reference to the Elimination of Chestnut (*Castanea dentata*) by Chestnut Blight. Unpublished Thesis, Department of Biology, Franklin and Marshall College

STEBBINS, G. L. (1966) Processes of Organic Evolution. Englewood Cliffs, N. J., Prentice-Hall, Inc.

STOCKNER, J. G. (1968) The ecology of a diatom community in a thermal stream. *Br. Phycol. Bull.* 3(3): 501–514

STRONG, D. R., JR. (1974) Rapid asymptotic species accumulation in phytophagous insect communities: the pests of cacao. *Science* 185: 1064–1066

STRUHSAKER, T. T. (1967a) Behavior of vervet monkeys and other cercopithecines. *Science* 156(3779): 1197–1203

STRUHSAKER, T. T. (1967b) Ecology of vervet monkeys (*Cercopithecus aethiops*) in the Masai-Amboseli Game Reserve, Kenya. *Ecology* 48(6): 891–904

STRUHSAKER, T. T. (1974) Correlates of ranging behavior in a group of red colobus monkeys (*Colobus badius tephrosceles*). *Amer. Zool.* 14: 177–184

SVERDRUP, H., M. JOHNSON, AND R. FLEMING (1942) The Oceans: Their Physics, Chemistry and General Biology. Englewood Cliffs, N. J., Prentice-Hall, Inc.

SWAN, L. W. (1961) The ecology of the high Himalayas. *Sci. Amer.* 205(4): 68–78

SWAN, L. W. (1963) Ecology of the heights. *Nat. Hist.* 72(4): 22–29

TABER, R. D., AND R. DASMANN (1957) The dynamics of three natural populations of the deer, *Odocoileus hemionus columbianus*. *Ecology* 38: 233–246

TEAL, J. M., AND I. VALIELA (1973) The living filter. *Oceanus* 17: 7–10

THEODORSON, G. A., ED. (1961) Studies in Human Ecology. Evanston, Ill., Row, Peterson and Co.

THORNTHWAITE, C. W. (1948) An approach toward a rational classification of climate. *Geogr. Rev.* 38: 55–94

TILESTON, J. V., AND R. R. LECHLEITNER (1966) Some comparisons of the black-tailed and white-tailed prairie dogs in north-central Colorado. *Amer. Midland Nat.* 75(2): 292–316

TODD, W. E. C. (1940) Birds of Western Pennsylvania. Pittsburgh, University of Pittsburgh Press

TUCKER, V. A. (1966) Diurnal torpor and its relation to food consumption and weight changes in the California pocket mouse *Perognathus californicus*. *Ecology* 47: 245–252

TURNER, R. D. (1973) Wood-boring bivalues, opportunistic species in the deep sea. *Science* 180(4093): 1377–1379

U.S. Bureau of Sport Fisheries and Wildlife (1973) Threatened Wildlife of the United States. Washington, D.C., BSFW Resource Publication 114

U.S.D.A. (1949) Trees (The Yearbook of Agriculture for 1949). Washington, D.C., United States Department of Agriculture

U.S.D.A. (1973) Air pollution damages trees. United States Department of Agriculture, Forest Service Publication (Northeastern Area)

VALLENTYNE, J. R. (1957) Principles of modern limnology. *Amer. Sci.* 45: 218–244

VAN DER PIJL, L. (1972) Principles of Dispersal in Higher Plants. Berlin and New York, Springer-Verlag

VANNOTE, R. L. (1969) Detrital consumers in natural systems. In K. W. Cummins, ed., The Stream Ecosystem, pp. 20–23, Michigan State University Institute of Water Research, Technical Report No. 7

VARLEY, G. C., AND G. R. GRADWELL (1970) Recent advances in insect population dynamics, *Annu. Rev. Entomol.* 15: 1–24

VERNER, J., AND M. F. WILLSON (1966) The influence of habitats on mating systems of North American passerine birds. *Ecology* 47: 143–147

VESEY-FITZGERALD, D. F. (1960) Grazing succession among East African game animals. *J. Mammal.* 41:(2): 161–172

VESEY-FITZGERALD, D. F. (1965) The utilization of natural pastures by wild animals in the Rukwa Valley, Tanganyika. *E. Afr. Wildlife J.* 3: 38–48

VESSEY, S. (1967) Effects of chlorpromazine on aggression in laboratory populations of wild house mice. *Ecology* 48(3): 367–376

VOGL, R. J. (1967) Controlled burning for wildlife in Wisconsin. *Proc. Tall Timbers Fire Ecol. Conference* 6: 47–96

WASHBURN, S. L., AND I. DEVORE (1961) The social life of baboons. *Sci. Amer.* 204(6): 62–71

WATSON, A. (1972) A current model of population dynamics in red grouse. *Proc. 15th Internatl. Ornithol. Congress:* 134–149

WATSON, J. P. (1967) A termite mound in an iron age burial ground in Rhodesia. *J. Ecol.* 55(3): 663–670

WATT, K. E. F., ED. (1966) Systems Analysis in Ecology. New York, Academic Press

WATT, K. E. F. (1968) Ecology and Resource Management. New York, McGraw-Hill Book Co.

WATTS, C. R., AND A. W. STOKES (1971) The social order of turkeys. *Sci. Amer.* 224(6): 112–118

WATTS, W. A. (1971) Postglacial and interglacial vegetation history of southern Georgia and central Florida. *Ecology* 52(4): 676–690

WEATHERLEY, A. H. (1966) Ecology of fish growth. *Nature* 212: 1321–1324

WEBER, W. J. (1972) A new world for the cattle egret. *Nat. Hist.* 81: 56–63

WEBSTER, D. B. (1965) Ears of *Dipodomys*. *Nat. Hist.* 74(2): 26–33

WELCOMME, R. L. (1964) The habitats and habitat preferences of the young of the Lake Victoria *Tilapia*. *Rev. Zool. Bot. Afr.*, 70: 1–28

WELLINGTON, W. D. (1964) Qualitative changes in populations in unstable environments. *Can. Entomol.* 96: 436–451

WELLS, P. V. (1965) Scarp woodlands, transported soils and concept of grassland climate in the Great Plains region. *Science* 148: 246–249

WELLS, P. V. (1970) Postglacial vegetational history of the Great Plains. *Science* 167: 1574–1582

WENT, F. W. (1955) The ecology of desert plants. *Sci. Amer.* 192(4): 68–75

WENT, F. W. (1973) Competition among plants. *Proc. Natl. Acad. Sci. U.S.A.* 70(2): 585–590

WETZEL, R. G., AND B. A. MANNY (1972) Decomposition of dissolved organic carbon and nitrogen compounds from leaves in an experimental hard-water stream. *Limnol. Oceanogr.* 17(6): 927–931

WHITE, L., JR. (1967) The historical roots of our ecological crisis. *Science* 155: 1203–1207

WHITFORD, W. W. (1969) Heart rate and changes in body fluids in aestivating toads from xeric habitats. In C. Hoff and M. Riedesel, eds., Physiological Systems in Semiarid Environments, pp. 125–134, Albuquerque, University of New Mexico Press

WHITTAKER, R. H. (1965) Dominance and diversity in land plant communities. *Science* 147(3655): 250–260

WHITTAKER, R. H. (1967) Gradient analysis of vegetation. *Biol. Rev.* 42: 207–264

WHITTAKER, R. H. (1969) Evolution of diversity in plant communities. In G. M. Woodwell and H. H. Smith, eds., Diversity and Stability in Ecological Systems (Brookhaven Symposia in Biology, No. 22), pp. 178–196, Upton, N. Y., Brookhaven National Laboratory

WHITTAKER, R. H. (1970) The biochemical ecology of higher plants. In E. Sondheimer and J. B. Simeone, eds., Chemical Ecology, pp. 43–70, New York, Academic Press

WHITTAKER, R. H., AND G. E. LIKENS (1973) Carbon in the biota. In G. M. Woodweil and E. V. Pecan, eds., Carbon in the Biosphere, pp. 281–302, Washington, D.C., U.S. A.E.C. Technical Information Center

WHITTAKER, R. H., AND W. A. NIERING (1965) Vegetation of the Santa Catalina Mountains, Arizona: a gradient analysis of the south slope. *Ecology* 46(4): 429–452

WIENS, J. A. (1966) On group selection and Wynne-Edwards' hypothesis. *Amer. Sci.* 54: 273–287

WILDE, S. A. (1958) Forest Soils: Their Properties and Relations to Silviculture. New York, The Ronald Press Co.

WILDE, S. A. (1970) Soils and forest growth: Their relationship in terms of regression analysis. *Bioscience* 20(2): 101–108

WILEY, R. H. (1974) Evolution of social organization and life-history patterns among grouse. *Quart. Rev. Biol.* 49: 201–227

WILLIAMS, P. K. (1973) Seasonal Movements and Population Dynamics of Four Sympatric Mole Salamanders, Genus *Ambystoma*. Ph.D. Dissertation, Indiana University

WILLIAMSON, P. (1971) Feeding ecology of the red-eyed vireo (*Vireo olivaceous*) and associated foliage-gleaning birds. *Ecol. Monogr.* 41: 129–152

WILSON, D. O. (1972) Phosphate nutrition of the aquatic angiosperm *Myriophyllum exalbescens* Fern. *Limnol Oceanogr.* 17(4): 612–615

WILSON, E. O., AND W. H. BOSSERT (1971) A Primer of Population Biology. Stamford, Conn., Sinauer Associates, Inc.

WISTENDAHL, W. A. (1958) The floodplain of the Raritan River, New Jersey. *Ecol. Monogr.* 28(2): 129–153

WOLLIN, G., AND D. B. ERICSON (1971) Amino-acid synthesis from gases detected in interstellar space. *Nature* 233: 615–616

WOLMAN, A. (1965) The metabolism of cities. *Sci. Amer.* 213 (3): 179–190

WOODWELL, G. M. (1967) Radiation and the patterns of nature. *Science*, 156(3774): 461–470

WOODWELL, G. M., C. F. WURSTER, JR., AND P. A. ISAACSON (1967) DDT residues in an East Coast estuary: a case of biological concentration of a persistent insecticide. *Science* 156: 821–824

WRIGHT, R. D., AND H. A. MOONEY (1965) Substrate-oriented distribution of bristlecone pine in the White Mountains of California. *Amer. Midland Nat.* 73: 257–284

WRIGHT, H. E., JR. (1968) History of the Prairie Peninsula. In The Quaternary of Illinois, University of Illinois College of Agriculture, Special Publication 14, pp. 78–88

WRIGHT, H. E., JR. (1971) Late Quaternary vegetational history of North America. In K. K. Turekian, ed., The Late Cenozoic Glacial Ages, pp. 425–464, New Haven, Yale University Press

WURSTER, C. F., AND D. B. WINGATE (1968) DDT Residues and declining reproduction in the Bermuda petrel. *Science* 159(3818): 979–981

WYNNE-EDWARDS, V. C. (1962) Animal Dispersion in Relation to Social Behavior. New York, Hafner Publishing Co.

WYNNE-EDWARDS, V. C. (1964) Population control in animals. *Sci. Amer.* 211(2): 68–74

WYNNE-EDWARDS, V. C. (1965) Self-regulating systems in populations of animals. *Science* 147: 1543–1548

ZEDLER, P. H., AND F. G. GOFF (1973) Size-association analysis of forest successional trends in Wisconsin. *Ecol. Monogr.* 43: 79–94

ZELENY, L. (1972) Can we save the bluebird? *Living Wilderness* Autumn: 24–31

ZERVANOS, S. M., AND N. F. HADLEY (1973) Adaptational biology and energy relationships of the collared peccary (*Tayassu tajacu*). *Ecology* 54: 759–774

INDEX